Samuel Wilks

Lectures on Diseases of the Nervous System

Samuel Wilks

Lectures on Diseases of the Nervous System

ISBN/EAN: 9783337173203

Printed in Europe, USA, Canada, Australia, Japan

Cover: Foto ©berggeist007 / pixelio.de

More available books at **www.hansebooks.com**

ON DISEASES OF THE

NERVOUS SYSTEM

DELIVERED AT GUY'S HOSPITAL

BY

SAMUEL WILKS, M.D., F.R.S.

LONDON
J. & A. CHURCHILL, NEW BURLINGTON STREET
1878

PREFACE

In presenting this work to my readers, I may fairly be expected to preface it with a few words of explanation. In the year 1868 a part of my course on "Medicine" delivered at Guy's Hospital relating to nervous diseases was published in a periodical form. Ever since that time the lectures have been frequently perused by my pupils, who have constantly demanded of me their reprint in a separate form. To this appeal, so often made, I now respond in the present volume. It contains my original lectures with the additional matter which a subsequent ten years has enabled me to accumulate. Much of this has already appeared in the 'Guy's Hospital Reports,' but its repetition in this volume was inevitable, since it is evident that the cases which have been thought of sufficient importance to publish separately would necessarily be the most valuable ones with which to illustrate my lectures.

The order of the subjects which I have found useful for lecture cannot be justified on scientific grounds, but I may console myself with the conviction that with our present existing nomenclature it is impossible to frame a systematic view of nervous diseases on any rational basis whatever, be it anatomical, pathological, or clinical. I have therefore endeavoured to make the best of a heterogeneous system.

Had I under other circumstances believed that there was room for such a work as this, I should have prepared to sit down and write a systematic treatise, which would thus have enabled me to omit many of the explanations now offered expressly for the instruction of students, and to add more precise scientific material than the present occasion demands. I might then also perhaps have attempted in my descriptions of disease to approach nearer to a scientific method. Time would then also have permitted me to make appropriate references to the various authors whose works would have surrounded me, and to mention more emphatically the original observers in this department of medicine. The form of a lecture, however, does not admit of a reference to the source of information from which the instructor draws. He offers what he has in his possession, but how much of his wealth has been acquired by his own digging, how much delivered over to him by his predecessors, and how much he has been unconsciously acquiring from his contemporaries, is unknown even to himself. From whatever sources my own knowledge has been obtained, I have endeavoured thoroughly to assimilate it before adapting it to the use of others; at the same time I have always endeavoured to remember the name of any observer who has added a new fact to the general stock. In saying this I cannot but feel how indebted we all are to such men as Hughlings Jackson, who are always pouring out their best thoughts before the profession, and who therefore assist us in a greater measure than we ourselves are aware of in forming our opinions.

As regards myself I have offered my contributions to the profession from time to time, and were I asked what amount of originality may have been displayed in the lectures when published ten years ago I should point to the general view I took of cerebral physiology and pathology, the truth of which all observations and experiments have since tended to confirm. Indeed, from my earliest days as a teacher, and whilst controversies were still warm as to the interpretation of a double brain, and the textbooks of physiology were throwing no light upon it, I taught that the two hemispheres were necessary complements to the separate ganglia with which they were associated, and that the distinction of the latter was a necessary accompaniment of the independent action of the limbs. The pathological facts of hemiplegia, of hemispasm, as well as physiological considerations, all combined to prove the correctness of this position.

I might also express a belief that my lecture on migraine or sick-headache gave at that time an impulse to the further study of that malady; at all events I was prompted to give the lecture from failing to find any description of this complaint in the best-known works on medicine.

Nor am I aware that any account had previously been written on alcoholic paraplegia, or the effects of alcohol in producing a myelitis or meningitis. Some other minor matters relating to nerve pathology in which I have been interested I need not mention.

Finally, I must express my fear that the somewhat disorderly arrangement of notes for lectures has caused the description of cases to be a little more informal than I should have desired.

SAMUEL WILKS.

GROSVENOR STREET;
January, 1878.

CONTENTS

PART I—BRAIN

PART II—THE SPINAL CORD

PART III—FUNCTIONAL AND GENERAL DISEASES

PART IV—NERVES

LECTURES

ON

DISEASES OF THE NERVOUS SYSTEM

PART I—BRAIN

INTRODUCTORY REMARKS ON THE PHYSIOLOGY OF THE NERVOUS SYSTEM

GENTLEMEN,—Those who attended here last year will remember
that no lectures were delivered on diseases of the nervous system.
The session, indeed, is too brief to allow of any approach to a complete course of medical lectures such as was attempted to be at one
time given. It is very true that our distinguished predecessors,
Bright and Addison, were enabled to take up the whole series of
subjects in their nosology, and, by the force of their genius, to shed
upon them a light which illumines us to the present day, but at the
same time it must be confessed that a quarter of a century has done
much to expand the brief system with which they dealt. We shall
see, for example, how the physical diagnosis of the chest has
wonderfully unfolded the subject of thoracic disease, and the
physiology of the nervous system enlarged our knowledge of
nervine ailments. It were almost impossible, therefore, to
embrace all subjects in one short course. At one time an inflammation of the brain or a paraplegia might be all that held
the attention of the lecturer and his class ; now, although nerve
pathology is still in its infancy, we have to dilate on a variety of
nervous affections. These are so highly important that I beg your
serious attention to the disorders which I may bring before your
notice, and must also insist on this because it is only by a close
observance of the symptoms that cases of nervous disease can afford
any interest to you. To many students, I know, they constitute the
driest and most repulsive forms of malady, and for this reason—
that, wishing rightly to store your minds with as many facts as

1

be productive of one set of symptoms, and in another part of the
brain of another set of symptoms.   Effusion of blood shows a
totally different class of phenomena as occurring on the surface or in
the centres of the cerebral hemispheres.   We must, therefore, of
necessity, consider the physiology of the nervous system before
we pass to its pathology, and you will, I know, pardon me by
recalling to your remembrance some of the main points which
physiology has taught us.   In possessing this knowledge we shall
be prepared for the occurrence of those symptoms which we are
sure morbid changes in certain regions of the brain must produce,
and we shall also save much time by the avoidance of their repetition when we come to treat of distinct diseases.

Look, for a moment, at the constitution of an animal body ; here
is the framework, or skeleton, held together by ligaments, and
covered by muscles for the purpose of moving one bone on
another.   The action of the muscles necessitates wear and tear, and
consequently a fresh supply of nutrient material.   With this object
there is blood sent to them through vessels and propelled by the
heart, also other vessels to carry away the débris ; the latter is
got rid of by lung and kidney.   For a fresh supply of blood the
abdominal organs are required to manufacture it from the food.
Since these organs have certain definite functions to perform, they
are regulated for the most part by nerves called the sympathetic,
which convey a power originating in certain bodies styled ganglia.
I apprehend that a creature so constituted might exist or live.
There is not such a one, I know, for the very presence of parts I have
mentioned would be useless without a higher organisation ; but
there are animals of the molluscous kind which seem to have very
little higher nervous development than this—an arrangement for
the regulation of their simple animal machinery.   Now, remember that what the lower animals possess, we have also, and that
which we possess in common with them is not to be regarded as
an inferior portion of our nervous system, but, on the other hand,
the more important.   I shall have to show you that our very
existence depends on the integrity of the sympathetic system of
nerves, but not necessarily on the brain.

The creature we have contemplated having muscles to be stimulated to action, requires nerves to proceed to them which
originate in ganglia, or centres of force.   These latter are collected
together in a chain constituting the spinal cord.   These centres
are excited through the influence of other nerves which have their
origin over the whole surface of the integument of the body, so
that if the surface be touched, a stimulus is conveyed along one
of these sensitive nerves to the spinal centres, a reaction takes

place, and a corresponding effect is seen on the muscles by an influence conveyed back by the motor nerves.

We have now a higher class of animal which can be excited to movement by an external stimulus. Now, bear in mind that we also have this system, the spinal or excito-motor, in common with the lower animals, and do not ignore it because we have still some further development of our nervous centres.

Thirdly, imagine the vertebræ developing into a skull case at the same time that the spinal marrow itself expands and terminates in two large ganglia, known as the central ganglia of the brain—the "head-centres," or the thalami optici and corpora striata. These are the termini which become subsequently influenced by higher powers acting upon them, and through them the whole spinal cord; but whether they possess other properties, independent of those of the cord, I cannot, from my own clinical experience, positively declare. It is believed, however, that an animal possessing these bodies has a sensorium, and a kind of perception, and thus, when these are stimulated, evinces what are styled emotions; that a human being in possession of these structures, without any further cerebral development, might display, for example, such emotions as laughing or crying. I think myself that these are merely reflex actions and that the ganglia are nothing more than the superior terminations of the cord which come immediately under the influence of the cerebral hemispheres.

Fourthly, suppose, proceeding from these bodies which constitute the upper part of the cord, a number of white fibres pass towards higher ganglia, and we have still another system. These ganglia in a mass form the cineritious part of the brain, a region where fresh sensations are received and fresh powers are developed. This is the largest mass of grey nervous substance in the body, and would form a layer as large as this table, were it not folded up so as to be packed in a small compass. Herein the sensations conveyed from the surface of the body, and from the special senses, become to the recipient as perceptions and ideas: the animal has become a reasoning being. The power here produced reacting on the body is spoken of as the will, and the animal has become a voluntary agent. The amount of reasoning power and the strength of volition depend, I have no doubt, mainly upon the development of the cerebral hemispheres—small but not absent in the lower animals, most perfect in such a splendid head as that of the late Surgeon Lawrence or the first Napoleon. The quality of the brain, however, or its texture, no doubt varies considerably, and thus some idiots have had enormous heads.

You perceive, then, that the machinery of the body is worked or ruled by the ganglionic nervous forces; you may witness the fact daily in the wards of the hospital, in patients whose brains and spinal cords are irretrievably diseased. You see also how many operations of the human body are due to forces residing in the spinal cord; and that this cord is again overruled by the large cerebral hemispheres. All these systems are intimately united; even the sympathetic is closely joined to the spinal to regulate the functions of the viscera, and therefore you might infer that some of the operations going on in the organism within might at times be made perceptible to us. This relationship between the two systems is so important that I shall have to refer to it again.

The nervous system, as generally described by physiologists, is made up of two parts, the force-producing elements and the conducting cords, analogous to a galvanic battery and its connecting wires. The grey or ganglionic substance is the potent material and the nerves are the conductors; the former is the seat both of the active power, which is exhibited as the motive force, and also the highly vitalized receptacle for impressions from without. As regards the nerves, it would appear, from experiments lately made, that those for motion and sensation are structurally alike, and therefore would convey equally well sensory or motory vibrations. The grey or potential matter in the spinal cord can be stimulated to action from any irritation on the skin, and a movement be reflected back to the part touched. This occurs equally well or better where the spinal marrow has no connection with the sensorium alone; but, of course, under these circumstances, the animal has no sensation (or feeling in the ordinary sense of the term), nor voluntary power over the body. The brain, when connected with the cord, receives impressions from it, and elaborates them in a manner so that they become to us as mental phenomena; but the latter do not free themselves from the impressions which caused them, for these continue ever after to be intimately associated or even composed both of sensations and movements. This is established by the late experiments of Hitzig and Ferrier, who found not only, as the older experimenters had already discovered, that irritating the corpus striatum and brain over it would produce movements, but that irritation of particular convolutions would cause uniform and definite actions on certain muscles of the animal, as, for example, movements of the front and hind leg or ear respectively, according to the parts operated on. This would lead us to the belief that the various actions of the body are not merely set agoing by the voluntary power exerted on the spinal cord by the brain, but that each portion of the brain has its own separate function in relation to the several movements, and even that the mental conceptions which we subjectively possess are intimately associated with, if not the actual products of, the material impressions which caused them. The material world is on one side of the mirror, and our feelings are like the reflected image which glances off on the other ; so that running parallel with the sensori-motor impressions on the brain are our own mental states.

This doctrine was pretty clearly seen many years ago by Dr Gull, and enunciated in his ' Gulstonian Lectures for 1849,' where he says, " The combination of sensation and voluntary muscular movements has by Volkman been shown to be the source of our knowledge of locality and direction. It is not my purpose here to consider how he has applied this to visual directions and positions of objects, but I may just mention that it has long seemed to me obvious, and I have so long taught it in lectures, that as it is by the muscular movements of the upper extremity that we test the direction of any force acting upon the sensitive senses of the fingers, so it is by the contraction of the muscles attached to the eye that we tell the position of any object which sends its rays to the retina." " There is a close subjection of motion to sensation in the action of the muscles of expression. The movements, which are called emotional, are as directly *excito-motor* as any, with this peculiarity, that emotion or sensation form a necessary part. If we style the spinal movements *mechanical,* they are *psychico-mechanical.*"

The intimacy of our mental relations with the external world is therefore of the closest ; mental operations cannot be dissociated from the material objects which originated them ; when the metaphysician attempts to do this he is only studying the human mind in part ; and his department of knowledge may be called the science of consciousness, but this is not coextensive with psychology. The doctrine of Locke had already been expressed in the Aristotelian maxim, " Nullum in intellectu, nisi in sensu ; " and the same idea may be frequently found floating in the mind of Shakspeare,[1] as, for

[1] A consideration of the genius of Shakspeare tends to confirm the belief in the doctrine that the brain can be operating upon phenomena derived from without, and, subsequently, and only when the thoughts are committed to paper, becoming conscious of them. I cannot otherwise understand what is meant by the inspiration of Shakspeare, or how he came to speak of the centre of gravity before Newton, or the circulation of the blood before Harvey, or the various doctrines of mental philosophy before Locke. The following quotation from Carlyle is much to my purpose :—" Shakspeare is what I call an unconscious intellect ; there is more virtue in it than he himself is aware of. His dramas are products of Nature, deep as Nature herself. It is Nature's highest reward to a true simple great soul that he gets thus to be a *part of herself.* Such a man's works, whatever he with utmost conscious exertion and forethought shall accomplish, grow up withal *un*consciously from the unknown depths in him, as the oak tree grows

instance, in the following: "Oh! now I see; Queen Mab hath been with you; she gallops, night by night, o'er courtiers' knees that dream of courtesies straight; o'er lawyers' fingers, who straight dream of fees; o'er ladies' lips, who straight on kisses dream."

You will see that the impressions conveyed from the cord, which is in connection with the outer world, produce these higher or more elaborate qualities in the brain. The brain, therefore, stands, as it were, superior to the spinal cord; the latter is its servant to supply it with wants, and to be governed by it. The governing power of the brain is well worthy of our consideration, for the general doctrine is one which is applicable to a great many instances of natural and morbid conditions, although perhaps not rigidly proved by a scientific method, experimenters having differed in their results. It was, however, long ago shown that the action of the spinal cord was greater when the brain was removed; and it has been said that if a pigeon's head be cut off, and the spinal cord be still active, enough to move the wings, its power is at once stayed by application of a galvanic current. We know, too, in the human subject if the foot be touched, there is not the same reflex action as in a case of paraplegia when the communication between the cord and brain is severed. And, moreover, the tendency to movement can be overcome by the will of the individual, as in the resisting the effects of tickling, or in the instance of coughing or sneezing, where the involuntarily act, arising from some irritant to the glottis or nose, is much more violent than could be produced by any effort of the will. There are cases where the larynx is paralysed, and the patient unable to cough when directed to do so, and yet sometimes a violent explosion will occur when some mucus or foreign body irritates the glottis. Darwin mentions how his friends at table declared they could not take a pinch of snuff without sneezing, but on the expression of his disbelief, and their consequent endeavour to sneeze, not one of them could accomplish it. It may be that this controlling power of the brain over the spinal cord is not only analogous to but is the very material foundation for the influence which the mind exercises over the body, or the reason of the superiority of the higher faculties over the lower animal appetites and passions. For example, cannot we say that the material facts and the moral teachings correspond when we see a parrot, whose first impulse is to bite, by a simple reflex act, control the impulse when educated to obey a higher law; and do we not see the same in the taming of all

from the earth's bosom, as the mountains and waters shape themselves. How much in Shakspeare lies hid—much that was not known at all, not speakable at all, like roots, like sap and forces working underground. Speech is great, but silence is greater."

animals, and in children who are taught to subdue their passions? If this be so, cannot we also say that the strongest minded men and women are those who govern their lower natures, and in forming a scale of mental capacity amongst human beings, be able to graduate it according to the influence which one mind has over another. An idea of this kind will throw a light upon such a disease as hysteria, where the cerebral power is so weak that the whole organisation seems for a time under the control of the spinal system: the will and regulating power being gone, reflex acts reach so high a pitch that the body is thrown into a paroxysm of convulsions. Again, in various other mental derangements, where we say the intellect is disordered, a close analysis of the case will often show that all the strange phenomena are at once accounted for on the supposition of the temporary abeyance of the controlling power. The strange acts and behaviour of the patient are not mental vagaries or new vices, but simply the animal instincts and passions in full swing, without any controlling power present to restrain them. In intoxication this is very evident, and the saying " in vino veritas " receives its interpretation from this view of the physiology of the cerebro-spinal centres, for in the inebriate the innate propensities appear full blown as soon as the brain has become paralysed. And what is even a sadder spectacle, when the brain decays in old age, many long hidden follies or vices again make their appearance. If we wished to take a higher flight of thought in this matter, we might argue that a man's responsibilities are in direct proportion to the power which his brain or higher powers can exert over his lower spinal or animal life, and try to understand the saying that "though the spirit is willing the flesh is weak." A man cannot raise laughter or tears at will, but he may go and see a comedy or tragedy which will produce them. I am not mixing up material, mental, and moral questions which have no relation, for I regard them all as intimately related, and that in fact the only true interpretation of man's higher nature must be found in the physiological basis on which it rests. Such questions as the freedom of the will, which have been argued from all time by divines and moralists, cannot be solved without a knowledge of physiology.

We are in the habit of saying the brain is the organ of the mind. Within this skull, " the dome of thought, the palace of the soul," sits the mind enthroned. Here are perception and thought and judgment. Here originates the will or volition which starts the levers for setting in motion many mechanical movements of the body, the vital processes proceeding under their own independent forces. Our movements appear like direct acts of the will, for we are unconscious of the machinery which intervenes, and the mental power seems to

originate in the brain. We have, however, only to feel ill—to know
that consciousness is associated with the whole composite being.
The mind, looking upon itself, cannot see the mechanism through
which it works, much less the mechanism of which it may be
the resulting force, and thus, of necessity, rational beings are
obliged to place the "ego" behind or anterior to all physical events.
The term "mind," or will, however, must still be used in medical
lectures as the only word in common use to represent the feeling of
effort on the part of the individual.   The term mind I use here as
intimately associated with the cerebral hemisphere and man's
organisation.   The question how far this mind, or what portion of
it, is a distinct immaterial principle, worthy to be called the soul,
cannot be discussed on scientific grounds, for we have no data to
work upon.   The arguments for there being a spiritual part of man,·
which can remain after the body decays, depend upon moral and
religious considerations; and there are some who think that these
convictions are almost equal in cogency to mathematical demonstra-
tions, since they depend, like the latter, upon axioms which the
mind cannot further analyse, but must receive as true.   Believing,
however, as most of us do, that there is an immortal spirit, there is
no reason to suppose that any such immaterial influence is operating
in our organisms, and this is the reason why I allude to the subject
and depart apparently from my province.   I mean as a perfectly
independent guiding principle, for the belief in a soul operating
through the brain of course can have no significance in our physiological enquiries, since its presence would be lost in the acts of the
physical organism. We may, therefore, if we choose, without
renouncing scientific methods, regard the brain as the soul's frail
dwelling-house, or the mind as the resultant force of the two instruments working together.

> "My brain I'll prove the female to the soul,
> My soul the father; and these two beget
> A generation of still breeding thoughts."

But there are some in our profession who, perplexed with the
various phenomena of the human mind, have considered it necessary in their lectures to introduce into the problem of life a
perfectly independent force, and to ask if we do not admit this how
can we, for example, account for memory, and how can the mind of
man, having a material source, gaze into futurity, and have thoughts
of infinite space and eternity.   Now, I mention this to warn you
against such a belief as this, or if you hold it, to give up the
study of physiology and the pursuit of medicine ; what purpose can
we gain by studying the phenomena of the nervous system if some
unseen or extraneous force is pulling the strings?   It is of no use

physiology informing us that the Welsh fasting-girl, in order to keep up her temperature 40° above the surrounding atmosphere, must be supplied with food, and if she does not eat she will die, if some subtle immaterial agency may come in and supply the force required. For once admit such an extravagant notion, and you may believe that an unseen power may carry you up in the air, that flowers may float in at your window, or you may read a newspaper by sitting upon it. If we once admit such an independent power we do not know where we are, and a study of the physiology of the animal economy becomes a useless labour. What we, as physiologists, have to do is to take the bodily machinery as we find it, and endeavour to unlock its recesses and discover how the hidden wheels work ; but because we have not yet arrived at this knowledge, it seems to me absurd to introduce some extraneous and independent agency. The doing so seems much on a par with the savage who, terror-stricken with the report of a gun, and not understanding its mechanism, declares it to be the work of his gods.[1]

When it is said that the human mind can grasp infinity, it is really only attributing a negative quality to the mind and declaring that we are incapable of conceiving any object without the idea of something beyond it, or of a particle of matter that is indivisible ; and when it is said we can grasp eternity, it means that the human mind cannot conceive of time without an end. For my own part, I do not see in these so-called vast conceptions of the human mind any other than the necessary attributes of a conscious being on an earth like ours. Then, again, memory is a quality of the human mind, which to some seems almost to necessitate the belief of a spiritual essence, for how, otherwise, they argue, can it stretch over years of time and hundreds of miles of space.

The mode of connection between mind and matter has by the world at large been generally regarded as an insoluble problem,

[1] In a review of these lectures I am "charged" with being a materialist. The word "charge," not being used in scientific phraseology must mean that I am guilty of an offence against morals. Those who, like myself, have been styled materialists have never denied the existence of a future world, or the continuance of our being in another state, and therefore the term can be applicable only to the question of the independent existence of an immaterial principle operating in our human bodies. That this is the issue, I think, is evident from the fact that one of the writers who so charged me explained to a quasi-scientific society, that people could do very well without brains, of which he gave examples. The same people, who, in their childhood of knowledge believe this, also naturally feel a horror at the word "automatism," which implies that the mechanism of the body has a power of working itself, whereas they consider it more intelligible or rather more moral to suppose there is some independent principle always pulling the strings.

for the very terms employed to designate the phenomena of each are as opposed as subjective and objective, and therefore not translatable into one another. The words expressive of our feelings being of so different a kind from those by which we describe outward objects, that the mere absence of a common language for the two, is not only indicative of the total inability hitherto existing to cross the gulf, but in itself contributes a sufficient barrier to any one who tries to make the attempt. All explanations involve us in contradictions in terms or in simple assumptions. In spite of this, if there be any truth in physiological researches, the intimacy after all between mind and matter is of the closest, and possibly the material process and the metaphysical expression may be more allied than we imagine. It is impossible for us to see a physical process, as of a motion in our brain cells, changing into a mental one, as the latter is altogether subjective, but it does not follow that there is not a material movement of the kind, which results in what we call consciousness. I think it possible to imagine a higher being looking upon the active brain of man, and seeing both the material changes in operation and the resultant forces which are mental, and regarding them as one. The difficulties are in part common to all questions of organisation and functional activity, quite apart from mind and brain. For example, no one, looking at the bile, could predicate the liver from which it came, or looking at a mass of liver, would know that one of the products of its activity was the biliary fluid. I cannot conceive by what method of the minutest microscopic examination such a knowledge of the function of the liver could be ascertained. If with material substances the connection between organism and function are difficult to ascertain, how much more difficult when the function is altogether subjective? All attempts, therefore, to explain the association of mind and matter must be given in words, which are in themselves as diverse as the subjects of which they treat; in fact, to some persons any expressions used to designate material changes by metaphysical names and the converse, are simply absurd. The attempt, however, I think, is good, if it be only to show that physiologists will continue to endeavour to explain mental phenomena by a regard to the material organisation, instead of leaving them equally inexplicable in the hands of the metaphysician. Instead, therefore, of saying that memory necessarily implies an immaterial agency in man's nature, see what speculation may do for us on the material side; and such speculation will go on with every fresh advance in scientific discovery, until the goal be reached. I quote from two American authors, of whom their country may be justly proud. Dr Draper says that impressions made upon the brain may lie like the photographic picture on glass, unseen or unknown, until developed. " Thus I have seen landscapes and architectural views taken in Mexico developed, as artists say, months subsequently in New York, the images coming out after their long voyage in all their proper forms and in all their proper contrasts of light and shade. The photograph had forgotten nothing. It had equally preserved the contour of the everlasting mountains and the passing smoke of a bandit fire. Are there, then, contained in the brain more permanently, as in the retina, more transiently the vestiges of impressions that have been gathered by the sensory organs ? Is this the explanation of memory—the mind contemplating such pictures of past things and events as have been committed to her custody ? In her silent galleries are there hung micrographs of the living and the dead, of scenes that we have visited, of incidents in which we have borne a part ? Are these abiding impressions mere signal works, like the letters of a book which impart ideas to the mind ? or are they actual picture images inconceivably smaller than those made for us by artists, in which, by the aid of a microscope, we can see in a space not bigger than a pin-hole a whole family group at a glance ? " " During a third part of our life in sleep we are withdrawn from external influences ; hearing and sight and the other senses are inactive, but the never-sleeping mind, that pensive, that veiled enchantress, in her mysterious retirement, looks over the ambrotypes she has collected, for they are truly unfading impressions, and combining them together as they chance to occur, constructs from them the panorama of a dream."

Oliver Wendell Holmes takes up the same illustration, and says, " I need not say that no microscope can find the tablet inscribed with the names of early loves, the stains left by tears of sorrow or contrition, the rent where the thunderbolt of passion had fallen, or any legible token that such experiences had formed a part of the life of the mortal, the vacant temple of whose thought it is exploring. It is only as an inference aided by an illustration, which I will presently offer, that I would suggest the possible existence, in the very substance of the brain tissue, of those inscriptions which Shakspeare must have thought of when he wrote—

" Pluck from the memory a rooted sorrow ;
Raze out the written troubles of the brain."

It must be remembered that a billion of the starry brain cells could be packed in a cubic inch, and that the convolutions contain one hundred and thirty-four cubic inches. My illustration is microscopic photography. I have a glass slide on which is a minute photographic picture, which is exactly covered when the head of a small pin is laid upon it. In that little speck are clearly to be seen by a proper magnifying power the following objects: the declaration of Independence, with easily recognised facsimile autographs of all the signers, the capital at Washington, and very good portraits of the Presidents of the United States, from Washington to Mr James Polk. These objects are all distinguishable as a group with a power of fifty diameters; with a power of three hundred, any one of these becomes a visible picture. You will see, if you will, the majesty of Washington in his noble features, or the will of Jackson in those hard lines of the long face, crowned with that bristly head of hair, in a perpetual state of electrical divergence and centrifugal self-assertion. Remember, that each of these faces is the record of a life. Now, recollect that there was an interval between the exposure of the negative of the camera and its development by pouring a wash over it when all these pictured objects existed potentially but absolutely invisible and incapable of recognition in a speck of collodion film which a pin's head would cover, and then think what Alexandrian libraries, what congressional document loads of positively intelligible characters, such as one look of the recording angel would bring out, many of which we can ourselves develope at will, or which come before our eyes forbidden, like "Mene, mene, Tekel, Upharsin," might be held in those convolutions of the brain which wrap the talent entrusted to it too often, as the folded napkin of the slothful servant hid the treasure his master had lent him. Memory may, therefore, be a material record, and the brain, scarred and seamed with infinitesimal hieroglyphics as the features are, engraved with the traces of thought and passion."

Of course, this is all speculation, and almost as unintelligible as any other theory which attempts to explain mental processes by physical changes, but if there should be any truth in it, we might further observe that there are reckoned 600 million grey cells in the brain, which would allow fifty every minute in a long life to receive impressions.

ANATOMY AND PHYSIOLOGY OF THE BRAIN

I will recall to your memories the main facts in the anatomy and physiology of the cerebro-spinal system, so that we may be assisted in determining the seat and symptoms of the various diseases of which we shall treat. There are the sensory nerves proceeding from all parts of the surface of the body towards the spinal cord; some of these clearly pass at once into the ganglia; others apparently proceed directly upwards through the bulb and crura to the thalami optici, whence fibres spread out again to the posterior parts of the hemispheres. In a corresponding manner, mutatis mutandis, their physiological direction being in an opposite course, run the motor nerves, connected with the grey centre of the spinal cord and forming strands of fibres, which pass to the corpora striata, and then again by the radiating fibres to the anterior portions of the hemisphere.

There appears to be, then, on each side of the spinal cord a column of sensory and motor fibres, the one terminating in the thalamus and the other in the corpus striatum. How far this is a true statement of facts can be best decided by the experiments made for us on the human subject, but at present it is not absolutely proved that the thalamus does constitute the final receptive seat of the sensory tract. The exact distribution of the fibres in the corpus striatum, and subsequently in the hemispheres, is beginning to be made out both by dissections and physiological experiments. It is clear that the anterior portion of the crus cerebri or crust passes to the corpus striatum, and thence by the corona radiata to the hemispheres; in the same way the posterior portion of the crus or tegmentum passes through the thalamus to the posterior portion of the brain. It is interesting to note that the grey cells of the anterior convolutions of the brain are very large, like those of the anterior cornu of the cord, and that they are different in the posterior part of the brain which is associated with sensation. You will see by diagrams, which give transverse sections of the brain through the ganglia, the true arrangement of these parts. The grey matter of the corpus striatum is divided into an intra-ventricular nucleus or nucleus caudatus and an extra-ventricular nucleus or lenticular ganglion; and this again is separated from the thalamus by some white fibres, called the internal capsule, and also from the convolutions of the surface by other fibres called the external capsule. In this last is a nucleus of grey matter called the tæniaform band.

It is evident then, that there are fibres passing upwards and downwards in the spinal cord, and that these are connected, as well as the fibres proceeding from the grey centre, with the spinal nerves. It is an important question, however, whether there be any direct connection between the higher ganglia in the cranium and these nerves, or whether the connection be only through the grey centres. It is evident, from the smallness of the cord compared with the aggregate of all the nerves of the body, that the former cannot contain any great number of the latter, and therefore, in all probability, the nerves proceeding from the cord have their origin principally within it, and the connection between the cord and the brain is by other and distinct fibrillæ. The general belief is that there are spinal centres which rule over sets of nerves for particular objects or complex operations, and the cerebral ganglia set these centres in action by connecting nerve fibres between them; for example, in breathing a number of parts all work together in unison, being supplied by several nerves proceeding from the respiratory spinal centre; and all that is required is, when a voluntary effort is made to breathe or to stay respiration, that this centre should be influenced from the brain by a single fibrilla. Inasmuch as we cannot direct so complex a process as breathing by any act of the will, there is no need for all the individual nerves to pass to the supreme central ganglia. In the case of the limbs, not only is there a prearranged order of cells whence the nerves arise, but a further blending of the nerves after they leave the cord, as seen in the brachial plexus, and this is no doubt for the purpose of associating together the action of several muscles for the production of the complex movements of grasping, pronating, supinating, &c. The consequence is that it is impossible to discover the origin of a nerve supplying any particular muscle, except in the case of the so-called cranial nerves. You will find that a nerve, for example, which, arising in the uper part of the spine, goes to form the brachial plexus, divides into a number of filaments, and sends its branches to various muscles of the arm. Or, on the other hand, a nerve traced upwards from the arm will proceed to this plexus, which then enters the spine at different intervertebral foramina. Consequently, although probably each fibrilla has its distinct origin in the cord, the fibres become so blended that injury to the nerve at its exit cannot answer to paralysis of one individual muscle. I shall presently show you that the nerves to the extremities are more especially associated with the upper part of the spinal cord, or the central ganglia in the skull; and here, again, the fibres are so blended, that disease of any one spot of the ganglia appears sufficient to produce paralysis of the whole extremity. Thus, a small spot of disease in the corpus striatum produces a weakness of the limbs but of no one set of muscles, and a further extension of disease a more complete paralysis. Whether every distinct fibre supplying its own portion of muscle may not have its origin in particular cells of the cord, and that disease in the latter may not affect that portion of muscle solely, is another question. The facts observed in progressive muscular atrophy would rather favour this conclusion. The arrangement I speak of has for its object, no doubt, the grouping together of certain movements; and thus, we cannot will the action of any one muscle separately. In walking, when we use the flexors and extensors of the leg, we use them as a whole; we have no separate power over any particular muscle. There is some centre therefore, which is controlled or put in action by the brain through our volition, and that stimulates the whole group of muscles by means of the nerves. An apparatus of this kind is already made or is born with us, and once set in action will continue in operation without fresh voluntary effort on our part; and in like manner a centre of force may become educated so as to perform a regular system of movements, as for example, in the playing an air on the piano, which may be performed when the mind is not " willing," showing that the centre has become educated to perform certain work, as in the simple act of walking; hence the meaning of the common expression that use or habit is second nature. Just in the same way as the organic system of nerves can keep the viscera in play, so the spinal cord, by means of the properties with which it is endowed, can, through its nerves, produce various complex movements, these either having been acquired by the cord or arranged by a natural organization. I need not mention the case of the jumping of the frog or the flying of the bird when decapitated, but will only remind you of the anencephalous or brainless infant sucking at its mother's breast.

It would appear from this that the nerves may have their origin in various grey centres, possessing their own functional peculiarities. For instance, if I use the facial nerve for talking or laughing, I am probably setting in operation different centres in the two cases, and, therefore, the nerve must have more than one origin, and dissections are showing that this reasoning is founded on fact. The nerve arrangement may be likened to an apparatus where a dozen bells are pulled by a dozen strings, but where each string does not ring a separate bell, but by the interchange and combination of the cords each string may ring a particular series.

We, as physiologists, have to deal with men as animals, and, in spite of the prejudices against the notion, man must be studied as an animal. Thus we see many of our actions are, in common with other animals, in some way dependent on the spine. They may be regulated, excited, or arrested by cerebral influence, and a hard case it would be for us if in dressing ourselves, for example, or in eating, every necessary movement were dictated by volition acting on a particular muscle. Now, the whole grouping is to a certain extent arranged, and what is not arranged is brought about by education. Thus, the spinal centres, like the brain, as I shall presently show you, become educated ; and, therefore, I verily believe that the spinal cords of two different persons, although apparently alike, are functionally very unlike, and that an adult man's cord is a very superior organ to a child's. So educated to a particular purpose may the spinal cord become that in using our microscopes we pass the slide from right to left, and *vice versâ* when we want to examine it at exactly the opposite end, and I have known the following circumstance occur :—A gentleman going up to his room to dress for dinner had forgotten altogether the purpose of his visit until he found himself in bed. I think it is Professor Huxley who relates the case of an old soldier being observed by a former companion crossing the street with his Sunday dinner ; his friend called out "attention," whereat the man's hands fell to his sides, and the mutton and potatoes into the gutter. No better example could be given of the character of our spinal cord. The word entered the ear, impressed the ganglia within, touching the old key, and the stimulus was carried down the arms before the cerebral hemispheres could bring their superior influence to bear on its arrest.

I am impressed with this idea of the automaton action of the spinal system almost every day when I come to the hospital by the train. I observe at the Charing Cross railway booking-office, the passengers asking for tickets for only two or three places, as the line is short, and almost before the name of the station is completed, the ticket is delivered to them by the clerk ; there is no time for thought, and often, indeed, he is conversing with his fellow-clerk. So instantaneous is the word and action between the passengers and the clerk, that I regard the whole operation as reflex, and I quite believe that if it were possible to remove his cerebral hemispheres the whole process of delivering tickets might go on as before. It is true that the process was first directed by the action of the superior cerebral ganglia, but now I have no doubt the impulse from his ear to his arm takes a shorter and more direct cut through the medulla. In the case of the French soldier, whose case occupied the attention lately of the savans in Paris, the man, after falling into a kind of torpid state, might be played upon like a musical instrument. The case is very interesting, and I give it as copied from the newspaper :

A Living Automaton.—A curious patient is just now an inmate of Dr Mesnet's ward at the Hôpital St Antoine. His profession was that of a singer at the Cafés Chantants. During the war of 1870-71 he was hit over the left ear by a musket bullet, which carried off about 2½ inches of the parietal bone, and laid bare the brain on the left side. This led to a temporary paralysis of the members on the opposite side, as is always the case; but he was eventually cured of this, while the tremendous wound on the skull began to heal, so that after a time he could resume his professional duties at the cafés to the satisfaction of the public. Suddenly, however, he was seized with nervous symptoms, lasting from 24 to 48 hours, and of such an extraordinary nature that it was considered safe to take him to the hospital. His malady is easier to illustrate by examples than to define. When he is in his fit he has no sensitiveness of his own, and will bear physical pain without being aware of it; but his will may be influenced by contact with exterior objects. Set him on his feet, and, as soon as they touch the ground, they awaken in him the desire of walking; he then marches straight on quite steadily, with fixed eyes, without saying a word, or knowing what is going on about him. If he meets with an obstacle on his way he will touch it, and try to make out by feeling what it is, and then attempt to get out of its way. If several persons join hands and form a ring around him, he will try to find an opening by repeatedly crossing over from one side to the other, and this without betraying the slightest consciousness or impatience. Put a pen into his hand; this will instantly awaken in him the desire of writing; he will fumble about for ink and paper, and, if these be placed before him, he will write a very sensible business letter; but, when the fit is over, he will recollect nothing at all about it. Give him some cigarette paper, and he will instantly take out his tobacco bag, roll a cigarette very cleverly, and light it with a match from his own box. Put them out one after another, he will try from first to last to get a light, and put up in the end with his ill-success. But ignite a match yourself, and give it him, he will not use it, and let it burn between his fingers. Fill his tobacco-bag with anything, no matter what—shavings, cotton, lint, hay, &c.—he will roll his cigarette just the same, light and smoke it without perceiving the hoax. But, better still, put a pair of gloves into his hand, and he will put them on at once; this, reminding him of his profession, will make him look for his music. A roll of paper is then given to him, upon which he assumes the attitude of a singer before the public, and warbles some piece of his repertory. If you place yourself before him he will feel about on your person, and, meeting with your watch, he will transfer it from your pocket to his own; but, on the other hand, he will allow you, without any resistance or impatience whatever, to take it back again.—*Galignani*.

The difficulty which many have in accepting these views is owing to the dislike they feel to the idea of the machinery of the body being able to act by itself; their inclinations leading them to the belief that the fact of consciousness implies an immaterial principle which is pulling the strings and regulating our bodily movements. Physiologists, however, are obliged to renounce this, since they know that acts are performed without consciousness, as when a person walks in his sleep or passes through the streets in a reverie. Certainly a large number of operations go on in our body without any knowledge or will on our part. The spinal centres and ganglia take cognisance of the proceedings of the viscera over which they rule, and this with so much method that, did the stomach exist as an independent animal, its operations would be regarded as due to instinct, or to its having an " unconscious will." Not only the spinal system but the brain itself will act when we are not conscious of its operations, as in sleep. Numerous instances could be quoted of both scientific and literary men who have discovered on rising the results of operations which had been performed silently during sleep. Sir Thomas Brown said, " Sleep is the waking of the soul; the ligation of sense, but the liberty of reason." If this be so, what becomes of consciousness as the basis of all mental philosophy? No one who has cared to examine the operations of the human body can hold to the doctrine " cogito ergo sum," which can have but one interpretation—that consciousness and existence are identical expressions.

Now, it is very clear that any injury to the cord which severs a part of the body from the sensorium above and the centre whence volitionary acts proceed must produce paralysis of sensation and motion. We look, therefore, to the spinal marrow as the seat of all paralysis,and it cannot be too distinctly remembered that the division into spinal cord and brain is not in correspondence exactly with that of the vertebral column and cranium ; for just as the bones of the face are transfigured vertebrae, so the spinal cord within the cranium is the same organ merely altered in form, and the nerves it gives off true spinal nerves. It therefore follows that in perfect paralysis of all kinds there must be some lesion of the spinal system, between the intra-cranial ganglia above and the termination of the cord below. You must not, therefore, use the term cerebral paralysis in the old-fashioned way, as if disease of the brain proper could produce paralysis of the limbs, for this has not yet been proved to be true. This, which I have always taught, needs but little modification from the recent researches of Hitzig and Ferrier, nor from the few facts which appear to show that an inflammation of the brain or injury of the cortex will produce weakness of some of the muscles. The latter, probably, is not a paralysis in the ordinary sense of the term ; and as regards the experiments made by the observers just named, although it is clearly shown that the various motor fibres of the cord proceed to the corpus striatum and again spread out to pass to the convolutions of the brain, so that irritation of particular convolutions will set in action a particular group of muscles, yet it has not been clearly proved that a lesion of a convolution will produce paralysis of any part of the body while the ganglion below is healthy. This statement is founded on the fact already well known—that portions of brain may be removed after fracture of the skull with impunity. The two views will probably be reconciled by shewing that the motor region is limited in extent, and therefore loss of motility can only be looked for in lesions of a circumscribed spot. It is a question, therefore, which will, no doubt, soon be solved.

I might take this opportunity of saying that these experiments of Ferrier have more fully confirmed the doctrine which I have for many years taught at this school relating to the interpretation of the double nature of the brain. You know that physiologists and metaphysicians have puzzled over the question, some thinking that the one hemisphere was appropriated for certain faculties of the mind, and the other hemisphere for other faculties ; some thinking the one side of the brain was the seat of the good qualities and the other of the bad ; some also, again, suggesting that one was the receptive part of the mind and the other the objective and active part. Now it has always seemed to me that there was no purpose in propounding these fanciful theories when the plain facts were before us. The body, you see, is made up of two halves, joined together in the mesian line, each supplied by its own nerves proceeding from their individual ganglia; now, as the trunk is made up of these halves closely approximated, the ganglia are fused together and act as one in the various movements of the chest and abdomen; thus it is that we cannot move one side of the chest independently of the other. But you perceive, also, that besides the trunk we have limbs which are used independently, as well, also, as the muscles of the face and the tongue; from this it necessarily follows that the ganglia ruling over them must be distinct and not fused together; the cord ought, therefore, to split or be divided into two halves in that region which rules over the limbs and those parts which have an independent action. This separation does take place at the summit of the cord after it passes into the cranium, where it terminates in two large ganglia, whose isolation is made more complete by the lateral ventricles of the brain. That the separation of the cord into its two distinct halves has this object is made clear by the fact that disease of either of them produces paralysis only of those parts on each side which have an independent action, and more especially of the limbs. This division of the cord, necessitated by or associated with the independent action of the limbs, requires also that the superior ganglia or cerebral hemispheres which govern them should be independent and distinct also. I say a consideration of the mechanism of the human body seems to necessitate the idea of a double brain. For if the movements of the trunk as a whole require a fusion of ganglia, if there be any independent movement of limbs on either side, a separation of the ganglia which rule over them is also necessary, and so, again, a separation of the higher cerebral spheres which govern these. The proof of the intimacy existing between the convolutions and the ganglia below, as I have been in the habit for many years of showing, lies in the results of experiments made for us by an injury or disease like syphilis, which involves a portion of the convolutions of the brain. In these cases it has long been observed that if the patient had fits the convulsive movement occurred on the opposite side of the body to that of the disease. A step further in this direction was made · by Hughlings Jackson, and proved to be true by the experiments of Ferrier, that irritation of particular convolutions would produce corresponding and special convulsive movements in the muscles of the face or limbs. Ferrier, by using electric stimuli, proved that the excitation of certain convolutions was followed invariably by the same movements, and concluded that the anterior portion of the brain was for voluntary movements and the active outward mani·

festation of intelligence.   He found that irritation of certain convolutions corresponded to definite muscular movements, and as these
parts of the brain are now all carefully mapped out and named, it will
be as well to remind you of those of the greatest importance and most
scientific interest.

You remember the division into the different lobes; also that the fissure of Sylvius, beginning at the base of the brain and running up the
side, divides into two portions.   Then, commencing above and running downwards and forwards, is the great central fissure or fissure
of Rolando, which divides the frontal from the parietal lobes.
Behind this again is the parieto-occipital fissure.   The frontal lobe
is that portion of the brain which lies under the frontal bone, and
herein are some of the convolutions which of late years have had
most interest for us.   In front of the fissure of Rolando is a very
constant convolution ; it is called the anterior central or ascending
frontal convolution.   Proceeding forwards from this, along the
margin of the hemisphere, is the superior frontal convolution. Below
this is the middle or second frontal convolution, and below this is the
third or inferior frontal convolution.   The latter forms the superior
boundary of the front of the fissure of Sylvius ; it is the well-known
convolution on the left side which is associated with speech and
called Broca's convolution.   Besides these convolutions there are
sulci, which have received special names at the hands of Ecker and
others.   As regards the parietal lobe, you may remark that behind
the fissure of Rolando is a convolution called the posterior central
or ascending parietal convolution.   From this convolution others
pass backwards, divided by the inter-parietal fissure ; one of these
is called the superior marginal convolution, and another, one of the
most important of this lobe, turns round at the end of the Sylvian
fissure, and is called the angular gyrus.   It is thought to be
peculiar to man.   Then there is the occipital lobe, which also has
its named convolutions and sulci.   Then, also, the temporal or
temporo-sphenoidal lobe, separated from the frontal and anterior
part of the parietal by the fissure of Sylvius, but less defined
posteriorly.

Ferrier found that irritation of the postero-parietal lobule caused
movements of the opposite leg and foot.   Irritation of the convolutions bounding the fissure of Rolando caused complex movements of
the arms and legs.   Irritation of the posterior extremity of the
superior frontal, at its junction with ascending frontal, caused
extension of the arm and hand. Irritation of the posterior extremity
of the middle frontal, near ascending frontal, flexion of the forearm;
irritation of ascending frontal convolution, movements of mouth ;
irritation of posterior extremity of third frontal convolution, where

it joins ascending frontal, opening of mouth and protrusion of tongue ; irritation of posterior half of superior middle frontal convolution, lateral movements of head and eyes ; irritation of ascending parietal convolution, movement of hand and wrist. Up to the present time no one has fixed upon the exact locality of the perceptive centres of the senses ; but Ferrier has found in his experiments that all these are probably situated within a circumscribed area, for he has observed that the senses are lost by the destruction of certain convolutions ; for example, destruction of the angular gyrus impairs the sight on the opposite side, and this he therefore is inclined to regard as the visual centre ; in the same way destruction of the superior temporo-sphenoidal lobe causes deafness, and this he would regard as the auditory centre. A destruction of the hippocampal region impairs the sense of touch or common sensation, and this he calls the tactile centre. As I shall presently tell you, the experiments made for us by disease on the human subject, tend to show that lesions external to the thalamus produce anæsthesia, but this is explained by Ferrier on the supposition that disease merely interrupts the path of transmission to the true centre. The lower part of the temporo-sphenoidal lobe he considers to contain the centres of smell and taste.

The intimacy between the convolutions and the motor tract, therefore, is very great, both anatomically and physiologically, as well as in a mental aspect. The pathological relation was shown several years ago by Flourens and others, who had remarked that spots of softening in the convolutions very often corresponded to similar spots in the corpus striatum. Although I have spoken only of the motor tract and the anterior part of the hemisphere which rules over it, the same laws prevail with reference to the sensory tract, the thalamus, and posterior part of the hemisphere. Here we believe sensations are received from the outer world ; but since the impressions must be very similar on the two sides of the body, the necessity for the division, except for giving rise to special motor impulses, is not so clear. Although it might be supposed that a double set of special senses might also necessitate distinct perceptive centres, and, therefore, a double brain. The uniformity in respect to sensations and perceptions may be the reason why a large portion of one hemisphere may be removed without any apparent destruction of any mental faculty. Indeed, it is clear that each hemisphere is complete in itself. Probably in accordance with anatomical relations the posterior hemispheres are receptive, and the anterior active ; and in reflex mental acts a process occurs between them corresponding to that in the spinal cord, the ganglia in the latter case being more developed and differentiated. It may be that in ordinary quiescence and contemplation some parts of the brain only, as the posterior lobes, are at work, while in active thought, especially in writing and speaking, the anterior portions are mainly in operation. We may obtain some kind of clue to the parts of the brain which are at work by an example like the following. Since aphasia is due to destruction of a certain convolution, and spasmodic affections of certain muscles may be produced by irritation of other neighbouring convolutions, we may conclude what parts of the brain are employed in speech and in particular kinds of movements; when, therefore, we observe a child in a picture gallery examining its contents and all the time speaking (or thinking) aloud and pointing to the objects which please him, we seem to see how the impressions are conveyed by the eyes to the perceptive centre, and so excite to action special organs of the brain. Even if impressions on the two sides of the brain are not alike, it does not follow that injury to one hemisphere would deprive the person of his mental faculties in any more evident manner than destruction of one eye would deprive a person of sight; for just as in vision the two eyes, by means of the optic chiasm, allow a more perfect and full appreciation of the object, as, for example, its solidity, so the two hemispheres, receiving different impressions, may combine them by means of the corpus callosum to form a single but complete mental picture. I might allude to the case of hallucinations of sight or hearing by one eye or ear only, as showing that one of the perceptive centres of a double sense may be affected and not the other, the perceptive centre being in the hemisphere. I believe that late observers have discovered a marked difference in the number as well as in the form of the cells in the grey substance of the anterior and posterior portions. Both Trousseau and H. Jackson have stated their belief that there is more mental disturbance when the posterior lobes of the brain have been affected, and they have found this associated with cases of hemiplegia, where, contrary to rule, the arm has recovered before the leg. I see no *à priori* objection to the belief that the higher intellectual faculties may be more associated with the posterior and receptive parts of the brain; for there is the striking confirmatory fact of the greater development of the posterior lobes in man, so as to cover the cerebellum, and I believe their size is always regarded as a sign of advancement in the scale of cerebral development.[1]

[1] A study of physiognomy would lead one to the belief that the higher faculties are as much associated with the posterior as the anterior parts of the brain. The best observers, I believe, are the artists of 'Punch' and the comic journals. In turning over the pages of these periodicals one will observe how the picture of a man intended to be foolish-looking is made with a good expanse of forehead, but with

You will see, then, that the duality of the brain is as much a necessity as the duality of the body, or rather the two are coextensive; the brain is one as far as the body is one, and double as far as the body is double. If the limbs could have been separately governed by the brain as a whole, it would probably have formed one mass instead of being double. The counterpart of this is probably seen in the optic nerves, which are in part united and in part distinct, an arrangement expressed in the idea of the two retinæ overlapping, so that an object formed on parts of both may be but one, while the remainder of each surface has its own independent action.

There will be less difficulty in embracing this view when it is remembered that the limb is an exponent of the character of the whole animal; that the movements which the hemisphere rules over are associated with its entire instincts and habits. The foot of the lion and its mechanism implies a certain condition of teeth or stomach, and the same applies to all other creatures in the world, so that the palæontologist can, from the small bone of a limb, build up the entire framework of the animal. Indeed, much of the naturalist's classification depends on the form of the extremity, implying that its peculiarity carries with it a corresponding set of functions in the whole body. Sir Charles Bell, in his work on the 'Hand,' dwells upon the fact that throughout creation the limbs and the general organization correspond; and then, as regards the hand of man, he says, "With the possession of an instrument like the hand, there must be a great part of the organization which strictly belongs to it conceded. The hand is not a thing appended or put on, like an additional movement in a watch, but a thousand intricate relations must be established throughout the body in connection with it, such as nerves of motion and nerves of sensation; there must be an original part of the composition of the brain which shall have relation to these new parts before they can be put into activity." We need not, therefore, say with some of the ancients,

no back to his head; and, on the other hand, the configuration of an intellectual cranium is shown by a curve sweeping out behind from the neck. We have only to look at the head of a person with his faculties well developed to see a considerable projection behind, whilst in a person of low development the neck and head are in one line. Agreeing with Herbert Spencer that a strict morality and power of abstract reasoning are intimately associated, there is no reason why a high forehead should not be associated with a very low morale. Thus, in the public gallery of Basle there is a picture by Holbein, in which he marks out Judas Iscariot by the most villanous formation of head which is conceivable. This is done by making it immensely high, in sugar-loaf shape, so that there is a lofty forehead, but, at the same time the back of the head is in a perfectly straight line with the neck.

" Quia manus habuit propterea est sapientissimum," nor with others,
" Quia sapientissimum erat, propter hoc manus habuit," but rather
declare with Sir Charles Bell, " that with respect to the superiority
of man being in his mind, and not merely in the provisions of
his body, it is, no doubt, true ; but we shall find how the hand supplies all instruments, and by its correspondence with the intellect
gives him universal dominion." We might even go further than
this writer, and, entering the domain of physiognomy in its larger
sense, show how the outward form of the hand itself corresponds to
the character, as, for instance, the small hand with delicate tapering
fingers which is seen in members of families who have done no
manual work, whether they be aristocrats or gipsies, and the
large hand with big fingers and thumb of the man who possesses
great manual skill and dexterity. The word I have just made use
of—dexterity—has become an epithet for special ability in manipula-
tion, and is generally associated with much force of will and energy
of character ; it is an attribute of some of our best surgeons.

I am sorry to say that my acquaintance with Mr Herbert
Spencer's writings has been comparatively recent, but since the
great pleasure has been afforded me I have been gratified to find
that his opinions are in general accord with the views I have propounded. He shows how the sense of touch is the only perfect
sense, and into this, impressions derived from other senses must be
translated. The feeling or handling is the sense which gives us
most information, and this is the reason why, I suppose, animals
like to touch us before they make friends, or little children like to
stroke a lady's dress before they can fully appreciate its value. This is
the common sentiment of mankind, for example, if a person had an
illusion and thought he saw a ghost, he would endeavour to grasp
it to make sure, and thus Macbeth, when he asks " Is this a dagger
which I see before me ?" and fails to clutch it; says, " Art thou not,
fatal vision, sensible to feeling as to sight?" and then knows it is a
" dagger of the mind, a false creation, arising from the heat-
oppressed brain," and declares his " eyes are made the fools of the
other senses."

Mr Herbert Spencer, I think, explains on the superiority
of the sense of touch the intellectual capacity of many animals,
and has afforded an explanation in this way of the great intellectual capacity of parrots :—" If we examine in what they differ
most from their kindred, we find it to be in the development of the
tactual organs. Few birds are able to grasp and lift up an object
with the one foot while standing on the other. The parrot, however,
does this with ease. In most birds the upper mandible is scarcely
at all moveable. In the parrot it is moveable to a marked extent.

Generally, birds have the tongue undeveloped and tied down close on the lower mandible. But parrots have it large, free, and in constant employment. Above all, that which the parrot grasps, it can raise to its beak, and so can bring both mandibles and tongue to bear upon what its hand (for practically it is a hand) already touches on several sides. Obviously no other bird approaches to it in the complexity of the tactual actions it performs and the tactual impressions it receives."

The Vascular Supply of the Brain.—In considering the anatomy of the brain it is important to bear in mind its vascular supply, since obstruction of any vessel produces necessarily disturbance or disease of the part to which it is distributed. The vascular distribution corresponds in some cases to distinct physiological areas, and thus disease of a blood-vessel is often attended with very definite symptoms. The supply of blood is by means of the carotids and vertebrals, which at the base of the brain form the circle of Willis; and it is worthy of note that the brain receives a disproportionate amount of blood compared with other structures, and that the grey matter is four times as vascular as the white part. This demonstrates the great activity of the organ and its power-producing qualities. The importance of a good supply is shown in the experiments of Sir A. Cooper, where, after tying the vertebral arteries in a dog, he placed ligatures on the carotids; on tightening these the animal fell senseless, but immediately recovered when the blood was allowed again to flow. Mr Bryant relates that Mr Key once put a ligature on the carotid artery, and the man died immediately on the operating table. It was afterwards discovered that the opposite carotid had long been occluded. You yourselves must have witnessed the result of tying the carotid in the production of a hemiplegia, which is sometimes speedily recovered from and at other times ends in softening. The varied results arising from ligature have given rise to the suggestion that they depend not so much upon the immediate deprivation of blood as upon the implication of the nerves in the sheath of the vessel, whereby the smaller arterioles become paralysed, and so devitalize the tissue ; I think, however, the proofs of this are wanting. You should know, however, that the effects of ligature on or obstruction of the cerebral arteries are sometimes very remarkable, for the fact that they do not exhibit anæmia, but congestion or sanguineous stagnation. It seems as if the sudden stoppage of the blood, by exhausting the part for a time of its fluid, caused it subsequently to suck back, as it were, blood from other channels until it became choked with it. This is constantly seen where a small artery of the brain has been blocked and the part beyond it is found congested, instead of being anæmic as might have been expected. A very remarkable case occurred here some time ago, where the carotid artery was tied and death occurred three days afterwards. The brain on the ligatured side was of a deep purple colour, whilst on the healthy side it was pale and natural. This colour was due to the plugging of the whole of the arteries in that hemisphere as well as the veins, so that on opening the sinuses clots were seen projecting into them. It seemed as if the blood had passed backwards along the sinuses and so filled the smaller veins and arteries. The contrast between the dark purple soft hemisphere and the healthy white one was most striking.

It is important, also, to remember the arrangement of the vessels in connection with the nerves, since the latter might be involved by aneurismal tumours.

In any anatomical work you will read how the circle of Willis is formed. From this, small vessels are given off which penetrate to a slight depth into the substance of the base of the brain, the pons Varolii being supplied by the basilar. The larger arteries which come off from the basal supply the mass of the brain and the convolutions. The anterior cerebral passes to the anterior lobes and to a small portion of the anterior surface of the corpus striatum. The middle cerebral supplies the corpus striatum by a vessel dividing into two branches, the one going to the external and the other to the intra-ventricular portion; it also supplies the anterior and outer part of the thalamus opticus, and then passes up to supply the outer side of the middle lobe. The posterior cerebral arteries supply the posterior lobe and mainly the thalamus. The cerebellar arteries, of course, supply the cerebellum. The posterior cerebral and superior cerebellar supply the corpora quadrigemina, a fact of importance in connection with the function of the optic nerves. Most interest is attached to the middle cerebral, because it is one very likely to be diseased or affected by embolism, and in consequence, by damaging the corpus striatum, productive of hemiplegia. The vascular distribution to the corpus striatum is, indeed, three times as likely to be affected in various ways as that to the thalamus.

Minute Anatomy.—The nervous substance is composed of fibres and ganglionic cells. The latter constitute the bodies whence are supposed to originate the nerve forces, whilst the latter are looked upon as their conducting cords. The grey cells are found on the surface of the brain, in various internal parts, and in the spinal cord. They are surrounded by a sheath which is connected with the nerves. They are of various shapes and sizes, being unipolar, bipolar, and multipolar. They are more numerous in the anterior than posterior parts of the brain, also larger and of a pyramidal shape ; in the posterior smaller and rounder. The nerve-fibres are composed of an axis-cylinder, which is looked upon as the conducting cord ; this is surrounded by a peculiar matter, called myeline, or the white substance of Schwann, and around this again is a sheath of fibres or neurilemma. In a large nerve many of these fibres are bound together by cellular tissue. They vary in different parts, and in the character of the investing sheath. The axis-cylinder is well shown by adding carmine to stain it. Uniting the cells and fibres together is the neuroglia or nerve-glue, of whose exact nature there is still a diversity of opinion. It is of interest to us clinically as being probably the seat of most of the inflammatory processes which go on in the brain, especially those of a chronic nature. It is a fibro-granular material, containing nuclei.

Morbid Anatomy.—The cerebro-spinal centres, together with the nerves, are all subject to various morbid conditions ; of these I shall have to speak separately, but they may be spoken of as a whole as follows :—There are the accidental affections of the brain arising from injury, and those due to rupture of blood-vessels, where the organ may receive a severe and fatal laceration. The diseases beginning within, of an inflammatory kind, are acute cerebritis, leading to softening, and chronic cerebritis, tending to the same result and sometimes to the formation of new products of connective tissue, causing hardening or sclerosis. There is also atrophy of nerve structure and the production of amyloid bodies. Then, there are special changes in the brain-cells, or vesicles, whereby they become altered in form, lose their connecting processes, and filled with a dark pigment, or even become chalky. Of late much has been said of the morbid changes in the brain in connection with the perivascular canals or spaces around the blood-vessels from their supposed identity with the lymphatics ; herein is the seat of tubercle and other inflammatory products. Some pathologists, however, attach little importance to the changes found in connection with the vessels, since it seems that long-continued congestions, arising from any cause, will tend to the congregation of leucocytes, with a wasting of the coats of the vessel and the tissue around, giving rise to the formation of the vacuoles which have attracted so much attention in various forms of disease, and thought to have a special pathological origin.

It is also very important to note the secondary changes which take place in the brain, spinal marrow, and nerves, for a knowledge of their occurrence may enable us to explain many otherwise obscure cases of disease. The fact to note is that the nutrition of nerves does not depend so much upon a vascular supply contiguous to them as upon the integrity of the root of the nerves, or some distant centre whence they spring ; so that a nerve, when cut, degenerates along its whole course. Close to the spine the nutritive centre of the sensory root appears to be the ganglion at its origin, whilst that of the motor is the grey centre in the cord where it arises. These facts have been taken advantage of by Waller in ascertaining the distribution of nerves. He found that after dividing a nerve it gradually became changed into a fibrous cord ; at first he found a mere granular fatty change, with the axis-cylinder remaining, but subsequently the whole nerve decayed, which he could trace through the muscle to which it was distributed. Bouchard found a somewhat similar change take place in the fibres of the brain and spinal cord, for example after an apoplectic seizure due to effusion of blood in the corpus striatum he observed a softening process which extended through the crus cerebri and anterior columns of the spinal cord as far as the dorsal region. Not only in a case of this kind, but in various other forms of lesion, inflammatory or degenerative changes may extend to a great distance, taking the course of the anatomical and physiological arrangement of the nerve structures. You must often have seen this occur in a rough manner in cases of injury to the cord from fracture of the spine, and where an evident softening process has extended a long distance downwards in that particular tract which had been injured. Bouchard made use of these facts for ascertaining the course of the fibres in the cord, for by injuring a certain portion of a given tract he could watch the direction in which degeneration travelled. He showed also an upward degeneration extending from the cord to the restiform bodies and cerebellum, and he believed by this method of research he had traced fibres passing directly to the brain without the intervention of the grey matter of the cord, and also, as shown by anatomical dissection, that some fibres of the crus pass down on the same side of the cord. All these facts show how degeneration, or some other subtle change, may rapidly take place in the spinal cord, and how disease may be propagated from the centres to the nerves, and from the latter to the cerebro-spinal centres. They also show that nerve-fibres are something more than mere conductors, that there is a principle of activity pervading them as a whole, and which may come and go under varying conditions.

Differential Diagnosis.—Having seen that true paralysis is associated with disease of the spinal cord, we may ask, what are the symptoms which would be connected with disease of the brain proper ? This organ being intimately associated with perception, with the will as influencing the cord and other mental operations, we

hemispheres are impaired. If seriously diseased, either by organic changes in its substance or by poisoned blood circulating through the hemispheres, all perception and voluntary efforts would be gone, and in course of time there would follow dementia. If affected by acute inflammation, as in meningitis, there would be delirium ; also, since irritation of the surface produces movements either directly or through the motor ganglia below it, there might be also convulsions. I think, therefore, the symptoms associated with affections of the cineritious substance may be said to follow a pretty general rule. As regards the other portions of the brain, it is remarkable how few symptoms are associated with diseases of the medullary matter, whether these be effusions of blood, abscesses, or tumours. Probably if a large number of the radiating fibres were cut through, some kind of paralytic symptoms would result, but as this does not occur none of these diseases which I have named may make themselves manifest, or if they do become known it is only by their encroaching upon parts having more defined functions, as the cerebral ganglia or the cranial nerves at the base.

As regards the cerebellum, the symptoms dependent on its diseases are not striking, although when these have been long existent we can generally infer their seat pretty accurately. In course of time there is a tottering in the gait when the patient attempts to stand or walk, which shows the organ is associated with the process of co-ordination. In these cases, too, blindness is very often noticed, although, of course, this is not peculiar to disease of the cerebellum, and it is remarkable that nystagmus is often present. This is interesting in connection with the experiments of Ferrier who found, on irritating the cerebellum, movements of the ocular muscles (as if through the third nerve) were produced. Dr H. Jackson states that rigidity of the limbs is often a symptom of cerebellar disease.

A very important question to solve is whether there is pain in disease of the cerebro-spinal centres. Of this there can be no doubt ; that most extensive disease may exist within them without the occurrence of any pain whatever, and it also seems to be clear that the instances where pain is the chief symptom the disease has involved a nerve after its exit from the centre. In the case of the head, we know that where the greatest pain exists the membranes rather than the substance are involved, the dura mater being supplied with sensitive nerves. The cases which most contradict this statement are those of cerebral tumours, where, no doubt, intense pain often exists, but at the same time it may be remembered that this comes on in paroxyms, as if the tumour merely induced it by setting up irritation elsewhere, and as a rule the pain is not situated in the region of the growth. It has been thought by some

that a part of the brain only is sensible, and this view is supported by showing that sensation is affected when disease
occurs in the neighbourhood of the thalamus and summit of the
sensory tract.   Dr. Copland attached much importance to an old
observation in the diagnosis of cerebral diseases—that patients involuntarily or instinctively place one hand over the genital organs.

Now, it is only right to inform you that the ordinary approved
opinions of the functions of the brain being located in distinct
regions, and that special forms of paralysis are connected, therefore, with defined lesions is discarded by so high an authority as
Brown-Séquard.   In some lectures lately delivered by him he has
attempted to show that our facts are not clearly based on such principles, as I have endeavoured to prove to you to be true, and that no
definite lesions have yet been found in correspondence with special
symptoms.   He denies that the brain can be played upon as we do
on the keys of a piano, but that its functions are scattered through
its substance as a whole.   He says, for example, that a small effusion
of blood in a well-defined spot may be attended by convulsions,
altered pulse, change in pupils, sickness, and a number of other
symptoms which cannot be connected with derangement of that
small spot only.   These symptoms must imply that an irritation has
been transmitted to far distant nerve-cells, and so an irritation of
one spot may affect a large part of the healthy structure to the production of an over-activity or paralysis.   "The latter appears in
brain disease, not because conductors or centres employed in voluntary movement are destroyed and lose their function, but because an
irritation starting from the diseased part or its neighbourhood goes
to a more or less large numbers of brain-cells, scattered in many
places in the nervous centres, and stops their activity."   In this way
he explains paralysis as arising, not so much from the lesion as from
the irritation set up in the healthy portion of the brain.   Herein he
finds an explanation for the numerous cases which he quotes of
paralysis occurring on the same side as the disease of the centre.
His arguments are not sufficiently convincing to allow us to discard
our old opinions as to the localisation of the functions of the brain,
although it must be admitted, and this has generally been done, that
an irritation of neighbouring parts must have its share in the production of the manifold symptoms which occur in very specialised
lesions.   In the case he gives of small effusion of blood the first
symptoms are due to a disturbance of the brain as a whole, but in a
short time the phenomena become very defined.   Brown-Séquard
even seems to deny the frequency of crossed paralysis, a fact taught
ever since the days of Aretæus.   This author says the body is
divided into two equal parts, and what affects one side does not the

ether. " If the cause of the paralysis should be below the head as in the spinal marrow, the parts below are paralysed, but above this the nerves are not in a direct line, but are inverted, passing to the other side like the letter X, and in this way one side or the other may be paralysed." This has been the uniform experience of all observers during the many centuries since this was written.

PARALYSIS OF ONE SIDE AND APHASIA

Paralysis of Motion and Sensation of one half of the Body.—A perfect hemiplegia, according to our present views of the physiology of the nervous system, can imply nothing less than a disease or temporary impairment of the complete half of the spinal cord, or of all the centres which rule over the movements of half of the body; or if not this, it must imply disease of some one centre which governs all those minor ones which do possess the motor power. In the same way a loss of sensation of half of the body would indicate disease of the whole of the sensory tract or of that region into which all the sensory nerves pass.

Now, as a matter of observation, I am unacquainted with any organic disease which can produce either form of paralysis, and as regards that of motion there is really no such affection as a perfect hemiplegia, either of the organic or functional form. The so-called hemiplegia is of various kinds, amongst which a functional hemiplegia as seen in hysteria is quite different from that arising from disease of the cord or the central ganglia. Also, a true and perfect hemianæsthesia certainly does exist, but, as far as I have seen, is nearly always associated with that form of hemiplegia which is styled hysterical. Loss of sensation on one side from an organic cause has never, in my experience, been complete.

Hemiplegia from spinal disease or injury would arise from lesion of one side of the spinal cord, all the parts below being paralysed. For example, a man had a complete severance of the restiform body of the medulla. There was a partial paralysis of motion of the arm and leg, but sensation was unimpaired. The origin of the glosso-pharyngeal and pneumogastric nerves were involved, causing complete inability to swallow. He also had constant hiccough, and died a few hours after the accident.

Hemiplegia from inflammation of one hemisphere of the brain.— This is a cause of paralysis which I was at one time reluctant to admit, but have not hesitated to do so after Mr Hutchinson's

observations on the subject. The facts I knew, but the difficulty of
the explanation arose in consequence of the inflammation or unilateral arachnitis having of necessity its origin in an injury. In
idiopathic meningitis both sides of the brain are affected, but
where one side alone is involved some injury from without must
have been inflicted to cause it, and thence the possibility of some
other lesion of the brain being present, which might have given rise
to the symptoms. Indeed, cases have been reported where such
injuries have been quite sufficient to account for the symptoms
without having recourse to the effects of the supervening inflammation. I now feel sure, however, after removing all these
possible causes that an arachnitis is quite capable of producing a
paralysis. In some cases sensation is impaired, and pain also may
be a possible symptom. Convulsive twitchings are well known as a
result of irritation of the surface of the brain. This form of
paralysis excites additional interest at the present time, when we
are beginning to regard the cerebral convolutions not as merely
having a governing power over the ganglia below them, but as
really being a further and higher development of these summits of
the spinal system, and consequently closely connected with the motor
apparatus. Although I cannot say from my own observation that a
loss of a portion of a convolution will produce a localised paralysis,
yet in our present physiological knowledge it would not seem impossible, seeing that a destruction of the whole hemisphere can produce what is equivalent to a complete hemiplegic condition. We
believe we see this in the cases of unilateral arachnitis, of which we
are now speaking and where the function of the convolutions is
for the time destroyed. The experiments of Ferrier, by destruction of portions of the cortex of the brain, gave a variety of results ;
in some animals the whole of the hemispheres could be removed
without any loss of motility, whilst in the higher animals, as
monkeys, distinct paralysis followed the injury of particular convolutions. The author believed this difference to be due to the
degree of development of the animal, for whilst in the very lowest
of animals, movements might be automatic ; in the higher classes,
owing to a more complete education, they would be more intimately associated with the cerebral hemispheres. The French
physicians are by the light of these experiments beginning to make
more accurate and closer observations with respect to the effects of
lesions of the cortex cerebri, and if it be true that the motor region
is limited to the antero-lateral portion of the brain, it is evident
that lesions of this part and those of the posterior lobes would of
necessity produce very different effects. It is firmly believed by some
that injuries to the motor region will produce corresponding para-

lytic symptoms. I myself could quote cases which might appear to corroborate this, but I am reluctant to make use of any of an equivocal nature; nevertheless, I believe the experience of surgeons could show that persons having received an injury to the head followed by paralysis have recovered after the removal of portions of fractured bone, where the surface only of the brain could have been affected. I will briefly relate three cases of paralysis accompanying an arachnitis which occurred to me before I was in a position to recognise the true relationship between the symptoms and the disease. They are already published in the 'Guy's Hospital Reports,' as well as the case which follows them. The latter is one of interest as bearing upon the possibility of a disease in the convolutions causing a paralysis.

Injury to Head; Unilateral Arachnitis; Partial Hemiplegia

CASE.—A boy had a blow on the head from a stone, causing a scalp wound. He had at first no symptoms, but subsequently he became very ill, with great febrile and cerebral disturbance. He gradually fell into a semi-comatose state, and was found to be partially paralysed on the right side. He was trephined, but only a few drops of pus escaped. The post-mortem examination showed that there was no fracture, but that the bone at the seat of the injury contained pus in the diploë. The left side of the brain was covered with purulent lymph, some of which was free on the surface, while some occupied the subarachnoid space; the cineritious substance was also infiltrated with inflammatory products of a pale yellowish colour.

Injury to Head; Unilateral Arachnitis; Hemiplegia

CASE.—A man fell from a loft several feet high on to a piece of wood, which produced a compound comminuted fracture on the right side, a piece of the bone being driven in. He at first suffered from concussion, but soon recovered; he remained in a doubtful state for a day or two, when arachnitis came on, and he died a week after the accident. Before death the right side was observed to be paralysed. The post-mortem showed a fracture, beneath which the brain was reduced to a pulp; and on the opposite side there was an arachnitis, the whole surface being covered with lymph.

That disease of the cineritious surface of the brain is adequate to produce paralysis is seen in the following case, which I published several years ago.

Disease of the cineritious surface of the Brain; Hemiplegia

CASE.—A woman, æt. 36, struck her head against a beam of the ceiling several months before death; she subsequently became ill, and had two or three attacks of vomiting. About three months before her death she became very ill and took to her bed; she had febrile symptoms and sickness, and was evidently suffering from cerebral disturbance. After the urgent symptoms had passed off she lay quiet in bed, and appeared quite sensible of what was said to her. She gradually became weaker in her limbs, and especially in those of the left side, until in three weeks time her left side was quite paralysed. After this she lay quiet, apparently sensible and able though with difficulty to answer questions and to put out her tongue. On being asked a question she would wait and deliberate, and then slowly answer. The most remarkable symptom was the fact that she never seemed aware that she had lost the power of the left side. On being asked about her arm she always declared that she could move it, although it lay helpless by her side. She gradually died. The post-mortem examination showed nearly the whole surface of the right hemisphere to be diseased; it was in a very peculiar condition, as if the whole of the grey substance was undergoing disintegration. This was not so apparent on the surface as on taking off a thin slice, when the cortical part of the brain was seen to be of a yellowish colour, soft and as if worm-eaten. The disease in some parts affected the deeper layers more than the external. The disintegration reached the medullary matter, but did not appear to penetrate it. This morbid condition extended all over the right hemisphere, with the exception of the base. The left hemisphere was quite healthy.

Dr Moxon had a case, in the clinical ward, of fatal erysipelas, where it was observed before the man's death that he was hemiplegic on the left side, and apparently had lost feeling on the same side of the face. There was no convulsion. After death a unilateral meningitis was found on the right side, lymph covering the whole of the right hemisphere. The reason of its occurrence was not made out.

Hemiplegia from effusion of Blood external to Dura Mater.—I believe some good authorities doubt the possibility of a paralysis occurring under these circumstances, but I cannot myself hesitate to admit the fact of a partial hemiplegia arising in cases where there has been an extravasation of blood on the surface. Whether this may be attributable to simple pressure transmitted to the central ganglia, or due to a direct impairment of the convolutions, may be a question. The cases I especially allude to are those where the middle meningeal artery is lacerated from injury, and blood is effused external to the dura mater, forming a circumscribed clot, which is found after death to have produced a deep depression on the brain. The only cases of the kind from which any inference could be drawn with respect to the symptoms resulting from this cause would be those where the brain was not injured, and where, indeed, no severe concussion had occurred. They would be those where, as a result of the hæmorrhage, coma had slowly come on ; and therefore, if in these cases anything like hemiplegia were observed, it could scarcely be attributable to any other cause than the pressure of the blood. I might say that the paralysis would never be anything more than partial, and it may be further noticed that the blood would always be situated over the motor region.

Ruptured Meningeal Artery ; Paralysis of Motion and Sensation

CASE.—A man, æt. 30, had a fracture of the left parietal bone, caused by a poker. This led to the effusion of a large clot of blood on the dura mater compressing the brain. When admitted he was almost in an unconscious state, with contraction of the pupils (particularly the right), and was very restless. Two days afterwards, although he had not spoken, he seemed to have some degree of consciousness, and it was found that he had loss of motion and sensation in the right arm and leg, and particularly in the former. Two days afterwards he had somewhat recovered from his lethargic state, and appeared to feel when he was touched. After this he again became comatose, and gradually sank on the eighth day.

Ruptured Meningeal Artery ; Hemiplegia ; Trephining ; Recovery

CASE.—Some years ago a man was admitted under Mr Cock's care in an insensible condition, having fallen from a height. He gradually recovered from the concussion, and remained sensible for some time ; but in the night he was found in a deep coma, with stertorous breathing and with insensible contracted pupils. The man seemed on the point of death, when Mr Cock determined to trephine, being guided in the choice of locality by the fact that the left arm and leg freely moved when they were pinched, whereas not the slightest motion could be excited in either of the limbs of the right side. A large clot of blood was removed ; the stertor almost immediately ceased, and on the following day the man could move his right arm and leg freely. He shortly after resumed his work, and remained well for thirteen years, except that he had some fits towards the close of his life. He at last died of apoplexy.

Ruptured Meningeal Artery ; Effusion of Blood ; Paralysis ; no loss of Consciousness

CASE.—A man was taken out of an area apparently in a state of intoxication and was put into the police station ; but not having thoroughly recovered himself he was on the following morning brought to the hospital. No fracture could be detected, but there was bleeding from the ear. The left arm appeared very weak and almost helpless. This partial hemiplegia remained, although he recovered his consciousness. He died ten days afterwards. A fracture of the base was found, and external to the dura mater there lay a large clot of blood which had indented the brain into a deep hollow. Outside the clot there was some fluid blood, which appeared to have been more recently effused, and to have been the more immediate cause of death.

Ruptured Meningeal Artery ; Effusion of Blood ; Paralysis ; Coma

CASE.—A man fell from a plank, striking his head upon a heap of bricks. He was brought in with incomplete paralysis of the right side. He was not uncon.scious, but was in a lethargic state, and when spoken to was very irritable. He passed his evacuations involuntarily. There was found a fracture of the skull on the left side, and a large clot of blood lay external to the dura mater compressing the brain.

Hysterical Hemiplegia.—We generally mean by this expression that form of paralysis of one side which depends upon a temporary absence of voluntary power whereby the patient cannot will to move one side of the body. It is evident that this form cannot resemble the ordinary variety of hemiplegia which depends upon an organic lesion at the summit of the motor tract, for in the latter there is an absolute falling of the muscles on one side of the face, with thickness of speech, and, at the same time, the patient's will is good to open the mouth, the eyes, and other parts which are not affected by the paralysis. In hysterical hemiplegia there is no falling of the face, as no want of will could possibly produce such a condition, and the active drawing up of the face on the other side would be but a poor imitation of palsy of the opposite seventh nerve. There is, however, very often an inability to open the eye, or rather the eyelid falls; at the same time there may be an inability to open the mouth, thrust out the tongue, or swallow; all of which symptoms are clearly due to a want of voluntary power. As regards the helpless limbs, it will be found by raising them, and by a little manœuvring, that they are not absolutely dead and powerless, as in a truly paralysed part, but some resistance is made, especially when the arm is placed in an awkward position. Very often, too, in such cases, there is a want of forcing power over the rectum and bladder. It will thus be clearly seen that in hysterical paralysis those parts are affected over which the will more especially exerts its influence, and if the patient is said to be paralysed on one side the distinction between her complaint and the commoner form of hemiplegia is obvious.

CASE.—Mary B—, æt. 30, was sent in as a sufferer from paralysis of one side of her body. She was a governess, and said that two years ago her left hand became numb, afterwards weak, and at the end of three months her whole arm was helpless. Subsequently the left leg became affected in a similar way, and after this the left eyelid drooped so that she was unable to raise it. She said also she had a fit, and after this was unable to open her mouth, and lived in consequence on liquid food. On admission she was found unable to move her left arm and leg; the arm when raised dropped lifeless at her side. She had ptosis of the left eye, but the eye was natural and movable. She was unable to open her mouth beyond half an inch. The tongue protruded straight. It was evident that these symptoms did not correspond to any which would arise from a known special lesion, and therefore the case was regarded as one of "ideal" paralysis. She was cured in five weeks by moral treatment. She was told to use an effort to open the eye and move the limbs, and that an improvement was expected daily. The cure was thus speedily effected.

Hemiplegia with Spasmodic Contraction

CASE.—J. B—, a young woman, subject to various hysterical ailments, began to have weakness of her left leg sixteen months before her admission and soon afterwards of her arm also. The limbs then gradually contracted until they reached the condition as seen on admission. The arm was flexed tightly across the chest and the fingers clenched. The leg was drawn up to the abdomen and not stretched out straight as usually seen in this class of case; when the limbs were forcibly straightened great pain was experienced, and they immediately returned to their original flexed position. There was no wasting, good reflex action with hyperæsthesia over the left side of the body as well as tenderness over the left inguinal region. There was always a bad odour emanating from the arm, the clerk said, like old boots. She subsequently complained of dimness of vision of one eye. She was ordered galvanism and faradisation, and told to move her limbs as much as she could. She was soon able to get up and walk about, and left well in four months.

Ordinary Hemiplegia.—The commonest form of hemiplegia is that which arises from disease of the corpus striatum, due generally to sanguineous effusion or softening. It is characterised by paralysis of the arm and leg, of the face in part and of the tongue ; that is, the lower part of the seventh nerve is affected and the ninth. The trunk and other parts of the body are not involved in the paralysis. These facts have opened up a subject of great interest both to writers on physiology and pathology. For since it is seen that in hemiplegia the will can still operate efficiently on all other parts but the limbs and a small portion of the face, it might be conjectured either that there was some special channel of communication between the nerves which supply these other parts and the cerebrum proper ; or that, the centres of these nerves lying, not apart, as those which rule over the extremities, but in juxtaposition, can still be stimulated by their neighbours. Be this as it may, we shall find that if special nerves, or any other than those which supply the extremities, are paralysed in hemiplegia, we shall be able to conclude that disease exists at their roots or at the very centre whence they spring, and thus we shall have a means whereby we can detect the exact seat of the mischief.

Owing to the intermixture of fibres, we find that a small spot of disease in either of the central motor ganglia will produce paralysis of the limbs on the other side ; a slight lesion, partial paralysis ; and a severe lesion, complete paralysis, and not, as might have been expected, had there been no blending of fibres in these ganglia or in the plexuses of nerves, a paralysis of particular muscles, according to the exact site of the lesion. When, then, disease of the corpus striatum exists, we have hemiplegia ; any disease will produce it, as it is a necessary result of the lesion. You must not, therefore, say, as I heard one of you the other day conclude, that a patient necessarily had apoplexy because he was seized with hemiplegia. The fact of hemiplegia occurring so often in brain disease is owing to these central ganglia being so frequently attacked, and this arises, I

apprehend, from their great vascularity ; consequently, their vessels
are liable to rupture or to become aneurismal, or, being diseased, to
lead to softening or to be plugged by an embolus. It is for the very
reason that lesions of particular portions of the brain are productive
of certain definite symptoms that I am pursuing this method of
passing in review the consequences of morbid change, in whatever
way produced.   It is quite impossible to describe the symptoms of
apoplexy, abscess, or softening, since the symptoms would be as
numerous as the sites of the disease might be different. I want you
strictly to bear this in mind, and for the present to remember that
paralysis of the limbs means lesion in the motor tract, since no
paralysis succeeds to injuries to the brain proper, unless the term
paralysis is used in the larger sense of loss of function, whether this
be bodily or mental. The simple powerlessness, however, of coma
or drunkenness is not what is generally intended by the term
paralysis.   What I am saying may appear self-evident to some of
you, but yet it is sometimes forgotten, for I have been present in a
court of law and heard it declared that the plaintiff could not have
injured his brain because he had no paralysis, implying an affection of his arm or leg, whilst all the while the medical man had
overlooked the failure of his mental activity, which would have been
the true indication of a cerebral lesion. Thus it has happened that
a man with a considerably impaired state of cerebrum has received
no pity because he could walk or use his arms, whilst another man
has been commiserated as a cripple, although still able to conduct
his business and provide comforts for his family. It is of no use,
therefore, learning in this room the physiology of the nervous
system unless you carry that knowledge into practical life.   With
the exception of the possibility of arachnitis of one hemisphere or
lesions of particular spots on the cortex producing a condition
resembling paralysis, you must understand that hemiplegia necessarily implies disease of some portion of the motor tract.

In hemiplegia arising from disease in the central ganglion you
will find those parts more especially affected which are under voluntary control, and thus the respiratory process is not interfered
with. The paralysis, in fact, affects only or mainly the arm, leg, and
face. The chest is seen to expand as before, and the abdominal
muscles move equally on both sides, the diaphragm acts equally, and
the patient appears to be able to turn the head. As regards the
chest there can be no doubt that it is not usually affected in hemiplegia, but I have seen a hemiplegic patient who, when requested to
take a deep breath, was unable to move one side of the chest as
well as the other.   And many years ago Dr Walshe, who was quite
cognisant of the opinion generally held as to the non-implication

of the chest, took the trouble to accurately determine this question
of relative expansion by measurement, and concluded that the side
of the chest corresponding with the hemiplegia did not expand
equally with the other.   I cannot but think that in these cases the
greater movement on the unaffected side was attributable to the
action of the auxiliary muscles which really belong to the upper
extremity.   It may, however, be taken as pretty certain that those
parts of the body which act together as a whole, and are ruled over
by nerve centres which are blended into one mass, are not affected in
lesions of the central ganglia; for I believe it is for the maintenance
of this independence of the extremities that the upper part of the
cord is bifurcated in the manner we see. The parts, then, which are
affected in hemiplegia are the arm, leg, face, and tongue. If a
patient be suddenly seized as from effusion of blood in the central
ganglion, he is speechless, and the face is seen to be fallen on the
paralysed side.   When the shock has passed it is observed that the
paralysis of the face is but partial, whilst the tongue is thrust out
towards the weak side.   The mouth drops, but not the eye, the corrugator and orbicularis being responsive to the will, as before. It was
this peculiarity, the fact of the seventh nerve being not completely paralysed, which led the late Dr Todd to seek elsewhere for
an explanation for the falling of the face.   He thought, as it was
generally taught, that the buccinator muscle was supplied with
motive power by the fifth as well as by the seventh nerve, and that
in a paralysis of this branch of the fifth nerve might be found the
source of the symptom. Such an explanation, however, did not at all
simplify the question, for it was as easy to believe in a partial
paralysis of the seventh nerve as a paralysis of a particular branch
of the fifth.   Of late, however, the matter has been set at rest
by the statement of anatomists, that the long buccal nerve is really
a nerve of sensation ; that it reaches the muscle only to penetrate
it ; and is then distributed on the skin.   Pathological facts also
have shown that, when the fifth nerve has been paralysed from a
growth pressing on its roots, although all the muscles of mastication
are paralysed, the buccinator has escaped.   Quite recently, also, Mr
Turner, the Professor at Edinburgh, has dissected a subject where
the long buccal nerve came off from the second division of the fifth,
this root, as you know, being altogether sensory. You must therefore understand that it is the facial nerve which is partly paralysed
in hemiplegia.   At the same time the tongue is involved, shown by
its being thrust out towards the weakened side ; it was at one time
said that the tongue might diverge in either direction, but this
certainly is not correct, its direction being always towards the paralysed side. If the organ is voluntarily thrust out of the mouth,

it can only move in a given course, when one side is in action and the other inert.

There is an additional symptom sometimes met with in the so called "lateral deviation of the eyes."[1] In a certain number of cases, and more especially when the paralysis is on the left side, the head is forcibly turned away from the paralysed side, and the eyes are also turned in the same direction and a little upwards. That is, if the patient be lying in a half conscious state with paralysis of the left side of his body his head will be found turned in the opposite direction, and his eyes also turned to the right. If the head be placed straight it will again resume its former position ; this, however, would not occur if the patient were asleep. In my own experience the pupils have also been of moderate size and insensible to light. This deviation of head and eyes is explained by Ferrier to be merely an additional symptom of the paralysis, for the healthy motor centre being no longer counterbalanced by the other, the lateral movement necessarily occurs ; it is an exceptional occurrence, but important as showing the greater severity of the lesion. What you have to remember is that in hemiplegia the parts paralysed are the arm, the leg, the seventh and ninth nerves. The parts involved are for the most part those over which we have voluntary control, and thus it is that the face, tongue and limbs are more especially affected, whilst the body is left free. As before said, the trunk is made up of two united halves, and moves as a whole, whilst the limbs act independently, due no doubt to the separation of the motor tracts at their upper part.

Owing to the paralysis of the mouth there is usually some defect in utterance, and this occurs equally whether the paralysis is on the right or left side ; but in right hemiplegia there is the very frequent additional symptom of aphasia, a condition in which language or the use of words is lost. This seems to be due to an implication of the third frontal or Broca's convolution, an implication likely to occur in any disease which attacks the corpus striatum, in consequence of the parts being supplied by the same blood-vessel. When, a few years ago, the fact was first promulgated I almost at once recognised its truth by referring to old reports of cases, where I found that it was in cases of right hemiplegia only that the speech was lost. One of these cases was that of the celebrated chemist Dalton, who was struck with paralysis of the right side, together with loss of speech.

[1] As long as I remember anything of nerve disease this was pointed out by Sir W. Gull. See his Gulstonian lectures, where he gives cases of lateral deviation of eyes, and in all "the eyes were turned from the paralysed side."—*Med. Times,* 1849.

Hemiplegia arising from Lesions lower down the Motor Tract, and implicating special Cranial Nerves,—If disease occurs lower down in the motor tract, we still have hemiplegia, but it is of a different variety, as it is associated with paralysis of special nerves, according to its seat; if, therefore, in conjunction with hemiplegia, we find special nerves are paralysed, we are sure that the grey centres of these nerves are involved in the disease, and we fix upon the seat of the lesion. Thus, in a very interesting case related by Dr Weber to the Royal Medical and Chirurgical Society, a man was seized with an apoplectic fit, and found to be hemiplegic, with paralysis of the third nerve on the opposite side—that is, he had paralysis of the right arm and leg, with ptosis and dilated pupil of the left side of the face. Dr Weber concluded that there was an effusion of blood in the left crus cerebri, which turned out to be absolutely correct.

If disease occurs lower down in the pons Varolii, we might expect that other nerve-roots would be implicated, as those of the fifth, sixth, and seventh; and this is the case. As we are speaking of the motor tract, I will first allude to the case where the seventh or facial nerve is paralysed. I have already told you that in the commonest form of hemiplegia—that arising from disease in the ganglionic centres within the brain—the facial nerve is only slightly affected, so that if you meet with a case where the face is completely paralysed on one side you may know that the lesion is not in the ganglia above named, but in the pons Varolii. This I have verified over and over again, and on several occasions seen a correct diagnosis made as to the exact seat of the hæmorrhage. In this very model before you, where you observe the pons Varolii cut through and an apoplectic clot in its midst, the true site of the lesion was anticipated long before the death of the patient. In these cases the paralysis is sometimes on one side of the face and sometimes on the other, according to the position of the clot, the explanation which is offered by Brown-Séquard being probably correct—that the fibres of the seventh nerve cross in the pons. It would therefore happen that if disease occurred on one side implicating the motor tract, and at the same time the origin of the seventh nerve, we should witness the case of hemiplegia accompanied by palsy of the same side of the face. If, however, the disease occurred somewhat lower down in the pons, the centre of the seventh on that side would escape, but the fibres crossing over from the opposite nerve would be involved; we should then witness a case of paralysis of the arm and leg on one side and paralysis of the face on the other. I had an example of this under my care not long ago, in which the man almost entirely recovered.

If effusion of blood takes place in the middle of the pons, we have before us one of the most difficult cases for diagnosis, since from both motor tracts being involved, and a general paralysis necessarily resulting, it is far from easy to distinguish this state from one of coma, or from one simply where the voluntary power is for a time in abeyance, as in stupor. You have, for example, a patient lying insensible in bed, and you lift the arms or legs, and they fall lifeless at the side. Does this result from paralysis or coma ? I have seen a most experienced physician mistake a complete case of paralysis for poisoning by opium, and another physician a similar case for uræmic coma ; and, on the contrary, I have seen a most careful practitioner regard a case of dead drunkenness as one of sanguineous apoplexy and paralysis. The diagnosis is most difficult. In the case of effusion of blood in the pons Varolii, where the patient is insensible and wholly paralysed, the condition is very like that caused by opium poisoning. The resemblance is the more striking when, as is often the case in effusion of blood at the base of the brain, the pupils are minutely contracted ; should, again, the respiration be lowered in number or laboured, the similitude would be exact. Thus it has often happened that patients who have been taken to hospital with apoplexy of the pons Varolii have had the pump employed for the purpose of emptying the stomach of supposed laudanum. I should state that the coma is due to the sudden nature of the attack, or to the large amount of effusion of blood, by which pressure is exerted on surrounding parts, but that unconsciousness is no necessary symptom of apoplexy of the pons ; and in cases of chronic softening of this part the intellect is in no way impaired. Not long ago a medical man was most unjustly censured by the magistrate because he had raised a suspicion of poisoning by opium in a case of this variety of apoplexy, and I believe there is scarcely an hospital in London in which a similar experience of the difficulty of diagnosis could not be given. Some of you may remember the case of a woman who was brought into the hospital, and when I visited her in bed found her in a state of quiet stupor, and her hands dropping helplessly when raised from her side. When the urine was examined it was found to be highly albuminous, and I then stated my belief that it was a case of uræmic coma. We were shortly, however, informed that she had been suddenly seized whilst in an omnibus, and then, of course, I changed my opinion to one of sanguineous effusion ; but without this history I should have formed an erroneous judgment. Here there was a clot of blood in the pons.

If the portio dura is affected, you may ask, Why not the portio mollis ? It is probable that this is much oftener the case than is supposed, owing to the impossibility of discovering deafness in many sudden affections of the nerve centres, but in cases of chronic disease both roots of the seventh pair have been observed to be affected at the same time.

Cases, however, occur where the diagnosis is difficult, as in the following, where it was at first thought that the lesion was in the pons, but afterwards another view was taken of it.

Paralysis of the right Arm and Leg; of the seventh Nerve, with loss of Taste; no Aphasia; Galvanism

CASE.—William W—, æt. 49, a clerk, has had gout which he strongly inherited; not very hale man but temperate and regular in his habits. A week before admission he noticed his speech was thick and his right eye felt strange; soon afterwards also that he could not move the right side of his face as well as the left, and that he was deaf in the right ear; these symptoms became aggravated and at the same time weakness of the right arm and leg appeared.

On admission the patient was seen to have complete paralysis of the right side of the face, he could not move the right side of the occipito-frontalis nor close the eye. When asked to do so the eyeball turned upwards and slightly outwards. The cheek was motionless and he could not move the angle of the mouth which was drawn up on the other side. The tongue and soft palate were unaffected; on galvanizing the affected side the muscular contraction was much less than on the sound side, whilst a weak faradization affected readily the right side and more powerfully than the sound one. There was also complete deafness on the right side, so that he could not hear a watch when it touched the skin. No paralysis of ocular nerves; no loss of sensibility. The right arm was weak but he could move it; his right leg was weak and he dragged it as he walked. He had no aphasia; continued pain in the forehead, otherwise sound.

This man remained in the hospital a long time and somewhat improved, his principal complaint being headache. He was tested several times with respect to the sense of taste; his tongue was found to be perfectly sensitive to touch, but he had lost the sense of taste on the right side towards the tip; thus anything salt placed on the right side was unperceived, whilst it was immediately appreciated on the left side. The sense of smell on that side also seemed somewhat impaired. At the end of three months he walked about without any very perceptible weakness of the leg, and the arm also regained most of its strength. The paralysis of the face was considerably less, as he could close his eyelashes. The deafness remained and the left ear appeared also somewhat affected. Taste still lost on right side of tongue. He remained in the hospital some time longer, with a still further improvement which included the sense of taste. The headache, however, was more or less constant, and constituted his chief complaint. No remedy cured it, but he always expressed himself benefited after any new medicine, as iodide and bromide of potassium; guarana and gelseminum. He said he was much relieved for a time by the guarana.

A case like this is worth a very careful study, both in reference to the diagnostic difference between it and ordinary hemiplegia, as well as for the positive points of interest contained in it. It must have been a very superficial observation which suggested, in the first place, that it was a simple hemiplegia, since there was loss of power in the arm and leg, and a complete paralysis of the face. You will remember that there is only a partial palsy of the seventh nerve, in the common form of hemiplegia, but here it was complete—a fact at once suggesting that the disease was not in the corpus striatum; nor was there any deviation of the tongue, nor was there aphasia. The case was one of hemiplegia, but not of the ordinary kind; and the only question therefore asked was, did the lesion involve the motor tract in the pons, near the centre of the seventh nerve, by which both portio dura and portio mollis would be involved, or was there a mere association of hemiplegia, with a disease of the trunk of the nerve in the temporal bone? I believe it was the latter, for several reasons. First, the paralysis was complete; for probably there is no one centre in the cord which, if destroyed, could cause complete powerlessness of the seventh nerve. In bulbar paralysis this is only partial. If complete, therefore, we should infer that the trunk was involved. Again, too, if the paralysis had arisen from central disease the chorda tympani would not have been affected, as was shown here, by the imperfection of taste. The deafness, too, could as easily be accounted for, from disease of the ear itself; and this view was supported by the constant headache.

All these reasons suggested that the trunk of the facial nerve was affected in the petrous bone; but why it was associated with a partial hemiplegia was not very obvious.

Hemianæsthesia.—It might be thought that if the physiological doctrine was correct, that the sensory tract terminates in the thalamus opticus, disease of this ganglion would produce loss of sensation in half the body. Observations, however, have not yet substantiated this; and very good practical physicians still deny that their cases in any way show such a connection between the symptoms and the portion of brain affected. A few well-reported instances, however, seem to prove that in those forms of paralysis which have had their seat near the thalamus, sensation is more affected. As a fact, in many cases of hemiplegia there is no loss of sensation, whilst in others it is impaired, and even in others altogether lost. This is scarcely explicable, except on the supposition that different nervous tracts must have been involved in the different class of cases. A great difficulty arises in ascertaining the degree of loss in persons who have been struck with apoplexy, and also from the fact that it appears to require so great a lesion to cause absolute anæsthesia, whilst, on the other hand, this is a sense so often lost from mere functional disturbance. Loss of sensation, for instance, is common enough as a nervous affection arising either from disturbance in the nerves or their centres; and, what is very remarkable, a complete anæsthesia of half of the body is always, in my experience, functional; it is too complete to have its origin in the thalamus, especially as it includes in it a loss of the special senses at the same time. Such cases show rather that the higher sensorium is at fault, and the whole hemisphere must be for a time in abeyance; unless, indeed, it is supposed that a certain region, viz. the temporo-sphenoidal lobe contains the centres of the special senses. To go back, however, to the question of whether a limited anæsthesia may be due to disease of the thalamus, there has been a growing opinion of late that pathological observation does show a connection between them. Türck made some very careful observations, and his conclusions were in favour of the view that sensation would be found impaired when a lesion existed on the outer side of the thalamus, involving some of the fibres of the corona radiata. In some cases of lesions of the brain in this region he found a loss of common sensibility, as well as electro-sensibility; also that reflex excitability was less than in ordinary hemiplegia; and all physiologists agree in stating that destruction of the thalamus produces no changes in motility as in the corpus striatum. In looking over my cases I cannot but think that there is some evidence to confirm this, and that there is some association between disease of the thalamus and loss of sensation. In two cases of effusion of blood into the thalamus there was hemiplegia of motion and sensation; also in another case where sensation was not tested motion was very little impaired. In a case of a growth in the thalamus, accompanied by fits and pain in the head, there was apparently loss of feeling in the limbs, more especially in the arm. Dr Crichton Browne gives in his adhesion to the view that sensation is associated with the thalamus opticus; his experience being that when this body is implicated in disease that sensation is destroyed on the opposite side, and that the limbs do not respond to the stimulation as in ordinary hemiplegia; in fact, that common sensibility and reflex excitability are abolished. Bastian, and, I believe, Broadbent, two of our best observers, do not consider the theory proved, and the former physician thinks that in all probability the parts that would be affected in loss of sensation would be those which appear to rule over this sense, as described by Vulpian and Longuet, viz. the upper and posterior strata of the pons, midway between the lateral border and the middle line. Bastian seems to think that disease in this region, or below it, might affect sensation, but that the parts above, towards the hemispheres, could not, and speaks of cases of extreme atrophy of the hemispheres and central ganglia, accompanied by loss of motion, but by none of sensation.

It being then somewhat doubtful whether we at present know of any local lesion of the brain which can cause a complete loss of sensation, we must look for the causes of anæsthesia elsewhere. The only complete cases of hemianæsthesia which I have met with have been clearly functional and mostly combined with a hysterical hemiplegia, and not only the loss of feeling but that of pain—that is, analgesia—has been associated with anæsthesia. In this paralysis of motion the parts which are affected do not follow any anatomical rule, but are merely those over which voluntary power is not exerted, and therefore if any portion of the brain is for the time in abeyance it must be the whole hemisphere, or, if not the whole hemisphere, that large portion which rules over motility, and the adjacent region in which is contained the tactile centre, and those of the special senses. This hypothesis would include in it a loss of perception of touch, as well as of all the special senses. It is possible that in some of these cases the loss of motion might have been more apparent than real, and due to want of co-ordination, but in most of them it is clear that for all motor purposes the anterior parts of the brain are functionless, and for sensory purposes the posterior. In hysteria, where there is want of will to move, and loss of feeling at the same time, we must suppose that the brain is healthy but is not at work ; in the same way that we see a watch not going, and we might, therefore, suppose it to be seriously damaged in its internal machinery, yet on looking into it we find a perfect instrument which only lacks winding up.[1]

In these cases of hemianæsthesia it may be remarked that the loss of feeling is complete. There is a total loss of sensation in the body as far as the median line ; the face, including the eye, is also senseless, and sight, hearing, smell, and taste are deficient. There is either amblyopia or a more distinct hemiopia. It would thus seem that the whole perceptive power is deficient. The first case is an example of the condition I have been speaking of. The next one, that of a young man, was very remarkable, from the paralysis coming on suddenly, and therefore there might have been some grounds of suspicion of an actual lesion. If so, it appears to me it could only have been of one kind ; that is, effusion of blood over the whole surface of the hemisphere, which might by its pressure have obliterated for a time the action of this side of the brain, and so produced the hemianæsthesia and partial paralysis of motion. It is remarkable, too, in this case, that there was spasm in the limbs, a condition which is so often purely functional and associated with

[1] Since this was written I have read Charcot's 'Account of Hysterical Anæs-thesia,' in which he generally finds one of the ovaries very sensitive on pressure.

the hysterical condition. I should say that I am merely recording my own experience when I state that a perfect hemianæsthesia has been always functional, for I am aware that cases have been recorded where from the suddenness of the attack a sanguineous apoplexy has been diagnosed. Inasmuch as in these the special senses have been affected also, a large region in the neighbourhood of the thalamus, as the temporo-sphenoidal lobe, has been supposed to be involved.

Hemiplegia—Hemianæsthesia

CASE.—Mary K—, æt. 24, was admitted under my care several months ago with what I thought at the time to be a serious disease of the brain. She had been out of health for some time; had inflammation of the eyes, and the right one had been removed at a provincial hospital; she also had amenorrhœa, cough and spitting of blood, sickness and severe pain in the head. After being in the ward for some time, she became delirious at night, and suffered much from frontal headache. Some time afterwards it was found that she had lost power in the left arm and leg. It was then thought that she might have a tumour in the brain, as many of the symptoms were present, such as persistent pain in the head, sickness, and hemiplegia. She daily grew worse, complained incessantly of pain at the back of the head, so that injections of morphia were daily used. It was said also that she had fits, and foamed at the mouth. I did not see her for some time, but the impression was that she was slowly dying with a tumour in the brain. When I took charge of the case again, nearly six months after her first admission, I altered my view of the case; her head was better, also the sickness, and she had grown fatter. She still said that she had lost all power over the left arm and leg; she had no falling of the face; she now also could not open her mouth, nor thrust out her tongue, this being no part of a hemiplegia; she moreover had loss of sensation of the left half of the body, or complete hemianæsthesia. She had lost feeling, not only in the left arm and leg, but in the trunk as far as the mesian line, also on the left side of the face, including the eyeball; she was also deaf on the left side, and said she could not see, nor smell, nor taste, on the left side; thus the hemianæsthesia was complete, as regards common sensation as well as the special senses. On raising her arm and diverting her attention by conversation, it was found to be not utterly powerless, and on moving it about, some resistance was offered, showing that volition was not altogether powerless over it. The eye examined by the ophthalmoscope was healthy.

She left the hospital and again re-entered, much in the same state, with paralysis of motion of left arm and leg, no falling of face, and spoke well, but could not open her mouth wide enough to put out the tongue; also complete hemianæsthesia of left half of body, including the special senses. I regarded the case as one of hysteria.

CASE.—H. A. T—, æt. 22, a Swiss; been attached to an equestrian circus, and at different times received injuries to his body and head. He had been out of occupation for a fortnight, wandering about the streets and with scarcely any food. On evening of admission, after having felt giddy for some time, he stopped to take some coffee at the corner of a street, when he suddenly felt a rush of cold in the right arm; he shook, and presently fell. He was picked up and brought to the hospital in a quarter of an hour, where, on his arrival, he was thought to be dead. When this was found not to be the case, he was put to bed, and in about an hour afterwards he regained his consciousness, and was able to give an account of himself. It was then found that his right side was affected in the manner in which I saw him on the following day. The right arm was spasmodically flexed, the elbow and wrist bent, and muscles rigid. The hand was turned out, the fingers separated, the thumb drawn in, and little finger tightly flexed, whilst the last phalanges of other fingers were extended. Thus the muscles supplied by ulnar nerve appeared especially affected. His foot was extended, and muscles contracted ; he could stand on leg, but used it as if it was a wooden one. There was no apparent paralysis of the face, but he said he could not whistle. Besides this spasmodic contraction of the right limbs, he had almost complete anæsthesia of the same side; he could not feel when touched, nor discern hot from cold. This included the whole of the face on this side, with the eyes and nose, the right side of the trunk as well as the extremities ; also lost the sense of hearing, of smell, and of taste on same side. He said he had a pain in the right temple and forehead. On testing with both forms of galvanism, it was found that the muscles did not react so well as on the healthy side.

At the end of a week sensation was found returning in the face and body, although still very imperfect in the limbs; hearing and taste also returned. The rigidity the same.

He gradually improved, and at the end of another fortnight, he could extend all the fingers but ring and little fingers, which remained flexed. Leg weak, although he could stand upon it. Sensation still much impaired. In the night he had some kind of fit, when all the old symptoms returned, including rigidity and anæsthesia, but on the following day was again better.   On the next day he left the hospital unexpectedly, after having stolen some clothes ; from this and the whole demeanour of the young man, it was evident that his " morale " was very bad.

APHASIA

I have already stated that in right hemiplegia we have aphasia. This is so important and interesting a symptom that I shall devote a great part of this lecture to the subject.[1] If you go into the wards you will see that loss of speech depends, in the majority of cases, upon three very different causes. First, on disease or disturbance of the whole brain, whereby the perceptive powers and intelligence are destroyed ; secondly, upon paralysis of the parts employed in vocalisation ; and, thirdly, on a condition in which the patient appears to have forgotten his language or the use of words. It is the latter to which the term aphasia is usually given, and this is associated with a very defined and positive lesion in the brain. It is difficult to define exactly what we mean by aphasia, as all are not agreed as to its exact significance ; and, therefore, Dr Hughlings-Jackson (to whom we are greatly indebted for having first drawn attention to its pathology) adopts the term " defect of expression."

[1] This has been already published in the ' Guy's Hosp. Rep.'

Others have avoided a strict definition by saying that in aphasia physical expression of thought cannot be rendered intelligible; the voluntary power of using words which express ideas has gone; a portion of brain is always in full activity for the production of outward speech, and this part is damaged. Whatever may be the definition of aphasia the clinical facts are very evident. You will see clearly by examining patients the essential difference between the forms of loss of speech which I have mentioned.

In bulbar paralysis, for example, where the muscles used in speech are paralysed, you will see the patient making a great effort to express himself; he speaks slowly, and the almost unintelligible words appear to come from him with the greatest difficulty. You will also observe in partial paralysis of the facial muscles that the patient speaks like a drunken man, as seen in the general paralysis of the insane; but in aphasia the attempt at utterance is very different. The patient, in his attempt to answer you when you put a question to him, sets up a meaningless gabble; his mouth and lips move quickly, but nothing else than noise results. Without looking at patients belonging to these different classes you would, by their manner of speaking, or attempt at speaking, be able to recognise their peculiar form of malady.

Dr Voisin has delivered a very interesting lecture[1] on the troubles of speech, especially in the general paralysis of the insane, showing the significance of the different modes of utterance. Amongst them he notices stuttering, drawling, hesitation, jabbering, stammering, and quavering. For the perfect faculty of speech, he says, there must be (1) soundness of the cortical substance of the brain, the seat of intellect; (2) of the nervous fibres passing from the cortical substance to the bulb, the conductors of the will; (3) of the bulb and the nuclei of the nerves animating the muscles which are called into play during speech; (4) of the nerves animating these muscles; (5) of the muscles themselves. It is not enough, therefore, that the organ of speech is right, but there must be a sound intellect, as well as a knowledge and memory of words. He says stuttering, hesitation, and drawling, show an embarrassment of speech occasioned by a disturbance of memory, and are often accompanied by a misplacement of words. The seat is in the brain and intellect, and is noticed in general paralysis and other diseases. Stammering, jabbering, and quavering, do not result from troubles of the intelligence, but from a want of harmony amongst the muscles, and show a change in the medulla. In cases where speech had thus been interfered with, degenerations had been found in the grey cells of the facial

[1] 'British Medical Journal,' June 19, 1875.

and other nerves. Cases of absolute muteness were due to a degeneration of the muscles of the tongue and pharynx.

It seems pretty certain that aphasia is anatomically due to disease of the third frontal convolution, and as this lies immediately over the corpus striatum, the reason is clear why these two portions of the brain are so often affected together, and that aphasia so frequently accompanies right hemiplegia; the reason being that the middle cerebral artery supplies the corpus striatum and the convolutions over it, and therefore disease of this vessel, or more especially an embolus plugging it, would impair the function of both these parts. This convolution, then, is the organ which we find damaged in aphasia. Speech, you may remember, might be accomplished if the patient knew the word to utter, and therefore he can often ejaculate and repeat a word suggested to him.

The clinical facts connected with aphasia have long been observed, but special attention has been given to it since the discovery by Broca of its anatomical basis.

Thus, Abercrombie says, "in regard to the paralytic state in general, we may notice that in some cases of palsy there is loss of motion without loss of feeling, in others the feeling is lost also; but in some singular cases on record loss of feeling took place without loss of motion. In one case there was loss of motion on one side, and loss of feeling, without any diminution of motion, on the other. Some interesting phenomena are sometimes presented by the conditions of the mental faculties; one of the most common is a loss of memory of words, and this is sometimes observed to be confined to words of a particular class, as nouns, verbs, or adjectives. The patient is frequently observed to have a distinct idea of things and their relations, as well as of persons, while he is utterly unable to give their names, or to understand them when they are named to him; or there may be a modification of the affection by putting one word or one name of an object in the place of another, and, very singularly, the patient may always apply the name in the same manner."

It is interesting, also, to observe, that in the phrenological system of Gall and Spurzheim the organ of speech was placed in the region which has now been discovered to be affected in aphasia. In an illustrative case the description of the symptoms is so good that I shall not hesitate to quote it. It is to be found in the 'Transactions of the Phrenological Society' for 1822. Mr Alexander Hood, who brought it forward, said he felt sure the man's mind was perfect, although he could not speak, and from this he argued that the brain was not a single organ, and that every part of it was necessary to each mental act; and he quotes from Spurzheim, who had recorded a similar case : ' L'homme comprenait tout ce qu'on lui disait, mais il ne pouvait pas trouver la prononciation des mots dont il avait besoin. Si on lui montrait une couleur telle que la verte, et qu'on lui demandait si la couleur était brune, jaune, ou toute autre que verte, il répondait que non ; aussitôt qu'on nommait la véritable couleur, il disait qu'oui. Ces phénomènes prouvent que les idées et toutes les fonctions des facultés intérieures doivent être séparées des signes arbitraires et qu'elles précèdent les signes." Also Dr. Gregory had met with a case where language was lost and all the mental faculties remained. Mr Hood's case was as follows :—

CASE.—" R. W—, a blacksmith, æt. 61. On the evening of September 2nd, 1822, in the midst of his family, he suddenly began to speak incoherently, and became quite unintelligible to all those who were about him. The complaint, in the first instance, appeared to be pretty much like delirium or the effects of liquor, with this remarkable difference, however, that the words which were uttered were unconnected with the significations with which they are generally associated. On the morning of the 3rd of September, when I first saw him, he was in bed, and seemed to be somewhat confused, for though he could speak, no general ideas could be collected from the words which were expressed, as he only rendered himself intelligible by signs. Being apprehensive of apoplexy, as there was some fever present, with a full strong pulse, upwards of ninety beats in the minute, I took fourteen ounces of blood from the arm ; but he having become faintish, the wound was bound up and leeches applied to the temples. A brisk purgative was also administered, and towards the evening the skin became cool and the pulse moderate, but the mental affection remained the same, and it was now discovered that *he had forgotten the name of every object in nature.* His recollection of *things* seemed to be unimpaired, but the *names* by which men and things are known were entirely obliterated from his mind, or rather he had lost the faculty by which they are called up at the control of the will. He was by no means inattentive to what was going on, and recognised friends and acquaintances, perhaps, as quickly as on any former occasion ; but their names, or even his own or his wife's name, or the names of any of his domestics, appeared to have no place in his recollection.

" Under the serious apprehension that this strange mental affection might probably be the harbinger of death, he was extremely anxious to settle his affairs and make his peace with the world. A gentleman, who had often suggested to him the propriety of making a testamentary settlement, now occurred to his mind, though he could not by any effort call up his name. He laboured with the utmost assiduity more than an hour to make his family understand what he wanted, and ultimately succeeded in directing them to the individual by depicting the number of houses and doors between his own house and that in which his friend resided. Thus directed, some one of the family asked him if it was such a one, naming the person that he wished to see. He seemed to be overjoyed, and signified by various gesticulations that this was the person, and that he was desirous his friend should be brought to him immediately.

" I was afterwards informed by the gentleman that my patient had succeeded completely, by means of signs, hieroglyphics, and a few explanations from one of the family, in making known to him the manner in which he wished his children severally to succeed to the possession of their respective shares of his property.

The same evening had been fixed for a commitee meeting of a friendly society, of which he was a member; but though he recollected the society, the meeting of the committee, the time and place of meeting, and other circumstances connected with it, yet he had forgotten *all the words* by which these ideas are expressed. He seemed to regret much his inability to attend, and wished to convey this idea to his family, but could make them understand what he referred to only by forming a circle with chairs, and placing one more conspicuous than the rest, indicative of the president's, by which his meaning was at last discovered.

"On the morning of the 4th September, much against the wishes of his family, he put on his clothes and went out to the workshop, and when I made my visit he made me to understand, by a variety of signs, that he was perfectly well in every respect with the exception of some slight uneasiness referable to the eyes and eyebrows. I prevailed on him with some difficulty to submit to the reapplication of leeches, and allow a blister to be placed over the left temple. He took also a full dose of calomel and jalap, which operated well, having elicited, besides feculent matter, a copious discharge of bile. From this time he declined all medical treatment, excepting taking occasionally a dose of salts. He was now so well in bodily health that he would not be confined to the house, and his judgment, in so far as I could form an estimate of it, was unimpaired; but his memory for words was so much a blank that the monosyllables of affirmation and negation seemed to be the only two words in the language the use and signification of which he never entirely forgot. He comprehended distinctly every word which was spoken or addressed to him, and though he had ideas adequate to form a full reply, the words by which these ideas are expressed seemed to have been entirely obliterated from his mind. By way of experiment, I would sometimes mention to him the name of a person or thing, his own name, for example, or the name of some one of his domestics, when he would have repeated it after me distinctly once or twice, but generally, before he could do so a third time, the word was gone from him as completely as if he had never heard it pronounced. When a person read to him from a book he had no difficulty in perceiving the meaning of the passage, but he could not himself then read, and the reason seemed to be that he had forgotten the elements of written language, viz. the names of the letters of the alphabet. In the course of a short time he became very expert in the use of signs, and his convalescence was marked by his imperceptibly acquiring some general terms, which were with him at first of very extensive and varied application. In the progress of his recovery time and space came both under the general application of *time*. All future events and objects before him were, as he expressed it, 'next time;' but past events and objects behind were designated 'last time.' One day, being asked his age, he made me to understand that he could not tell, but, pointing to his wife, uttered the words 'many times' repeatedly, as much as to say that he had often told her his age. When she said he was sixty he answered in the affirmative, and inquired what 'time' it was, but, as I did not comprehend his meaning distinctly, I mentioned to him the hour of the day, when he soon convinced me that I had not given him the proper answer. I then named the day of the week, which was also unsatisfactory; but upon mentioning the month and day of the month he immediately signified that this was what he wanted to know, in order to answer my question respecting his age. Having succeeded in getting the day of the month, he then pointed out the 'time' or day of the month on which he was born, and thereby gave me to understand that he was sixty years of age and five days or 'times,' as he expressed it."

Mr Hood, who commented upon the case, expressed his belief in the phrenological doctrine, that the organ of language was situated above and behind the eye, and that this man had disease in the anterior lobe of the brain, and probably on the left side, as it was here that he had complained of pain. In a later volume of the 'Transactions' of the Society there is a further account of this case, in which it is said that the patient was subsequently seized with a fatal apoplexy and paralysis of the right side, when a cavity was found in the anterior part of the left lobe of the brain, in the spot suggested during life, besides a recent clot of blood in its neighbourhood.

In order that you may more fully understand the subject, I will relate to you another case, and then make my comments upon it.

CASE.—Elizabeth H—, æt. 24, a domestic servant, admitted July 4th, 1871. A fortnight previously she went to bed perfectly well, but not rising at the usual hour, her room was opened, and she was found, as when admitted, paralysed and speechless. There was no previous history of any illness whatever.

She was a well-grown woman, very pale, and with a vacant expression of countenance. The right arm and leg were completely paralysed, the mouth slightly drawn up towards the left side when she used any muscular effort. She could move her mouth well in eating, and could use her larynx; there appeared some little difficulty in protruding the tongue. There was also considerable loss of sensation in the arm and leg, also some loss of power over sphincters of rectum and bladder. On examination of the heart no bruit could be detected. On speaking to her she appeared to understand all that was said, but could not answer a word. Her mouth moved, and she uttered a senseless jabber. She was put on good diet, tonic medicines, and, after a time, Faradization to the right side.

In about a fortnight's time her general health was improved, although the paralytic symptoms remained as before, and then I was better enabled to test her knowledge of language. She appeared to understand written and spoken language perfectly; she read books and the newspaper, she received letters from her friends, and on one occasion a telegram; all of which the sister of the ward said she perfectly understood by her actions and gestures, since she prompted what answers to give. When any object was held before her, and the name demanded, she merely moved her mouth, or uttered an unmeaning sound, but immediately assented by a nod when the correct appellation was given, just as she would shake her head when any wrong name was purposely uttered. At the end of another month the power of the leg slightly returned, so as to enable her to sit up, but the arm remained paralysed, and I waited with expectation to see if, with a slight recovery of the limbs, the speech would also return. There did not, however, appear to be the slightest indication of improvement; at the same time, she seemed as rational as any other person in the ward. It was then attempted to teach her afresh, and the method, up to the present time, has been eminently successful. A box of letters of the alphabet, with pictures upon them, was brought her, and she was taught in the same manner as a child, or as a person learning a foreign language. On repeating to her several times the name of one of the objects on the cards, she would at length articulate the name herself. On the following day she would say more of them correctly, and forget others, or use the wrong name, just as a child might do. On one of the cards was a picture of an umbrella, and she evidently knew it was a long and tiresome word, and immediately exclaimed "butterfly," but in a moment shook her head to express her error. A butterfly was on one of the cards, and she had been taught the name. She was also taught to say other words, as "good morning," "Guy's Hospital," and her own name; all of these she would suddenly bring out when I visited her on my rounds, being evidently delighted with her improvement. Although it was true that she could move the muscles of the face, yet every word appeared to be formed with an effort, as if she had never before put her mouth into shape, and much as a stammering person does when trying to give utterance to a word. That an actual difficulty existed was seen in the dissimilarity to the correct sound on the first attempt to speak, although afterwards the words would be properly formed; in fact, if one has observed a child attempting to utter a hard word, and mixing the syllables together or skipping a letter or sound entirely, the exact condition of this woman may be understood. On being told to say "seventeen," the number of her bed, she said "eventeen," but then being requested to make a hissing noise to precede this, she immediately did so, and so produced the correct sound. A very similar performance may be seen going on at any school where the master is endeavouring to teach French; strange grimaces and many feeble attempts are witnessed before the children can say, for example, "Donnez-moi du pain, monsieur." In this way the woman is gradually learning to talk, and, as far as can be ascertained, she has not used a single word which has not been taught her quite recently; she does not appear to have used any expression which might have cropped up from recollection or from any return of memory.

As regards agraphia, the ability to write was rather a difficult faculty to test, seeing that few persons can write at all legibly with the left hand, but I apprehend it is the same faculty which is put on trial when a number of letters of the alphabet are placed before the patient, and she is requested to arrange them into words. I doubt whether this woman would have voluntarily put letters together to form words, for she never attempted to do so; but when told to spell a common word, like "horse" or "cow," she immediately endeavoured to accomplish it, but generally spelt it wrong; there was a tendency for the letters to come right, but they were seldom placed together quite correctly. When wrongly placed she shook her head, to express the error, and if the word was made right would accord by a nod.

I should have said that in learning to speak she was not guided by the movements of the mouth or larynx, as are the deaf and dumb, but by the sounds through the ear; this was known by speaking the word we wished her to utter behind her back, when she was found to copy it just as readily as when she saw the face.

The patient was slowly improving, when she was removed to the north of England by her friends.

I relate this case because it presents, in a well-marked degree, those characteristics which have of late years rendered aphasia so interesting a subject of study. The various forms of the affection have been fully dwelt upon by writers, and distinctions made according to the amount of loss which the faculty of communicating ideas has undergone in different cases. In some there has been a total forgetfulness of words; in other cases words could be written down, but not spoken; whilst in others the exact converse has occurred. The true aphasic condition is seen in the case under consideration, and it is this which is generally supposed to exist when the term aphasia is used. Some writers have spoken of this form as not common; but I should regard it rather as the typical state of what has of late been understood by the term, and it is the exact condition observed in this patient which so perplexed many of the older medical authors who have described cases of loss of speech. I am referring now to the case where a person understands what is said to him, and apparently appreciates what he reads, but at the same time is unable to express his ideas in words, either spoken or written. Lesser degress of this aphasic state may be met with, as, for example, where there is not a total inability to speak, but a partial forgetfulness of some words, which, however, if remembered, are sadly disfigured in the attempt to enunciate them or to write them down. Another form is said to occur where the patient can speak, but cannot write, and to this failing the name "agraphia" has been given; whereas, if he can write, but cannot speak, the term "aphemia" would be used. In cases like the one reported, where both the power of speech and writing failed, the term "aphasia" is the one usually adopted. In all these forms faculty of language is not entirely lost, but only the power of expression, and then the term "ataxic aphasia" has been used in order to distinguish it from "amnesic aphasia," where the memory of words is altogether lost. In the case related by Dr Bristowe, at the Clinical Society, where the patient was taught to speak after he had ceased to be able to do so for some time, the ability to write had never failed, and, therefore, it was clear that the power of expressing himself in language had not entirely departed, but merely the faculty of knowing how to use the organs of speech. He so quickly regained the use of language that his ataxic condition can scarcely be regarded in so formidable a light as one of true aphasia. In the case, however, which I have related the power of expression, both in writing and speaking, was entirely gone, whilst a knowledge of language remained, and it is this kind of case which creates so much interest as regards the question of localising speech in a particular part of the brain, and, indeed, as to the nature of language generally, together with the mental state of that person who has lost the power of communicating his ideas by signs.

Where the power of expression is entirely gone, as in many cases of disease, no explanation of the phenomena is needed, for whether we believe that speech is situated in one spot, or is intimately associated with the working of the whole hemispheres and the entire mental faculties, it is not difficult to comprehend how a severe lesion might deprive a person of speech, so that he should not be able to speak or write, or even know the meaning of words. What his mental condition is under these circumstances may, of course, be very difficult to discover. But the case is altogether different and far more perplexing when he has an appreciation of language, but yet is unable to communicate his ideas by writing or speaking. Here we have an intricate problem, and one which some of the best minds in the profession have not yet been able satisfactorily to solve.

The difficulties of explanation are in some measure removed and the way made more clear when we attempt to analyse the faculty of language. For its production three, if not four, different processes are employed. First, the mind is impressed with ideas through the eye by means of written signs, that is, impressions formed on the retina pass through their own perceptive centres (to adopt Dr Bastian's expression) to the cerebral hemispheres, where they are further developed or compounded with other perceptions. Secondly, impressions made on the ear by sounds are carried through the auditory perceptive centres, also to the hemispheres. Now, these signs conveyed by the eye and ear have nothing in common; there can be no connection between the letters which spell, for example, the word "field" written on paper, and the sound which we attach to this word; the association in our mind is one of our own constructing, and a perfectly arbitrary one. We are compelled to this association from our earliest infancy after the following manner:—When a child is shown a certain word in his book, a certain sound is made to represent it, and thus the visual impression and the auditory impression become intimately blended in his mind. In all probability a picture of the object (say, a dog) is shown to the child, to which these visual and auditory signs are always to be attached; but this is not all, for the sound made by the master is to be imitated by the child until the latter can use its organs of speech in a particular way when the said object is presented to him. The whole complex faculty of language is thus taught through signs by the ear, by the eye, and by the organs of speech, together with a representation of the object itself. One would naturally ask what idea of language would exist if it had been taught by means of one or more of these processes, and I think we may get some answer to the question. Let any one who has acquired a knowledge of a language by means of a dictionary ask himself if he need have the slightest comprehension of it when spoken; not having acquired it through the ear, it would be, under these circumstances, quite without meaning to him. Thus the common remark made by many persons, that they can read French or German, but cannot speak those languages, is, of course, true, for they have never tried. The processes of learning through the eye and through the ear are, in fact, totally distinct, although by education we may have intimately blended them together. Then, again, when we speak, we are employing a third faculty; we are not then gaining ideas through the ear or eye, but we are communicating ideas to another person through a totally different channel. If a child were taught the name of an object by a visual or auditory sign, and that child did not learn to make the customary sound which belongs to it, he would understand language by reading or hearing, but he could not speak it. He might gain ideas through the eye or through the ear, but he must learn to transmit them through the brain to the vocal organs in order to communicate them. In illustration, is the following. A little boy, the son of French parents, born in England and going to the town school, speaks English like other boys, but at home the conversation is carried on by his parents in French. They speak to the boy in French, and he readily obeys all that is demanded of him, so that there is every reason to believe that he understands the language perfectly; but up to the present time he has never been known to utter a single syllable of it. He cannot yet read easily, and, therefore, what information he might obtain from books I cannot say. Here is a case of partial education which much resembles that of our aphasic patient. We might illustrate this, although somewhat imperfectly, by the following:—An engine driver has learned the meaning of certain signals, so that he could explain their object to another person, but it does not follow that he could take the signalman's place. He might know that for a certain purpose a certain sign was to be shown, and the instrument for making the signals might be perfect in its working; but he would be quite helpless, because he had not been taught how to use it. He would be much in the position of the French child just mentioned, who understood the language, but could not use it. If he had once learned the use of the signal instrument, and then forgotten it, he would rather resemble our aphasic patient. The man who sits in a tunnel, and works his instruments without being able to see the effect on the distant signals, is like the deaf and dumb child, who has learned to talk by imitating the movements of the mouth of another person, but hears no sound. Or you can imagine a piece of machinery made for some special object, such as turning, planing, or boring; a motive force alone would not produce the result, but a complex apparatus must intervene to convert

the original power into the end desired. In like manner, between
the brain proper and the vocal apparatus there is a portion of
brain where the ideas are put into form necessary to operate on the
larynx.

That this compound of language, made up of spoken, auditory,
and visual signs, may be thoroughly broken up, we see to a limited
extent in every-day life. For example, a considerable amount of
attention, and expenditure of vis nervosa are required in continuous
speaking; therefore, in an exhausted state of the nervous system,
this faculty may be virtually paralysed. Let a person be prostrated
from want of food or over-exertion; he cannot talk, he forgets
what he wants to say, he uses one word for another, but he does
not forget the meaning of words which are spoken to him or which
he may read. Even under ordinary circumstances, in perfect health,
we may lose the memory of words, and are for the time exactly
like our aphasic patients; we cannot speak them nor write them,
but we recognise them when written or spoken, or if a long list of
names be given we immediately assent when the right one is
arrived at.

It has long been discussed, what amount of intelligence can
exist without language, the opinion being that the cultivation of
the mind, and the perfection of language, are coextensive; take,
for example, the small vocabulary of words found in savage nations, compared with the vast number of words used by civilised
people, to say nothing of their wonderful inflections as seen in
the tenses of the Greeks. This opinion implies that every idea
necessitates the conjuring up its appropriate sign in language, and,
therefore, probably cannot exist without it. A very prevalent opinion
is, that when we are thinking we are really mentally speaking,
and the fact that some persons not having sufficient control over
themselves do think aloud, tends to corroborate the opinion. If we
remember, however, that language is a complex faculty, and acquired
by different processes, it may be true that, in thinking, we must
conjure up an appropriate sign, but it does not follow that this is by
mental or imaginary speech; and herein we may, perhaps, find a
clue towards a solution of some of the difficulties met with in our
aphasic patients. If a person can speak a language fluently, and
is recalling to his mind some sentiment heard in conversation, he is
no doubt thinking of words as they are uttered, but if he had never
learned to speak the language, and had acquired it only by reading,
that person would, when recalling the sentiment to memory, have before his mental vision a particular page of a book with a certain line
towards the top or bottom. It seems, therefore, almost true of necessity that a person may know a language but cannot speak it, just as

another person may understand it when spoken who cannot read or write it. One person does not know language by sight because he has not seen it in print ; another does not know it by ear because he has never heard it pronounced, and a third cannot converse in it because he has not learned how. Then again, if it be true that in thinking we must have a mental object before us, it does not follow that it must take the form of language, for I suppose the child may think of the picture of the dog in its lesson book without necessarily attaching to it any spoken, auditory, or visual sign, and I apprehend that the child could then go on to think about various actions of the dog, as running, eating, &c., without the idea of words or any artificial signs. One boy at school, in thinking over his proposition of Euclid, will have the figure in his mind's eye, and so go through the proof, whilst another who has learnt it by heart, will be impressed by the letters, and these he will repeat. I apprehend that in this way two totally different methods have been adopted in learning the problem. If the boy, again, is working an equation in his head, and mentally says $a^2 + b^2$, he may be silently articulating the letters, but not necessarily so, for if he has not been in the habit of doing his sums aloud, it is probable that the mere mental vision of them is sufficient for the operation. If this be true, it shows that the idea may take the form in which it has been acquired. In the case of the little boy who understood French, but had never spoken it, we should consider that if he were thinking or dreaming of his mother, he would not conjure up language by any imaginary movement of the vocal organs, since he had never used them for the purpose, nor by the mind's eye, but simply think of her words as they had impressed him through the ear. It follows then, that even if it be true that we have no thoughts without some corresponding remembered sign, this need not be a spoken sign, for in our aphasic patient in whom speech was lost, the remembrance of a spoken or written word still remained. There may be, therefore, different interpretations put upon the meaning of such an expression as the following, used by a person who has gone to reside abroad :—' I soon learned to speak the language, and after a time to think in it ;' or as a foreigner in England will say, ' I have never been so familiarised with your language as to be able to think in any other than my native German.'

But, as before said, even without language, pictures of objects may be brought before the mind's eye, and thus mental operations may still proceed. How is it with the deaf and dumb ? It is said they think with their fingers, since they are seen to move them in their dreams. Or how with the blind who have learned to read with various kinds of raised letters, but without speaking aloud ?

For the same reason as with the deaf and dumb, their fingers
should be seen passing over imaginary embossed books. How is it
with animals who are said to dream? They probably do no more
than picture to themselves a series of dissolving views, accompanied
by the music of familiar sounds.   Probably, little more is going on
in the human being when he is in a state of reverie, or in the mind
of the ploughboy, who whistles as he goes for want of thought. Most
persons will state that they cannot accurately think out a subject
without speaking or writing, showing that the ideas must have expression in form.   What condition the mind of the truly "amnesic"
patient is in is very difficult to know; but if in ordinary life most
persons' thoughts are not worth much unless put in consecutive form
by means of spoken or written language, it would follow that the
mind of man without language would be a blank. It, however,
could scarcely be this, for with mental vision of objects before him,
together with remembered sounds and odours, he would be as well off
as the dog. Whether really any deeper intellectual processes can go
on without language is very difficult to ascertain. It is possible that
some obscure, dreamy, and pleasurable condition may exist without
any consecutive current of thought passing through the brain, and,
indeed, nearly every poet speaks of abstracted states of this sort.
Wordsworth might often have been in this state.   In speaking of
the 'Wanderer,' he says :

> " Sound needed none,
> Nor any voice of joy ; his spirit drank
> The spectacle ; sensation, soul, and form
> All melted into him ; they swallowed up
> His animal being ; in them did he live,
> And by them did he live : they were his life,
> Thought was not; in enjoyment he expired."

The American poet Poe says of himself :

> "Not long ago the writer of these lines
> In the mad pride of intellectuality
> Maintained the power of words, denied that ever
> A thought arose within the human brain
> Beyond the utterance of the human tongue."

And then he goes on to say how untrue this is, for he is the
subject of—

> " Unthought-like thoughts that are the souls of thought
> Richer, far wilder, far diviner visions
> Than even the seraph harper, Israfel,
> Could hope to utter."

"I cannot write—I cannot speak or think—
Alas! I cannot feel, for 'tis not feeling
This standing motionless upon the golden
Threshold of the wide-open gate of dreams,
Gazing entranced adown the gorgeous vista."

The question is a very interesting one, but I should not have dwelt upon it had it not a practical bearing. We are sometimes consulted by lawyers as to the intelligence of patients with regard to the disposal of their property when in the aphasic state. There are some writers who maintain that you cannot separate entirely the office of speech from the thinking process which suggests it, that the workings of the brain can never assume a definite shape without the use of signs or words, and that our mind can never at any time undergo the process of thinking without certain definite terms already made appropriate to the thoughts coming before the mental vision, or without the larynx and muscles of vocalisation passing, in the imagination, through a kind of inarticulate speech. "Names are impressions of sense, and as such take the strongest hold of the mind, and, of all other impressions, can be most easily recollected and retained in view. They therefore serve to give a point of attachment to all the more volatile objects of thought and feeling. Impressions that when past might be dissipated for ever are by their connection with language always within reach. Thoughts of themselves are perpetually slipping out of the field of immediate mental vision, but the name abides with us, and the utterance of it restores them in a moment. Words are the custodians of every product of the mind less impressive than themselves. All extensions of human knowledge, all new generalisations, are fixed and spread, even unintentionally, by the use of words."

Dr H. Jackson, who has given considerable attention to the subject, is inclined to the opinion that loss of speech is a mental as well as a bodily symptom; for, he says, we speak not only to tell other people what we think but to tell ourselves what we think, and a proposition is not only the ending of a mental operation but the beginning of another.

Professor Sayce also says thought precedes language, which represents our ideas, and not until we have clothed our thoughts have we truly got them. They must be in a definite shape for further thinking. Thought creates language, but in its turn language creates thought. Speech is necessary for conversation, but also from the intellectual development of man himself. For ideas to be kept in the mind they must be imagined or pictured there. Without thought there is no language, but equally without language there is no thought, in the true sense of the word. Those must admit the force of this who have found how difficult it has been to map out clearly any subject of thought until they have sat down and commenced to put their ideas on paper, when the whole matter has gradually unfolded itself before them.

Dr Jackson has also clearly shown that the simple utterance of a word may not be expressive of any idea, and thus it happens that aphasic patients will frequently ejaculate in monosyllables, and this at a time when they are thought to be speechless. They may use words, he says, but cannot propositionise. He was impressed by this some years ago by the fact of a patient in the next bed to a sufferer from aphasia informing the doctor that he believed his neighbour was an impostor, for he had heard him swear. Where-upon Dr Jackson showed convincingly that the mere exclamation, as in uttering an oath, is not using language as an intellectual operation in the manner already described, and as a system of signs corresponding to ideas, but a mere emotional or automatic action, and thus affords an additional proof that the organs of vocalisation are not at fault in aphasia. He further showed how his clinical observations corroborated the statements of Horne Tooke and other writers, that swearing is not an intellectual process, but exhibits merely a state of feeling. A man who swears much shows an exuberant amount of sentiment, but detracts so much from the real force of his language. Dr Jackson relates the case of a patient who attempted to write with his left hand, and, when given a slate and pencil, wrote all over the former the word " damn."

These observations tend to throw light upon those very painful cases which we sometimes have to witness, where old and bed-ridden persons, whose minds have gone to decay, and thus incapable of any mental effort, distress their friends by their repeated exclamations in oaths. Persons who have led the strictest and most pious lives, and who have never been heard to swear in the whole course of their existence, will now adopt the most horrible expressions. This simply means that the intellectual part of the nervous system has gone, while some of the emotional remains, and thus, without a thought, such expressions as " Oh, dear ! " " Good God ! " and so on, until the foulest words in the category of oaths, may be uttered. I have seen an old man in this state who could use the most san-guinary and condemnatory expressions, but could not utter a syllable for any intellectual purpose.

There are some great authorities who regard the faculty of speech as the great characteristic of man, and especially belongs to him as an instinct. The physiologist is not, however, content with this bare assumption, but is endeavouring to discover the meaning and origin of language. Hobbes' definition of language, according to

Mill, is unexceptionable, "A name is a word taken at pleasure to serve for a mark which may raise in our mind a thought like to some other thought we had before, and which being pronounced to others may be to them a sign of what thought the speaker had before in his mind." Professor Whitney says, thought may be regarded as an act of the mind, but every word is an act of the body, and of the body only, performed under the direction of the mind, but not different in kind than beckoning with the fingers, brandishing of an arm, or kicks with the foot. The apparatus of thought has no more immediate connection with the muscles of utterance than with those of facial expression or of gesture. Talking is just as much thought as dancing is, and not one whit more. He thinks that language might begin as a cry or a growl, just as an animal makes when pleased or displeased, and men also make use not of only words, but grimaces, gestures, postures, &c. Captain Burton says there is a tribe in North America who possess such a scanty vocabulary that they can hardly converse with one another in the dark. Professor Key holds much the same view, and looks upon language as the mode only by which thoughts are conveyed from one person to another ; but this might be done through other senses. The voice and ear are used in oral language, but there is a visible language in writing and hieroglyphics.

Some of the best known writers on language have failed to see this, and also that language is not commensurate with speech, but something much more complex. The written signs being arbitrary, must, no doubt, have been invented last, so therefore it only remains to inquire whether the spoken names applied to objects arose from an intuitive feeling of correspondence between the ejaculation and the object, or whether these were mere imitations of natural sounds. The first theory implies that, in accordance with some intuitive mental perception, there would be some natural cry indicative of fear, anger, or love. This has been styled the interjectional or "pooh-pooh" theory. If, on the other hand, language has been acquired, not from within, but from without, and is a mere copy of sounds, the theory has been called the imitative or "bow-wow" theory. If the idea be carried still further, and it be supposed that every object in motion produces a sound (as the "scratch" of a pen upon paper), we have what has been called the "ding-dong" theory. I should think that physiologists would incline to the theory called ironically the "bow-wow" theory.

This consideration of the nature of our present language, and that it is a mental production formed by several processes, enables us to separate these analytically, and to perceive how one of the parts may fail whilst others remain intact. If language is regarded as a simple faculty bound up with thought, it is difficult to understand how a knowledge of it can exist, and at the same time be forgotten; but if looked upon as a complex process, we can see how words can be intelligently perceived through the ear or eye, and yet cannot be spoken by the tongue; the organs of speech, notwithstanding, being all the while intact.

As there are direct channels from the senses to the cerebral centres, by which impressions become converted into ideas, the one leading inwards from the gateway of the ear, and another also inwards from the gateway of the eye, so it may be presumed that there is a third, which leads outwards from the brain, for the purposes of spoken thought; or, as there is an auditory perceptive centre and a visual centre, by means of which sensations are carried to the hemispheres to be converted into ideas and where the various mental processes are performed, so between the intelligence and the vocal organs we may presumably suppose there is a centre intimately associated with the production of speech. For the same reason, as when a portion of brain is injured and vision lost, and yet the eye itself is not affected nor the intelligence dimmed, so we might suppose an injury to another part of the brain and power of speech lost, whilst the vocal organs and the mind still remained entire. If the up line can be damaged or cut, so can the down line. That there should be a portion of the brain whose especial duty it is to rule over speech seems less remarkable when we remember that the tendency of physiological opinion of the present day is to map out the cerebro-spinal centres for various and distinct purposes, and to believe that the different complex movements of the body are prearranged and regulated by certain dominant points, either for respiration, speech, or motions of the limbs; and, moreover, that influences pass by special channels from the cerebral hemispheres to these points. It is clear, in the case of aphasia, that some powerfully presiding influence over the organs of speech has been lost, for not only is the remembrance of words gone, but the organs themselves, without being paralysed, appear to have quite lost their habit of accommodating themselves for talking.

That the loss of speech is mainly physical, and not intellectual, is shown by the case of a girl who, after recovering from scarlatina, became aphasic, so that she could not say a single word, nor attempt to read, but she sat down to the piano and played the various old tunes which she had learned before her illness.

Considerations of this kind on the analysis of speech and the compound office of the brain tend to elucidate some of the difficulties presented by such a case as that I have related. But there are still further points of interest in this case not yet alluded to, and one is the reacquisition of language after it had been lost. I have no doubt the usual tacit explanation in such a case has been—that language has returned with recovery of the injured brain; but before this can be satisfactorily determined some more rigid observations are required to show if the facts answer to the explanation. In bringing to my recollection several cases of right hemiplegia with aphasia, where there was no recovery of the limb, the speech also appeared to be irretrievably gone, and I know more than one case where patients, under these circumstances, remain absolutely dumb. I have just now under my care a sailor, who was the subject of this affection seventeen years ago; he partially recovered, so as to be able to resume his employment, but his speech even now is most imperfect. In the case under discussion the woman recovered in part the use of her leg, but not at all that of her arm, and at the same time there was not the slightest appearance of the return of speech. The few words she was acquainted with at the time of her leaving the hospital she had altogether newly acquired, and there seemed no reason why she should not have learned as much French, or any other foreign language, in the same space of time. It becomes, then, a question whether this fact be not an argument in favour of the theory that speech is located on one side of the brain, and that when language is relearned the other side is being educated for the purpose; in fact, whether the same process is not going on with language as with the left hand when it is learning to write and do what the paralysed one has been accustomed to. If after a violent concussion all idea of language was knocked out of the brain, no argument could be founded upon the recovery of it; but when the mind is entire, language understood, and yet the power of speech gone because one part of the brain is damaged, it seems to follow that, if language again return, it must come by re-education, and what more likely than that the part corresponding to the damaged one should be the seat of the training—that this should take up the lost function of its fellow-convolution? If speech was originally learned in a special way, it must be regained by the same method.

Believing as I do that the aphasic state is intimately associated with destruction of a convolution on the under surface of the left anterior lobe of the brain, as stated by Bouillaud, Broca, and Jackson, I have come to consider that the reacquisition of language by an aphasic patient is an additional fact in confirmation of it. I may say that I have always expressed my adhesion to the theory first propounded by my colleague, Dr Moxon, in explanation of the

localisation of the organ of speech on one side ; that it is entirely
owing to the education of the two different sides of the brain, as
this affords, to my mind, a far more probable explanation than
any other.　Those who for a moment have rejected the idea of the
lateral localisation of speech as *primâ facie* untenable, or even
absurd, have not, probably, considered that the two hemispheres
must have different functions, and that these have been produced
mainly by education.　When they see this, the difficulties regarding
a local organ of speech will be much lessened.　And not only by
personal education, but by a long usage of one hemisphere through
many generations, for all observers are agreed that the left hemisphere is larger than the right ; and Dr Bastian asserts that
the posterior lobe projects slightly further back than the right,
and Broca, that the convolutions of the left hemisphere are more
complex.

We believe, for example, that the cerebral hemispheres stimulate
the large ganglia lying below them to move the limbs at the time
when we will to do so, and we know that if the ganglia on one side
are diseased, the limbs are also paralysed; the will is good
to move them, but the line is cut.　The brain as the seat of will
remains healthy as before, but it has no influence over the weakened
limbs ; whence it follows that these central ganglia rule the limbs
and prompt their action on one side only.　Then comes the question, when an arm is educated for any particular movement, as, for
example, in playing a musical instrument, what is intended by the
term education ? or rather, what part of the body is educated or
trained ?　Is it the muscle or nerve of the arm, or is it the brain
which rules over them ?　I think there can be but one answer to
this, that it is the nerve centre which is educated; whence it follows
that just as the two arms are trained to different movements, as
the one to handle the bow, and the other the strings of the violin,
so must the two sides of the brain have been educated to regulate
their action.　Since the two hands cannot be interchanged in playing, it shows that each side of the brain must have been specially
educated for their particular movements.　They have thus become
physiologically different.　At birth the two sides were alike, or
only so far differed as hereditary transmission had made them to
do, but they have soon become functionally unlike ; they may be
employed equally in larger operations, but in matters of detail
each is performing its own work.　Let us suppose the case of a
person who communicated his thoughts by certain movements of
his right hand instead of by his larynx, and these movements were
called speech, it would follow that if he was struck with hemiplegia,
speech would be lost from disease of one side of the brain.　It is

true that such a case is not identical with that of aphasia, since the organ which we suppose is employed in communicating ideas is paralysed; but they may, nevertheless, be conveniently compared, for if what we understand by language is associated with the operations of the whole of the brain, how is it that the left hand was not ready to take the place of the right. The case shows this, that it is possible to have educated a part of the body for the purpose of communicating ideas, and that one side of the brain becoming diseased, the art which has been acquired is lost. Such a case is not identical with that of aphasia, and no case exactly like it can be imagined, so it must stand alone; but the supposed case shows how easy it might be for some equivalent of speech to be destroyed through a softening of one side of the brain, and that always on the same side. A conception of this kind tends to make the lateral localisation of speech less remarkable than might appear on first consideration. The reason why the left hand did not immediately follow the dictates of the mind, was simply because, being uneducated, it was unable to do so. It is thought by some most unlikely, and I agree with them, that language should reside in one spot of the brain, seeing that it is associated with every faculty of the mind; it can reside in no special place, seeing that it is everywhere; but in most cases of aphasia, as in the one reported, language is not lost, but only the faculty of speech. Now, if this language was expressed through the right hand, and the left side of the brain was diseased, we should undoubtedly have an instance of loss of speech from a local lesion. The loss would be of a different kind, it is true, but we should, nevertheless, be familiar with the fact that disease of one side of the brain would prevent the ordinary intercommunication between people. We should also find that so far from language being gone, the other arm could soon be educated for the purpose. Take, again, the case of music; this is intimately associated with the higher operations of the intellect, so that a genius will clothe some of his subtlest thoughts in the harmonics of his own creation; these he expresses in part through the left side of his brain when he produces them with his bow in the right hand: let a clot of blood form in the brain, and the power of performance is gone. If every faculty of his mind, if his whole soul be imbued and penetrated with music, we know that the concord of sweet sounds came through the small channel of the ear, and therefore it is not remarkable that it should again flow through a channel of equally small dimensions. So ordinary language, gained from impressions which have passed by the narrow inlets of eye and ear, becomes intimately associated with all the operations of the mind, and yet must again be concentrated towards one spot in order to make its exit in the shape of winged words.

The difference between the case of aphasia and that of hemiplegia, just mentioned, is, that in the latter case the organ is paralysed, but in the former this is not the case; but if we suppose that some spot in the brain which lies in mid-channel between the brain proper and the nerve-centres of speech is affected, a somewhat similar result would ensue. This leaves only the final difficulty that the organs of speech are double, composed of structures belonging to both sides of the body, and, therefore, not acted on by one side of the brain only. This difficulty may, perhaps, be overcome by having recourse to the theory of Dr Broadbent, who shows that although the nerve-centres which rule over the limbs are separate, yet those which regulate the body act in unison; they are cemented together, and are both affected by the same stimulus, so that if one channel of motor power be destroyed, the other will transmit an influence to the common centre, and thus no paralysis will occur. He thus explains why in hemiplegia there is no paralysis of one side of the abdomen or chest. Although I believe his theory to be sound, I do not know why, in the case of hemiplegia, it should be assumed that half paralysis of the body would occur as the necessary consequence of disease of the central ganglia, did not the law just mentioned come in to prevent. I am not sure that it is necessary to assume the dependence of half of the body on these ganglia, seeing that it is not wholly paralysed when they are diseased. For if we acquire a knowledge of the function of a centre by the amount of paralysis which its destruction causes, we have only a right to say that disease of the central ganglia causes paralysis of the arm and the leg together, with the ninth and lower portion of the seventh nerves. But as regards speech, the theory of Dr Broadbent would explain how the larynx could be stimulated to action, should its nerve-centres receive an influence from one side of the brain only. Seeing that the larynx, though having a double set of nerves, is a single organ, it is most probable that a stimulus acting on one nerve-centre would be sufficient to cause its movements. Seeing, again, that nearly all the voluntary movements of the body do take place through the sole action of one side or the other of the brain, it is not unlikely that the side which is most active is that which should rule over the organs of speech. The fact of several cases of left-handed men having been recorded who, when paralysed, did not become aphasic, tends to corroborate this view. In any double piece of mechanism of man's contrivance in which the two portions had to be worked in perfect unison, it would be almost impossible to attain this end if each half was regulated by a separate machinery. In the larynx, which must move as a whole, although the stimulus to harmonious movement is acting through separate nerves on the several parts of each side, yet it would almost of necessity follow that the centres whence these nerves proceed must be intimately united, and if so, an impulse falling upon them from one side of the brain would be amply sufficient for their excitation. If this were so, and the larynx had to be trained to any special use, it is most likely that one side of the brain would alone be employed for the purpose, and thus, if this were diseased, although the organ could be stimulated to simple movement by the other side, all those modulations necessary for speech could not be accomplished.

Simple Aphasia.—I have already explained to you the reason why aphasia so often accompanies right hemiplegia; it is that the same blood-vessel supplies the corpus striatum and Broca's convolution over it, and therefore, the probabilities are in favour of the same lesion involving both organs. This association, however, does not always occur; and therefore, just as you may sometimes meet with hemiplegia without aphasia, so you may sometimes have aphasia without paralysis. This may arise in some instances from actual lesion of the convolution; in others from a mere temporary abeyance of its function in common with a similar loss in other portions of the brain. It is, therefore, met with after typhoid fever and other severe illnesses, which may affect the integrity of all the organs of the body. One may see, indeed, how in a weakened brain loss of speech or power of utterance may be the predominant nervous symptom. It is a faculty which is soon lost, and the first to depart after an exhaustive disease. We may all of us understand this by remembering the condition of a man who is exhausted from fatigue; how his intelligence might in no way be affected, since he could understand all interrogatories, and could read what was put before him, yet he might scarcely be able to utter a word, or, if he did so, the word might be wrongly placed, and he would, in all probability, be altogether forgetful of proper names. You will see that some effort is required to speak, to collect the thoughts, as we say; whilst none, or but little exertion, is wanted to understand what is said by another. It is not, therefore, difficult to see how loss of words or language occurs after a severe fever. In some cases, I have no doubt, there is an actual tangible cause for it in the brain, and of the same kind which occurs in other organs, but made more manifest because situated in a highly specialised part. For example, in some forms of blood-poisoning and embolisms, there has been reason to think that some of the smaller vessels have become plugged; but no effects might be manifested if the obstruction was in any other part of the hemispheres than the anterior portion, and then aphasia would be evident. In cases of epilepsy, where one side of the brain is especially implicated in the "discharge," and the other side of the body convulsed and temporarily weakened, a brief aphasia is likely to occur.

I need not detail to you instances of temporary aphasia which follow febrile disorders, but draw your attention to those comparatively rare cases in which it occurs as a primary affection, and is probably due to a slight hæmorrhage, or a lesion of the same nature as takes place in other parts of the nervous centres. It is important to recognise it, because the case is not so easy to diagnose as you might suppose. It is easy enough to make out aphasic symptoms when the patient is lying in bed before us, and we have time to analyse the case, but let a person be suddenly struck speechless, his strange condition will soon become evident to bystanders, but what may have happened to him is very difficult for them to say.

CASE.—I will give you an example in the case of a man who was lately brought to the hospital by the police from the railway station; whilst about to start for his destination, he suddenly felt strange sensations, and was unable to express himself; when spoken to, he could not give any account of himself, and was brought in here. He walked into the ward and was put to bed, he looked confused, and could only say a few words which meant nothing. His gabbling noise and his manner generally were so strange that it was thought his mind was deranged, and the nurses were ordered to keep a strict watch over him. On the following day, when I saw him, and carefully went into the case, it was clear that all his actions were those of a sane man, and that he simply had aphasia. He rapidly improved, his speech returned, and in a few days he was able to give a rational account of himself. He said he had never lost his reason for a moment, he knew everything which happened at the railway station, he knew where he wanted to go, but could not remember the place, nor indeed find words for his own name or address. He might have shown some exasperation at being regarded as mad. To use his own words, he said " he could think well within himself, but he could not say it." He soon left almost recovered, and after repeated conversations with him as to every particular of the attack, I could make out no other symptom than that of aphasia.

CASE.—A medical man informs me that on three or four occasions he has suddenly become aphasic, and he has not quite recovered for some days. On all the occasions he has had numbness and a strange feeling in the right arm. It is clear that whatever has disturbed the equilibrium of the convolution has also troubled the corpus striatum, and the case may have been a form of epilepsy.

As another proof that the functions of the brain are localised, and that aphasia is due to lesion of a particular spot of the brain, I might mention cases of injury to the organ. A case ready at

hand I will read from the journals as occurring at St Bartholomew's Hospital. A boy, æt. 6, had a heavy piece of wood fall on
his head, towards the front and left side, making a wound and exposing the bone. He had feverish symptoms for six days, when he
was seized with paralysis of the right arm. A piece of fractured
bone was removed by Mr Smith and pus escaped. He remained
paralysed, and was quite unable to say a single word; he only
made noises when spoken to. He so continued for a week or
two, when the power began to return in the limb, and at the same
time the power of speech came back. He then rapidly recovered, and
left at the end of two months, the arm only slightly weaker than
the other; his speech had quite returned; he was very irritable,
and swore with great volubility.

Simple dumbness or mutism in children.—You will be frequently
consulted about children who cannot speak, and you will often find
that the difficulty is purely a mental condition; that the children
are in fact idiotic. More rarely you will be asked about children
who are speechless and yet perfectly intelligent. In these cases
speech is acquired slowly, although I cannot say that this has
been so in all my cases, as I have, unfortunately, not been able to
follow them out.

A little girl, æt. 4, who has never spoken, has fits, and, therefore,
in her case, there may be some actual lesion of the brain present.
She is, however, perfectly intelligent in every particular; she obeys
all that is said to her by her parents, plays with her brothers and
sisters, and shows no peculiarities whatever.

A little boy, nearly 4 years of age, has never spoken. He is
well grown, shows no peculiarity in his formation, is perfectly intelligent, takes a watch, puts it to his ear, examines it, and in every
respect appears as sharp as other children.

A boy, æt. 8, I have been consulted about who has never spoken;
his hearing is perfect, and in every way he is as intelligent as
other children.

A little boy, æt. 6, was brought to me because he had never
spoken. He was well grown, with a good proportioned head, and
intelligent looking. He understood everything that was said, and
did all that was told him. He could make noises, but no intelligible
sound.

I am under the impression that in all these cases speech will
eventually come, but it is remarkable that so long a time should be
required to rouse the governing organs of speech into activity. I
apprehend they are merely dormant, much in the same way as those
of the little French boy just mentioned, who, being born in England,
talked English well with other boys; at the same time he had

never expressed himself in French, although he understood perfectly what his parents said to him in that language. It is not uncommon for hysterical women to remain speechless for a long time, as if the volitionary effort could not be roused to stimulate the organs, and maniacs will sometimes remain absolutely silent for years.

A friend of mine, Mr Frank Campbell, has ingeniously suggested in such cases that one of the parents might have been left handed. He has remarked that, whatever may be the cause of right-handedness, it has become in part a matter of hereditary descent, but that where one of the parents has been left handed the children have not so readily adopted the usual custom; their brain has shown no proclivity to action on either side. If in the same way speech belongs to one side rather than the other by the long-continued education of the same side which rules over the hand, a similar perplexity would arise in the speech convolutions of the child who came of right and left-handed parents.

APOPLEXY

What do we mean by the term apoplexy? I am sorry to say that I am much in the same difficulty as I am with some other terms in use, scarcely able to define it, for the simple reason that the expression has had a different value at various times, the probabilities being that it is used in a very different sense nowadays from what it was fifty years ago. Then the term apoplexy was applied in its more strict signification to a malady characterised by certain symptoms; of late years it has been used to denote a particular pathological condition. Then it was applied to the case of a person falling in a fit; now to a hæmorrhage in the brain.

The word ἀποπλήσσω referred of course to the fact of a person being struck down, and the old definition of apoplexy implied that a sudden seizure had taken place, with loss of consciousness, from which recovery might soon take place unless death occurred. Thus, Cullen speaks of apoplexy with effusion of blood, or sanguineous apoplexy, serous apoplexy, hydrocephalic apoplexy, apoplexy from poisons, drunkenness, epilepsy, &c. All this is rational, but by using the term in this extensive sense, it comes to signify little more than insensibility, and thus wholly loses its value as a technical expression, the same word being adopted for the case of one who has extensive effusion in the brain, and for another who is simply dead drunk. I should say, however, that there are still some physicians who use the term in this extended sense, and speak of any one who falls in a fit as apoplectic. This is, however, I believe, not the usage of the profession, for I have taken the trouble to inquire of medical men whom I meet what they imply by the term, and I should say that the word apoplexy is restricted by the majority to cases of effusion of blood in the brain; our experienced resident medical officer, Mr Stocker, always uses the term in this sense. If any doubt existed as to the usage of the profession in this matter, we might recall to mind that effusions in other parts of the body are called apoplectic; and thus we speak of apoplexy of the lung, of the spleen, &c. Of course, if you consider the etymological meaning of the word, no expressions could be more absurd than these, but they show in what sense the term apoplexy is generally understood. I am under the impression that nine medical men out of every ten would imply by the term apoplexy sanguineous effusion. If this be so, we cannot use the term in its original sense, as was done by Abercrombie, and the writers who have followed him; for, by so doing, we shall have to include epilepsy and many other diseases which have already their distinct appellations affixed to them. If, on the other hand, we use the expression for effusions of blood simply, we are departing from the original meaning of the term. Some of you may say, why not limit the expression to those cases of sudden fits which are due to effusions of blood on the brain? The answer is this—that effusions of blood do not necessarily produce sudden fits of insensibility, or those symptoms which were described formerly under the term apoplexy. We are on the horns of a dilemma. The fact is, that pathology has completely upset our notions about the disease. Apoplexy, or "the being struck down insensible," as formerly understood, included such a variety of complaints that the term could be of little value. After some years it was limited, as now, to certain cases where blood was effused in the brain; but it unfortunately so happens that these are the very cases in which apoplexy, strictly speaking, is not present. I should have liked to renounce the word, like many other medical expressions, but as it must be retained we will imply by the term the case where blood is effused, and forget altogether the etymology. This we do in the case of lungs, where the term is altogether out of place.

I am compelled to make these remarks because I believe that although there is now a pretty general consent amongst members of the profession as to the use of the term, much confusion is introduced into the subject by the adherence of writers to the expressions used by our ancestors. Thus, by some apoplexy is defined as being "a disease characterised by the sudden loss, more or less complete, of volition, perception, sensation, and motion, depending upon sudden pressure upon the brain, the tissue of which may be morbid, originating within the cranium." This is so wide a definition that it almost ceases to be a definition at all; but it certainly implies that the term apoplexy is to be used for cases where a certain set of symptoms are present. Another writer defines it as loss of consciousness, with feeling and voluntary motion impaired, or a suspension of the functions of the brain. I have no objection in the abstract to these definitions, if we only agree to use the term in the same sense. I cannot but think, as I just now said, that if the opinion of the profession was canvassed you would find the term apoplexy was not applied nowadays to those cases where there was simple loss of consciousness and no other evidence of effusion of blood on the brain.

Most writers, I say, have followed Abercrombie, who strictly kept to the true signification of the term. This capital observer says the attack occurs under three forms. In the first form the patient falls down suddenly, deprived of sense and motion, and lies like a person in deep sleep, face flushed, breathing stertorous, pulse full; in some cases convulsion occurs, in others rigid contraction of muscles. This may pass off. Such cases as these he calls *primarily apoplectic*. I think there is very little doubt that some of these cases were instances of Bright's disease, many of them epilepsy and various other diseases, which would at the present day be recognised as distinct affections, but certainly not be styled examples of apoplexy. Abercrombie, pursuing his theme, and adapting his expression apoplexy to the etymological signification, then speaks of the second form of apoplexy, or that *not primarily apoplectic*, where there is no loss of consciousness, but syncope, with temporary recovery and subsequent sinking into an insensible or apoplectic condition. In these cases, he says, blood was often found effused, so that the very cases which we at present style, *par excellence*, apoplectic—those cases where a vessel is ruptured and blood effused—are those which Abercrombie styled *not primarily apoplectic*. He used, as you see, the terms apoplectic and insensible as convertible terms. His third form is that where the patient is suddenly deprived of power of one side of the body and of speech without stupor. These he styles the paralytic cases, and says the symptoms were due to effusion of blood or softening. The issue is this : a man who suddenly falls down insensible and struggles, presenting the symptoms which Abercrombie describes as being primarily apoplectic, we should now say has a disease bearing some other name, whereas the very patient who, according to this author, is not truly apoplectic is the very one whom we should nowadays declare to have apoplexy. It is a hard thing to have to perplex you in this way, not about matters of fact, but about difficulties of our own making, and due more especially to authors using the term in one sense when writing, and in another in speaking or in conversation.

In giving, therefore, the symptoms of apoplexy in the modern sense, you will see that they must be of the most varied character, since effusion of blood taking place into different parts of the brain would produce of necessity very various effects. It would be absurd, therefore, for me to detail to you the symptoms of apoplexy as formerly given in the books, since these have reference alone to a condition which, on altogether other grounds, received the name. These symptoms often, indeed, denoted epilepsy or Bright's disease, although very often, too, an effusion of blood. If, for example, you had been called to a patient who had been suffering for some hours from increasing effusion of blood in the brain and you had present the symptoms described, the case would have been apoplectic in every sense of the term. The general idea of a person taken with apoplexy is that he is struck down senseless, having, perhaps, previously had some headache, or felt sick and faint. He may, in fact, have been first collapsed; then, a reaction having set in, his skin has become hot, his face red, and his pulse throbbing or labouring. He will be lying quite insensible, the face drawn up, showing some paralysis, or with the limbs on one side hemiplegic. If you are called in to see a patient in this condition, you may call him apoplectic if you will, using the term in its original sense; but I believe, unless you had the history of the invasion, it would be impossible to say whether or no that man had extravasation of blood in the brain. As I cannot describe every possible variety of apoplexy, we had better divide them first into the cases where effusion of blood occurs on the surface; secondly, where it occurs in the substance; and, thirdly, where it takes place in the central ganglia.

Beginning with the last, which is the most common—the patient, say a man, and generally somewhat advanced in life, has an extravasation of blood occur in the corpus striatum or its neighbourhood. He experiences suddenly a strange feeling in the head, a giddiness, he may fall, he turns pale, his pulse goes down, and he is sick. He shortly recovers, and, perhaps, with help, is able to walk a short distance, when it may be observed that one side is weak. These were the symptoms of a man who was so seized whilst working on the hospital premises, and, being brought to the ward, we had an opportunity of immediately seeing him. An hour afterwards he was lying in bed completely hemiplegic, having lost power of his right arm and leg; the face also somewhat fallen on that side; and,

being thus paralysed, he was not able to speak. When asked
to put out his tongue it protruded towards the paralysed side.
These were all his symptoms. The earlier symptoms, due to the
shock which the brain had received, had passed off, leaving merely
those due to the local lesion. It was a clear case. He had suddenly had some injury to the motor tract, which, no doubt, was due
to effusion of blood ; and, as no special nerves were affected, I said
it was in the corpus striatum or its neighbourhood. At this time his
difficulty of speech could not be decided as attributable to aphasia.
Now mark, this man had not lost his consciousness, and therefore
was not apoplectic in the old-fashioned use of the term ; in fact, as
Abercrombie would have said, he was not primarily apoplectic.
Since, however, we nowadays should call such a case, *par excellence*,
one of apoplexy, and the clot which we found after death an apoplectic clot, it certainly implies a contradiction in terms to say a
man is apoplectic who has no unconsciousness, and yet we must
adopt this term.

That, in such a case of effusion, paralysis is the main symptom,
is seen in the fact that the extravasation may occur during sleep,
and the patient wake as usual with his intellect unclouded ; he then
finds that on attempting to get out of bed he falls to the ground,
and afterwards discovers that he cannot articulate, or that his
speech is thick. He is hemiplegic in the manner in which I told
you. If the clot be circumscribed in either of the central ganglia,
no worse symptoms need arise, but the patient gradually recovers,
first getting the use of his speech and then of his extremities. But
it is only in a certain number of cases that the effusion is so slight ;
and thus, as in the case of the man just now mentioned, whom I
saw after the occurrence of the fit, I said that the blood was then
limited to the ganglia, but whether it would there remain or extend
still further I could not say. I said that if it burst through into
the ventricles, then he would fall into a state of coma and
have the usual so-called apoplectic symptoms. I came in to see
him in the evening, and the event had occurred—the blood had
broken through into the ventricles. A great pressure was taking
place on the surrounding parts, squeezing the convolutions out flat
against the skull, whilst passing downwards, the blood had filled
the fourth ventricle, and so involved the respiratory and other
nerves. He was then in the truly apoplectic condition as originally
understood—in a perfect state of insensibility or coma. The whole
body was paralysed, the limbs dropping when raised ; the rectum
had lost its power, and there was a fæcal escape into the bed.
The breathing was irregular, mucus collected in the tubes, and in
a few hours he died.

There are many symptoms of the true apoplectic condition which it is important to notice : I will first mention stertor or snoring. Snoring is a temporary, whilst stertor is a real, paralysis of the palate, whereby it flaps to and fro as the patient breathes. At stertor occurs during inspiration, so there is a peculiar puffing-out of the cheeks during expiration from paralysis of the buccinator and other muscles, occurring more on one side than on the other, according to the side affected. As the chest and the lungs are paralysed, so the mucus collects from an inability to cough, and thus it flows from the mouth, or is blown out on one side by force of expiration. The respiration is often altered in rhythm. It will sometimes get slower and slower, and cease for half a minute, and then go on again for a few seconds and again cease. Or some-times the expiration is extraordinarily prolonged, and then a deep short breath is taken. Dr Stokes has described a somewhat similar kind of expiration in cases of diseased heart, where the patient for a moment ceases to breathe, then short and superficial respirations commence, which gradually deepen, until an extreme one is taken, when they gradually diminish until the arrest recurs again. The surface of the body is congested and the face livid. The heart is also partially paralysed, and thus decreases in frequency, the pulse coming down, perhaps, to 50, and may be irregular. In this condition you will often find the pupils contracted. Much has been said about the condition of the pupils in brain disease, so as to assist us to diagnose between concussion, compression, apo-plexy, and drunkenness, and on more than one occasion we have had letters in the newspapers informing us that mistakes would not occur if we had recourse to the state of the pupils. But, as I shall again have to remark I think there is no rule in any of these con-ditions, and that the state of the pupils will depend much upon the part of brain affected ; but you may be sure that either an extreme dilatation or contraction denotes disease. In that severe form of apoplexy where blood is poured into the ventricles, and presses on the base of the brain, the pupils are very generally minutely con-tracted. Where a large effusion has occurred into the substance of the hemisphere the pupil on that side will often in the first in-stance be larger than the other, although subsequently it may be found smaller. Sometimes both eyes are forcibly turned away from the side that is paralysed, as I have mentioned under hemi-plegia. In a case I saw lately of a man who died in profound coma from an apparent apoplectic attack, and in which no special paralysis could be made out, the eyes, which were contracted, con-tinually rolled from side to side. Slowly and methodically they first deviated to one side and then to the other ; this continued for

some considerable time whilst I watched him. Sometimes the paralysed side is rigid, which was regarded by Dr Todd as an evidence
that the effusion had occurred in a healthy brain, and that the
paralysis was not due to softening. I cannot say that this explanation is correct, but that an irritation is set up in the corresponding
ganglion on the other side you may know by the twitching or convulsive movements which are frequently observed in the healthy
limbs. Sometimes it is merely seen as a constant restlessness and
pulling up of the bedclothes by the hand ; therefore, unless carefully watched, the patient may wriggle off the bed. The convulsive
movements are sometimes so severe that the case is regarded as
one of epilepsy ; and thus, just as epilepsy accompanied by unilateral paralysis is sometimes mistaken for apoplexy, so I have on
several occasions seen apoplexy called epilepsy. When there is
much pressure on the brain, the surface in a few hours becomes
inflamed, as I have frequently seen, and this is, perhaps, another
cause for special symptoms. Sometimes the limb is painful when
moved, and this, I believe, augurs badly for the recovery of the
patient. Whether the pain is centric or is due to a neuritis in the
limb, has not been made out. In a case lately in the hospital, of a
man who had from infancy a weak arm and leg the former became
rather suddenly flexed across the chest, and any attempt to move
it caused most exquisite pain and almost threw him into convulsions.
After death a most careful examination failed to show any recent
inflammatory process in the brain, cord, or nerves.

I should say that this which I have described is the commonest
form of apoplexy, and, when witnessed from the commencement,
is the easiest to diagnose. There is first the shock, then the
hemiplegia and loss of speech, followed by such symptoms as would
denote the spread of the blood through the substance of the brain,
or its bursting out on to the surface, or, as more frequently
happens, into the ventricles—viz. coma, stertor, and all those
symptoms which are usually described as constituting apoplexy.
It often happens that some hours may have elapsed before the
doctor is called in, and then all the earlier symptoms have passed.
He finds the patient in a perfectly insensible condition, with
contracted pupil, a slow labouring pulse, flushed face, profuse
perspiration, stertor, froth issuing from the mouth, and other signs
which "foretell the ending of mortality."

Having cursorily regarded the physiology of the nervous centres
as far as can be elaborated from pathology, you will perceive that
the symptoms of apoplexy will vary according to the seat of the
disease and the amount of the effusion. The commonest form is
that which I have mentioned where the effusion takes place into

the central ganglia. If the blood be restricted to these parts, the symptoms proceed no further than the hemiplegic stage. Let the blood exceed these boundaries, a truly apoplectic state results, and death usually occurs within twenty-four hours. But it is clear that if a large vessel gives way in the first instance, the symptoms of the first stage, the primary shock and the hemiplegia, may soon pass off, and the blood pouring out into the ventricles may so compress the vital parts that death will rapidly ensue in a few hours. If blood be effused to a small amount in any other part, the symptoms depend upon the site. If in the pons, and interfering merely with the motor tract, it is not necessarily fatal; but if in larger quantity, so as to compress neighbouring parts, then, of course, fatal symptoms rapidly ensue.

I need not detail any particular case to you, but if you take up a volume of our Inspection Books for any year, you will find plenty of cases in illustration, of which the following are outlines.

CASE.—A cook, æt. 52, whilst employed at her occupation in a gentleman's house, was seized with apoplexy. A medical man, who was speedily in attendance, bled her and sent her to the hospital. She was hemiplegic on the left side, and retained her consciousness, which had never left her. She died on the following day, when a large clot of blood was found in the right corpus striatum, tearing up the substance of the brain; also some fluid blood in the ventricles.

CASE.—A man brought in perfectly insensible; stertor; quite powerless; contracted pupils. The ventricles found full of blood, which passed down through the third and fourth to the base. The corpus striatum was only superficially ruptured, so that the blood had flowed at once into the ventricles. There was no clot in the substance of the brain.

CASE.—A woman of loose character took up her night's lodging at a coffee house. On the following morning she was found perfectly insensible. The police who were called in believed she was drunk, and the doctor that she had been drugged. She was sent to the hospital in a perfectly unconscious state, with stertor, minute contraction of the pupils, and limbs quite powerless and flaccid. After death a large clot of blood was found occupying the right hemisphere. This had burst into the ventricles, which were full of coagula; blood also had passed through to the base, and then travelled up to the sides of the brain.

I have already spoken of the results of effusion of blood in large quantity into the pons Varolii, and the difficulty in diagnosis. When the effusion is less, there need be no loss of consciousness, and, if on one side, the seventh nerve is often implicated in the paralysis.

CASE.—A man, æt. 50, who was said to be well in the morning, was seized with a fit. When brought in, he was conscious, and spoke, though hesitatingly.

6

The left side of the body was paralysed, and the right side of the face. Effusion in the pons was diagnosed. On removing the brain after death nothing was noticeable, except a swollen condition of the pons Varolii. This, on section, showed a clot on one side, just breaking into the fourth ventricle.

Dr Moxon had lately the case of a man under his care with Bright's disease. One day he suddenly became dull and half conscious; he muttered and spoke with difficulty. The pupils were contracted. He seemed also to have paralysis of the left facial. On the next day his consciousness had returned, and he moved his limbs. He subsequently fell into a state of coma, and died. A clot of blood was found in the pons.

I had lately an opportunity of seeing a case which you may regard as a type of the ordinary form of apoplexy. A man, whilst engaged in his business, was seized with violent pain in the head. He then began to talk incoherently, and was assisted to his room, when he fell down, and had a convulsion. The doctor found him collapsed and cold, and with a pulse scarcely to be felt. He gradually came out of the syncope, and then passed into the apoplectic state, with stertor, great rise of temperature, throbbing pulse, left side paralysed, right constantly moving, and hand pulling the bedclothes. Death in a few hours.

The collapse is sometimes very remarkable. A few months since I was called, with Mr Stocker, to see one of the treasurer's servants. She had fallen in a fit in the yard, and when brought indoors was cold and pulseless. Brandy was administered, and it was doubtful whether she would survive the shock. She gradually revived, when the side was found paralysed, and she has now nearly recovered.

Considering the different causes which may give rise to effusion of blood in the substance or on the surface of the brain, the symptoms must of necessity differ in detail. If the case be fatal, and a careful examination be made after death, the symptoms and post-mortem appearances can generally be made to accord, but it is not to be argued from this that such symptoms of necessity denote that particular form of disease. Very difficult of diagnosis are aneurisms of the larger arteries of the brain, in which there have been symptoms indicative of pressure on some particular region, and subsequently those of apoplexy, due to bursting of the sac and flooding of the brain with blood.

Cases are constantly occurring in which effusions of blood may be conjectured to be present, but cannot be absolutely diagnosed, as the symptoms are not sufficiently distinctive. For example a lady goes to bed well; wakes with an intense pain in the head, which continuing the medical man is sent for; just as he arrives she sinks into an apoplectic condition and is collapsed; she gradually recovers, and when I see her some time afterwards she has perfectly regained her consciousness, but has complete paralysis of the left third nerve. Here is a case where you might form several conjectures as to the cause and site of the lesion. The subjects of apoplexy, as a rule, have the premonitory symptoms just mentioned, and then fall into a state of insensibility ; then a gradual recovery takes place with some form of paralysis. Sometimes the premonitory symptoms are of longer duration, exemplified by pains or strange sensations in the limbs, headache, giddiness, &c. In other cases the vessel which ruptures is a large one, the blood bursts out in large quantities, coma comes on, and death ensues in two or three hours. These are the cases which the late Dr Addison was accustomed to style a "smash into the ventricles."

Sometimes loss of sensation accompanies loss of motion in the hemiplegic, or, to use technical phraseology, " anæsthesia often accompanies akinesia." I have already alluded to this when speaking of anæsthesia, and suggested that the thalamus was the seat of the lesion. Much of our difficulty arises from the doubt as to whether there be any loss of feeling or not. There is often some numbness, and yet the patient feels when pricked. He will also often not know in what position his arm is lying, and feel for it with the other.

In the rapidly fatal cases the blood has most usually had its source in one of the central ganglia, then ploughed up the substance of the hemisphere, or burst into the ventricles. In those cases where no further symptoms occur after the first shock and the hemiplegia, recovery takes place. If at a subsequent time the brain is seen the remains of the effusion or its consequences are found to be limited to the spot where the rupture took place.

In the common form of ingravescent apoplexy Dr Broadbent thinks that the symptoms are due not only to the slowness of the extravasation, but the seat of it ; that it does not take place in the corpus striatum proper, but between this and the island of Reil ; nor does the blood flow into the Sylvian fissure.

Occasionally blood is effused into the *white substance* of the brain, and then, as you might suppose, the symptoms are most obscure. In some cases where such effusions have been found, the symptoms, when sought for, have been of the vaguest character—so wanting in preciseness, that when occurring I could not expect you to diagnose apoplexy of the medullary matter of the brain, although you might suspect it. There have been symptoms which denote cerebral disturbance, as pain in the head, giddiness, sickness, faintness, followed by some reaction and labouring pulse, but nothing more distinctive.

Somewhat more definite are the symptoms which result from *meningeal apoplexy*, that is, the case where blood is effused on the surface of the brain ; when the fact is known the symptoms are quite appropriate, but I cannot say, on the other hand, that the symptoms are sufficiently distinctive to warrant the diagnosis of meningeal apoplexy, although they may be suspected. You might imagine from what I have already told you what the symptoms might be—no paralysis, but mental confusion or delirium, ending in coma, and perhaps convulsions. The great importance of having a knowledge of such cases is in a medico-legal point of view, seeing that a hæmorrhage on the surface may be either spontaneous or arise from injury. I will not enter upon this subject, as I have done so elsewhere. If blood be poured out in large quantity and quickly, coma and death rapidly ensue. In less amounts the symptoms are those of irritation. These are best seen after injury, but occasionally also in the medical wards. Thus there may be convulsive movements, or even attacks of a distinct epileptiform character. The majority of such cases in my experience have occurred in Bright's disease, and have been attributed to uræmia.

CASE.—A little girl was under the care of Dr. Pavy with renal dropsy, when one day, and a few hours before death, she was seized with a screaming fit, followed by convulsions. A layer of blood was found covering the brain.

CASE.—A young woman came into the hospital suffering from pains in the joints and a mitral bruit. Being soon after her confinement, the condition was undetermined between rheumatic and puerperal. She soon afterwards complained of great pain in the head, and became perfectly maniacal, in which state she died. There was found a diseased mitral valve and embolic concretions in kidneys and spleen. The anterior part of the brain was covered with blood, effused under the arachnoid, and which passed down between the convolutions. There had evidently been a recent rupture of a vessel in the pia mater, although, as usual, undiscovered.

CASE.—A woman, aged 58, having previously felt giddy, fell down in her room. She was taken up insensible, but recovered in about ten minutes, and spoke. In twenty minutes she again became unconscious, and, remaining in this state all day, was brought to the hospital. She was in a perfect state of coma, with stertorous breathing ; limbs rigid, with occasional twitching, and drawn up when touched. Pupils continually varied in size, and the face appeared paralysed, sometimes on one side, and then on the other. The diagnosis was uræmic poisoning. After death an examination showed the whole surface of the brain covered with blood, and which had evidently proceeded from a large vessel in the pia mater.

CASE.—A young man fell and struck his head on the pavement. After recovering he went home, but appeared in a half-stupid state, and on the following day he had a fit, which was called epileptic, and on the succeeding day another. When taken into the hospital his condition was like that of a man who had delirium tremens. On the following day he had another epileptic fit, which left him in a state almost maniacal. He continued thus for nearly two weeks, having fits, and being in the interval in a condition much resembling that of delirium tremens. The post-mortem examination showed a fibrinous clot of blood closely adherent to the cineritious surface of the brain.

I have on more than one occasion seen a patient trephined for meningeal apoplexy, showing how difficult the diagnosis is. A man fell down whilst at work, and was immediately brought into the hospital. He was pale, comatose, with stertorous breathing, and had rigidity of his limbs, with jerking movements. Blood was found effused over both hemispheres.

Another man fell downstairs, and was brought here insensible, with a scalp wound, and soon afterwards had a fit. He was trephined, and nothing found under the bone. After death effusion of blood was found on the brain, and some old local disease. It was then learned that he had been subject to fits.

In some cases of meningeal apoplexy an old disease of the brain has been found, as if this had been the source of the haemorrhage, just in the same way as we believe it sometimes occurs in the spinal cord.

In a case which occurred lately of a man who had fallen and cut his head the ventricles were found full of blood, but a most careful examination failed to discover the source of the haemorrhage.

In the case of a man who was brought into my ward in an apoplectic condition the symptoms were of the most heterogeneous kind, and were found after death due to effusion of blood on the under surface of the brain and into the fourth ventricle, the brain itself being healthy. He was taken, a few hours before his death, with giddiness, pain in the neck, inability to swallow, closure of left eyelid, right pupil strongly contracted, left dilated, but could move his arms and legs. The symptoms were due to an implication of the nerves rather than to any affection of the brain itself.

Concussion.—The effects of concussion were well seen in a case under Mr Durham, where a young man fell down stairs on his head. He was insensible, with pupils dilated and a marked rigidity of his arms and legs. He continued in this state for nearly forty-eight hours, when his temperature rose very high, and he died. There was no fracture, no laceration of the brain, nor effusion of blood. Its whole surface was very red, a section showed it everywhere studded with red points, and the substance of the brain everywhere was of a pinkish yellow colour. The surface of the ventricles was covered with minute ecchymoses, as well as fornix, septum lucidum, &c. The central ganglia were also similarly dotted.

Effusion of Blood on Surface from Injury

CASE.—A young man, aged 18, was admitted into the hospital with cerebral symptoms of a not very severe nature and died on the following morning. Very little history was procurable, but it appeared that eleven days before, he had been fighting in a barge with some other men, and received a blow on the head. He did not suffer much in consequence, and continued his employment during the next ten days; on the following day, feeling unwell and his head ache, he came to the hospital; he walked up to the ward, appeared quite rational, his only complaint being pain at the back of the head. There could be little doubt that he had been more ill than he chose to admit; he was depressed and wished to conceal his condition. Body that of a strong muscular young man; the head presenting no signs of injury. On lifting up the dura mater, the surface of the brain on the right side was seen to be covered with blood; the blood was mostly fluid, and about four ounces in amount; a few coagula remaining adherent to membranes. The whole of the brain on the right side was surrounded by blood which compressed it, and somewhat flattened the pons. The left hemisphere was untouched. The brain was carefully washed in order to discover any breach of surface or rupture of blood-vessels, but it appeared perfectly sound in every part. A close examination of the blood showed that a part of it might have been due to a recent effusion, but there were thin layers of coagulum which had evidently been there for some days; these were assuming a brownish or ochrey colour, and were becoming very closely attached to the dura mater. No crystals of hæmatoidin discernible.

Effusion of Blood into the Cerebellum.—Dr Broadbent has lately published two very interesting cases of this form of apoplexy, and in neither were there any characteristic symptoms.

One was that of a girl, æt. 20, who had an effusion of blood in the left lobe of the cerebellum, and died after a few days. She lay on her side, unwilling to be moved or spoken to. She had pain in the head, and seemed as if she suffered from great weariness, and wished to be left alone. She at last died suddenly from rupture into the ventricles.

The other case was that of a girl, æt. 16, in which there was effusion of blood bulging into the fourth ventricle. She was seen leaning against a wall, complaining of pain in her head. She was taken to St. Mary's Hospital, and was then insensible. She moaned, and could move all her limbs, but resisted all movements made by others, and gave evidence of pain. She died in two hours.

You will observe from what I have said, although I should think most of you have seen a sufficient number of cases to render the statement unnecessary, how erroneous is the opinion that apoplexy (that is, effusion of blood in the brain) is suddenly fatal. This is a popular opinion, and exists, I believe, even somewhat extensively, in the medical profession. This is evidenced by the fact that in stories and theatrical pieces the characters are made to die suddenly of apoplexy; and as regards a very prevalent opinion amongst ourselves one may read in the daily papers the accounts given of inquests held on persons who have died suddenly, where the medical man has attributed death to apoplexy. I will just say, once for all, that apoplexy does not cause sudden death—a popular mistake, and one not yet eradicated from the mind of the profession. The case of shortest duration of which I know is where an effusion of blood occurred on the brain, and the patient being dragged through the street, survived only an hour.

This opinion has been promoted by another delusion—for I cannot help calling it so, although still held by some members in the profession—which is this: that persons of a certain configuration are prone to apoplexy. It is said that the pattern of body which is most prone to apoplexy is denoted by a large head and red face, shortness and thickness of the neck, and a short stout squat build. This remark is as old as the time of Hippocrates. Hippocrates and those I quote are very good authorities, and it might appear presumptuous to differ from them; but the difference of opinion lies probably in the explanation of the cause of death. What did they observe? That such persons as just described died suddenly. True, and, according to my showing, the very proof that they were not apoplectic. The mistake has arisen for two reasons—first, the error as to the cause of death, more extended observations telling us that the suddenness of the death must have reference to the heart; secondly, the error arising from mere vague impression. Thus, a man with a red face is thought to have more blood in his head than a pale one, and therefore it is always ready to burst out into his brain. You know very well that a man with a red face has no more blood in his brain than another: it is a mere idle fancy; it is the associating two things together, in our imagination, which have no real connexion. It is like the association of hydrophobia with the dog days, these being named after the star Sirius, which is to be seen in our winter nights; or the erroneous belief that fever is most prevalent in summer. The fact is, that blood is poured out in the brain because a vessel has burst. The person in whom the vessels are diseased is consequently he in whom apoplexy is most likely to occur. Such a person is often pale and thin, with a long neck. I knew a gentleman some years ago who had such an extraordinarily red face that some young friends disliked to walk in the streets with him lest he should die of apoplexy; his face was of a deep purple hue, like a ripe gooseberry, ready it was thought to burst and let out the contents. This gentleman died of heart disease.

If a small amount of blood be effused, producing hemiplegia, and no further result follow, the patient gradually recovers. The paralysis of the face quickly passes off, and if it be on the left side the patient is soon enabled again to speak, although the paralysis of the arm and leg remain. I should think that, in many cases, the effusion of blood can do little more than compress and temporarily suspend the function of the part, for otherwise we could scarcely account for the complete recovery, which may sometimes ensue in a few days.

In most cases, however, a very lengthened time is required for recovery to occur, and this is only partial ; as the blood is absorbed the parts again come into order, and their function is resumed. But since some of the conducting fibres are absolutely severed, it is impossible that motion will ever be perfectly restored. You will find, as a rule, that the leg recovers before the arm, but at the end of some months nearly all hope is gone of either limb permanently recovering, if not restored by that time. You will see, therefore, how absurd it is for us to assent to the suggestion of a patient to do something for a paralysed limb after years of its existence. A week or two ago a lady asked me if something could be done for her arm and leg, which had been paralysed for twenty-five years. She was really asking for the removal of a cicatrix from the brain, and a restoration of the original tissue. This would be strikingly absurd if asked in reference to a scar on the skin.

During the paralytic state the muscles are flaccid, and respond readily to either form of galvanism, and reflex excitability is retained. The temperature is often slightly increased. Sometimes, after a few hours, the limb becomes rigid and painful when moved. This symptom, in my experience, is unfavorable. When recovery takes place, the speech becomes distinct, the tongue is put out straight, and power returns in the limbs, beginning in the leg. If the arm recovers first, Trousseau considers the case a bad one, and that the mind will suffer. With this view Dr H. Jackson coincides. Dr Gull, in his Gulstonian lectures, gave two cases where the arm recovered first, and the disease was situated in the inferior part of the posterior lobe of the cerebrum. If there has been a lateral divergence of the eyes, a diplopia or hemiopia may exist a long time during convalescence. This is so now in one of my patients, and also an inequality of the pupils. If further observations tend to confirm the views of physiologists as to the localisation of functions in the hemisphere, it may perhaps be shown hereafter that these additional symptoms are owing to a wider implication of the surrounding convolutions ; and that hemiopia, for example, is due to the injury of a part situated more posteriorly, which Ferrier has shown to rule over vision.

Now, what becomes of the clot ? The blood disintegrates, it becomes yellow, ochrey, or coffee-coloured ; the corpuscles break up, and out of the colouring matter crystals arise ; these take at least a fortnight to form. Whether a true cicatrix may follow is doubtful ; more generally some inflammation occurs around the clot, lymph is poured out, which hardens, and thus a cyst is often produced. In a person who has been long hemiplegic a small cyst containing fluid may be found in the corpus striatum, or a brown spot may be seen looking like a cicatrix, the remnant of the dried-up clot. In some cases of effusion of blood in the brain it may be useful to ascertain the time of the seizure ; the presence of crystals may assist us in the inquiry, as they seldom are found before two or three weeks. These crystals, which spontaneously form, are called hæmatoidin, and differ from the hæmine crystals which are formed artificially by acetic acid. The latter are small rhombic plates with acuter angles, whilst the former, which form spontaneously, are much larger and broader crystals, and of a deep ruby-red colour.

If this process of recovery do not ensue, a softening may gradually go on until neighbouring parts of the brain are involved, and then further symptoms arise, of which I shall speak when I come to softening. You may often find that the emotional powers are readily excited, as if less under the control of the patient ; but whether the disease has or has not in these cases proceeded beyond the region of the ganglia I am not quite certain.

The contraction of the limb in some long-standing cases of paralysis is very remarkable. I have already alluded to a rigidity of the limb which is sometimes observed at the commencement of the attack, but I am now speaking of the permanent contraction, with a certain amount of withering of the limb, which gradually comes on after the paralysis has existed for some time. Dr Todd believed it was due to an excitation continually sent down to the nerves from the brain, and which, under these circumstances, always presented the condition of a cicatrix, the irritation caused by which was the origin of the phenomenon. I have always known that this could not be the explanation, seeing that it occurs under a variety of morbid conditions of the brain ; indeed, the most marked cases of contraction I have witnessed have been those where, so far from there being a cicatrix, the brain tissue had become altogether destroyed. In fact, it was long ago maintained by Lallemand that contraction more frequently occurred in softening. I remember the case of a man who had had his arm rigidly flexed across his chest and his fingers contracted for nearly two years, and whose brain presented little more than a hollow space on one side. It is for such reason that I have always maintained the impossibility of the rigidity being due to any influence propagated downwards, the nerves being in a state of mere negation as regards cerebral influence. An explanation first offered by Dr Charcot, and confirmed by Dr Béhier, the clinical professor at Paris, was that the paralysed limb had undergone a great change in its tissues; the muscle was atrophied, and the nerves had become enlarged and indurated through the addition of a connective tissue. The nerve filaments had become reduced in size, whilst the neurilemma was thickened In this condition the term "cirrhosis of the nerves" had been applied. The change might have occurred by a descending neuritis from the cord. An extension of disease along the motor tract from the c. striatum had already been demonstrated by Bouchard. The muscle itself is altered, and does not become flaccid when the patient is under the influence of chloroform. At other times a slight œdema may occur in the paralysed limb. In the early rigidity there is often pain on moving the limb; and it may be a question whether this is due to a commencing neuritis or to an extreme sensibility of that part of the cord near the seat of disease. Certainly this may be suspected when the limb is seen to move or jump on the patient gaping or sneezing. Sometimes the limb is constantly moving, as in a limb affected by chorea, but this is generally at a later stage, when the arm is permanently rigid. If we suppose a local inflammation of the brain in early rigidity, and a chronic inflammation in late rigidity, we should have the same pathology in the two cases. Rarely a subacute inflammation of the joints occurs which would account for the pain. The skin also of the palm of the hand becomes soddened, and emits often a very disagreeable odour, showing that the cutaneous secretions are morbid. The occurrence of these and other symptoms of the kind, indicating that nutritive changes are going on, probably implies that the morbid process has descended to the cord; for, as a rule, no wasting or any other marked alterations occur in a hemiplegic limb.

Causes of Apoplexy.—According to the old definition, these might be numerous, the simplest being that, where a ligature is placed around the neck, producing temporary congestion of the brain; but restricting ourselves to the definition of effusion of blood arising from the rupture of an artery, the main cause would be disease of the blood-vessels, hence the reason of increasing liability to apoplexy with advancing age. In younger persons a cause might sometimes be found in aneurism of the cerebral vessels, in all probability much more frequently than is now done if they were carefully looked for. I remember thirty-five years ago, seeing the late Mr Wilkinson King carefully washing a brain so as to expose the blood-vessels, as he expected an aneurism, at the same time informing me that he had discovered a ruptured sac on three several occasions in the midst of an apoplectic clot. Further researches of late have shown that these are far from uncommon, and that many may be found in the same brain. These are called miliary aneurisms. Since in apoplectic patients there is disease of the blood-vessels, you might suspect that they were often the subjects of Bright's disease. You are constantly being told of the various degenerations which occur in the tissues in morbus Brightii, and more especially in the blood-vessels, one evidence of which is witnessed in the amaurosis and effusion in the retina, which is a frequent concomitant of Bright's disease. In most cases you will find the cerebral vessels evidently diseased, as seen by the naked eye; in other cases you find the smaller arteries thickened, and the more minute ones, when placed under the microscope, are seen to have undergone a fatty degeneration. You know also how frequently hypertrophy of the heart is found in chronic morbus Brightii, and in connection with this circumstance is the interesting fact that this state of heart had long been observed in fatal cases of apoplexy, and a theory was held that the effusion of blood was due to the increased pressure on the vessels in consequence. The observation shows that these were really cases of Bright's disease in which apoplexy occurred, although, most remarkably, the kidneys were never examined. It is important to remember the connexion, because the discovery of albumen in the urine of a person lying in a lethargic condition might suggest uræmia, and lead us away from the idea of apoplexy, whereas it would only be another evidence in favour of diseased blood-vessels. This hypertrophy of the heart led to the notion that increased pressure on the blood-vessels had much to do with their rupture. Now, as a matter of fact, I believe they must be actually diseased before they can give way. But supposing the moment has all but arrived when the catastrophe is to happen, you may imagine that a very little increased force acting on the vessel will cause it to rupture. More recently Dr G. Johnson has enforced this doctrine—that the first change in Bright's disease is on the blood-vessels, which become hypertrophied, and subsequently the left ventricle; that the compensatory thickening of vessels is not uniform, and therefore the cerebral and retinal vessels, not partaking of the change, more readily yield to the increased action of the hypertrophied heart.

It is said that increased tension in the arterial system may cause rupture of a vessel, or delay in the venous system do the same, but in a healthy person strangulation does not produce apoplexy, and in those frightful cases of congestion of the head seen in bronchitis apoplexy is not an ordinary result. The fact is, the character of the circulation of the blood in the head is not yet quite understood; it appears that it can never contain more than a certain amount of fluid. Then, again, the anastomoses are so remarkable that we may have such a state as I one day met with in a case of ligature of the carotid, which certainly would never have been anticipated. The common carotid had been tied, and when the patient died a few days afterwards the whole of the cerebral vessels, as well as those of the pia mater, were filled with solid coagulum. It is said that the immediate causes of apoplexy are over-exertion, straining at stool, or any undue excitement. I cannot say from my own experience that such causes are effectual. It appears to me that it has been too readily assumed that during exertion or movements of the body there is greater stress thrown upon the blood-vessels, and that they are then more liable to rupture. The presumption has been taken from the fact of hæmorrhage on the surface of the body being arrested by pressure and quietude ; but no inference can be drawn from this as to the amount of force exerted on the blood-vessels during rest or movement. You know Mr Durham has made experiments whereby he has shown that the circulation of the blood through the brain appears less active in sleep than waking. However this may be, I cannot disregard our experience as to the occasions on which hæmorrhage is likely to occur from the bursting of a diseased vessel. In the case of a diseased cerebral artery I am not at all sure that the pressure is greatest upon it when the circulation is active, for how can we account for the very frequent occurrence of apoplectic seizures in the night when the patient is asleep ? Take, again, the case of hæmoptysis in connection with tuberculous disease of the lung. I believe it is rather the rule than the exception, when this occurs, for the patient to wake in the night with blood in the mouth ; in the day-time there had been no sign of its occurrence when the patient was actively engaged. In spite of this fact, when we are treating a case of hæmoptysis, we insist on the patient being absolutely quiet, to lie in bed, to make not the slightest movement of the arms, such as putting on a coat, which we suggest might seal his fate. Believing that such practice has been determined entirely on *à priori* considerations, I have for some time ceased to adopt it, and certainly with no ill consequences to the patient. I am inclined to believe, although I must not teach this as a proven fact, that the pressure exerted on the blood-vessels is greater during sleep, and is lessened by exercise. The case of sudden rupture of the heart will also bear out my statement. In several cases which I have seen where the patients were found dead in bed a fatty heart had suddenly ruptured, and this in persons who had been at work, or at least walking about, all day. I believe there are cases of heart disease where a little gentle exercise will promote a more active circulation through the system, and thus quiet the palpitating and irritable organ. These remarks arise out of the question as to the exciting causes of apoplexy, and therefore I express my suspicion that those usually given have been concocted by writers in their study, and not adopted as the result of experience.

I have said little of the premonitory symptoms of apoplexy which are often spoken of, as numbness and tingling of the fingers. I shall again allude to these, but regard them as of little value in the diagnosis of approaching apoplexy. Certainly they are of little importance when we remember their excessive frequency in gouty and nervous persons who again get better of them. Headache in like manner is of no diagnostic value. It is true that pain in the head sometimes, though exceptionally, precedes an apoplectic attack, and remains after the vessel has ruptured, but in such cases some special or accidental cause must be in operation.

Diagnosis of Apoplexy.—According to the old definition, this is easy. If we go back to Cullen, it is simply synonymous with insensibility, and would be applicable to cases of poisoning or drunkenness as well as to cases of disease. If, however, we mean by apoplexy effusion of blood into the brain, how do we then frame our diagnosis? If the attack be of the more ordinary kind mentioned, this is easy. If there be the sudden seizure of pain in the head, giddiness, &c., followed by thickness of speech and weakness of limbs, the case is clear. If, however, you are called to see a patient at a still later period, when he is quite insensible, you will have to inquire into the history. You may find him in the ordinary condition which is known as apoplectic, with stertor, coma ; but you may clearly make out from the face and movement of one side that a paralysis exists, and therefore you may diagnose apoplexy. If the coma is profound, the chance of his ever coming out of it again is most remote. If you see a patient immediately he has fallen, and you find him in a state of deep coma, you may be pretty sure the case is not one of apoplexy. If you hear that he has struggled, it is in all probability epilepsy, and the fact of his having apparently some hemiplegia is no argument against it. In this case it would probably be not one of simple epilepsy, but one of disease near the corpus striatum. When a medical man is called to see a person who is lying insensible with apparent paralysis of one side, the case presents considerable difficulty. I have seen an epileptic patient under these circumstances regarded as apoplectic ; and, on the other hand, I have, on several occasions, seen the case of fatal apoplexy looked upon as epilepsy. This was no doubt owing to convulsive movements. The most difficult case which the house-surgeon of an

hospital can meet with is that of a man brought from the streets
in a perfectly insensible condition. He may have fallen and have
a wound on the head. The questions the surgeon asks himself are
—Is there an injury? Is the case one of apoplexy? Is the
patient poisoned or drunk? In complete insensibility there may
be effusion of blood in large quantities in the ventricles, at the
base of the brain, or in the pons Varolii. In these cases the pupils
may be contracted, and thus the resemblance to poisoning by
opium. The probabilities would be in favour of the latter if the
breathing were very slow. You have heard Mr Stocker express
an opinion in favour of apoplexy if, on undressing the man, there
is found a fæcal discharge in his trousers.

If the patient be in a semi-conscious state there are three conditions which are much alike—concussion, drunkenness, and uræmic
poisoning. In none of these are there any very characteristic symptoms; the patients move their limbs and perform automaton-like
movements, and the pupils are of the ordinary size. The so-called
serous apoplexy of older writers was probably Bright's disease.
Two cases, which much resemble one another, are those of ingravescent apoplexy and effusion of blood from rupture of the middle
meningeal artery. In both there is the shock and the collapse, with
the reaction, coma, and paralysis; the latter, however, is only partial
in the case of compression. About two years ago we had two cases,
admitted to the hospital in the same week, in which the diagnosis
was most difficult, owing to the erroneous history. A man was
found lying insensible in his workshop, and brought to the hospital,
his friends saying that he had previously had fits. His head was
shaved and most carefully examined, and not the slightest trace of
injury could be discovered; consequently the case was regarded as
one of disease; yet, when examined after death, a crack was found
through the temporal bone, and blood effused on the dura mater. The
other case was that of a man who, after reaching home, said he had
received a blow on the head, complained of pain, and gradually sank
into a state of insensibility. He was brought to the hospital with
apparent paralysis of one side. The house-surgeon was about to
trephine, but, deferring the operation for consultation, it was not
performed. There was found after death a clot of blood in the
thalamus opticus, and not the slightest trace of injury.

Time will not permit me to endeavour to unravel every possible
case of the kind, and indeed I could not, for I believe, when you are
called in to see a patient lying insensible, it is often perfectly impossible to form a diagnosis, although you may make a good guess.
I ought to have said that the suddenness of the paralytic attack
does not preclude the possibility of the case being one of soften-

ing. This process may go on gradually, and then suddenly some connecting fibres will be severed, and the consequent paralysis ensue.

Diagnosis from Embolism.—This is generally made out from the circumstances under which the attack occurs; for example, if a patient with known heart disease be attacked with hemiplegia, we naturally look to the heart for the cause, and much more if the patient were still a sufferer from a recent endocarditis; or if we are called to a patient previously unknown to us, who has had a fit of paralysis, and we find on examination that he has a cardiac bruit, we premise that the cause of the attack may be embolism; but other circumstances, as well as the nature of the attack, also assist us in the diagnosis; an old person, the known subject of Bright's disease, would, in all probability, be a sufferer from a change in the blood-vessels, whilst this would be less likely to exist in a younger person. Then, again, the suddenness of the attack would be in favour of embolism rather than sanguineous apoplexy, for the fibrinous clot carried into the vessel and plugging it, immediately renders the part of the brain which it supplied functionless, whereas effusion of blood, as a rule, takes place more slowly, and the effects on the brain take a longer time to be produced. Then, again, it is found by experiments that the left middle cerebral artery is more liable to be plugged than the right, consequently a right hemiplegia would more likely be due to embolism than a left hemiplegia. Again, this artery not only supplies the greater part of the corpus striatum, but the convolutions over it, and therefore, if plugged, the additional symptom of aphasia is likely to arise. Thus it is that this symptom is so often met with in embolic hemiplegia. It might be thought that unconsciousness would be less likely to occur in embolism than in sanguineous effusion; but this is not always the case, for although it is true that there is often complete coma in large effusions from rupture of small vessels, as in ingravescent apoplexy, the unconsciousness and mental confusion are often less than in embolism. There may be some difficulty in explaining this, although the fact appears to be as I have stated. It may be that the circulation through a larger part of the brain than is supplied direct by the plugged artery is for the moment disturbed, or that in consequence of the sudden obstruction it undergoes some kind of œdematous swelling. You will see, however, that the very hasty opinion sometimes expressed that a paralytic attack is due to hæmorrhage on account of its suddenness will not hold, since the very suddenness is rather a sign of its embolic nature.

You must remember that the effects on the brain by the plugging of a blood-vessel are very remarkable. I have just told you of a case where the carotid was tied, and the brain on that side, so far

from being anæmic, was intensely congested, and the same occurs
when a smaller vessel is blocked by a fibrinous embolism. If the
brain be seen a short time afterwards, the portion of it supplied by
the blocked artery is red from stagnant blood, It seems almost
as if the vessels beyond the point of obstruction, becoming for a
moment emptied, subsequently, by a kind of suction or exhausting
action, drew the blood back from the veins into the affected part,
and there retained it in a stagnant state. In these cases of embolism sometimes a rapid recovery takes place ; this is no doubt due
to the circulation of blood being restored by means of anastomosing
vessels, sometimes by the channelling out of a new path through
the obstructed vessel. If the recovery do not soon occur, *i. e.* in two
or three days, a softening may take place in the affected part of the
brain. But, very remarkably, the result is not always a softening,
but rather a hardening, for, from the blocking of the vessel, a firm
mass may be formed, as in the embolic impactions of the spleen.
Since these results have not always been followed by an obstruction
of a vessel, as in ligature of the carotid, it has been thought that the
softening did not arise so much from the diminished blood-supply
as from an injury to the vaso-motor nerves which accompany the
blood-vessel, but this idea is quite conjectural.

I have been speaking of embolism of the middle cerebral artery,
but, of course, the symptoms would vary according to the particular
vessel which is blocked. Thus, if the posterior cerebral was the one
involved, sensation might be affected as well as the sight, since this
vessel supplies the corpora quadrigemina. A few such cases have
been recorded, and a temporary blindness has resulted.

The effects of obstruction of vessels by an embolism elsewhere are
worth observing, as showing what may occur within the skull. In
a limb, embolism, as you know, may cause gangrene. In two cases
where the femoral was blocked the patient was seized with sudden
and intense pain ; the leg got cold and almost powerless. After
some days the circulation returned. In a case of obstruction of the
aorta the patient was almost paraplegic.

Treatment.—Now, first of all, as regards bleeding. I will make
some general remarks with respect to this, as it will save me much
repetition. This was once the universal remedy ; now it is all but
discarded. How do you account for this? You must ask the
older members of the profession, who at one time practised one
method and now another. If you do, you will receive for answer
that disease has changed its type ; that fever was different half a
century ago from what it is at the present day ; that pneumonia
was different then from what it is now. But how about this
disease which we are considering—apoplexy ? Is the bursting of a

blood-vessel different now from then? In answer to this it will be said that the type of disease not only has changed in its own inherent nature (whatever a disease is *per se*), but that the patient has also changed; there is no longer that vigour of constitution which formerly existed amongst us, and therefore, although apoplexy cannot have altered its character, the sufferer is a different man. You have often heard me express my opinion about this doctrine of change of type: do not believe it. I have read descriptions of disease in the works of our older authors, and I fail to discover the difference between these and those of modern writers. Then, as regards any impoverishment or deterioration of the human race, there is no proof of this. When a handful of men the other day reconquered India, how could we believe in the diminished prowess of our countrymen? You may remember that in Shakspeare's Henry V, the king, speaking on the day of the battle of Agincourt, exclaims—

> This day is called the feast of Crispian.
> He that outlives this day, and comes safe home,
> Will stand a-tiptoe when this day is named,
> And rouse him at the name of Crispian.

Now, it is very remarkable that on that very day of November, more than four centuries after, there was fought the battle of Balaclava; and I would ask whether our soldiers showed they were the degenerate descendants of the conquerors of Agincourt when came the cry—

> Forward the Light Brigade!
> Was there a man dismayed?
> Into the valley of death
> Rode the six hundred.

As the question turns so much upon the subject of bleeding, I have taken the pains to inquire whether the results were different some years ago than at the present time, but I cannot learn that they were. I know that people receive severe injuries, and lose pints or quarts of blood, and the surgeon treats the matter with indifference. And, as regards the effects of bleeding formerly, I have often asked the late Mr. Monson Hills as to his experience when persons came here, especially at the "spring and fall," to be bled by the dozen or twenty every morning. I had supposed that they would walk in and as quietly walk out after the operation, but he would answer, "no such thing;" they very commonly fainted, and they might be seen lying in rows on the surgery floor like so many slaughtered sheep.

I have read much of what has been said about this change of

7

type, and I am still waiting for a single good fact to substantiate its truth. What are the probabilities in favour of a change of type rather than of a slight error in observation, and what are the arguments used in support of it? They are these: When one of our profession is said to treat his cases differently now in his old age from what he did in his youth, he exclaims, "My principles are not changed; nature has changed! I used then to take a few ounces of blood from my patient with fever or pneumonia, and now I give him a few ounces of wine instead; the majority recovered then, and only a few die now. Surely disease has altered its type." The fallacy of this lies in the undue preponderance given to the influence of medicines; and shows, therefore, to us how important it is to study the progress of disease uncomplicated by our interference, and to pass unheeded the contempt occasionally thrown on the truly scientific procedure of the study of the natural history of disease.[1]

The question of bleeding, then, I discuss with no reference to any fanciful opinions about the change in type of disease. I believe an apoplectic man, if it be thought useful, will bear venesection at the present time as well as he could half a century ago; and what is fifty years in the world's history? If I can show you that patients benefit now by bleeding, I suppose we must conclude that the type has altered once more. I am not, however, going to do this, for I have little experience to guide me as to its use in many cases in which it was formerly practised. If you read the older writers, you will perceive that they did not discriminate between the general effects and the local or mechanical effects of bloodletting. They had a notion, which is opposed almost universally to the doctrine held at the present day, that inflammatory diseases, including fevers and the pyrexiæ, were due to increased vital actions going on in the body—that the blood was too much in quantity, too rich, or too stimulating. Consequently disease was to be "knocked down" by bleeding, purging, blistering, &c. When, therefore, a man was bled on suspicion of an inflammatory attack, and this did not reveal itself, the conclusion was that the disease was arrested. It is difficult to form an estimate of such cases; but when you read, on the other hand, of the doctor being called in to a patient sitting up in bed or in his chair, purple in the face, and gasping for breath as if every respiration would be his last, and you read how the doctor took out his lancet and bled the patient *pleno rivo*, the face meanwhile resuming its natural colour, and

---

[1] So difficult is it as years increase upon us to give a fair judgment on the past and present, that we see one of our most esteemed physicians, now retired from practice, declaring that the treatment of disease has been retrograding during the last twenty years.

the breathing becoming tranquil, you can have no doubt of its efficacy. In this case you will perceive that the lungs were gorged, the right side of the heart loaded, and that the lancet came just in time to disencumber the overburdened organs, so as to allow them free play again for their functions.

The effects for evil or for good of venesection, having for its object the arrest of inflammatory processes, have not been ascertained with sufficient accuracy to warrant me in offering an opinion as to its value. I have seen, however, a sufficient number of cases bled to know that it is not a very fearful measure. In fact, if I were obliged to adopt one method only—that of venesection, or brandy-giving—I know which I should prefer.

It has fallen to my lot to have seen three patients with typhoid fever bled, and it is remarkable that they all got well. It is very extraordinary what importance we attach to the artificial withdrawal of a few ounces of blood; and yet we see our patient with typhoid fever have a large hæmorrhage from the bowel, and we have but little fear. The surgeon also treats his cases of fractured skull and fractured leg without much regard to the pints of blood which have flowed away.

When I refer, however, to bleeding as a means of relieving engorged lungs and heart, I can speak with some certainty and authority. I have no hesitation in saying that I have saved patients' lives by this treatment, the diseases in which I have adopted it being bronchitis, heart disease, apoplexy, and epilepsy. Now, you must remember that the indications for bleeding in these cases are very different or very opposite to those which would suggest its employment in the class I have mentioned, and it is owing to this being not rightly understood that the practice is not more frequently adopted. A patient was formerly bled because his pulse denoted an inflammatory condition—it was full and hard. The tradition has descended to us; and thus at the present day, when, meeting a medical man in consultation over a case of chronic bronchitis, and I have suggested the propriety of bleeding, he has often said, "I dare not; the pulse is scarcely to be felt, and the patient is dying. The removal of a few ounces of blood would be the finishing stroke." Herein lies the error. If bleeding is of use in the cases I name, it is to relieve congestion of the lungs and heart. And it is especially in severe cases of bronchitis, such as we have seen this last winter, that I would recommend it. You see your patient sitting up in bed, face, tongue, and lips blue or purple, and the jugular veins starting out of the neck, and often visibly pulsating; the heart beating quickly, and perhaps with a tricuspid bruit, indicating the gorged right heart and obstructed lung; the veins in the body are full to bursting;

the heart can scarcely work longer, as it cannot get rid of its blood, and yet all the while little passes into the left ventricle which again meets with an obstacle in its action by the engorgement of the entire capillary system. The pulse is consequently, as you might expect, very small, very weak, or scarcely to be felt. Because this is its character, the doctor dares not bleed. He thinks he is to feel the state of the artery, in order to know if the venous system wants emptying, and rejects the proposal for bleeding because the pulse is of the character I state. It is exactly the pulse, joined with the other indications, which points to the remedy.

In short, without discussing it further, if you take away blood from such a person, you relieve the heart and the lungs, the circulation becomes freer, and the pulse improves in fulness. Many years ago I was asked to see a little boy suffering from bronchitis after hooping-cough. He was lying half raised on a pillow, gasping for breath, his face livid, eyes starting out of his head, and the superficial jugular making itself most apparent. The medical attendant and myself looked at this vein, and, in spite of the protestations of the father, opened it, and let out a few ounces of blood. The lividity passed off, the child sank back on the pillow into a tranquil sleep, and from that time recovered. You must make no comments upon the valves at the commencement of the jugular veins, for, in spite of these, the blood will run out in abundance.

Many of you in this room saw the case of epilepsy which one of my clerks bled a few months ago. The man was a strong agricultural labourer, and came here suffering from severe epileptic fits. One day, on going round, we were informed that he had never been out of a fit for four hours. We found him lying in bed with constant convulsive movements, but the most striking condition was the engorgement of his lungs, his labouring heart, and increasing lividity of the surface of the body. I requested that he should be bled, and one of you, wishing to do well at his first attempt, or from the lancet being over sharp from disuse, fulfilled my object to the utmost; the blood poured out in a torrent; the face rapidly became pale; the man opened his eyes and spoke for the first time since the morning, the interval having been a blank to him. The effect was one of the most striking that I have ever witnessed, and it is worthy of note also that he had no fit for a long time afterwards. In heart disease, the expectoration of blood when the lungs are apoplectic affords often the greatest relief to the patient. The practice has been too often to give remedies to arrest it, but the beneficial effects to the sufferer, in spite of our vain attempts at prevention, have been so striking in two or three instances which I have witnessed that I cannot do otherwise than direct you to let nature take her course.

Finally, I come to the subject which suggested these remarks, the treatment of apoplexy by bleeding. It was recommended, and is so still by some, for the reason that it diminishes the amount and force of blood in the system, and so tends to lighten the pressure within the cranium, also because it prevents the tendency to subsequent inflammatory action. I have always felt some difficulty in accepting this explanation, because, if of any use towards the object named, it ought to be performed at the onset of the attack ; this, however, is the time when the patient is collapsed, and you are recommended to give a stimulus, which, indeed, often appears to be absolutely necessary. After reaction sets in, I believe bloodletting is often useful ; whether it acts by diminishing the flow of blood to the brain is doubtful, but by relieving the congested lungs, it is often highly beneficial. The patient often dies directly through the lungs ; if, therefore, you can relieve them and give him a few hours' respite, you might just get him over the critical moment. The cases in which we read of immediate cure by bleeding I should think were really those of epilepsy. Once more, then, bleeding in order to relieve congested lungs is highly useful, but whether of advantage as one of the old antiphlogistic remedies to combat inflammatory disease I am uncertain. I could not inform you of its exact value unless I had statistics before me.

It is one thing to recommend you to adopt a method because it forms a part of a given routine plan, and another to exhort you to the use of particular remedies, because experience and a discriminative trial have found them useful. Thus, if you are called to see a person seized with an apoplectic fit, and you find him collapsed, you administer stimulants, which, under the circumstances, are necessary. If you see him at a later stage, when the coma and stertor are present, I have no objection to your bleeding him if other circumstances permit. At the same time we endeavour to withdraw the blood to distant parts, and thus mustard poultices are placed around the calves, and a blister is applied to the back of the head. I order these because others do, but I cannot demonstrate to you their exact value. The next thing is to give a purge, such as calomel or croton oil. I think all are agreed as to the beneficial effect of aperients ; indeed, in a large number of cerebral troubles, purging is of the utmost value. Then, if the patient recover, it is the fashion with some to give mercury to prevent inflammation. I have seen this adopted on several occasions, and I cannot say that I have seen the slightest good accrue ; on the contrary, it has tended to weaken the affinities of the cerebral tissues and

promote their disorganisation. I do not know that mercury has
any effect in arresting inflammation. It is a drug which appears
to act on all the secretory organs of the body, and thus promotes
various physiological processes. Consequently, it will often cause
the absorption of inflammatory products already poured out. During
convalescence the question of lowering or supporting treatment is
all-important, but must depend on circumstances. I lately saw a
man lying in bed totally hemiplegic two weeks after the seizure,
and pursuing the old-fashioned antiphlogistic plan. His condition
suggested an opposite method, and being put on a mutton chop
and pint of porter for dinner, he began rapidly to improve. The
recuperative process required further aid from nourishing diet. It
is a common practice to apply cold lotions to the head, but
then the same thing is done in arachnitis, in fever, or delirium
tremens—indeed, in all cerebral disorders. It is impossible
that a universal plan can be either rational or useful. Galvanism is beneficial after some weeks have elapsed since the attack.
The continuous use of this and faradization for a lengthened
period preserves the nutrition of the muscle, and is therefore often
of great service. A large number of cases of hemiplegia pass
yearly through the electrifying room, and Mr Sandy thinks that
about half of them receive considerable benefit. I have never more
than once ordered a shock through the head, as I consider the experiment a hazardous one. Many years ago Marshall Hall
made some observations with respect to the effects of galvanism in
various forms of paralysis, and stated that this agent was powerful
in cerebral paralysis but not in spinal paralysis. Although in one
sense both forms are spinal, there is a great difference between
that resulting from disease in the corpus striatum and that originating in the grey centre of the cord. In the former case, as in
ordinary hemiplegia, the nutritive centres of the nerve are not involved, and the muscles react to both kinds of galvanism; it is
very different, however, in spinal paralysis.

Heat Apoplexy or Sunstroke.—Opinions have varied considerably as to the nature and causes of coup de soleil, but of late much
weight has been attached to the observations of those who have
taken the trouble to more closely watch the phenomena, and more
especially to note the increased temperature of the body. This rise
of temperature has been so repeatedly observed, and the symptoms
accruing from excessive heat are now so well known, that there
seems little doubt that all the phenomena of sunstroke are due to
the effect of its sudden increase. It must be remembered that one
of the most remarkable facts in the animal economy is the constancy
of its temperature, even to the fraction of a degree. Whether we

eat much or little, take exercise or remain at rest, live in a tropical or temperate climate, the result is the same—our temperature is uniform. It is clear, therefore, that as we are producing more heat at one time than another, and losing unequal amounts at different times, there must be a regulating power in the body which preserves the normal standard of warmth. The balance is supposed to be regulated by the spinal cord. Injuries of this part destroy the equilibrium, and cause irregularities of temperature. In various morbid conditions, such as rheumatism, a hyper-pyrexia may suddenly occur, and which can only be accounted for by some sudden change in the nerve centres. This high temperature is incompatible with life, as the blood and muscles undergo a change, and the capillaries become filled with the débris of the disintegrated tissues. Exactly as the process may be watched in the human body in disease, so in the same way it may be closely imitated in animals by exposing them to great heat. The temperature of the body is in health kept down to the standard mainly by the evaporation from the surface by the perspiration, and we can thus easily see the danger to which we are exposed should the atmosphere be unable to carry off the superfluous heat from the body.

In the coup de soleil, as it occurs in India, the patient often suddenly falls down in a kind of syncope; this is probably due to the effects of the heat acting directly on the spinal centre, and so giving a shock to the heart. A reaction then occurs, and all the symptoms of pyrexia develop themselves, hot skin, red face, quick pulse, and rise of temperature from 108° to 110°. As all these symptoms are the result of increased temperature, a heat stroke may occur in the night as well as the day. Half of the patients die who are struck down by the heat, and those who recover often remain for ever after the victims of an impaired nervous system. They suffer in various ways from cerebral symptoms or become confirmed epileptics. The true nature of the malady is confirmed by the efficacy of the remedy. Just as cold affusions save the life of the patient who has a sudden pyrexia from disease, so in sunstroke the same means have been found equally efficacious. Quinine has been said also to be a most effectual remedy in India, when speedily used, either in the ordinary way or by the hypodermic method. This I can conceive from its known apyretic properties. Since the thermometer has come into use as a diagnostic agent, and when it has been at hand to apply it, several cases of heat stroke have been observed in London. In one of the late hot summers a lad wheeling a barrow down Fleet Street was taken suddenly ill and carried to St. Bartholomew's Hospital, where he died in an hour. His temperature was ascertained to be 110°. A

case also was related by Dr. Thompson of a man who having been out
on a very hot day, went home giddy and sick with a pain in his head.
He was delirious in the night, but arose on the following morning
and went out, when he fell unconscious, and was taken to the
Middlesex Hospital.   He was in a state of perfect coma and died in
an hour.   His temperature was 107°.

If heat may produce sudden death, there is every reason to
believe that it might cause injurious effects of a less severe nature,
or such as might lead to a fatal result in a few days rather than
in a few hours or minutes. There has been a very ancient and general
belief in this noxious influence of heat in the case of children,
although the subject never seems to have been scientifically studied.[1]
I have always been impressed with this since my earliest childhood,
on account of a little brother of mine dying of head disease, the first
symptom of which was severe headache after playing in the garden
on a summer's day, and from the distinct remembrance I have of
my father likening the case to that of the Shunammite's son, of
whom we read that—"When the child was grown, it fell on a day,
that he went out to his father to the reapers. And he said unto
his father, My head, my head. And he said to a lad, Carry him to
his mother. And when he had taken him and brought him to his
mother, he sat on her knees till noon, and then died."

Lightning Shock.—In sudden death from this cause no marked
changes are found ; the disturbances produced in the cerebro-spinal
centres are too subtle to be manifest to the eye. In some cases, however, positive injuries to the body are inflicted in the same way as
when a tree is rent by the stroke ; and it has been stated that the blood
is found coagulated and that there is no rigor mortis. In a case lately
reported (the name of the author I forget), the body was found
scorched and blistered on the surface, corresponding to holes
burnt in the clothes. The examination was made eighteen
hours after death, and the rigor mortis was well marked. The left
ventricle contained a small quantity of dark fluid blood, and the
veins contained a similar kind of fluid. There was no coagulum
anywhere. A woman was once brought to the hospital, collapsed
and insensible from the effects of lightning ; after three days she
began to speak, and on the fifth day was again quite well. In a
similar case of which I heard, a woman was struck down and
found lying cold and speechless. She soon afterwards came to,
and then some febrile action occurred. In less severe shocks
patients have experienced pains and strange feelings in the body
and limbs.

[1] Dr. Gee has lately studied this affection and given it the name of *phrenitis
æstiva.*

INFLAMMATION OF THE BRAIN

The next subject we come to is inflammation of the brain. A difficulty arises at the onset as to the interpretation of the term as used by authors and by medical men generally. It must be self-evident that inflammation of the brain cannot be treated after the simple manner of inflammation of the lungs, for, in so complex an organ as the brain, the symptoms must vary immensely with the part affected, as also with the cause. In other organs we make a division into the inflammation of the substance itself, or of the viscus and its covering—for example, pneumonia and pleuritis ; but an inflammation of the membranes of the brain alone, without involving the cerebral structures, is almost impossible. Such a term is a mis-nomer ; the symptoms, indeed, which are ascribed to it imply an implication of the brain itself. Since, however, there is an affection where the membranes seem to be especially or primarily involved, to this the general term meningitis can be applied ; of this several varieties may be noticed. On the other hand, there is an inflammation of the brain proper, followed by softening and, in certain instances, by abscess : to this the term cerebritis or encephalitis can be ap-plied. What I should have preferred would have been to take distinct pathological processes, examine them separately and the symptoms accompanying them, and subsequently discuss those cases where no distinct morbid changes have yet been discovered. As, however, we cannot ignore the terms in common use, I am forced to treat the subject in a manner different from what I should have desired.

Cerebritis and Softening.—As one of the chief results of inflam-mation is softening, and as this may arise under a variety of circum-stances, I think it will be as well to allude to this first.   Softening —or *ramollissement*, if you prefer the French term—is used in a very vague manner.   Generally, when we say softening is present, we mean a chronic change has taken place in the brain substance, whereby it has become disintegrated, and its function lost.   But softening, as a result of inflammation, may be acute, and be developed in a few days.   Such a case we ought not to designate by the name of softening, but by that of cerebritis or encephalitis.   In all acute inflammations the tissues become soft, but we should not, therefore, name the disease after one of its effects. We should not, for example, call pneumonia a case of softening of the lung. In the case of the brain, however, we are often compelled to speak of the result as if it were the disease itself, being altogether ignorant of the cause, not only during life, but even after death ; a doubt even then existing whether the softening be due to inflammation or be the result of a chronic disintegration from a change of nutrition. Then, again, besides these actual and tangible forms of softening, we are using the term in the vaguest possible sense as applicable to a great variety of symptoms. Thus, when a person becomes a little feeble in his mind, and has some slight paralytic symptoms, we often say the patient has softening of the brain, intending only to express the idea that some impairment has taken place in the cerebral structure, and not necessarily a change like that of softening, which is visible to the naked eye.

There are, no doubt, a large number of changes going on in the living brain whose effects are at once perceptible by some alteration in the working of the machine, expressed by some physical or mental failing, which in any other organ would not be manifest. A slight structural change, for example, in the liver would not be apparent, except perhaps by some general feeling of *malaise*, but in the brain this would at once evince itself. What these changes are and how associated with distinct symptoms, we have yet to learn. I hear sometimes the remark made that morbid anatomy has taught us enough, and that all we want is some medical philosopher to arise to generalise from our facts and supply us with theories; but I think I am in a position to say that our facts are meagre or scanty, and that we are only in the infancy of the science of cerebral disease.

As regards the softening—this tangible softening, with its evident symptoms—we have been in the habit of expressing the difference between a chronic softening resulting from decay or degeneration and that arising from inflammation, by styling them white and red softening. You are familiar with the terms red softening as denoting inflammatory, and white as meaning a more passive or atrophic change. If, however, red softening does result from inflammation, then it would be more desirable to at once designate it inflammation of the substance of the brain, or cerebritis, or encephalitis; but the reason, as I before said, why we cannot do this is that it is only in exceptional cases that this inflammatory process is evident as an acute and idiopathic process, and thus we are obliged to speak only of the effects. In the majority of cases the softening is chronic and associated with other disorders. If the softening be of a red colour, we call it inflammatory, the redness being due to the greater vascularity. We are influenced also in our decision by the age of the patient, and by the circumstances connected with the illness. In older persons, and especially where the arteries are diseased, as in morbus Brightii, we expect to find rather the white or non-inflammatory softening.

Suppose we make a post-mortem examination and find softening, how does it display itself ? In some cases, when you make a section through the organ, you see the hemisphere presenting a peculiar appearance in the medullary matter ; a certain portion, more or less circumscribed, looks and feels pulpy, resembling somewhat a piece of blanc-mange. As a rule, however, it does not look smooth, but it is disintegrated, and thus, if a section be made, it shows a broken surface. When you pass the knife through the substance it sticks to the blade, and if you stir it about you can make it into a pulp or paste. If merely a number of softened spots were present, these would be apparent, when you made the section, by an equivalent number of broken surfaces.

If the softening has proceeded a stage further, then the brain matter may be quite broken up, or be semi-fluid, and a portion of this running off, a depression is left. If a stream of water be allowed to trickle upon it, the brain matter may be washed away, and a distinct hole be left corresponding to the softened part. Sometimes even during life a disintegration and absorption occur, so that, when you make a section of the hemisphere, you find a large hollow space filled with a fluid like lime-water and the *débris* of brain substance. All these cases where there is actual loss of substance come under the category of white softening. They arise in connection with diseased vessels and general decay. In the red or inflammatory softening the disintegration is not so great. Besides these two kinds, some authors have spoken of a yellow softening, which they surmise to be of a peculiar kind, and due to a chemical change going on in the fatty acids of the brain. In some of the best-marked cases, however, which I have witnessed, I have considered that the yellowness is due merely to an altered condition of the colouring matter of the blood which has been present in it.

Then, again, showing how difficult it is to decide by the mere colour whether the softening is inflammatory or not, if we take the case of acute hydrocephalus or tubercular meningitis, we know that there is an inflammatory exudation into the ventricles, and that the central parts have undergone a remarkable softening ; the septum lucidum and fornix and adjacent parts are broken down and diffluent, but they are perfectly white—milky white. So marked is this that those who maintain that a structure must be red to indicate inflammation would say that this central softening of acute hydrocephalus was due to a simple death or atrophy of the part, or had occurred from the presence of so much fluid, which had, as it were, melted it down. Of this there is no proof, but, there is evidence that the change is inflammatory. You no doubt might think that the

microscope would positively inform us as to whether a softening
was inflammatory or not, but I am sorry to say it does not do much
for us in this respect ; for when the cerebral structure is broken up,
and a number of new products are present, it is extremely difficult
to say whether inflammation has anything to do with the process or
not. The microscope is extremely useful in proving the fact of
softening, because, besides the broken nerve tubules, it displays a
quantity of new formations, such as granule masses, which, to say the
least of them, are morbid. It often happens that we wish to know
whether a part of the brain has undergone a morbid softening or
not, and by using the microscope and finding these bodies we are
sure of the fact.

Softening is most commonly localised, whether it be due to an acute
or chronic cause, although occasionally we find large portions of the
cerebral structures affected. In cases where there is much disease
of the blood-vessels, spots of softened tissue may be found throughout the whole brain, and in the much rarer cases of acute encephalitis,
nearly the whole cerebral structure may be found to be undergoing
disintegration. In cases of this kind destruction of so large a part
of an important organ will, of necessity, very speedily bring about
a fatal issue, but in instances of local inflammation and softening,
life may be prolonged for many months, and ulterior changes result ;
one of the commonest is for the brain tissue to perish until a mere
vacuity is left, containing a whitish fluid with remnants of blood-
vessels. In some cases this cavity is lined by a smooth and tolerably
thick membrane. Should the inflammatory process proceed to the
stage of suppuration, then an abcess is formed. This may or may
not be contained in a cyst. The latter, under these circumstances,
is not merely a thin delicate membrane, but a thick, firm bag, composed of tough lymph. A very important question, whether idiopathic inflammation of the brain ever ends in suppuration, you have
heard discussed on other occasions—whether, indeed, a cerebral abcess does not signify either that the morbid process has been started
in the cranium, or that it is pyæmic. The question has a very wide
pathological significance, referring, as it does, to the mode in which
the various tissues of the body undergo their own peculiar modification in disease, but to this I shall have to refer to again.

For convenience sake inflammation of the brain may be divided
into a *general* cerebritis and a *local* cerebritis, and these again into
the *acute* and *chronic* forms. In all these varieties the usual result
would be a softening of the tissue, but in some cases the inflammatory
products are organised and an induration takes place ; this receives
sometimes the name of *sclerosis*. Of course, there is no reason why
induration and softening should not be associated, as they often are.

Patches of sclerosis are often also designated by the term *grey induration* or grey degeneration. If circumscribed, the inflammatory product is scarcely distinguishable from a new growth or tumour. The softening, as before said, is usually styled red and white, and these terms are often considered to be synonymous with inflammatory and non-inflammatory softening. This is not, however, always evident, since the redness is due only to a stagnation of blood. The yellow softening is, in all probability, only a further stage of the same process, where the red corpuscles are broken up and some fatty degeneration has taken place in the substance of the brain. In chronic degeneration of brain tissue we find broken-up nerve tubules, granules, masses, and amyloid bodies. Now, this might have been inflammatory in the first instance and, therefore, there are undoubtedly many cases where the red, yellow, and white softening are merely stages of the same process. In all probability the local softenings in old age are due to disease of the blood-vessels arresting the flow of blood or causing the plugging of an artery by means of fibrin formed at some atheromatous spot. The condition of brain is, therefore, analogous to a limb in senile gangrene. As I have already told you, although a vessel may be quite blocked, the brain tissue beyond it is of a red colour, owing to the reflux of blood into it, and sometimes a hardened mass is formed instead of a softening. Not only is there a stagnation, but small effusions of blood sometimes take place into the tissue. Although the red softening occurring under these circumstances may not be clearly distinguishable from the acute inflammatory form, as the redness in both is due to hyperæmia, I would not say that they were identical, seeing that the causes which are instrumental in producing them are different. Besides the well-known effects of emboli in the larger vessels, there can be no doubt that the smaller ones become plugged, giving rise to less defined symptoms, as delirium, wandering &c. In cases of ulcerative endocarditis I have seen aphasia without paralysis, suggesting an implication of special convolutions; and the same also in typhoid and other diseases. The plugging of the vessel is due in the cases first named to a coagulation of blood on a roughened surface of the artery, and this leads to a complete blocking of all its branches with fibrin; this must be distinguished from the case of embolism, where the fibrinous mass is carried from the heart. The former is merely atheromatous, and we endeavour to judge between it and embolism by the presence of endocarditis, the age of the patient, and the suddenness of the symptoms. When the smaller vessels are plugged, leading to a more diffused softening or spots of softening, the case is very obscure, but it is very likely that many anomalous cerebral symptoms are due to a plugging of the smaller arteries by embolic particles, which are carried into various parts of the medullary and grey substances of the brain. Thus, a girl was under my care for rheumatism, she having had several previous attacks and also mitral disease, when she, one day, was found insensible, and during the following four weeks had paralysis of the third nerve, of the seventh, congestion of the optic discs, and a variety of other nerve symptoms of a most anomalous character. After death there were found ecchymoses of various parts of the brain and scattered spots of softening. The basilar artery was completely plugged by an embolus and the superior cerebellar and posterior cerebral were almost obliterated. I might also mention here other and more minute changes which are sometimes met with in the cerebro-spinal centres and whose nature has been explained by Dr Lockhart Clarke. They result from a long-continued congestion of the nerve substance and are met with in tetanus, hydrophobia, diabetes, and some other diseases. He finds a dilatation of the blood-vessels, with a degeneration of the nerve tissue around them, which produces a kind of excavation. Cavities are thus formed, which are visible to the naked eye, and contain extravasated blood, pigment, and products of nerve decay. If these are absorbed simple vacuities remain. This condition is found both in the brain and spinal cord, but more especially in the pons and medulla oblongata.

A general cerebritis, although not so evident in a post-mortem examination, is probably not infrequent; it would be associated, in all probability, with an inflammation of the membranes, and therefore a meningo-cerebritis would be the more appropriate term. The membranes are seen to be thickened; in the meshes of the pia mater hard inflamed lymph may be found; at the same time the ventricles are distended with fluid, and the whole of their surface is in a granular condition. The brain substance throughout has not a healthy appearance or consistence, and the microscope shows inflammatory products and degenerative changes. I have found this condition in connection with cirrhosis of the liver, granular kidney, and chronic pleuro-pneumonia in cases of alcoholism, and therefore I judge the meningo-cerebritis has exactly the same pathology as they have.

I have met with three or four cases where the inflammation has not been confined to the brain and membranes, but has implicated the skull.

CASE.—A man, æt. 40, was admitted into the hospital with a history of headache, loss of memory, difficulty of speaking, &c., of nine months' duration. A general torpidity came over him until he became quite unconscious and generally paralysed. A conjectural diagnosis was made. The inspection showed that an inflammation had attacked not only the brain, but the membranes and skull itself. The whole internal surface of the latter was covered with bony granulations, with a roughening of the corresponding surface of dura mater. Arachnoid surfaces adherent in places; substance of brain soft, with an inflammatory cyst in anterior lobe.

A young man, æt. 21, had a very similar disease, where the whole of surface of skull was scabrous and adherent to dura mater. Brain affected throughout, and cyst in cerebellum. He was in bed for months with blindness, deafness, and a general weakness of the limbs. This probably arose from injury.

Symptoms.—In speaking of the symptoms of softening, if you take the case of acute general cerebritis, they may be of the most obscure character. There would, of course, be severe pyrexia, with delirium, dulness of intellect, and final coma, but perhaps no other symptoms especially referable to the brain. I believe cases of simple general acute cerebritis are not very common. In those which I have seen, the nature of the disease was by no means evident during life. A young man was admitted in a condition like that of fever, it being said that this torpid state had gradually come upon him. He lay perfectly motionless in bed, and never spoke, although his eyes were open, and he appeared to understand; he did not live long. After death the brain was found softened throughout; the whole was pulpy, and some parts were actually semifluid. I have seen one or two similar instances. In cases of less rapid progress, the cerebral symptoms would be more marked, whilst the pyrexia would be less, and there might be headache, sickness, a slow labouring pulse, constipation, delirium, and gradually approaching coma. In two cases I have seen lately, the patient lay in a listless state, answering when spoken to, and then falling into sleep. This sleepiness is especially noticeable. So wanting in character are the symptoms, that the most fatal of disorders has been regarded as a simple functional disturbance; and I have seen the case of a lady in which the phenomena were ascribed to hysteria. According as particular parts of the brain are involved, so would special paralytic symptoms be present; if any portion of the motor tract, a hemiplegia; if the base of the brain, paralysis of the cranial nerves. If the softening should be confined to one spot, then the symptoms would be proportionally limited, just as I told you in the case of sanguineous effusion; if to the corpus striatum or thalamus, a simple hemiplegia; if to the medullary substance, symptoms of the most indefinite character. If involving the cineritious substance, more remarkable symptoms denoting mental disorder.

Meningo-cerebritis

CASE.—A young woman, admitted February 18th. Father and mother healthy. One brother has had rheumatic fever twice. No history of tumour or consumption.

Personal history.—The first symptom noticed was seven months ago, when she laughed without a cause at her mistress and was discharged for it. She was a nursemaid, but could never bear the noise of the children. On leaving service she kept company with a young man who left her, and this caused a mental shock. Since then she has been strange in her manner. Four months ago she began to suffer from incontinence of urine; this has continued since and the urine is now offensive smelling. Two months ago she lost the power of walking; she used to be helped up stairs at first, and then carried up. About this time she lost the power of talking, though she could occasionally speak perfectly well. When placed upright in a chair she would bend over till her head reached the ground. When spoken to she generally laughed. Appetite was voracious six weeks ago; since this she has eaten in the way described under present condition. She has complained of pain and aching in her limbs, more especially when warm in bed. She has menstruated regularly till the last three months; since then only once. No history of discharge from the ears, blow on the head, or severe pain in it.

On Admission.—She has taken a journey of thirty miles. Patient is a healthy looking woman with a good colour in her face. She has a peculiar stupid expression and when spoken to generally laughs. When asked a question she seems willing to answer, but unable to do so; she can, however, speak in a slow deliberate tone of voice and very low. When she eats she keeps the food in her mouth for an hour at a time and does not seem able to masticate her food. She can understand perfectly any questions put to her. There is no paralysis of either side of the face, and sight and hearing are unimpaired. She can protrude her tongue slightly, and it is indented at the edges and covered with a white fur. Pupils generally dilated and equal, contract with light. There is a want of will in all her actions. Right side of body: When her right arm is placed vertically to the bed she retains it in this position for a great length of time, and can grasp firmly your fingers but has a difficulty in unloosening her hand. She can write words with her right hand, not with her left. The right side of her body has altogether more power than the left; the leg responds better to reflex action by a prick on the sole of the foot; the left leg also responds but to a less extent. She cannot hold her left arm up in the same way as the right. The right leg is generally moved over the left when the left is pinched. She has fits of jogging in her right leg; she can feel a pinch anywhere over the body. No heat of skin. She occasionally heaves a deep sigh. Bowels generally confined. She has had no motion while in the hospital. Heart: There is a thrill over the apex of the heart, but the chest walls are very thin. No bruit. Chest: Sounds and resonance normal.

February 20th.—She takes hardly any food while in the hospital. There is a bedsore coming over the sacrum. She is menstruating.

22nd.—Died this morning at 6.20 a.m. Yesterday she could not swallow. In the evening her temperature was 102 deg.

Post mortem.—Head: The falx cerebri was rather adherent to the hemispheres, and on its separation it was seen to be, especially on its right face, thick and finely tuberculated by a yellow gelatinous granulation like a layer of new tissue, and on section of this layer and the sublying falx, it appeared that the opinion formed as to the nature of the thickening was correct. The falx was unchanged, but lying upon it, more on the right side than on the left, was a layer of granulation tissue. The membranes of the brain in this region were yellow looking, thickened, and adherent to the substance of the organ. This state was most decided anteriorly, but extended back on the right side to the posterior part of the corpus callosum. The arteries of the brain looked quite normal. Elsewhere the membranes were perfectly healthy. They were neither greasy, nor tuberculated, nor opaque. Taking the right frontal lobe between the finger and thumb, it was felt to be decidedly indurated on comparison with the other side, and still more so when compared with other parts of the brain. On section both sides, but especially the right, cut in the frontal lobes with that peculiar gristly resistance found in cases of sclerosis. Throughout the whole brain this was more or less the case, but it was especially so at the anterior parts. On removing the upper part of each hemisphere, nearly to the level of the corpus callosum, the cut section showed a peculiar condition. The whole surface of both grey and white matter from the longitudinal fissure outwards as far as, but not involving, the grey matter of the external convolutions, and from the front to a little behind the fissure of Rolando posteriorly, was of a brick-red colour ; this was most intense as far as the anterior part of the longitudinal fissure on the right side, *i.e.* where the granulation tissue on the falx was most decided. The left side, in addition to the red tint, was also soft, but on the right side it was hard. It was also noticeable that the grey matter of the various convolutions jutting into the brain from the longitudinal fissure was nearly twice its proper thickness, swollen out, and it became quite impossible to define the outline of the convolutions, so perfectly did their outer border blend into the red tint of that in the white matter. On both sides the red tint was noticed from the surface down to the roof of each lateral ventricle, and on the left side the roof of the ventricle was very soft. The corpus callosum was soft and the fornix also. The ventricles and other parts of the brain appeared perfectly healthy, the substance was firm, the cut section purplish in hue from engorgement of the vessels.

Examined microscopically the parts over the left lateral ventricle had the usual appearance of soft brain tissue (red softening), *i.e.* numerous granule masses were seen and a general corpuscular state of the brain substance. The tissue from the right side (red induration) shewed a remarkable absence of anything like nerve structure, but appeared to consist mostly of vessels and a dimly fibrillated substance.

The medulla, pons, and spinal cord were perfectly healthy.

The parts that were most diseased corresponded with the first and second frontal convolutions. The third was healthy on both sides. The disease extended further back than this in the white substance of each hemisphere, but the greater part on each side would be included in the anterior part of a brain divided vertically across from the commencement of one Sylvian fissure to the other, that is, behind the third frontal convolution on each side.

Weight of brain 45 oz. Venous blood in excess. Other organs healthy.

When this girl was admitted the first impression made on the mind of the clinical clerk was that it was a case of hysteria. This idea was, of course, soon dismissed when the history was taken and the symptoms were more clearly revealed. The first impression, however, was very natural and gives a clue to the whole character of the case, for it indicates that there were no striking paralytic symptoms present, and that the girl's manner was both lethargic and emotional. In fact, she presented the symptoms which we sometimes see in extreme forms of hysteria. The history, however, showed that her illness began in a tolerably definite manner some weeks before, until she had reached the feeble state in which she was admitted. Then, as the report says, although there was some difference in the degree of power in certain parts of the body, there was no distinct paralysis as the term is usually understood. The inability to move appeared owing merely to a failure of the volitional act.

In true paralysis the spinal system is affected, whilst the will is good, hence the patient is seen to make the greatest effort to move a leg or arm, although the result may be ineffectual. In disease of the brain proper, or during its functional abeyance in hysteria, it is the will itself which fails; now it was evident that in this girl the power of acting or willing was gone. She made little effort to move, to speak, or even to masticate her food, and she allowed the bladder to empty itself in the bed. This latter fact removed the case from simple hysteria where the patient never allows herself to be inconvenienced in this way. She had not paralysis of the bladder, but simply, as in mad people and in the lower animals, she exerted no intellectual control over it.

The patient had clearly, therefore, some progressive disease of the nervous system, and the question was where was its seat and what was its nature? The absence of true paralysis shewed that the spinal system was not affected and by the spinal system is meant, not only the cord which is contained in the spinal canal, but the medulla oblongata, crura cerebri, and central ganglia within the cranium. The disease, therefore, was in the brain proper; a conclusion quite compatible with the symptoms, which were rather mental than physical. The question then arose whether it was a tumour or a diffused inflammation. The former, as is known, although localised, may produce in an inexplicable manner a disturbance of the whole brain with a corresponding insanity, but at the same time it is usually accompanied by well defined symptoms, as severe pain, sickness, convulsions, and amaurosis, due to atrophy of the optic discs. It was, therefore, concluded that the disease was of a more diffused nature and largely involved the cineritious matter; it could be none other than inflammation or encephalitis, usually known by the name " red softening." It turned out to be an inflammation, but only a portion of the brain was soft, the greater part of the affected hemispheres being indurated from the effused lymph within them. A considerable part of both hemispheres were structurally destroyed.

In our present state of knowledge it may be affirmed, when a patient has been ill only a short and well defined time with symptoms denoting a deep implication of his nervous system, and no paralytic symptoms are present, that the spinal system cannot be affected, but the brain proper, viz. the hemispheres. He would of course lie in bed in a lethargic condition, having no will to move. Some time ago a young man lay in Stephen Ward for several weeks in a kind of torpid state; he never spoke, but his eyes wandered after objects, as if he comprehended every occurrence which took place. After death his whole brain was soft or pulpy.

Cerebritis

CASE.—Henry A—, 48, admitted in December. He was a painter and had suffered from the effects of lead, he had also had once a severe fall on his head and he had been a hard drinker. His history was, that in June, whilst painting, the brush fell out of his hand several times; he went down stairs, and then fell on the ground insensible. He soon recovered consciousness and walked two miles home. About two weeks afterwards, whilst going to his work, he felt pains in his right leg; these gradually extended up to affect arm and head. He soon became insensible, and remained so for half an hour. He afterwards walked home, when he found he could not speak and was rather deaf. After being in this state a week his speech and hearing returned. The doctor believed he was going out of his mind. He went to work for five weeks, when a tingling of the left arm came on, and after being at home a month, again went to work and remained at it for seven weeks. He then began to have much more severe symptoms—pains in his limbs and head, swelling of the ancles, dimness of sight, and failure of memory; also great sleepiness. These symptoms continuing he was brought to the hospital in December.

He was in a semi-conscious state, objected to being moved or roused, sight almost completely gone, and could only just discern light; there was extensive hœmorrhage into both retinæ, obscuring the optic discs. There was a slight blue lead line on the gums. He was delirious at times and often lay quite unconscious. The kidneys and other organs healthy. There was no paralysis, as he could move his limbs freely. On January 12th he was much the same, lying in bed in a half conscious drowsy state, sometimes talking, but quite incoherently; the ophthalmoscope shewed that central absorption was going on, and the discs had a woolly appearance. At the end of the month he was in an unconscious state out of which he could be roused with the greatest difficulty, and bed sores had appeared. He thus continued and six weeks afterwards he was still lying in an apparently unconscious state, and quite blind. When roused would sometimes say a word and put out his tongue; he had a fresh hemorrhage in othe retina. No special paralysis of any part. During this time he had been taking the bromide occasionallly.

At beginning of March he began to improve; he answered more rationally. Quite blind, pupils dilated and insensible, except when asleep; they were then naturally contracted.

On the 18th of March he was much better and began to eat. From this time he slowly improved, so as to be able to sit up in bed, and answered questions put to him quite rationally. On the 26th he got out of bed and walked along the ward. The report of the examination of the eye was that the optic discs were anæmic, veins few and varicose, all blood spots absorbed. It was thought that this was due to a blocking of cavernous sinus. He continued to improve, and walked about the ward and apparently had quite recovered his intelligence On April 27th he had a convulsive attack, was almost insensible, and his arms and legs twitched. On May 10th had been still improving and sat at table to dinner. On May 11th he had a fit, was convulsed for an hour, and then died.

On opening the skull the convolutions were found flattened and much compressed; this was not due to an internal pressure from fluid, but to an increased size of the hemispheres from adventitious matter. A section of the hemispheres showed the medullary substance infiltrated with a new material, and more so on the right than the left side. Its colour and consistence distinguished it from the cerebral tissue, being firmer and of a slight grey colour, but in certain spots it gradually passed off into the natural brain substance. In parts where the new material was massed together it could be felt by the finger as tolerably defined from its hardness. It spread through the hemispheres and encroached on the cineritious substance, pushed the convolutions aside, separated them and absorbed some of them. Many convolutions were destroyed and they thus appeared to be fewer in number, with long distances between them. Until it was evident that a new material had been formed, the first appearance might have suggested an hypertrophy of the white substance of the brain, with an absorption of the cortical part. The great mass of new matter, amounting almost to a tumour, was in the right anterior lobe, encroaching on the Sylvian fissure to outside of corpus striatum. The central ganglia, however, were untouched. Cavernous sinus healthy. A section of Pons varolii shewed some minute and recent extravasations into its substance.

It was a question whether in this case the disease should be regarded as a cerebritis in which the inflammatory product had become isolated and hard, or whether it was of the nature of a new growth or tumour around which an inflammatory process had taken place. The most remarkable circumstance connected with the case was that the symptoms bore an inverse relation to the amount of disease—that as the one developed the other decreased. The state of brain found after death might well account for his earlier condition when he lay senseless in bed, and it can well be supposed that, though the central disease was progressing, the neighbouring healthy parts of the brain were at the same time recovering from some disturbance which temporarily paralysed them.

CASE.—A little girl, æt. 14, began to ail about a year before death with symptoms denoting a cerebral cause and yet of a very undefined character. It was said that she one day fell down and lost her sight, subsequently had headache, and was sometimes sick. She came to the hospital as an out-patient on account of her failing powers both in mind and body. Those who casually saw her believed her to be an idiot. She then came into the hospital, where she lay two or three months until her death. Her symptoms were almost entirely of a negative character; she had ceased to be able to stand and had very little power in moving her legs. Her arms she moved but feebly and slowly. She had a vacant stare with dilated pupils. She generally lay quiet making no complaints, and when spoken to smiled. When asked her name or a simple question she answered sensibly but remarkably slowly, so that it was not apparent for some time whether she understood the question, and was able to express herself. It evidently took her a long time and caused her a great effort to collect her thoughts. The only difference perceived from week to week was that the bodily and mental powers grew feebler. As regards the former she had no paralysis in the usual sense of the term, but she failed in the ability to attempt any movement. When food was put in her mouth she would cease chewing and swallowing, so that it would remain in her mouth until removed. She could evidently see, and on examination by the ophthalmoscope nothing abnormal was discovered on the retina. The skin appeared sensitive as far as could be ascertained. During the last few days of her life she lay with her eyes open; she looked at the nurse but could no longer speak, although by a movement of the lips appeared as if endeavouring to do so. She then became cold and her feet very livid, and so quietly died.

The diagnosis in such a case was difficult. There was nothing to warrant any other opinion than that the brain as a whole or the hemispheres were at fault. As, however, many cases have occurred where the whole brain has been affected in connection with a tumour, either functionally in a reflex manner, or organically by a secondary inflammation, especially of the ventricles, it was thought possible that a tumour might exist in the hemispheres or cerebellum. There was, however, the absence of the usual symptoms of this condition, viz. fits, violent headache, frequent sickness, and optic neuritis.

Post-mortem Examination.—There was a general meningo-cerebritis. No tubercles discernable in the brain or any part of the body. The character of the inflammation was not of the kind seen in tubercular meningitis. The whole surface of the brain had lost its transparency; the arachnoid was thick and opaque, and in the meshes was a considerable quantity of fluid; this was mostly serous and escaped when the pia mater was torn. In the meshes, however, there was some exudation of a firmer character. This appearance was universal, but rather more on the surface than at the base. The ventricles were very greatly distended with clear fluid; no signs of any inflammatory action except on the fourth ventricle, which was slightly granular. The brain as a whole was firmer than is usual 24 hours after death; it felt hard to the finger, and the septum lucidum could be stretched out firmly without laceration. No tumours or deposits in any part. The most striking morbid condition was the firm adhesion of the pia mater to the surface; in spite of the presence of fluid it was difficult to remove it on any part, and on the surface of the convolutions any attempt to do so tore off the brain substance. Generally the outer cortical layer of the grey substance came off with the membrane, so that the latter when removed was covered with a thin granular layer of cerebral matter. This condition reminded us very much of the similar fact on attempting to tear the capsule from a granular kidney. The bones of the skull were perfectly healthy, nor did there appear any cause for the occurrence of this general chronic inflammation of the membranes and substance of the brain.

In cases of *local softening* the symptoms would depend much upon the seat of the disease; for example, when, as is often the case, the parts supplied by the middle cerebral artery undergo decay, either from disease or plugging of the vessels, the symptoms would be those of a slowly progressive hemiplegia.

CASE.—A lady on rising from her bed found herself a little weak in her left arm and leg; being alarmed, she took a journey to rejoin her friends; when arriving she could scarcely walk up stairs, and her speech was observed to be somewhat thick. On the following day she could not get up, as her limbs were still weaker, and on the next day after that the left side was completely powerless, and her speech so indistinct as to be scarcely intelligible. She then became bodily ill from the onset of feverish symptoms, the skin being hot, tongue furred, and pulse quick. During this time we supposed that an inflammatory process was going on in the brain. After a few days the feverish and constitutional disturbances passed off, so that we concluded that the inflammation was confined to the region supplied by certain vessels. She subsequently got into a fair state of health, having paralysis of the right side and being highly emotional.

CASE.— *Softening and atrophy of the brain.* A man was admitted into the hospital in a fit, which was styled apoplectic, and it was thought that he would shortly die. He, however, gradually recovered from the attack, but for the remainder of his life he was paralysed. After he left his bed he was put in a chair, when it was observed that he gradually grew weaker in his powers of mind and body. He sat motionless the whole day, having a vacant look, and with his arm drawn up across his chest. He never spoke but could protrude his tongue. He continued in this sort of vegetative state until his death, fifteen months after the attack. The autopsy showed that nearly the whole of the left hemisphere was destroyed, its place being taken by a bag of fluid, which was confined only by the pia mater and arachnoid, and which burst when touched. It contained at least half a pint of fluid like lime water. The softening process had destroyed a large portion of the left corpus striatum and thalamus as well as the convolutions on the surface, but the fluid had not broken into the ventricles.

Chronic cerebritis or sclerosis, general and in patches.—There may be yet other conditions which may some day find a place in the category of inflammations ; for example, children die with brain symptoms, and having very large heads, water in the ventricles is suspected. This is, however, not always found, and then an opinion may be entertained that an actual hypertrophy of the brain has occurred : opinions at present vary as to the possibility of such an event. Then there is an occasional induration or sclerosis of the whole of the brain, as described by Bright in a very remarkable case in his ' Medical Reports.' It was that of a little girl, who for a year before her death, lay in a perfectly motionless and senseless state, with her limbs stiffly extended and without the possibility of her making the slightest movement. The white matter of the brain was found after death as hard as soft cartilage, so that the grey substance could be peeled off it, leaving the mould of the convolutions in the white substance. A stream of water washed off the grey matter, leaving the convolutions below on the hard white substance, giving it the appearance of a wax model of a brain. The ventricles looked as if they had been modelled in wax. The white matter passed in streaks into the grey convolution around. The cord was also hardened. Dr Bright regarded the case as one of chronic inflammation of the cerebro-spinal centres. The *sclerosis in patches* is met with generally in connection with a similar affection of the spinal cord, to which I shall have hereafter to refer. They are seen as isolated patches, scattered through the brain substance, and are composed of a grey adventitious inflammatory material.

Extension of disease from ganglia to surface of brain.—As I shall have occasion to mention again, when speaking of diseases of the spinal cord, it is especially worthy of notice that not only are cases of inflammatory disease and softening marked out by the distribution of blood-vessels, but they are dependent upon the anatomical and physiological arrangement of the nerve fibres. Not only is the nutrition of the latter dependent on the integrity of the

grey centre, but morbid processes run along them in the direction
of their physiological action. This is very evident in the case of
portions of the spinal cord, where a spot of disease proceeds in a
definite course. I shall have to tell you how the health of a muscle
is dependent on the nerve which supplies it, and this, again, on the
grey centre whence the nerve proceeds. From this centre, again,
changes may be found continuous with similar ones in the corpus
striatum ; and here once more I must tell you how morbid
changes in this large ganglion may include similar changes in the
convolutions with which the ganglion is connected by means of the
corona radiata. Thus it would seem that the ultimate relation
between the convolutions and every muscle of the body, as shown by
physiological experiments is corroborated by pathological processes.
The point is this—that associated with local softening in the central
ganglia of the brain there are often found corresponding atrophies
or degenerations of the convolutions with which they are associated.
I have often seen this myself, but the fact has been observed long
ago, as you will see by reading the cases of brain disease described
by Foville, Bright, and others. In the case just described, the
wasting was too great to allow of its being used as an example, but
cases constantly occur when, from disease in infancy, a child has
grown up paralysed or crippled on one side, and the hemi-
pheres and ganglia on the opposite side have been found atrophied.
A well known case is described by Schrœder Van der Kolk of an
idiot girl, æt. 27, who was paralysed on the right side from infancy.
The left hemisphere was found much smaller than the right, and
the space was occupied by fluid. The corpus striatum and thalamus opticus were much smaller than natural, especially the latter.
The convolutions were also small, as well as the left crus cerebri, the
left side of the pons, and the pyramidal body ; and what was very
remarkable, the right lobe of the cerebellum was smaller than the
left. The spinal cord was also smaller, especially in the dorsal region,
as well as the nerves which were given off from it.

Dr Ogle described the case of a man, æt. 62, who was of weak
intellect since childhood ; he had contraction and atrophy of the
right arm and weakness of the right leg. After death there was
found a large cyst in the brain, which occupied and destroyed about
the posterior half of the corpus striatum on the left side and the
outer part of the optic thalamus, as well as a portion of the outer
wall of the ventricle.

Other forms of degeneration.—I would warn you against con-
founding the term softening in a strictly pathological sense with
the *popular meaning of softening*. We intend by it a distinct
localised process accompanied by special symptoms, whereas the

extra professional signification is synonymous with a weakened brain, which might have its origin in a hardening as well as a softening.

In cases where a slow degenerative process has been going on throughout the brain, the symptoms would be of a less special kind, and indicative of a general decay of the organ. Indeed, it so happens that in the well-marked cases of softening of the brain the diagnosis is often wrongly made, whereas the term is applied to examples of disease which indicate decay, but not necessarily softening. Amongst our patients we constantly have persons come before us presenting a variety of symptoms indicative of an impaired brain, and, for want of a better term, we declare they have softening. They evince a paralytic condition both of body and mind.

Probably one of the first phenomena you notice in such a patient is his manner of speaking. It is not so much that he speaks thick or indistinctly, but he answers in a manner which shows that his mental vigour is departing; so far from hesitating in his speech, he answers rapidly or curtly. When a person speaks slowly, it may be from a careful consideration of what is to be expressed, whilst a mere hasty flight of words shows a want of this power. Thus, your patient answers you in a good-natured way, "Yes, yes, yes," and on questioning his friends you will find such an answer is not correct; it has been little more than the result of an excito-motor action. Of course, if there is much decay in the central parts of the brain, the speech may be indistinct or thick. Then, also, there may be slight paralysis of the limbs. The patient totters into your room, as if he had lost the use of his legs. In a hemiplegia, you see patients make an effort, but here he tumbles into your presence. So with the arm, you see him endeavour to take up an object, and his hand falters. Ask him to write his name, and his hand shakes. There may be also altered sensations, as a numbness of the hands and feet, or a complaint of tingling, &c. The patient cannot button his clothes or tie the strings of his dress. There may be also some difficulty with his bladder or rectum. As a rule, the emotions are soon disturbed; the patient laughs and cries, as in an hysterical condition. Such a state may go on until one particular part of the brain is destroyed, and then marked paralytic symptoms, such as those of hemiplegia, set in.

It may happen that the friends keep back these symptoms, which they do not care to recognise; or the patient, coming alone to your study, is unconscious of them, and suggests complaints to your mind altogether of a different kind from the real ones, or alludes to certain symptoms connected with his stomach or bowels. He complains perhaps of what he calls bilious attacks or constipation. This is important to remember, because, as age advances and the brain is liable to change, the earliest symptoms may be those of gastric disorder. I have more than once been asked to prescribe for sickness which has been in all probability cerebral, from the occurrence of subsequent brain symptoms. In these cases the vomiting is often very characteristic, being little more than the stomach evacuating its contents without any spasm, pain, or effort on the part of the patient. He eats his dinner, and then, whilst sitting quiet or lying down, the whole of it will return. The diagnosis of cerebral vomiting is aided by the presence of other cerebral symptoms and the absence of any gastric disturbance, as furred tongue, &c.

Meningitis.—Now, what is meant by inflammation of the membranes of the brain? What is meant by arachnitis or meningitis? You will find that many writers and most members of the profession use these as convertible terms, and thus I well remember, when a student, how sorely puzzled I was to unravel the mysteries of such a . subject as inflammation of the brain. I believe the term arachnitis was adopted by those who wished to compare this inflammation with that of other inflammations, and thus would imply that the exudation proceeded from the arachnoid. Those who did not pretend to assert the exact seat of the process were content with the expres·sion meningitis. I think, however, we shall find some broad distinctions in the seat of the inflammation, and with corresponding pathological differences. The first main distinction is between the cases where the inflammatory product is found in the meshes of the pia mater, or is subarachnoid, and those where it is in the arachnoid cavity. The difference is highly important pathologically. The latter always arises from without, the source of the secretion being the dura mater, whilst the former is the only variety which is idiopathic, although it may arise also from injury. Where the effusion is found in the arachnoid cavity, I have adopted the name arachnitis as a term equivalent to the inflammation of other serous membranes; and, having no term of strict application to the case of idiopathic inflammation where the effusion is subarachnoid, I have used the old expression meningitis. The latter may be traumatic as well as simple or tubercular.

I should like to be clearly understood on this point, as it is one of great practical importance. If I asked any novice what part was affected in inflammation of the membranes of the brain—as in meningitis, arachnitis, or by whatever name you choose to call it—he would, no doubt, remembering that in pleurisy or pericarditis the inflammatory exudation was contained in the serous sac, state that the same was true of the brain. But it is not so; the exudation does not lie in the serous sac, but altogether beneath the

serous covering of the brain. There is such a case as you picture
to yourself, but this arises altogether in a different manner, and has
its origin from without. The simple explanation is this—that in
inflammation of other serous membranes the character or seat of
the inflammation may be twofold, although you cannot distinguish
between them; but in the case of the brain you can. Take the
instance of pleurisy, where you find serum in the chest or lymph
covering both pulmonary and costal pleura. This may have had
two sources, and may imply two very different pathological processes. One patient, for example, has a pleuro-pneumonia, implying
a general and severe disturbance of the whole constitution, whilst
another in perfect health receives a stab in the chest (not touching
the lung), and has a severe pleurisy as the consequence. In the first
case, if the lung recovered itself it might be impossible to say
whether the lung or the thoracic walls were the source of the exudation in the pleura; but supposing the lung had a wrinkled or
convoluted surface, then the lymph would be discovered beneath
the serous membrane which stretched over it, and its source would
be evident.

The same difficulty exists in pericarditis and peritonitis. In the
case of the brain, however, it is different. The pia mater, stretching down among the convolutions, is the source of the exudation in
idiopathic inflammation, and there being much space beneath the
smooth surface of the pia mater, commonly called arachnoid, the
exudation remains for the most part beneath it. In the case,
however, where an injury has been received to the skull, or where the
bones are carious, so as to implicate the dura mater, an inflammation may be set up in the latter membrane, and an exudation
poured out from its smooth surface (also styled arachnoid), so as
to fill the cavity. If, then, on making a post-mortem examination,
and removing the dura mater, purulent fluid trickles down, and
you find both serous surfaces covered with the inflammatory product, you may be sure that this is not an idiopathic arachnitis, but
has its origin in the dura mater, and you at once look to the bones.
The idiopathic arachnitis or meningitis, you see, is not analogous
to pericarditis or peritonitis, for the serous surface shows scarcely
any exudation, the lymph being for the most part in the pia mater,
and beneath the arachnoid. Remember, then, simple arachnitis, or
intrarachnoid inflammation, arises from without; whilst subarachnoid inflammation is idiopathic. You ask, no doubt, if the converse
of the first proposition is true. No, it is not. Injury may give
rise to subarachnoid inflammation, as well as interarachnoid.

In the latter or simplest form it is possible that the brain is
only finally affected, and thus the reason why the symptoms are so

obscure. For example, after injuries, signs of cerebral disturbance are often altogether wanting until the final coma appears. It is possible for a large amount of lymph or purulent matter to be poured out into the arachnoid cavity, and for the pia mater, with the adjacent cincritious substance, to hold out against the inflammatory process for a long time, and thus it is only at last, and just before death, that any well-marked symptoms arise. On post-mortem examination surprise is expressed at the large amount of effusion with so slight and recent symptoms. The explanation is to be found in the fact that simple arachnitis arising from external causes is altogether a local affection, and also because the inflammation is often unilateral—that is, one hemisphere is found covered with lymph whilst the other is perfectly free. This never occurs in idiopathic inflammation. You must understand that although this simple arachnitis of which I have been speaking always arises from without, yet that the same form which constitutes the idiopathic variety may also arise from injury. A blow on the head, may, indeed, set up an inflammation of the brain and its membranes not to be distinguished from an idiopathic inflammation, but a simple arachnitis always arises from an external cause.

The inflammation of the dura mater is technically called a *pachymeningitis,* whilst that of the pia mater and appendages is called a *leptomeningitis.*

Although we are in the habit of speaking of an inflammation of the substance and of the membranes as distinct affections, and no doubt they may occasionally be found separated, especially if the inflammation of the brain be local, yet I think there is every reason to believe that in very many cases of meningitis there is also a cerebritis, and that the more correct term would be meningo-cerebritis.

That the brain to a certain extent is involved all must allow, for in inflammation of the pia mater the implication of the grey matter produces many of the most marked symptoms ; there are others, too, which are evidently due to the state of the ventricles. In fact, a meningitis is characterised by marked cerebral symptoms.

It will be observed that the idiopathic forms of inflammation of the membranes are mostly met with in children. In adults the affection is more rare, and is found generally in connection with various diseases or injuries of the skull and brain. It may be as well, therefore, for practical purposes, to consider these acute cerebral affections of children together, although they may not strictly come under the head of meningeal disease. You know that tubercular meningitis is often styled acute hydrocephalus, in consequence of the effusion in the ventricles being one of the most striking changes found in the brain. There is, however, a not infrequent fatal form of disease in children, accompanied by cerebral symptoms, where an increased ventricular effusion is found without any marked inflammatory products; so that, although in such cases we attribute death to the brain, it is very probable that this organ has merely participated in some other more general changes in the body. These cases may be called simple hydrocephalus, in contradistinction to the tubercular. We have, therefore, amongst the more rapidly fatal forms of brain affection *acute simple meningitis, tubercular meningitis, or acute hydrocephalus,* and, thirdly, *simple hydrocephalus.*

In *simple meningitis* the whole surface of the brain is found covered with a thick layer of lymph, and dipping in amongst the convolutions with the pia mater; there may be some opaque fluid in the ventricles, and a slight softening. This simple acute affection is not very common, but is sometimes met with in connection with scarlatina and the other exanthemata. Occasionally, also, children will die with all the symptoms of meningitis, where the post-mortem appearances do not reveal the characteristic conditions of either the simple or the tubercular form of the disease, the effusion in the ventricles and marks of inflammation on the surface being but slight. As these, however, correspond to those found in a greater degree in tubercular meningitis, it may be a question whether they may not be of a like kind with it, seeing that in the latter the tubercle is often little more than a granular lymph.

CASE.— A little boy, æt. 7, was admitted into the hospital on March 6th. The mother stated that he was well on the morning of the 4th, but during the day became feverish and drowsy; on the 5th he became almost insensible, and on the 6th he was brought to the hospital. He was then in a high state of fever, with quick pulse and contracted pupils. He had no convulsions, but remained in a state of coma until the afternoon of the 7th, when he died. The mother could give no cause for his illness; said he was a delicate boy, and as an infant had had inflammation of the lungs, with convulsions. *Post-mortem examination.*—No mark of injury could be found on scalp or bones. On removing the dura mater the whole surface of the brain beneath the arachnoid was seen to be covered with a thick layer of yellow lymph. This was in such abundance that the hemispheres appeared of a light colour, the cortical substance being quite hidden. The sides and base in like manner were covered with the effusion, though to a less amount than in the hemispheres. The inflammatory exudation proceeded downwards to the corpus callosum and between the convolutions, so that sections of the brain showed the lymph existing in large quantities in the pia mater between them. The ventricles contained a slight increase of fluid, and this was opaque and turbid, as if some inflamatory exudation had mixed with it. The whole cerebral substance was soft, and this softness very different from that due to ordinary post-mortem change, for the brain-matter was tenacious and sticky, adhering to the knife and fingers when touched; it was at the same time dry, giving out no watery exudation. This condition of the cerebral substance was seen more markedly in the attempt to tear off the pia mater when it stuck to the latter, so that they could not be clearly separated from each other. The parts forming the ventricles did not appear softer than the rest of the brain. The whole brain was of rather a dark colour from hyperæmia, A careful examination was made for tubercles, but none were found. All the other organs of the body were perfectly healthy. The lungs also were carefully examined for tubercles, but none were discoverable.

CASE.—A child, æt. 3, was brought to the hospital from the workhouse, and no other history could be obtained than that she had been ailing for three weeks. She was then brought merely for advice, but appearing extremely ill, she was taken in. She was in a dull, drowsy condition, resembling that of fever, and it was not quite certain that this was not her complaint. She only lived 24 hours, having a severe attack of convulsions shortly before death. On making inquiries as to any injury the child might have received, it was stated that, when 15 months old, she had a fall and struck her head, but suffered no manifest symptoms in consequence. *Post-mortem examination*—On removing the dura mater, the surface of the brain was seen to be covered with a thick layer of green-coloured lymph; this entered deeply between the convolutions, so that only small portions of the grey matter were here and there apparent amongst the effusion; this was altogether beneath the arachnoid, the surface being quite free. This exudation, which was to a great amount, covered also the sides of the brain and continued round to the base, where it was very slight, the under part of the organ being comparatively unaffected, compared with the surface of the hemispheres. The ventricles contained a slight excess of fluid, and this was clear. The walls of the cavity were natural, not being softer than other parts of the brain, which as a whole was rather soft. No tubercles could be discovered in any part. On examining the septum lucidum, fornix, &c., by the microscope, nothing more was found than numerous granules surrounding the capillary vessels. The whole of the veins of the brain presented a remarkable condition, from being closed by blood, which had evidently coagulated before death. The smaller veins of pia mater on surface were much distended by clots, the larger branches were in like manner occluded, and the longitudinal sinus itself was completely filled by a firm clot of blood. This was slightly adherent to the walls, and the centre was whitish and soft. The lateral sinuses were in like manner filled, and in these the clot was more broken down and softer than in any other part; the coagulum ceased with the sinus, the jugular vein being quite natural. The bones of the skull were carefully examined, in order to discover if any disease existed in them which could have given rise to the meningitis, but none was found. The temporal bones were healthy, as well as upper cervical vertebræ and other parts. *Lungs*—At lower parts there were many condensed portions of tissue; these were of dark colour, sank in water, airless, and dense; they were at the same time granular when incised, and showed by the microscope exudation-corpuscles. Heart had firm, white clot on right side. The stomach showed post-mortem solution, with contents of the organ in the abdomen. No tubercle discoverable in any part of the body.

Tubercular Meningitis.—After removing the calvaria and taking off the dura mater, you might for the moment doubt the existence of inflammation, as the hemispheres may exhibit no evidence of exudation; but on placing the finger on the surface a slight greasiness may be felt, and on looking at the sides towards the temples some streaks of lymph may be observed in the sulci. It should also be noticed that the convolutions are flattened, showing a pressure against the cranium from effusion of fluid within the ventricles. On removing the brain from the skull a large patch of lymph is seen at the base, covering the optic commissure and neighbouring parts, occupying, indeed, the large subarachnoid space situated there. Along the margin of the Sylvian fissure also lymph may be seen, and on lifting up the middle lobe the lymph is found to proceed quite into the fissure. The ventricles contain a large quantity of opaque serum, and the adjacent parts are very soft. The septum lucidum may be broken down, while the fornix is semi-fluid and the walls of the ventricle all around are softened. On stripping off the pia mater from the convolutions, and more especially in the fissure of Sylvius, tubercles may be found. These are not seen on the surface of the brain, but in the meshes of the pia mater which dip down between the convolutions.

From the position of the lymph, this form of inflammation is sometimes called *basilar meningitis*. It is important to observe that in the simple and tubercular, as well as in the other forms of meningitis, the ventricles are most usually involved in the inflammation, from which it would appear that the affection known as meningitis, so far from being an inflammation of the membranes only, unquestionably involves the surface of the brain to a certain depth, and at the same time the walls of the ventricles. It may therefore happen, and in all probability frequently does, that the brain throughout its whole substance is affected. We witness the effects on the surface without and the surface within, but there can be little doubt that very often these are the mere outward manifestations of a universal affection. Thus the term meningo-cerebritis would be the more correct appellation. The time will, I hope, soon come when we shall possess a more accurate knowledge of these different pathological conditions of the brain, so that we shall be able to give them their appropriate names and their corresponding symptoms. Since post-mortem examinations have been more carefully conducted, we have not infrequently found tubercles extending over the spinal membranes, and at the same time evidences of a slight inflammation thereon. This may have complicated the symptoms.

Symptoms.—For the sake of saving the memory, it is as well always to have a rational knowledge of symptoms, and not merely to hold in possession a meaningless category; from what I have already told you, therefore, of the functions of the brain, you might infer the existence of many of the phenomena which occur in meningitis. For instance, from the surface being involved, you would suspect delirium, passing on to a torpid state of mind, and finally unconsciousness; also, from what I have said, you might expect convulsion, either in its extreme form or a rigidity of particular muscles. From the centre or ventricles being involved with a large secretion of fluid, you might expect coma, or, owing to the softening of the ganglia on either side, right or left hemiplegia. This, of course, is towards the termination of the case. Then, from the inflammation at the base, with exudation and implication of cerebral nerves, you might not be surprised to find strabismus, inequality of pupils, and their inaction to the stimulus of light, or, from implication of the pneumogastric nerves, an irregular respiration, a slow action of the heart, and much disorder of the stomach. Now it so happens, as you might suspect, that many of these symptoms need not necessarily be present, and thus, if you abstract those which are not universal, you are left with symptoms which have no characteristic features about them. The brain, indeed, may become inflamed throughout (usually styled meningitis) ; the patient falls into a drowsy condition, which ends, in a day or two, in fatal coma. These may be the totality of the symptoms of inflammation of the brain, as I have witnessed over and over again, both in the traumatic and idiopathic variety. For instance, the surgeon will have a case in his ward of fracture of the skull, and the patient is apparently doing well, when he becomes listless, then drowsy, insensible, and dies. The dresser will inform me that he can give no explanation of the cause of death, for there were no symptoms of inflammation of the brain ; but on making the *post-mortem* examination, its whole surface is found covered with lymph. Then, again, cases are brought to the hospital under various names of delirium tremens, drunkenness, &c., which turn out to be examples of meningitis. For example, a sailor goes to a lodging-house, and on the following day is found in bed, talking strangely, or in a stupor, and is brought to the hospital with a suspicion of foul play, and then arachnitis is discovered. Only this day you have seen an example of this kind. A man was brought in from a neighbouring linendraper's, because for two days he had been acting strangely, in consequence, it was thought, of intoxication. He was placed in the spare ward, as on admission the case looked like one of delirium tremens. He would not speak, but was constantly moving in bed. There was not one single other nervous symptom. I suggested that he had had a blow on the head, and probably some effusion of blood on the brain, as I have seen exactly such symptoms from this cause. In a few hours he fell into a state of coma and died, and we have found an acute arachnitis, probably of a tubercular character. If you meet with an obscure case of recent disease to which you can only apply the term cerebral, without being able to declare the existence of any special lesion, it will generally turn out to be a case of meningitis, or rather cerebro-meningitis. You must, remember, then, the tendency to the final coma may be the only symptom in some cases ; in others, the symptoms may be many and marked, and in others such a combination of characteristic phenomena that the nature of the disease cannot be mistaken. These are mostly seen in cases of tuberculous meningitis, which running a course often of a fortnight, the symptoms may be made the subject of careful observation. But even here all the usual symptoms may be absent. A child may be seized with sickness, and then lie for a whole fortnight ejecting immediately everything it takes ; then falling into coma and dying. In such a case I have seen the membranes stuffed with tubercles. This disease, or acute hydrocephalus, as it is called, occurs in by far the majority of cases in young people of the tuberculous diathesis, is not common in infancy, is seen more in growing children, and not unfrequently in young adults, and not so seldom, as is thought, in middle or advanced age. I have several cases in individuals above fifty years of age.

Now, in a large majority of these cases the earlier symptoms of illness are associated with the inflammation, but previous to this it is believed that tubercles are present, and certainly in the lungs and other parts of the body they already exist. It may be, therefore, that in some cases the premonitory pyretic condition is due to the formation of these bodies. The principal objection to this is that in cases of general tuberculosis latent tubercles are not found in the brain, showing they are formed only just antecedent or during the time of the inflammation. Those who hold this view speak of granular inflammation of the membranes. In those cases, if a vessel be removed, the so-called perivascular space is seen around it, and in this may be perceived nuclear masses forming the foci of tubercles.

It is convenient to divide the disease into three stages. You will observe that a division of this kind is made in several other affections, as it is to a certain extent founded on a natural separation of the symptoms which occur in inflammation of special organs. First of all, there is the general constitutional disturbance denoting the onset of a severe disease ; secondly, there are the symptoms which indicate the disturbance of a particular organ ; and, thirdly, those which mark the paralysis or arrested function of that organ.

Thus, in inflammation of the brain there would be—first, the premonitory symptoms ; secondly, the invasion of the disease, with its special symptoms ; thirdly, those of profound cerebral mischief, or, in other words, a general constitutional disturbance ; then marked brain symptoms, and, finally, signs of collapse or paralysis of the brain.

In the first place, the child may be fretful, listless, have stomach disturbance, and other troubles, which bear no especial characters; afterwards feverish or pyretic symptoms, marked by hot skin, quick pulse, furred tongue, and headache. At this period the case is generally, I might say always, styled one of typhoid or gastric fever. It is not the exception, but the rule, to make a mistake; the stomach disturbances are so common, and these cerebral affections commence so insidiously that the severe nature of the case is not suspected. I believe, in the majority of cases to which I am called, the case is regarded as one of fever in which cerebral symptoms have become developed, and I think, with the utmost acumen which you can use, you cannot positively determine the disease at this early period. Some time ago I was asked by a more than ordinarily acute practitioner to see a young man who, having been ailing a few days, was causing some anxiety to his friends. He had febrile symptoms, and the question, as in all such cases, was whether these arose from a local inflammation or not. We could discover no cause, and were therefore obliged to style the case one of fever. We were both alive, however, to the possibility of tubercle formation, and of early meningitis, but failed to discover any proof of either. On the next day there was some suspicion of head affection, and on the following day there could be no doubt of its existence. You will see that the diagnosis of continued fever is as often made for negative reasons as for positive ones. A person is in a highly febrile condition, and if a characteristic rash is at the same time present the nature of the case is apparent; if there is no rash we look for some local inflammation to account for the symptoms; if none be discovered we have again recourse to the term fever.

Since, then, at the commencement of tubercular meningitis the head symptoms are not more marked than in ordinary fever, a positive diagnosis cannot be made. I am justified in this opinion by having to-day looked over my note-book, used when I was a student, and reading the particulars of the case of a child who was in the hospital under Dr Addison, to which are appended some notes of his clinical lecture. He alluded to the difficulties attendant on the diagnosis of tubercular meningitis and gastric fever, and after commenting on the case stated his belief that the symptoms present denoted a cerebral affection which would be necessarily fatal. It so happened that on the very next day these head symptoms passed off, and the child got rapidly well. I know that Cullen and some of the old writers speak of meningitis being ushered in by well-marked cerebral symptoms, as a violent convulsion or a fit, to which they could give no other name than "apoplexia hydrocephalica." Now I believe these were not head cases at all, for you

will find that such sudden cerebral symptoms as delirium and convulsions do not attend the onset of inflammation of the brain.
They denote merely a poisoned blood acting on the brain, and
generally imply that an exanthem is about to appear, that the
kidneys are at fault, or that there is some exciting cause in the
coming teeth or in the intestine.

In *the next stage* and subsequently to the fever the more characteristic symptoms appear: the child is obliged to lie down, complains of severe pain in the head, rolls it about, often moans and
screams out suddenly, and in the night grinds its teeth. The
headache is often now too severe to be the accompaniment of a
mere febrile disorder, and the distressing rolling of the head is
never seen in simple fever. At this time lymph is forming at the
base of the brain, and effusion is occurring into the ventricles. A
strabismus may occur as a very early symptom, and various organic
functions are now interfered with, probably through the pneumogastric nerve and its associations. The heart is lowered in number,
and from being 90 or 100 during the onset of the fever, is perhaps
only 60, although sometimes, if disease approaches insidiously, it
is slow from the beginning. This is one of the most important
symptoms to note, and on this alone I ventured a diagnosis the
other day, which turned out to be correct. I was asked to see a
young man with fever. I found him in bed, almost unconscious,
and therefore not able to make any complaints, and the febrile
symptoms ran very high. I confessed my difficulty in being able
to state positively whether the patient had continued fever or
meningitis, but, on feeling the pulse and finding it 58, I decided at
once on the latter. This turned out to be correct. Two days afterwards it was 120. It is usual for the pulse afterwards to rise.
Then the character of the respiration is to be noted. This may be
variously altered in number and rhythm by diseases involving the
mechanism of breathing; but an irregularity in the process, or a
sighing, may be usually referred to a nervous cause, and this in a
child is strongly indicaative of a cerebral affection. The temperature is never very high, and in this respect meningitis contrasts
with cerebral rheumatism. In connection with injuries an inflammation of the brain, however, has been accompanied by as high a
temperature as 106°. That the worst cerebral symptoms are due
to high temperature rather than inflammation, was well seen in the
case of a little girl who was in the clinical ward. She was at home
sitting over the fire in a listless state, when she suddenly had a fit
and was brought immediately to the hospital at 6 o'clock. She
was insensible with temperature of 100°, and very rapid breathing.
In the night she had several fits, and the temperature rose to 104°,

pulse 156, and respiration 54. At 11 o'clock the temperature was 105°, and at 12 o'clock 108°, and at 1 o'clock 109°; she soon after died. On post-mortem examination no morbid condition whatever was found. She died of hyperpyrexia, but the cause of this was never ascertained. Then again, and produced in all probability through the pneumogastric, there is the vomiting ; this, if marked, is one of the most important symptoms in head affection. It does occur occasionally in gastric fever, and sometimes very severely in typhoid ; but these are exceptional instances. At the same time, as the stomach nerve-supply is affected, so is that of the whole intestinal canal. The latter undergoes violent spasmodic contractions, so that after death intussusception is frequently found. With this narrowing of the intestines, the abdomen is much flattened or hollowed, and the bowels are confined. Here, again, is a point for diagnosis. If the question be between fever and meningitis, the state of the abdomen will enable you to distinguish : in the one it is full or tumid, with gurgling on pressure ; in the other disease it is flat or scooped out. In some exceptional cases, however, for reasons I cannot explain other than that the nerves are paralysed, the abdomen is tympanitic. You may remark, also, as a help to the diagnosis between these disease, an extreme irritability of the patient and sensitiveness of the skin. This is made an especial circumstance in the aid to diagnosis by Mr Stocker, who says that if you wish to examine the patient's body for the presence of a rash, you will find the fever patient assist in raising his nightdress, whilst the cerebral patient will whine and resist ; for you know that the patient suffering from fever is in a semi-conscious state, his blood is diseased, he lies in bed in a torpid state, and with a confused state of mind, but has sufficient sense to discern your wishes, and assists you in your effort to examine him, whilst the patient with head disease is suffering pain, and is unwilling to be disturbed.

The *final symptoms* appear in a few days by the child or young person being still worse. No food is taken, there is wasting of the body, repeated sickness, delirium, pupils disposed to dilate, or, if coma have come on, widely dilated, and scarcely at all acted on by the stimulus of light. On raising the eyelid you may often notice a mucoid film on the eye, which, I apprehend, implies that the fifth nerve is losing its function or becoming dead. Before the coma comes on, or accompanying it, there are often convulsions, sometimes of a general kind, with great distortion of the face, and, if the chest be involved in the spasm, much congestion of the skin and violent perspiration. A modification of the convulsion is seen in the rigidity of a limb or the turning in of the thumb on the palm of the hand. At this time, if you make a mark down the skin with your nail, a red stain remains. This was first mentioned by Trousseau as a diagnostic mark of cerebral disease, and is styled " tache méningitique." I never saw this well shown except in the later stages, where no doubt existed as to the nature of the disease, and you may sometimes observe it in other febrile disorders. I therefore do not attach much value to it as a diagnostic sign. Towards the close of the case it is not unusual to find the patient hemiplegic ; at this time the ventricles have become distended with fluid, and the surrounding parts softened. I apprehend this symptom may be accounted for by the central ganglia on one side being involved in the softening process. If tubercles exist on the choroid they may be sometimes observed by the ophthalmoscope. I have seen them on the eye, removed after death, in such numbers that I have no doubt their presence could easily have been ascertained during life if looked for. It is said that an optic neuritis often exists.

Towards the close of the case, when the more marked symptoms have passed off, some hope is expressed of recovery, but this is what so often occurs in other inflammatory diseases.

To summarise.—1. The child is getting out of condition, is wasting and fretful, with loss of appetite and constipation. 2. Pyrexia, but tongue not much furred, nor temperature high, headache, delirium, crying out, throwing head about or rolling it on the pillow, restlessness, pupils contracted, strabismus, vomiting, constipation, with retraction of belly, general hyperæsthesia, respiration irregular or sighing, face flushed, tache cerebrale. 3. Insensibility, pupils dilated, grinding teeth, convulsions or twitching, pulse perhaps slow, face sunken, skin clammy, paralysis.

It has been said by some foreign writer, whose name I forget, that many of these symptoms are due to the part of the brain affected, and in making an analysis of all cases of meningitis, the following symptoms may be regarded as especially denoting an inflammation at the base of the brain, viz. intolerance of light, eyes red and covered with mucus, turned up, strabismus and pupils irregular, headache, convulsions or spasms, and vomiting. I need do no more than give one illustration of the disease.

Case.—A child. æt. 4. When first seen had been ill a fortnight, and now suffering from well-marked symptoms of cerebral disease. During this period he had complained of pain in the back of the head, sickness, screaming at night, &c. Now he lies on his back and is in constant fear of being moved, as it increases his headache; constant sickness as before. He occasionally draws up his legs and arms, and then extends them ; but no decided convulsions ; thumbs turned in on palms. He afterwards sank into a state of coma, with pupils widely dilated.

Post-mortem examination.—Surface of hemispheres almost dry, or only a little greasy exudation to be scraped from them. At base a considerable quantity of inflammatory effusion within the pia mater ; the optic commissures, and parts around covered with an exudation of lymph. Upon stripping off the pia mater from the convolutions, tubercles were found in the membrane, especially in the fissure of Sylvius. Brain itself flat on superior surface from the pressure of fluid within the ventricles, and corpus callosum bulging out for the same reason. The fluid within amounted to about three ounces, and appeared to consist of little else than water, having a specific gravity of 1001, and becoming only slightly opaline by heat. The septum lucidum, fornix, and the ependyma, very soft, being in fact semi-fluid. Examined microscopically, these were found to contain exudation-globules, and fatty granules surrounding the capillary vessels. The cerebellum contained three or four hard scrofulous masses in the cineritious substance, each being about the size of a pea. The *lungs* contained miliary tubercles. Bronchial and mesenteric glands also. Spleen and kidneys contained a few tubercles. There was also post-mortem solution of the stomach, and discharge of contents into peritoneal cavity.

Tubercular meningitis in adults.

—You must not suppose that this disease is confined to children. It is by far more common in them, but I have frequently met with it in young men and women, and even in two persons between the ages of fifty and sixty. I make especial mention of it because it is more likely to be overlooked, and other complaints to be suggested as a cause of the symptoms, as, for example, fever, or in one case uræmia on account of albumen being present in the urine, or in another delirium tremens from a consideration of the habits of the patient. It would seem also that there is less likelihood of finding tubercles universally spread throughout the body than in children, in whom they are all but invariably found in every organ, as well as in the brain.

CASE.—A young man, who had always lived in the country, came to London three months before his illness to be an apprentice in the city. Sixteen days prior to his death it was observed that his manner was odd, he spoke in a strange style, and was unfit for business. He soon became very ill, but it could not be learned that he had any well-marked cerebral symptoms, and the case was considered to be one of typhus fever. It was principally from fear of contagion that he was moved to the hospital. This was on the twelfth day of his illness and three days before his death. He was then all but unconscious; he would put out his tongue when requested so to do, but was too ill to make any rational complaint. He was very restless and continually screaming out, generally moaning in a low tone during the day and being very noisy at night. His tongue was furred and brown; pulse above 100; skin hot; bowels confined for a week; and he had retention of urine. His chest was examined, and the lungs appeared healthy. There was no trace of maculæ, and it was very evident that the case was one of meningitis. In the absence of all direct exciting cause, such as injury or disease of the temporal bone, this was believed to be tubercular as its most probable character. The patient remained much in the same state, unconscious but continually raving. His bowels were not moved by repeated doses of croton oil. Finally, the pupils became dilated, and the breathing laborious, death ensuing through congestion of the lungs. He had no convulsions. His friends denied his ever having had a cough, or being delicate, and repudiated the idea of any of his family being scrofulous when a tuberculous disease of the brain was suggested. On this account they allowed a post-mortem examination.

Autopsy.—Body extremely wasted. The brain presented in a well-marked degree all the features of tubercular meningitis. The superior surface of the hemispheres was somewhat flattened; the arachnoid greasy to the touch. At the temporal regions some lymph was seen in the sulci. On removal of the brain a large patch of tenacious lymph was seen to occupy the base, forming a covering to the pons Varolii and optic commissure. The fissure of Sylvius was filled with lymph, and the pia mater covered with tubercles. The pia mater, when stripped from other parts of the surface, was also seen to contain tubercle. The ventricles were occupied by a large amount of fluid, and the surrounding walls were very soft. The lungs were much congested, with tubercles scattered throughout them. The liver, spleen, and kidney were full of tubercles.

CASE.—A young woman, æt. 18, was admitted on May 15th, and died on May 25th. She was a dressmaker, and had been in apparent good health until about a week before admission, when she began to complain of her head, and soon became very ill. When taken in, the case was looked upon as one of fever, and wine was given to her. It was soon apparent, however, that the case was a cerebral one. She lay coiled up on her side in bed, disliking to be touched or moved, and complained much of feeling cold. Although sensible she was very drowsy; one eyelid fell, and both pupils were dilated. This dilatation and inactivity of the pupils when they were approached by a candle remained constant throughout. During the last two days she became unconscious; finally, the breathing became difficult, and the skin intensely congested. She had no convulsions nor paralysis.

Autopsy.—The surface of the brain was slightly flattened, and the arachnoid was greasy to the touch. On each side of the temporal region there was a little lymph in the course of the vessels of the pia mater. At the base there was a patch of yellow tenacious lymph, involving the optic nerves and surrounding the crura cerebelli and covering the upper part of the cerebellum. The Sylvian fissure was filled with firm lymph, in which abundance of tubercles were embedded. The pia mater of the hemispheres was also full of tubercles. The ventricles contained a considerable excess of fluid, but the centre parts were not much softened. The lungs were very carefully examined in order to discover the presence of tubercles, but none could be found, nor were any to be seen in other organs of the body.

Simple hydrocephalus.—Under this name are included cases where the sole observable morbid appearance is an increased effusion in the ventricles. This is very often found to be of a purely passive character, as it is met with in impoverished and anæmic children, and the symptoms dependent thereon are relieved by stimulants. These cases were first distinguished from the inflammatory by Dr Gooch, and subsequently by Dr Marshall Hall, who gave them the name of *hydrencephaloid*. The children have been delicate and exposed to various debilitating causes. The physician finds the child lying on its nurse's lap, unable or unwilling to raise its head, half asleep, one moment opening its eyes, and the next closing them again, with a remarkable expression of languor; the tongue is slightly white; the skin is cool or cold. If depletory measures are used all these symptoms increase, and coma with dilated pupil come on. On the other hand, they are often relieved or cured by nourishment and stimulants. In some of these cases no morbid condition of the brain may be found, whilst in others there is an increased amount of fluid in the ventricles.

Convulsions.—I have examined the heads of several infants who have died of convulsions, and have failed to find any marked morbid appearances.

Simple arachnitis.—I have already informed you that in idiopathic meningitis the inflammatory product is found mostly beneath the visceral arachnoid, and although such an inflammation may arise from an injury, yet, whenever the exudation is found in the arachnoid cavity, or is interarachnoid rather than subarachnoid, the morbid process has begun from without, and you must look to the dura mater and bones for the cause. The following is an example:

CASE.—A boy, æt. 7, had a poker thrown at his head; the point struck the left side of his head, producing a small hole. No marked disturbance attended the injury, but soon afterwards cerebral symptoms came on, and he was admitted two days after the receipt of the accident. He was then evidently suffering from arachnitis, being in a feverish state, drowsy, and occasionally shrieking out. A small piece of bone was removed, corresponding to the spot which received the blow. The symptoms continued, and he died five days afterwards.

Post-mortem examination.—On the left side of the head was a small opening, through which a probe could pass, and in the corresponding dura mater also. On turning back the latter the most acute arachnitis was seen, the purulent effusion being in so great an abundance that it poured down on the ground in a stream. The inflammation had extended all over the brain, but was more on the left side, the inner surface of dura mater being coated with a thick layer of soft lymph. On washing off the effusion from the free arachnoid, there was found also to be some inflammatory exudation in the subarachnoid space, though this was comparatively little. This, however, was seen to pass down in the course of some of the pia-mater vessels, along the sides of the brain. The brain opposite the fractured bone was slightly contused.

Meningitis from special causes.—I would by no means say that because I have spoken of tubercular meningitis and a simple meningitis in children that I have included in them all forms of inflammation of the membranes. We occasionally meet with a meningitis, both in children and adults, which would not come into either category, viz. a rapidly fatal meningitis where the signs of inflammation are slight, and yet where no tubercles are discoverable. In such cases we are at a loss to discover a cause; in some there may have been an unknown injury, and in others an unrecognised blood-poison in the system; an extension also of inflammation from the brain to the cord and *vice versâ* is not uncommon.

A man had a compound fracture of the skull injuring the brain; an inflammatory softening took place, which reached the ventricles. This extended through the third and fourth to the spinal canal, so that on post-mortem examination lymph was found surrounding the cauda equina. A blow on the head had given rise to an inflammation, which reached the sacrum.

A young man lying in bed with caries of spine and lumbar abscess died from an extension of the meningitis of the cord to the brain.

A woman with bedsore after fever suddenly became ill and died in a few days. The cord had become implicated in the disease of the sacrum; the canal was laid open, and the inflammation had extended to the brain.

Treatment.—Now, as regards treatment, there is no use in endeavouring to evade the only conclusion which can possibly be arrived at, after fully considering the subject in all its bearings—that we in all probability have no means of arresting inflammation of the brain when its nature has become manifest. It is easy to talk loosely about inflammation of the brain, and confound together cases of convulsions, cerebral symptoms of the exanthemata, and true meningitis. I have already said that the violent onset of cerebral symptoms in a child previously well generally denotes anything but a head disease, and yet it is to such cases that remedies are actively applied, and to them the recovery is credited. It is a question whether an inflammation of the brain or of the meninges, when once it has evinced its characteristic symptoms, can ever be recovered from. It is a fair question, however, to ask whether, at the very onset, such inflammation might not be arrested. But then, again, we must inquire whether we are speaking of simple or tubercular meningitis. In the latter case the body is full of tubercles before the final inflammation sets in, and the beginning of the end has arrived before the doctor has even diagnosed the case. It may be a question whether, even if recognised at its very earliest stage, it could be arrested, seeing that at the very next moment it would be ready to be lighted up again. I therefore regard the diagnosis of tubercular meningitis as equivalent to pronouncing the child's doom. I have no knowledge of the instance of any person recovering from it. I do not know that this should in any way tend to weaken our faith in our art, because we are unable to withstand disease when the mischief is done; for this would be like saying that a fire-engine is of no service in saving a house from destruction because it is rarely known to be required until the whole premises are in a blaze. The fact is, it is becoming more and more the duty of the medical man to be the custodier of the public health, and thus to endeavour to divert the morbid processes at a time when they are recent, and prevent their development into disease. I believe, as regards pulmonary consumption, attempts in the direction of prevention would be of far more service to the public health than the adoption of fanciful remedies when the disease has set in.

We might fairly ask, however, whether simple meningitis (as there appears to be such a disease) is not amenable to remedies, and to this I would answer in the affirmative. In the first place my own experience and that of others is corroborative of this opinion, and, secondly, I have met with appearances on the post-mortem table which seemed to leave little doubt of the existence of previous inflammation long cured. About two years ago a little child was in one of my cribs, partially amaurotic and deaf, together with other cerebral symptoms, the result apparently of an acute meningitis following scarlatina. In the course of time, and whilst taking the iodide of potassium, the child gradually recovered, and at the end of three months had regained its sight, and was running about the ward. From some unknown reason severe cerebral symptoms suddenly set in, accompanied by convulsions, and in a few hours the child died. On post-mortem examination I found a quantity of lymph in the subarachnoid space, covering the optic commissures and neighbouring parts. This was hard and tough, and evidently of some age.

If you ask how this inflammatory process is to be arrested, I am looking forward to the time when the use of such remedies as I have before mentioned as having a control over the vaso-motor nerves may be found of practical service, but in the mean time I should recommend the old-fashioned methods. Having witnessed such marked subsidence of symptoms after the use of bloodletting, I would adopt it, and I should be strengthened in its recommendation by such an authority as the late Dr Alison, who, I remember, related some striking cases in its favour. I believe also antimony is an anti-inflammatory remedy, or one possessing an influence over the capillary or minute arterial circulation. I believe, therefore, that at the present day, if you think a child has acute meningitis setting in, you will have no better treatment at hand than the application of leeches and the saline with antimony.

As regards calomel, I cannot speak with any satisfaction as to its administration. One of the very few cases I have seen of salivation in children was that of a little girl with tubercular meningitis. To her was given a grain of calomel every four hours, and this was continued for above a week, but without the slightest influence on the disease. In such cases we usually apply cold to the head, and have assumed that, because inflammation exists within, it is the correct remedy. I should judge myself by experience, but even then it is unfortunately very difficult to test its value. I believe myself that heat and cold are very important agents in controlling or modifyiug morbid processes, and therefore the subject is deserving all our attention. I think it very probable that the piece of wet rag which generally adorns every one's head who has cerebral symptoms is not very potent in its action; and if we eventually discover that such means as I speak of are serviceable, we shall apply them in a much more effective and rational manner. As regards any criterion to be gained from the feeling of the patient, it has rather been in cases of purely functional brain disturbance that I have seen relief by cold afforded. We hear, for instance, of persons exposed to the heat of India plunging their heads in cold water, and we find women with hysterical headaches seek the application of cold. Then with sick headaches or headache following a debauch cold is used with relief. One such case I had an opportunity of witnessing not long ago in the person of a medical man. He often suffered with headache, he had been overtaxed with work, and often had recourse to drops of brandy. Under the combined influence of these three causes, he, one evening, became almost raving mad. I found him rolling about in bed, complaining frightfully of his head, and asking for cold to be applied. On putting cold water on his head, he called out for more, so that we hung his head over the side of the bed, and, standing on a chair, let it fall in a stream upon him. He said it was most delightful, prayed for its continuance, soon said his head was better, and sank into a sleep. He subsequently confirmed his statement as to the value of the remedy. I believe, however, that putting a drunken man's head under the pump is an old and approved popular remedy.

As regards medicines, I have seen many given, but without much success, such as colchicum, digitalis, and iodide of potassium. I am constantly asked to see cases of acute hydrocephalic disease or tubercular meningitis, and have largely prescribed the iodide of potassium, but I am sorry to say that it is the rarest possible exception to see any good result. I have seen one or two cases of children with head symptoms recover after its use, but in all probability they were not tubercular, even if cases of meningitis.

My experience is this—that active treatment is never adopted in the most usual form of meningitis, simply because the disease is not recognised, and even if it were adopted early I am not aware that it would be of any use. The cases in which active measures are used are those in which cerebral symptoms appear at the very onset of the illness, and in such a recovery often occurs. Now, as I before said, in many of these cases there is no inflammation of the brain at all—it is a state of "irritation," to use the ordinary expression, from remote causes. There is, however, a residuum in which such well-marked cerebral symptoms are present, indicative of inflammation, that I cannot but think that the remedies are sometimes useful. Idiopathic simple meningitis is not a common disorder; when it does occur it sets in with violence, and is over in a few days; in such cases I have every reason to think that the old-fashioned so-called antiphlogistic method was eminently useful.

My experience has been mostly with the unfortunate cases of tubercular meningitis which are invariably fatal. I have seen all the medicines, I have named given without success, as well as sundry local applications, as blisters and iodine to the head. In a large majority of instances, when a post-mortem examination is made, tubercles are not only found in the meninges but scattered through the body; our helplessness is evident, and that we have only been called in for advice at the beginning of the end.

It is a curious circumstance, as showing how little expectation we have of curing certain complaints like meningitis, that if the patient recover we think it more reasonable to suppose that we were mistaken in the nature of the case than that the medicine has arrested the disease, and on this idea there is a plan of treatment often adopted which has a very ludicrous side to it—that is, we treat a patient for a particular disease in the most orthodox manner, hoping that he is not the subject of it.

I think I have told you that tubercular meningitis is not often a disease of infancy, and therefore you have much more hope of recovery in a child a few months old suffering from cerebral symptoms, since these may be set up by external causes. You may remember, also, that in simple exhaustion or anæmia of the brain convulsions may occur, and that such must be combated by very different remedies from those I have been mentioning. The child may be atrophied or in a depressed condition from many causes, and the head may grow large from a passive collection of fluid in the ventricles; at all events, you may find a child in a very feeble condition, with a sunken fontanelle, and yet having convulsions. In this case a stimulant is the remedy. I saw a case only lately where a child was in constant convulsions, and these had been aggravated by depleting measures. An immediate change for the better occurred after the administration of a few doses of brandy and ammonia. I might here say that there is a state of atrophy in children where alcohol will cure when all nourishment and cod-liver oil have failed. I used to keep for the purpose at the Infirmary for Children a mixture made of rectified spirit. I should also say that the constant convulsions of children are much relieved by the inhalation of chloroform. In those cases, where the whole body is constantly distorted to the great distress of the mother, chloroform will cause the movements instantly to cease, and sometimes with their arrest the child may fall into an apparent sleep. I have given it in many cases with the greatest advantage and relief.

Results of inflammation.—It is a very important question to determine, if it can possibly be done, what are the changes which permanently remain after recovery from different forms of inflammation of the brain. It may be that instances of impaired mental and even bodily development, which we meet with in adults, may have originated in an acute inflammation, and that cases of so-called idiocy, especially when combined with crippling of the limbs, are only the results of an infantile cerebritis or meningitis.

You must know, however, that in many cases of enlargement of the head the effusion is a merely passive condition. An infant inherits a sickly constitution, or becomes cachectic from deficient food, and a change in the framework takes place—the chest falls in over the non-expanded lungs, whilst the belly and the cranium give way, the one being distended with gas and the other with fluid. This fluid may again be absorbed, or the head may remain permanently enlarged. I know now a highly intelligent youth with a good-shaped head, in whom the skull became much enlarged after a severe illness when he was about two years old. I am also seeing a child two years old with an enormous head ; she speaks well, and is more than ordinary intelligent. In two cases were the head was excessively large I had recourse to tapping, but without any success, so that I think I shall never do it again. It is remarkable, however, that no ill results followed.

CASE.—A little girl, now ten years of age, was well until three years ago, when she had scarlatina, and whilst in bed severe cerebral symptoms came on. She was unconscious, was continually screaming out, and had all the symptoms of acute head disease. She remained very ill for several weeks, and then gradually lost the severe symptoms ; she was, however, perfectly helpless, and for a very long time had difficulty in standing, holding, or eating, and there was slight strabismus. She gained a partial amount of strength, and then passed into the state in which she now remains. She totters in her walk, is feeble in her limbs, and is dull in intellect.

CASE.—A boy, æt. 13, was very well until three years ago, when he had a severe illness, the nature of which was not very clear, except that it was attended by cerebral symptoms ; subsequently his arms became rigid, so that he could scarcely move them to his head, and the legs contracted with the feet turned inwards, so that he has not stood since. He can scarcely speak except in monosyllables, and then his words are suddenly jerked out. His appearance now denotes one of congenital idiocy, with paralytic contraction of the limbs, and suggests the question whether cases of this kind observed in infancy may not have had a definite cause at a previous period, and not, in fact, truly congenital.

When recovery takes place after acute head affection, it is impossible to know what amount of inflammation may have occurred or what may have been its exact site. In the following case, although the boy was admitted for meningitis, his complaint might have been a local one confined almost to the ear.

CASE.—A boy, æt. 12, was admitted under my care extremely ill with cerebral symptoms. He was in a high state of fever, was in a drowsy state, continually throwing his head about, and very deaf. He had frequent vomiting, bowels costive, tongue thickly coated, &c. After lying in a most precarious state for some days the fever passed off; he appeared to recover his consciousness, but he was perfectly deaf. He then rapidly improved, and left the hospital absolutely deaf. Both tympanic membranes perfect. I saw him eight months afterwards and he was "stone deaf."

CASE.—A child, æt. 7, was lately brought to me perfectly deaf, not hearing apparently the slightest sound. A year and half before, she was quite well, pleased with music and sang airs; she then became ill, very irritable, and often screamed out as if in pain. When she recovered from this state she was found to be absolutely deaf.

CASE.—Thomas M—, æt. 18, a very intelligent young man, was at work during the hot weather in the summer and thought the heat affected him. He went home complaining of his head, and very soon acute symptoms set in, which the doctor attributed to inflammation of the brain. He was delirious, sometimes almost maniacal, and unconscious of everything around him for nearly a month. After the urgent symptoms passed off, he appeared perfectly idiotic. He very slowly improved, and when I saw him, six months afterwards, he had a vacant look indicating an almost blank state of mind, like that of an idiot. He was too weak to stand and sat up in bed; he understood what was said to him and he did as he was bid, but acted as an automaton or a cataleptic. When told to put out his tongue he would thrust it out, and keep it out; he would also keep his arms in any position in which they were placed. He had a book before him, at which he looked, but it was evident that he was not appreciating a single word. He resembled a person recovering from a fever or anæmia of the brain. A month afterwards he was walking about and his intelligence returning.

The most evident result of a morbid action in the brain is the persistent hydrocephalic condition, either shown by the large head during life, or by the increased quantity of fluid found in the ventricles after death. In these cases it may be difficult to pronounce upon the exact importance of this undue quantity of fluid in the brain, or how far it may be merely the most striking feature of a more general and less evident atrophy of the whole cerebral structure. In a very large head, even with this excess of fluid, we can suppose the brain to be of normal weight, but in an average-sized skull a large ventricular effusion must necessarily imply a diminution of brain substance. Mr Hilton thinks that a local inflammation might give rise to the dilatation of the ventricles by a closure of the iter ad quartum ventriculum, and that all the subsequent symptoms complained of by the patient might in this way be readily accounted for; that the relation between the amount of subarachnoid and ventricular fluids and the blood in the vessels being most inti-mate, any impediment to the flow of the serum into and out of the cranium caused by congestions would necessarily be productive of important, serious consequences. For instance, if the ordinary causes productive of increased vascularity of the brain were present, which would in health necessitate a removal of part of the ventricular fluid, and the fourth ventricle were closed, the fluid could not escape, and serious cerebral symptoms would result. It is certainly true that the most appreciable condition resulting from infantile cerebral disorders is an increased ventricular effusion, and therefore it may be allowable to associate the symptoms with this alone. The following cases show on what a precarious tenure these patients hold their lives, how a slight disturbance is enough to arrest the impaired machinery of the brain, and, therefore, with what con-sideration they should be treated.

CASE.—A gentleman, æt. 34, had always had delicate health; the nails of his fingers were peculiar in form; as a child he was active but very irascible, as a man very spare and delicate. He had a fancy for turning and gardening, but disliked the excitement of London. At the age of sixteen he had a severe nervous illness, with great depression, brought on by application to business in the city; this business, however, was not heavy, and would have been thought nothing of by ordinary youths. His food was of the simplest kind; even tea deranged his stomach. Winter and cold always affected him injuriously; he got torpid in the winter, and it was difficult to rouse him, when he scarcely knew where he was. He was subject to headache, derangement of stomach, and occasional deafness. He had a peculiar restless look of the eyes and a stare, became feeble, and stooped in his gait. A few months before his death he had a severe attack of vomiting, with great prostration, without any apparent cause. The last month or two were marked by a morbid activity and restlessness. On the day of his death he had been to the Crystal Palace; whilst there he vomited. He walked home, some distance, and when he entered the house he staggered and said he felt giddy and oppressed. He was placed in bed, but very shortly stertorous breathing came on, and he soon afterwards died.

On exposing the brain the convolutions were found flattened, and were appa-rently large and few; the brain structure appeared healthy. On opening the ventricles they were found to contain about 4 oz. of fluid. The ventricles were greatly enlarged; the foramen of Monro was large and rounded; the fourth ven-tricle greatly dilated. The cerebro-spinal opening between the under surface of cerebellum and the upper surface of medulla oblongata was completely closed by a tolerably dense membranous structure, which formed a kind of pouch, project-ing downwards, and showed the direction of the fluid tension upon it to have been from above to below. The other organs were healthy.

CASE.—A gentleman, æt. 50, had been of feeble intellect since childhood, and which incapacitated him from earning his livelihood. He was taken charge of by a domestic, who assisted to dress him, and who accompanied him in his walks. He had a very large head, which was noticeable by strangers, especially as he wore his hat at the back of it; this, with a peculiarity in his gait, made the state of his mind at once apparent. After having been ailing for a week or two with headache, he one day returned home from a walk, about 2 o'clock, not feeling well. He, however, partook of a hearty dinner, and then retired to his room. He was seen by the servant about an hour afterwards sitting on the stairs, when he was conducted to a sofa, and he sat there some time in a half conscious state; he shortly fell back insensible, was carried to bed, when stertor came on, and he shortly died.

On post-mortem examination the ventricles were found of great size, and holding half a pint of fluid. The brain was thought to be healthy, as well as all the other organs of the body. The medical men had some doubt as to the ventricular effusion being sufficient to cause death, and an inquiry was instituted.

CASE.—A young man, æt. 23, an engineer, was said to have had good health until a year ago, when he began to have numby and other strange feelings in his limbs. Whilst at work he would suddenly have to stop, but without losing consciousness; once he stood a whole hour without moving. Later on he had fits, in which he lost his consciousness. On admission, we were immediately struck by his large head and cerebral aspect. He had a vacant look and slight strabismus. When spoken to he answered slowly and coherently, but it was found that his statements were often incorrect. He had general muscular debility, moved his arms slowly, and had difficulty in chewing his food; he passed his urine in bed, but whether this was due to actual paralysis or not was uncertain. The pupils were dilated and the optic discs ill-defined, and there was some hæmorrhage in the retina. He remained a few days in this condition, indicating an extreme feebleness of body and mind, when he had a slight fit and died.

The post-mortem examination was made by Dr Fagge. The body was well-nourished and muscular, the bones of skull exceedingly thin and having no diploë, like a child's skull; dura mater easily removed. The brain was flattened, owing to the presence of a large quantity of fluid in the ventricles; this was clear, and scarcely altered by boiling. The ependyma thick, and in places granular. All the ventricles were dilated; the third formed a wide cyst-like expansion at the base of the brain, on which the optics nerves were stretched. The iter a tertio ad quartum ventriculum was large enough to admit a pencil. The fourth ventricle was enlarged so as to excavate the cerebellum. The medulla oblongata looked soft, but showing no very evident disease; the cord below was firm and healthy.

I have already said that although a ventricular effusion may be the most manifest result of a morbid action of the brain, it is far from certain that this is the only effect, as it may be only a part or consequence of a general morbid change throughout. It is difficult, therefore, in many cases to give the correct proportion of credit to the inflammation for the changes found in the membranes, the substance, or the ventricles, respectively. The term hydrocephalus is used when we find that the main results are in the ventricles.

CASE.—A boy, æt. 17, was said to have been lying ill at home for several weeks with fever before he was brought to the hospital. He was then lying on his side, coiled up, sensible and complaining severely of his head, and in a few hours he died. The body was much wasted. On opening the skull the surfaces of hemispheres were seen to be flattened and sulci almost obliterated. There was no effusion on the pia mater and no tubercles; the ventricles were distended with several ounces of fluid. The serum was slightly more opaque than the natural ventricular fluid on boiling. The foramen of Munro was very large; iter a tertio large, and fourth ventricle much distended. Surface of all the cavities was granular.

CASE.—A young man, æt. 20, had all his life suffered with his head, having occasional pain, oppression, and at times appearing as if his mind was enfeebled. During four months all these symptoms increased, and at last assumed an inflammatory character. After death the only morbid appearance discernable was an immense effusion into the ventricles. There was no opportunity of measuring it, but the quantity was thought to equal a pint.

CASE.—A few years ago, a schoolmaster was put on his trial for causing the death of a pupil by flogging him; he defended himself by describing the boy's peculiarities; the boy, he said, was wilfully obstinate, determined not to learn, and at the age of sixteen did not know or pretended not to know a sixpence from a shilling. It was also stated that he was a nervous, timid boy, was frightened to go over a plank by himself; he also suffered from chilblains in the winter. His head was large, and he had the appearance of one who had water on the brain. He died almost suddenly after the beating. On post-mortem examination several ounces of fluid were found distending the ventricles, the arachnoid at base somewhat thickened, and it was thought by some who were present that this might have obstructed the lower opening of the fourth ventricle. The pons and medulla flattened and compressed by the central effusion. Brain and other organs healthy.

Chronic Hydrocephalus from Injury

CASE.—A man, of middle age, was brought in in a fit; after this he lay in a drowsy state with a slow labouring pulse, so that it was thought he was suffering from concussion, as he had severely struck his head. It was learned that he had several fits previously; also that he had a severe fall on his head some years before, but it was not known that he had any cerebral symptoms. He never got out of this drowsy state, but remained lying quietly in bed until his death. His pulse was very slow, sometimes 40, but he had no fits nor any other marked cerebral symptom. After death an old fracture of the skull was found at its posterior part and base. The brain appeared structurally healthy but the ventricles contained fourteen ounces of fluid. This was of the ordinary natural kind, and showed no evidence of its having had an inflammatory origin, and there was no apparent obstruction in the veins of Galen or at the fourth ventricle.

Dr Fagge had also a case under his care of a man who evidently was suffering from some obscure cerebral disease, but he had suffered only from headache, and for many months constant vomiting. He finally had convulsions and died comatose. The principal condition found was a most enormous distension of the ventricles with fluid.

Chronic hydrocephalus.—What is generally meant by chronic hydrocephalus is the affection seen in infants, coming on soon after birth, and where the head may reach an enormous size. The patient may live and appear little the worse for having his brain expanded by this large quantity of fluid, as in the case of Cardinal, whose skeleton is in the museum. In delicate children an effusion of this kind is of a passive nature, and is again removed as they grow stronger. This I have several times seen. When all other remedies have failed, an operation has sometimes been had recourse to. It came somewhat into favour on account of its alleged success in the hands of Dr Conquest, who was said to have cured four cases out of nine in which he performed paracentesis. I have myself tried it in two cases, but should never propose it again—not on account of the severity of the operation, but from the want of possible success. In my first case it could not be said that the child suffered in any way from the operation, for no apparent effects were produced by it: there seemed no danger from any entrance of air or collapse of the brain. The child eventually died, but never had any marked brain symptom. In the second case the difficulty was with the after-treatment. You see, if fluid is rapidly formed in a skull which cannot give way coma ensues. Now, in these young children the bones separate, and no injury to the brain results. When the fluid is drawn off it will again form, and in order to prevent this, if a bandage be applied symptoms of compression follow. Our difficulty was so to regulate an elastic bandage as to exert pressure sufficient to prevent expansion, and yet not to cause undue compression. This was a problem we could not solve, and the child left for the country unrelieved.

Arachnoid and blood cysts.—I might here allude to the cysts which are sometimes found on the surface of the brain, and which have given rise to much controversy as to their origin. They have for the most part been met with in lunatic asylums, in connection with old cases of dementia or general paralysis ; the opinion being held that they have resulted from the organisation of blood clots which have had their origin in diseased vessels of the pia mater. If after an effusion of blood on the brain an organisation should take place a cyst will arise which, by involving the surface of the brain, gives rise to a variety of symptoms. In childhood such an occurrence would be the cause of a structural change and functional derangement. I mention this that you may be prepared for such a discovery in cases of obscure cerebral disease. I have described this disease, with a good specimen of it found in our dissecting-room, in the 'Journal of Mental Science.' You can easily see why these cases have been met with in lunatic asylums, and been associated with insanity. An effusion of blood organising on the surface of the hemispheres would necessarily lead to disturbance of the intellectual functions ; and, again, effusions of blood would be more likely to occur in those who had previous disease of the softer cerebral membranes. Thus, it has been frequently observed that in the general paralysis of the insane apoplectiform and epileptiform attacks are very likely to occur, and it is in this class of patient

10

especially that the membranous exudations have been found. Dr Hodgkin believed that they were caused by inflammation, but it has been shown in many instances where effusion of blood has resulted from injury, and experiments on animals have also proved it, that blood may organise in the manner mentioned; and it is very remarkable to notice how membranes forming on each surface of the arachnoid in this manner may produce in time distinct cysts. A very old specimen in our museum of this nature, which came from an epileptic patient, was thought to be formed by a separation of two layers of dura mater, but it is clearly a cyst formed from an old effusion of lymph or blood. This explanation of their formation does not preclude the possibility of fresh effusions of blood occurring from the vascular membranes already formed. It was Mr Prescott Hewett who first showed how they were formed from blood. Foville had also observed them, and stated that all his patients were remarkable for their dulness or stupidity; that they resembled statues, with this difference, that when pushed they walked, when set upright they kept their place, and when food was put into their mouths they swallowed it. The case I have alluded to, and elsewhere described, was one which came from our dissecting-room, and therefore there was but little history attached to it.

CASE.—The body was that of a young man who had died of phthisis, and had come from the workhouse. He had there been regarded as half-witted, and quite incapable of following any occupation. He was fond of frolicking and tumbling about. When spoken to he answered with a rude sharpness. During the seven years he had been an inmate had never had any illness. On removing the calvaria and touching the dura mater it felt boggy; on opening this there proved to be a cyst closely incorporated with the dura mater above and the surface of the brain below. It looked as if a cyst had been formed in the dura mater, that is, by a splitting of its layers, if this had been possible. It could, however, be completely stripped off the dura mater as well as the brain, leaving the cyst perfect. Its shape was that of the right hemisphere, to which it was applied, and adherent by tough fibrous tissue. The interior surface was smooth, contained three or four ounces of an opaque-white, glistening fluid, and which, on standing, deposited about half an ounce of cholesterine.

THROMBOSIS OF THE CEREBRAL SINUSES

This is a condition not infrequently met with in connection with disease of the bones of the skull, especially that of the petrous bone, and occasionally associated with various inflammatory affections of the brain. It is, however, also, but less frequently, found as an apparently primary state, and where no local morbid process exists from which it may have proceeded. I have met with it two or three times in children as apparently the main disease. Various cases of this kind, which have been reported, tend to show that it has occurred under conditions of great debility, suggesting that a mere retardation of impoverished blood may have caused the coagulation in the sinuses of the brain. This, however, is somewhat conjectural, and, therefore, this form of disease must be put in the same category with the so-called cases of phlebitis occurring in the limbs, and whose pathology and causes are in very many instances very obscure. There might be instances where there was evidence of blood disease in other parts of the body, and others where an excessive hæmorrhage or anæmia might be regarded as the cause. There may be also states of the brain itself where a stasis commencing in the tissue might be propagated to the larger vessels. Most of the evidence, however, at present tends to show that the cause is to be found in an impoverished state of blood.

CASE.—The following case has lately come under my notice:—A lady, below middle age, had been for some time a little out of health, when she was seized with headache, a complaint to which she was not used. When it had existed a week I saw her. The pain was all over the head, and especially at the back; it was constant, but with paroxysms of greater severity. She was anemic, had a weak pulse, temperature normal, no fever, no rigors, and, in fact, no feverish or other symptoms besides the headache. On the following day she was exactly the same, but in addition had occasional sickness. On the next day she thought she had a little weakness in right arm and leg, but this passed off. During the next two days the headache continued, but there was no fever, and the pupils were rather contracted. Her intelligence remained intact. During this time she was seen by several of the most distinguished physicians in London, who, individually and separately, expressed their opinion that she was not suffering from any organic affection of the brain, as there was not a single symptom to indicate it, nor any acute inflammatory affection; in fact, that she was the subject of a functional disturbance only. On the day after this she became very sick, and towards evening sank into a half-conscious state. On the following day her limbs became rigid and limbs flexed, then convulsive twitchings came on, and she died. She had thus been ailing for about twelve days.

Post-mortem examination.—The surface of the brain was turgid with blood; the vessels of the pia mater were everywhere tightly filled with coagulum, and felt like so many worms under the fingers. On opening the longitudinal sinus this was found to be completely closed by fibrin, which was adherent to the walls, and in some places softening in the centre; the lateral sinuses in the same manner were filled with fibrin; this was traced as far as the jugular vein, and there ceased. In the meshes of the pia mater there was a little yellowish serum, and in one or two places a patch of recently-effused blood. There was no disease of the bones of the cranium. The body was healthy, and there was no sign of coagulation in any of the vessels.

A case like this is in the highest degree obscure, both as to its pathology and diagnosis. No cause for so sudden a morbid state being lighted up in the brain was apparent, and as regards any suspicion of such a condition existing during life I have already said that the complaint was regarded by all who saw her as functional,

The error appears at first sight very grave, but a little consideration will show that this opinion is not a just one, and that the mistake really occurred in consequence of the advanced knowledge which we have of late made in the pathology of cerebral disease and its diagnosis. There was a time when a case characterised by so long continued headache would have been construed into an evidence of inflammation of the brain or similar severe disorder, but such an opinion would have been founded in ignorance and on a false pathology. We are now in a position to state that none of the important centres of the nervous system are involved in disease if paralysis is altogether absent, and also that the hemispheres are not the subjects of disease whilst the mind remains unclouded. Again, we deny the existence of any inflammatory disorder of the brain whilst there is an absence of fever or mental disturbance. In such a negative condition as here existed we could not declare more than that there was disturbance of the cerebral hemispheres. Now, in all probability, at the onset of the symptoms, no more derangement existed than an increased vascularity, and therefore the diagnosis was so far correct. That this stasis of blood should go on to a coagulation and blocking up of the vessels, ending in death, of course, was not suspected. A careful consideration, therefore, of the case clearly shows that, under an apparent error, a remarkably true and scientific diagnosis was made.

The two following cases, taken from the tenth and sixteenth volumes of the 'Transactions of the Pathological Society,' somewhat resemble the one which I have related. They are described by Dr Crisp and Dr Andrew respectively. In the second case it will be seen that there was evidence of blood coagulation in other parts of the body.

CASE.—A girl, æt. 16. A month before her death she went into the country, and whilst there was seized with pain in the head, confusion of intellect, vomiting, and other symptoms indicating cerebral disturbance. She was sent home, and then found to be suffering from some paralysis of right side, with loss of speech and inability to protrude the tongue; there was pain in the head, especially of left temporal, occipital, and post-aural regions, extending down the side of the neck, and aggravated to extreme suffering on motion. There was frequent screaming and moaning, and the left hand was frequently applied to the left side of the head and neck. Perception of external objects existed, expressed by signs. Pulse never above 80. No fever, pupils natural. Death occurred rather quickly, but before this it seemed as if the power of the limbs had become restored, as she moved them and afterwards spoke. Her symptoms had lasted altogether a fortnight.

Post-mortem examination.—Body well developed. On opening the head the superior longitudinal sinus was found filled with congulated blood, interspersed with portions of fibrin, closely adherent to the walls of the sinus. The superior cerebral and cerebellar veins were found to be distended and of a cord-like form portions of straw-coloured fibrin could be readily seen through their thin walls. The superior cerebellar veins were all plugged with fibrinous coagula, and all the cerebral veins entering the longitudinal sinus were firm and cord-like from the same cause. The cerebrum apparently healthy, and without any preternatural vascularity of its substance.

CASE.—Eliza S—, æt. 20, had suffered from anæmia and amenorrhœa for a twelvemonth, when, about ten days before her death, she complained much of headache, which afterwards became most intense. This was her principal symptom until delirium and vomiting came on, and at last coma.

Post-mortem examination.—Lateral ventricles contained an excess of blood-stained fluid; optic thalami were filled with firm dark clots, and left corpus striatum also. The veins of choroid plexus, and velum interpositum, together with venæ Galeni, were distended by firm, partly yellow and fibrinous clots. These extended continuously along the straight sinus, and for about one inch into the commencement of the lateral sinus, the latter not entirely filled with them, but the smaller veins were greatly distended by them. Several of the branches of the pulmonary artery had old clots in them.

Abscess of the brain.—This is not one of the usual results of inflammation, or cerebritis, and therefore its presence is not expected in cases where idiopathic inflammation has been known to exist, but only anticipated as a probable consequence of injury or caries of the bone, or as a secondary formation in pyæmia. This was the experience of Sir W. Gull many years ago, and I quite accord with him in this opinion. It is pretty certain, too, that an abscess may remain latent in the brain for a very long time without the production of symptoms, and it is only after death, when its existence is actually known, that we alight upon the cause in some long-forgotten injury or hidden disease. So long antecedent often is the occurrence of the injury that we are forced to believe in the possibility of the dormant presence of an abscess in the brain for months or even years. Indeed, I know of one case where the rupture of an abscess was suddenly fatal, and where fits had occurred ever since an injury to the head seven years before, so that the question was raised of the possibility of the abscess having been present in the brain during all this time. I am not aware that a dried-up abscess in the brain or pus concrete has ever been met with.

One reason for the absence of symptoms in such cases is due to the fact that abscess is more generally found in the white substance, and more especially in the middle hemisphere, and very rarely occupying the central ganglia. In these long-standing cases, too, the abscess is often encysted, and in this way it is cut off from the surrounding brain substance. As an example of how rapidly the cystic abscess may form, I may mention the case of a little boy, the specimen from whose brain is in the museum, who had fracture of the skull, and died eleven weeks afterwards. We found a cerebral abscess contained in a cyst whose walls were an eighth of an inch thick, and so strong that it could be taken out and held up by the forceps without rupturing.   In other more recent cases the abscess is not encysted, but is surrounded by softened brain, and the pus, very often of a remarkably greenish hue, mucoid and fœtid.   The first class of cases where the abscess is circumscribed may arise from injury or disease, as in the one just mentioned, but yet need not be in immediate contact with the bone, as a portion of healthy mem-brane as well as brain may come between them.   In other cases the abscess is in direct communication with the carious bone and the sloughing dura mater, as is so often seen in cases of disease of the petrous bone and other parts of the cranium.   The abscess in connection with this form of disease depends for its site upon that of the bone, and therefore may be either in the cerebrum or cerebellum.   The dura mater covering the bone is usually found green and sloughy, and the bone beneath it dark coloured and carious.   The lateral sinus is sometimes also involved in the inflam-mation.   It does not necessarily follow that abscess arises direct from disease of the petrous bone when the ear is affected, for I have seen an abscess with a perfectly healthy portion of brain outside of it under these circumstances, and it was therefore supposed, in this case, that purulent inflammation had extended to it from the internal ear by means of a vein in the aqueductus vestibuli.   In these cases of abscess their formation is quick and their existence generally surmised by the acute cerebral symptoms which precede death.   For example, in a long-standing case of disease of the in-ternal ear, when inflammatory symptoms come on with severe pain in the head accompanied by rigors, vomiting, convulsions, and other cerebral symptoms, the formation of an abscess may be suspected, although it is true that if attention had not been directed to the ear the interpretation of the brain symptoms might not have been clear.   I have seen one case of the kind styled ague, on account of the severe rigors which were present, and two others called fever, because of the severe pyrexia and torpor.   You may remember that I told you torpidity or lethargy may be the only marked symptom of inflammation of the brain or its membranes.   It does not follow, however, that these acute symptoms need be present, as an abscess in connection with caries of the bone may come on most insidiously.

I have seen more than one case of cerebellar abscess where there were no nerve symptoms whatever except pain—no paralysis, and the patient was able to get out of bed and move about.   You may remember that disease of the petrous bone may lead to death otherwise than through the brain; for by involving the lateral sinus a phlebitis may be set up which may extend to the jugular vein, and so infect the lung and the system at large. It is very remarkable that just as cerebral abscess has been present without any symptoms, so in cases of disease of the internal ear a meningitis or abscess may be diagnosed when the brain is healthy. I have seen cases recover where the cerebral symptoms have been so severe that an inflammation of the brain was thought without doubt to exist.

Now, besides these acute and fatal abscesses, there are those which quietly form and lie latent for a very long period. During this time there may be no symptoms, or, if any, not of so serious a kind as to suggest abscess. A young man, for example, was admitted into the hospital on account of an intense frontal headache ; it occurred in paroxysm, and had existed for four or five years. One day he suddenly died, and there was found an abscess in the middle lobe of the cerebrum, in connection with old disease of the temporal bone. Occasionally there have been symptoms somewhat similar to those met with in tumour — as fits, pain, stupor, but it is remarkable how seldom such symptoms occur in cerebral abscess, compared with that of tumour. This may be in part owing to their usual position being in the medullary substance.

CASE.—The following case related by Gull, is very remarkable as showing how few may be the symptoms. A gentleman had an encysted abscess in the posterior lobe of the left hemisphere of the brain following a chronic disease of the lung. He had no nerve symptoms until three weeks before his death, when, one day whilst writing, he was surprised by noticing a violent clonic spasm of the right arm which lasted several minutes, and which obliged him to support it. He soon felt quite well and went on with his duties as before. After some hours the movement occurred again, and in the evening a third time. On this occasion it affected the leg and face also. After this he seemed quite well again. On the following day he had another attack of clonic spasm of arm and face of right side. Subsequently he had a regular epileptic fit, with loss of consciousness. After two days he had another fit followed by partial paralysis of the right arm and leg. Besides these fits he had constant slight clonic convulsions of the right side without loss of consciousness. Sensibility of the limb was not much diminished. The fits increased, together with the paralysis of right side; he became aphasic, incoherent, and so died.

CASE.—Quite lately a man was brought into the hospital in a state of semi-coma and paralysis of the left side and discharge from the left ear. He had been in this condition for about two weeks, but had been ailing some time before, and a year previously he had had a severe injury to his head. On post-mortem examination the dura mater was found closely adherent to the brain over the right temporal region, and beneath it lay an encysted abscess the size of an orange encroaching on the central ganglia. The cyst was composed of tough thick walls, and was evidently of some age. On the left side the petrous bone was carious, and a large abscess was beneath it, passing down into the pharynx. It seemed, therefore, as if the first mentioned abscess in the brain had been the result of the injury which occurred a year before.

Where the cerebellum has been the seat of the abscess there

may have been no more definite symptoms than in abscess of the hemispheres; and no more than would warrant a diagnosis of meningitis or some obscure affection of the brain.

CASE.—A boy æt. 12 was admitted under my care looking very ill and thin. He was quite intelligent, but spoke sharply as if he did not want to be disturbed. He had been suffering for some weeks with intense pain at the back of the head, which had been called tic douloureux. He was quite deaf on one side and had a discharge from the ear; he was also sick and his respiration was irregular. He was so feeble that he could not sit in bed. He gradually died, when a large abscess was found occupying the right lobe of the cerebellum.

CASE.—A woman, æt. 26, stated that, eight days before admission, she caught cold, and this was followed by rigors, languor, and nausea. She was sent to the hospital as a case of fever, but she had no eruption, and her abdomen was shrunken and bowels confined. She was quite rational; she gradually got lower, having no marked symptoms until two days before death, when she became drowsy and finally comatose. A large abscess was found in the cerebellum in connection with caries of the temporal bone. It had burst into the ventricle.

I have seen several cases of pyæmic abscesses in the brain and where the symptoms of cerebral disturbances occurred only a day or two before death. I remember two, if not three where the abscesses in the brain occurred during convalescence from empyema and suppuration in the lung. I must admit that I have met with an abscess in the brain without any apparent cause for it, but so exceptional an occurrence would only suggest that the primary source of it had escaped notice.

An interesting case of abscess resulting from injury, and cured by an operation, has been reported by Mr Holden.

Abscess from Injury—Cured

CASE.—The patient was a young man, æt. 18, who was struck on the head, with a piece of iron and received an injury over the left parietal bone, so that small portions of brain escaped from the wound. The latter soon healed and the patient left the hospital; subsequently some portions of bone came away, but he was at work five months after the accident. Twelve months afterwards he had a fit and then several others. They then ceased and for six months afterwards he was fairly well. He then began to have other symptoms, as inequality of pupils, deafness, and attacks of shivering; presently he became drowsy and took to his bed. He was taken to the hospital, when his lethargy passed into coma and he appeared to be dying. This was one year and eight months after the injury. Mr Holden determined to trephine and without chloroform. During the operation his face was livid, the pulse reached 160, and then became too rapid to count; the respiration shallower and shallower until it almost ceased. The dura mater was now exposed and seen bulging up, whereupon a bistoury was inserted, and five ounces of pus of a very fœtid character spurted out. The relief was immediate; the breathing, which had almost ceased, recommenced, the lividity of face passed off, and the heart's action fell; the pupils also became equal in size. On the following day the patient knew his mother; the pulse was 68, and he had no bad symptoms. The wound healed and he left the hospital perfectly well with his intellectual faculties in no way impaired.

Tumours of the brain.—The symptoms connected with tumours of the brain are very obscure, although their presence after a time may often be pretty confidently anticipated. Abercrombie's knowledge of the morbid anatomy of tumours of the brain was very inferior to the clinical knowledge which he possessed indicative of their presence. " They are distinguished," he says, " by long continued headache, the pain varying in its seat and severity, and one very remarkable character of the affection is that the pain sometimes occurs in regular paroxysms, having intervals of comparative or complete relief. It is sometimes referred to a particular spot, as the crown of the head or the occiput. The diagnosis is difficult, but the duration and violence of the pain leads to a suspicion that the complaint is something more than common headache. Sometimes the paroxysms are accompanied by vomiting. In other cases the organs of sense become affected, as the sight, the hearing, the taste, and smell, and occasionally the intellect. The loss of sight generally takes place gradually, being first obscured, and after some time lost. In other cases there are paroxysms of convulsion, which may occur with some degree of regularity like epilepsy or only at particular periods. There is nothing which enables us to explain the diversities of symptoms in the three classes. In some cases there were blindness and convulsion, in others blindness without convulsion, and in others, pain alone without either of these affections."

The symptoms of tumours will necessarily vary according to their seat, but as they are often situated in the depth of the hemisphere without involving any of the central ganglia, the symptoms may be of the most indefinite character. Thus, a patient may have pain in the head, and gradually become listless and torpid, until death at last ensues without a single other special symptom. Tumours even of some size may be found in the brain accidentally, when there has been nothing to indicate a suspicion of their presence. If they grow to any size and involve the surface convulsive fits are liable to come on ; if they penetrate towards the base, special nerves may be implicated, or as is very often the case amaurosis ensues. But even if on the surface they need not necessarily produce symptoms. But lately I have had the case of a woman in Mary ward who had a tumour on the surface of the hemisphere, but she had no symptoms but lethargy and general weakness.

The earliest symptoms are, as a rule, those of headache, or strange feelings in the head or giddiness ; subsequently, weakness may come on, and if all these continue, they point to some cerebral mischief. After a time the pain in the head may become most severe, and what is remarkable in paroxysms. It is an intense fixed pain in the head, lasting for some time and then passing off. It would be an important matter to determine, if possible, the cause and seat of headache generally, as it might aid us very much in diagnosis. There can be no doubt that, as a rule, disease of the brain, or of the cerebro-spinal centres generally, is unaccompanied by pain, as witness abscess, softening, and the instances of injury to the brain where a portion has been removed without giving rise to any sensation; whilst on the other hand, when the outside of these centres is affected as in diseases of the membranes and nerves, the most painful affections are set up. Still, we know that a non-sensitive part under abnormal conditions, such as stretching, pressure, &c., may become the seat of the most exquisite pain; witness the alimentary canal in gastralgia and enteralgia arising from colic; or the senseless muscle when in the condition of cramp. It may be therefore that the pain in cases of cerebral tumour is due to pressure on the neighbouring parts, but whether this be so or not, we must admit that this form of disease is often accompanied by pain of a most severe character. As regards the seat of the pain in reference to the position of the tumour, there seems to be no relation between them; for in the last case in the hospital, where the growth was in the anterior lobe, the pain was fixed in the back of the head in the course of the occipital nerves; and even in the same patient the pain may shift its position.

I have said that in a case of long-continued headache and sickness a cerebral disease might be suspected; this opinion would be much strengthened if after a time convulsions or epilepsy came on. If during the fit any special part of the body was always affected by the spasm, we might infer the possible seat of the disease. If subsequently amaurosis ensued it would render the diagnosis almost certain, since in a large number of cases absolute blindness is a symptom of cerebral tumour. This had long been observed, especially in cases of tumours of the cerebellum in children, but we are indebted mainly to Dr Hughlings Jackson for having shown the reason of it in the fact of the optic disc becoming perfectly white and atrophied in consequence of spreading optic neuritis. He has pointed out that even before the patient has complained of any defect of sight, neuritis may have been set up, shown by the swelling and inflammatory change in the nerve. If this neuritis exists as well as the previous symptoms mentioned, we are almost justified in speaking with certainty of the existence of a tumour. The disc is observed to be red, swollen, and prominent; the edges indistinct, the arteries obscured by the swelling, and the veins tortuous and dark. The disc subsequently becomes more confused, the arteries not traceable, the veins larger, and sometimes effusions of blood are present. Finally, the disc merges into the fundus, and becomes permanently atrophied and white. The neuritis in these cases is usually double. In other cases of nerve blindness, and more especially when connected with spinal disease, the atrophy of the disc appears as the primary affection, there being no evidence of its having been preceded by active changes. I do not, myself, profess to be very ready in appreciating all these early stages of neuritis, but the resulting white disc in the complete cases is very striking. Allbut says this descending neuritis must be distinguished from true optic neuritis; in the latter there is the appearance to which he gives the name of "choked disc;" there is venous turgor, with swelling and venous infiltration. The disc is prominent, outline dimmed, and retinal vessels distended, dark, and tortuous; the affection is limited to the discs. The cause of the neuritis is not explained, it being difficult to say whether it be due to vascular changes in the brain or whether the optic tract is more directly involved. It does not imply that any part of this well-known path is affected, and therefore we cannot from the fact of the amaurosis make any prognostication of the actual seat of the tumour; the impairment of vision gives us no clue to the locality of the disease, as would be the case if the hearing, smell, or any cranial nerves were affected.

The prevailing symptoms, then, of tumour are giddiness, headache, sickness, convulsion, and blindness. The intellectual qualities may be in no way affected, as seen in those cases, mostly of children, where the tumour is situated in the cerebellum. Paralytic symptoms should be connected with disease of some special portion of the brain, and in the case where the cerebellum is involved there might be only a general weakness of the body or staggering in the gait when walking. Dr H. Jackson has observed in some cases of cerebellar disease a tetanic condition, the legs being stretched out and arms and hands flexed. In one case there was frontal headache, sickness, and drowsiness. Sometimes great wasting is observed in tumour of the brain, as if some trophic centres were involved. Sometimes the symptoms attending tumour of the brain are almost negative, or show themselves only by some slight aberration of mind or change of temper. Such symptoms in an excessive degree may amount to mania, and necessitate the removal of the patient to a lunatic asylum. Perhaps this is the most curious and interesting circumstance connected with tumours of the brain, that the mental disturbances constitute their predominant feature, so that they come

to be treated by the alienist physician. As far as my own experience goes, and from what I can learn from others, the mental
symptoms show no special characteristics, and therefore patients in
whom tumours have been found, on post-mortem examination, have
represented all classes of lunatics; they have been demented, had
delusions, been dirty in their habits, &c., and sometimes been classed
with the general paralytics.

The diagnosis, therefore, of tumour of the brain is by no means
always easy, since the combination of symptoms most frequently
met with resembles very much what is often seen in various diseases
accompanied by cerebral disturbance. For example, in Bright's
disease we meet with headache, convulsions, amaurosis, &c. I had
long under my notice a man whom I thought had Bright's disease,
and then I changed my opinion to tumor cerebri.

CASE.—A man was admitted on account of pain at the back of the head in
course of the occipital nerve. After some time his manner appeared strange, and
he became very morose; he then ceased to walk about, but sat quiet in a chair in
a lethargic state. He subsequently took to his bed, and remained lying there in
a simple passive state until he died. He had no paralysis, and except the pain at
the early part of his illness his only symptom was a gradually increasing dementia.
On post-mortem examination there was found a tumour the size of an egg in the
anterior lobe of the left hemisphere. It occupied the medullary matter, and the
convolutions were stretched over it.

I need not describe to you the various kinds of tumour which
are met with in the brain, as you will find a description of them in
most works on pathology. The most characteristic tumour is the
glioma, composed of a delicate fibre originating in a growth of the
neuroglia. It slightly projects when a section is made, and resembles
in appearance and colour the structure of the brain itself, so that it
is ill-defined, and appears to run gradually into and infiltrate
the cerebral substance. In one case of glioma of the pons the
cerebral structure was so infiltrated with the new tissue that
it had not so much the appearance of a growth as of the pons
gigantically enlarged. You may also occasionally meet with myxoma,
the firmer fibroids, and more rarely carcinoma. Hydatid is rare,
but sometimes met with; also large scrofulous masses in the cerebellum and cineritious substance of the brain. I have seen several
cases of tumour of the pons, and in these were present, as you might
imagine, various lesions of the cranial nerves, as well as general
paralytic symptoms. Syphilitic disease of the brain ought to require
separate notice, but as it is an affection so frequently affecting the
surface of the brain and producing epileptiform attacks, I shall
defer its consideration until I come to epilepsy.

If the tumour be at the base of the brain, so as to involve the
cranial nerves or the spinal tracts, then of course special symptoms

exist. In the following case the diagnosis was simply tumour, until it was evident that the fifth nerve was involved, when its locality became apparent.

CASE.—Wm. B., æt. 5. For some weeks past had had pain in the head and sickness, the latter being the most prominent symptom. On admission he looked ill, and it was evident that his sight was bad, but he had no marked symptom except the vomiting. On examination of his eyes it was seen that he had commencing optic neuritis. When he attempted to walk he could not keep a straight line, and he carried his head stiff. He generally lay quiet in bed, resting his head on his hands, but was quite intelligent when spoken to. His sight gradually became more dim, with dilatation of the pupils, and constant oscillation of the eyeballs. At the end of four months after admission he lay quiet in bed, drowsy, and often complaining of headache; when held up on his feet he fell down, and the optic discs had become white. At the end of another two months he had a fit, and about this time the right eye was observed to be inflamed. It then discharged, and subsequently the cornea became involved and sloughed out. On testing sensation it was found that the right side of the face was anæsthetic. He had two more fits, and gradually sank, seven months after admission.

Post-mortem examination.—Convolutions flattened, corpus callosum bulging outward, and when encised eighteen ounces of fluid escaped. The right lobe of cerebellum appeared much swollen by presence of a soft gelatinous tumour, which flattened the pons Varolii, crura and optic tracts. The Casserian ganglion was stretched on both sides, but more on the right. All the interior parts in like manner stretched, but the corpora quadrigemina and other structures not diseased.

Our records contain several cases of a similar kind. A little girl came as an out-patient with gradually increasing paralysis of the facial nerve on one side and of the sixth nerves on both; after death a tumour was found in the pons. Another child with exactly the same symptoms with the addition of partial paralysis of the limbs, also had a tumour in the pons. Also a very similar form of disease where difficulty of talking and swallowing was more marked. In a case quite lately under Dr Habershon, of a young man who had general tuberculosis, there was a hard yellow mass imbedded in the left half of the pons which caused great stretching of the roots of the nerves on that side. His first symptoms of this were a slight falling of the face about a year before; afterwards he found he could not eat so well on that side, and then his hearing became deficient. When admitted, he had paralysis of the right seventh nerve and left fifth with absence of taste; also some weakness of both sixth nerves; gait unsteady, no optic neuritis, but retinal veins large.

Scrofulous Tumour of Cerebellum

CASE.—Boy, æt. 4, of whom the mother gave the following account :—That fifteen months before the time he was first seen he had an attack of measles, and

that soon afterwards the abdomen swelled. In six months' time the abdomen
decreased in size, and then the head became affected; it grew large, and. there
was an alteration in the boy's manner; he was slow, and hesitated in answering; after this he began to lose power in his left side. When first seen he was
scarcely in a sensible condition; he lay in bed, unable to lift up his head, which
was very large, and he often applied his right hand to it; pupils dilated. At the
end of the month his state was much the same, but the left side was completely paralysed, the pupils were dilated, and the child, judging from his
vacant stare, appeared quite blind. In the night he often screamed and started.
Afterwards a watery discharge was observed flowing from the ear. Subsequently
the other side of the body became paralysed, as well as the rectum and bladder
The child .often had violent fits of screaming, but never any convulsions. He
lived six months after he was first seen, and nearly two years from date of the
first illness.

Post-mortem examination.—There was a bed-sore on the sacrum; the head was
as large as that of a child of ten years, and from the occiput the hair was
rubbed off, and an ulcer existed. The fontanelles were not closed; the surface
of brain healthy, but hemispheres burst open from the weight of the fluid
within. This could not all be collected, but was reckoned to amount to a pint;
it was clear, like water. The ventricles were thus of enormous size; the septum
lucidum was entire, but like a piece of tissue-paper, and the corpus callosum
much resembled it. The ependyma were not softened, as in the white softening
of acute disease. On attempting to remove the brain, the cerebellum was found
adherent to the dura mater, owing to the presence of two scrofulous tubercles
in the cerebellum; one was very large, the size of a small egg, and to this was
attached a smaller one; the larger one was two inches long and one inch broad;
the right lobe of cerebellum was thus destroyed, and neighbouring parts pressed
upon. The mesenteric and other lymphatic glands contained tuberculous deposit;
also lungs, liver, kidney, and spleen,

The following case of cyst in the cerebellum has just been published in the Guy's ' Gazette.'

"A young woman, æt. 21. About four weeks before her death she began to
vomit every morning on rising, and it was observed that her manner was altered,
being excitable and hysterical. She was obliged to leave her situation, and
go home, where she remained in a very listless and irritable condition, the
vomiting continuing, the food regurgitating as soon as swallowed.

On admission, had listless vacant expression, complained of constant headache,
referred to left temple, occiput, and back of neck, and sight was double. She felt
better sitting up than lying down. Answers all questions intelligibly. Slight
amount of optic neuritis in left eye.

On next day she was found sitting up, and on asking her to walk she rose
deliberately from the chair, then seemed to wait to steady herself, and placing her
hand over the left eye, started off, holding out the right hand to grasp any object
should she fall. Her gait was unsteady and shuffling, especially with right leg, and
she looked whilst walking as if every moment she were about to fall. Sickness
continued.

On following day headache so bad that she could not get up, and turned about
the bed with pain. Then came on a fit of screaming, and she fell into an unconscious state and died, respiration ceasing five minutes before the heart. No
convulsions.

A cyst was found occupying the left cerebellar lobe, outside the corpus dentatum, and nearer upper than lower surface of cerebellum.

Hydatid in the Brain

Eliza S., æt. 9, admitted in January into Clinical ward, and died two months
afterwards. About six months before she began to complain of pain in her head,
until this became constant; she lost her appetite, and was often sick. After a
little while she became somewhat better, and was able to leave her bed, but it
was then observed that her left side was weak; her appetite returned, and sometimes she was actually ravenous. A month before admission she completely
lost her eyesight. On admission she was quite blind, with pupils widely dilated;
she seemed quite intelligible, understanding all that was said to her. She was
very fretful and irritable, constantly asking to be moved or crying out for food.
She only occasionally complained of headache, and if so, this was situated in the
forehead. The motions were passed involuntarily. Both legs were weak, but
the left more so; the left arm was contracted and rigid. The right arm she
could move, though imperfectly; neither sensation nor excito-motility were impaired. When moved she called out as if it gave her pain, and the muscles
became more rigid. Mr Bader examined the eyes and found nothing very
noticeable, nothing, indeed, to suggest a cerebral tumour. When she was asleep
the pupils became contracted as in health.

The child gradually grew worse, was very fretful, very sharp in her answers,
and continually crying out for the nurse. The left leg became flexed like the
arm, whist the right remained extended, but powerless. The right arm appeared
disposed to be rigid. Respiration irregular. Death took place quietly.

Post-mortem examination.—When the calvaria was removed the convolutions
were found flattened, and on cutting through the brain a large hydatid cyst was
seen situated in the right hemisphere. It was large enough to hold half a pint
of fluid, occupying the middle and posterior lobes of the covering, within three
lines of the surface and separated from the ventricle by a very thin membrane.
There was a small hydatid cyst in the liver.

Tumour in medulla oblongata

CASE.—A girl came in complaining of pain at the back of the head which had
existed only a few days, and having no other symptom, the house physician
thought it was functional, and her condition due to hysteria. She had only been
in the hospital two days, when he was called to her to find her dying, and very
shortly she ceased to breathe. He kept up artificial respiration as long as the
heart beat, and this continued for eight hours! There was no attempt at spontaneous breathing, and the body was cooling down all this time. At the end of
this period the artificial respiration was stopped, and then, most remarkably, the
heart continued to beat for another twenty-five minutes.

The post-mortem examination showed a glioma in the medulla oblongata in-
filtrating and encroaching on all the important centres, including the fourth
ventricle. It was dilated with fluid and the surface granular.

The following case of Dr Goodhart was very interesting as being
apparently one of tumour which had been dispersed or shrunken.

Tumour cured

CASE.—Ada P— was attacked with severe headache and vomiting on Dec. 20, and this continued some days, when Dr Goodhart saw her. The headache and vomiting continued, and her sight was impaired. The optic discs were hazy grey, and swollen; the pulse 52. She was ordered bromide and other remedies with temporary relief when, the headache again returned in a most agonizing form, and if anything was given her she immediately began to retch. Sight very dim. Pulse 48. During next few days she looked very ill, a little delirious at night, and pain down right side of head. She subsequently became unable to distinguish forms, and see only a glimmer of light. Pupils dilated and external recti paralysed. Mr Higgens then saw the patient and found double optic neuritis, disc swollen, edges blurred, &c. On following day she was worse, only, half understanding what was said to her, but crying out with pain in the head, and she appeared to have some weakness of the left side. She was ordered bromide and iodide together. She thus continued until a month after the time when Dr Goodhart first saw her, and then she began to improve; all the urgent symptoms passed off, and the sight began to return. A careful examination of the eyes was continually made, and changes were seen. In March she was able to get out of bed and walk with assistance, and in May she was fairly well, with exception of dimness of vision. When Dr Goodhart reported the case a year and a half after this date, she was well with exception of imperfect sight and occasional sickness.

Aneurism of the brain.—Aneurisms of the larger vessels on the exterior of the brain have always been recognised, even occasionally in the substance of the organ itself, and, as I said when speaking of apoplexy, they have of late years, especially by French observers, been found to be far from uncommon. When very small and numerous they have been styled miliary aneurisms. As regards those of the larger vessels there may have been no symptoms indicating their presence until they ruptured, the brain became flooded with blood, and the patient fell into an apoplectic state. As an aneurism is as likely to occur in a young person as in those of more advanced years, it should always be suspected when any one of tender age dies with the symptoms of apoplexy, and its presence may be almost regarded as certain if, after death, on removal of the membranes, the brain is found covered with blood. In some cases there may be symptoms of long standing, which create a suspicion of a local growth, if not of aneurism, as, for example, those which would be due to pressure on some important region of the brain, by which particular nerves are involved. Several cases of the kind published in the 'Reports,' were collected by Sir W. Gull, and the specimens may be seen on the shelves of our museum. In some of these the aneurisms were situated on the vessels constituting the circle of Willis and its branches, and no symptoms of their presence existed. In other cases of aneurism of the basilar artery pressure was exerted on the pons Varolii, and the patient had a gradually increasing paralysis of the limbs, followed, after a time, by weakness of the muscles of the face, of speech, and deficiency of hearing.

In a case, of which we have the specimen, where an aneurism the size of a grain of wheat broke into the substance of the pons, the principal symptom had been a constant pain in the back of the head and down the neck, with an inability to bend the head forwards. Next to aneurisms of the basilar I think the most frequent are those of the middle cerebral artery. In aneurisms of other vessels the symptoms vary with the locality and the parts which are pressed upon; thus, in a fatal case of aneurism of the posterior communicating artery described by Mr France in the 'Reports,' the patient had for some time been an inmate of the eye ward for paralysis of the third nerve.

In this case and in another, where an aneurism existed in the middle cerebral artery, the patients were young, they had suffered from rheumatic endocarditis, and after death embolic infarctions were found in their spleen and kidneys. The combination of these conditions has been found so often that no doubt can be left that aneurisms in young persons have had their origin in embolism.

Such an origin for aneurism is not yet described in the systematic works on medicine, but the cases taken to the Pathological Society and described occasionally in the medical journals can leave no doubt as to its correctness. An embolic mass of fibrin sticks in a vessel, becomes adherent to it, undergoes a softening process, and carrying the coats of the artery with it, an aneurismal dilatation or pouch is at last produced.

Many years ago I exhibited a case of aneurism of the axillary artery in a little girl, in whom there was endocarditis, and one of the mesenteric artery in a young man, who also had a similar affection of the heart; and some other instances. Only just now we have seen an aneurism of the thigh at the division of the femoral and profunda, exactly at the place where an embolus would lodge; and in this case there was extensive disease of the aortic valves. I have had also two cases of embolic infarction of the brain where the aneurisms may be seen in process of formation.

Chronic meningo-cerebritis and general atrophy.—I have spoken of acute general cerebritis with softening, as well as a meningo-cerebritis of a subacute character, where the main symptoms are evidently due to a progressive morbid change in the brain; now, besides these more evident forms, there is a slow degenerative disease, not marked by any distinctive features, but only apparent, after a lapse of time, by the failure of brain power. Just as there is a phthisis, which may be compared to pneumonia, and a cirrhosis of the liver or granular kidney, which may be compared to acute hepatitis or nephritis, so there is a degenerative disease of the brain, which is a counterpart of these other chronic affections. In cirrhosis, or Bright's disease, the great fact which comes before us is a spoiled

or degenerated organ, whose commencement is obscure ; the pathological question whether either of these diseases is due to a simple
atrophy or to a change of a chronic inflammatory nature being still an
open one ; so likewise in the case of the brain where we find a wasted
organ with thickening of the membranes, the mode of production
may be equally obscure, and therefore is altogether comparable to
that of the granular kidney. In nearly all these diseases the change
in the organs is only a part of a more general one, existing throughout the body, there being some common cause productive of the
universal tissue alteration.

Under many and varied conditions we find the brain shrunken as
a whole, the convolutions gaping, leaving spaces occupied by fluid,
the arachnoid everywhere thickened and sometimes adherent to the
surface, the pia mater vessels also diseased, the ventricles distended with fluid, sometimes granular on the surface, the central
ganglia having lost their roundness, and the structure of the brain,
when examined by the microscope, often having undergone degenerative changes, especially in the grey cells. With this state of
brain it can be imagined that there will have been observable
during life failure of the powers both of the body and mind. In
old age such a condition is often met with when the patient has
become childish and garrulous, totters in his walk, writes with a
trembling hand, and everything betrays an increasing decay of muscular and nervous power. Now, a very similar condition may be met
with in earlier age in imbeciles and the demented. It may also
arise from injury and shock to the nervous system, and is most
markedly seen in alcoholism. The inveterate drunkard is, in fact,
prematurely old, and has literally been living too fast. Then also
from various unknown causes of a moral kind a degenerative change
may arise in the brain, as is found in the general paralysis of the
insane. From whatever cause it results such a brain must imply
the existence of a *dementia paralytica*, although, according to the
mode in which the change has come about, so would the symptoms
which lead up to the final issue differ. We can see how this would
be from taking the analogous case of the diseased kidney, and considering that, when this organ has undergone a structural change,
certain symptoms, such as uremia, would necessarily ensue ; but
according to the mode in which this change had occurred so would
the whole history of the case be peculiar, varying in acute
nephritis, granular kidney, or in suppuration. In like manner,
probably, according to the character of the morbid process going on
in the brain, would the symptoms during its progress be of a special
kind, although they would be all leading up to the final catastrophe,
the destruction of the organ, when they would all become as one.

It may appear to some of you that passing from inflammation of the brain to insanity was a long stride, and that I am going now out of the wards of the hospital to the madhouse; but, in fact, I am following the natural order of disease, for the forcible separation of many cases found in lunatic asylums from those which we have to treat here is artificial. For not only pathology but clinical facts unite them ; over and over again have I seen a patient sent here, and it has been a question whether he has softening of the brain, whether he be suffering from the effects of intemperance or he have dementia. Not long ago I had a man in my ward who was labouring under cerebral symptoms in consequence of an injury, and one of my clerks, who saw him first, finding his articulation defective, believed it was a case of disease of the medulla oblongata, but after going over all the man's symptoms I concluded it was a case of general paralysis of the insane. Thus you see in actual practice there is not that line of demarcation between cases which you think ought strictly to be found in lunatic asylums and those which we have to treat here in a general hospital.

You might very fairly argue that this doctrine would apply to all cases of mental disease, for all have their foundation in an alteration of the brain. But this is not really so. Insanity is not a pathological condition, as we understand the term in our post-mortem room. I believe all mental peculiarities are associated with a certain organisation, but that peculiar organisation is not of a kind which can be appreciated by our eye. We know that cerebral development is intimately connected with mental power, and that no doubt a peculiar formation exists with eccentricity of character. But if this eccentricity reaches that height to which we give the name of insanity, we should not expect to find any structural change in the brain, even if there was a visible peculiarity in its conformation. I would have you here remember, once for all, that the changes which our eyes or microscopes discern are either of a degenerative and destructive nature or of that formative kind which encroaches on the healthy structure, and therefore, under both these circumstances, there is a loss of good substance, with a corresponding failure of function. We know nothing of those alterations which are associated with a change of function or exalted function, if there be such a thing. If, then, a person be not born fatuous, but, having possessed the ordinary mental faculties of man, should lose any of these, and at the same time his bodily powers, we should expect some degeneration of the structure of the brain. If a patient becomes feeble or paralysed in any part, we ought to be able to discover the cause in his nervous system, or if he become demented, a lesion should be made manifest; but such a case is clearly separable from one where the patient has spent the whole of his life in an asylum.

I believe it is true that alienists are now making this distinction, and if imbeciles and old people are excluded, they are able to divide all cases into the functional and curable diseases, such as mania and melancholia, on the one hand, and into the organic and fatal, on the other, and which in an asylum seem to be called the general paralysis of the insane.

In speaking, therefore, of chronic inflammatory processes in the brain, we of necessity touch upon the subject of mental disease, and in this form of malady to which I now allude, and which may be compared to the atrophic form of Bright's kidney, the cerebro-spinal centres become wasted. You would expect, of course, a paralysis of body and mind—a "dementia paralytica." This is what you have, and is, in fact, the name of the disease. It is a disease in which the patient finally possesses little more than vegetative life. He sits in a chair or lies in bed perfectly helpless, and the mind gone. Now, you might suppose that this disease progresses slowly, and thus there is a gradual decay both of mental and physical power; which of these begins to fail first is a disputed question among alienists. Inasmuch as the membranes are often affected and the adjacent cortical structure, you might infer (from what has been already taught) that convulsive movements might be present, and so they are, epileptiform attacks being very frequent. As also the change is not simply a degenerative one, as seen in old age, but is a destruction due to a more active process, you might suppose that the dementia was ushered in by peculiar mental phenomena. This is so. The mental peculiarities are almost characteristic of the disease. The patient's mind is excited to the formation of the most extravagant ideas; he believes himself to be some exalted personage, to be inhabiting a palace, and to be possessed of enormous wealth. Thus it is that this disease is styled by the French "paralysie ambitieuse," or "folie ambitieuse."

Here, then, is a chronic disease of the cerebro-spinal centres having the same relation to acute inflammation as phthisis has to pneumonia, or a chronic rheumatic arthritis to an acute synovitis, or, more appropriately, a chronic Bright's disease to an acute nephritis. The disease has a duration of about two or three years, and then terminates fatally. After death in many cases there is no change evident to the eye; in others there are microscopic changes; and in others, as Cameil described, a very evident meningo-cerebritis, and with this a marked degeneration or destruction of the most important parts of the brain, the grey substance more especially, but no portion of the whole cerebrum is excluded from the process. The membranes —that is, the visceral arachnoid and pia mater—may be found thickened and closely adherent to the surface, so that on an attempt to remove them the cineritious structure is torn. In this grey substance the ganglionic cells would be found to have completely degenerated and altered in form and colour. Associated with them are amylaceous bodies and a quantity of new connective tissue, which binds the whole together and hardens it. Sometimes numerous blood-vessels are seen, and small extravasations of blood ; also considerable extravasations are often seen to have occurred on the surface, and organised into membranes.

In some cases, one of the most striking alterations is to be found in the blood-vessels ; thus, in one specimen in the museum, when a section was made, the vessels stood out like so many bristles. They could be pulled out of the brain to the length of several inches, and had undergone a most remarkable calcareous change. The patient was comparatively young, and such an alteration I have never met with in older persons with diseased cerebral vessels. This was an evident and marked change, but besides this it is said that the microscope is able to exhibit even more distinct alterations in the vessels. Dr Sankey thought that the vessels become tortuous, and put on a varicose appearance from protuberances on their surface, owing to a thickening of the walls, and regarded the disease in a pathological point of view as a case where there was hypertrophy of the connective tissue in the small arteries and veins of the pia mater of the cortical portion of brain. At the time these observations were made, the anatomy of the blood-vessels of the brain was not so well known as it has been of late, and therefore it would be necessary to compare these diseased vessels with those which may be regarded as normal as delineated by Professors His and Bastian. These gentlemen have shown that the blood-vessels of the brain are normally surrounded with a case, or rather they are contained in sheaths styled perivascular canals, and which some have regarded as lymphatics. With this new light our specimens must be studied afresh. As you might suppose, the disease does not end here, but affects the brain as a whole, and at the same time the spinal cord. Several cases have been recorded where the medullary matter was so hardened by the chronic inflammatory process and the production of adventitious matter that the grey matter could be scraped off, leaving the form of the convolutions in the white matter beneath. The spinal cord is also atrophied. In a spinal cord sent to me from an asylum I found grey degeneration of the posterior columns with abundance of amylaceous bodies ; but the history was not very good. Theoretically there may be no necessity for supposing that the cord shares in the disease, for, besides those in the central ganglia, the changes in the cortex might be considered sufficient to account for the symptoms, now the motor properties of this region are better known; more than this, it has been suggested that the peculiar movements of the patient may indicate the exact convolutions which are affected.

In the cases where there are no very evident changes to the naked eye in the form of inflammatory adhesions and wasting, the microscope may display profound alterations in the shape and conditions of the ganglionic cells. The ventricles may also show atrophied patches on their walls, and the latter may be granular. From such cases it can be seen that the marked alteration in the blood-vessels sometimes found cannot be regarded as the essential pathological condition, since they are not always present. A theory, therefore, exists which converts the degeneration into a malnutrition due to a prior derangement of the vaso-motor or trophic nerves. I believe, I heard this theory first propounded by Dr Davey. It has since been advocated by Dr Bonnet, who after some very careful investigations comes to the conclusion that the origin of the disease is to be found in the sympathetic system; he declares that visible changes are met with in the whole chain of ganglia, but more especially in those of the cervical region. In these glands he found the nerve cells atrophied and destroyed, and their place taken by cellular and adipose tissue. In the ganglia of the spinal nerves, and even in the cranial, there was pigmentary degeneration of the grey cells. To these alterations he attributes the changes of nutrition in the cerebro-spinal centres. In the brain and cord he does not find actual disease of the blood-vessels, although there may be some thickening of the external coat. The radical change is the deformation and fatty degeneration of the grey cells of the ganglia; these are found containing granular and pigmentary matter, diminished in size or destroyed, or their place occupied by adipose tissue; there may also be an increase of connective tissue. According to these views the primary disease is in the sympathetic system, and the original cause of the disease is not in the encephalic centres, which are only secondarily affected. I give this as the last new theory, but I confess to a prejudice against it, as the vaso-motor pathology is now the one in fashion.

We have, then, this disease styled "dementia paralytica," or the "general paralysis of the insane;" by the French "folie paralytique," or "paralysie ambitieuse." Pathologically it is a degeneration, and by those who insist on its inflammatory origin would be styled "meningo-cerebritis chronica" or "peri-encephalo-meningitis." It is a case of paralysis of body and mind, commencing with a physical weakness or some alteration in the manners or character of the patient. It is a question which takes the lead, the psychical or physical derangement. Those who have recorded most cases agree that the mental aberration is the first symptom discernible. I think it is Dr Blandford who has insisted on the mental alienation being the first and most important symptom, and that even the defect of speech does not at that time prove a necessary motor lesion, since by a strong effort of the will the stammering may be overcome. As regards the bodily symptoms, they are those of gradual progressing paralysis; there is a tottering in the gait, a want of power in articulation, and an evident paralysis of the muscles of the face. The commencing paralysis of the muscles of the face produces a vacant expression, which is at once recognised by the experienced medical man, and as soon as the patient talks his mode of articulation at once characterises the disease; he speaks thick, and clips his words like a drunken man. There is no necessary connection between paralysis of the face and insanity, but still, in a case where the brain is approaching decay, it might be the first symptom recognisable. Thus, from the earliest ages, a madman was characterised by an altered mode of speech and by an inability to hold his saliva. In this disease there is a vacant expression, the articulation is indistinct, and the tongue is protruded with difficulty or wavers. Afterwards the paralysis extends to all parts; there is a weakness of the limbs, and the patient walks unsteadily, like a drunken man. When it is said that the paralysis begins above and then affects the body, this is often only apparent, and because a change in the muscles of the face is more easily distinguishable. Thus, in a man who was in the hospital some time ago, there was the tremor of lips and tongue, as well as hesitating speech; but the man could grasp with his hands most powerfully, and when asked to walk would step out with vigour, and said he could continue for miles. It might thus have been inferred that he had perfect control over his limbs; but I caused him to write, and the tremor of the hand was evident, as you will see by this specimen, where every letter is formed by a dozen or twenty different strokes; and when he walked his gait was seen to be unsteady, although we had no opportunity of testing the amount of control over his limbs. If ever you go to a lunatic asylum, you may observe persons with this disease playing at cricket, and then you will perceive that, although they run and hit the ball, they do it in so awkward and grotesque a manner that it looks like a burlesque of the game. The appearance of a drunken man will give you some idea of a patient thus affected—in fact, a drunken man might be said to have an acute general paralysis of the insane. I should have mentioned also a fact which is very common in many brain-diseases—that the pupils are, as a rule, unequal in size, and both contracted more than natural. I had lately an opportunity of proving the importance of this symptom. A gentleman died, of what was evidently general paralysis, about two years after his life was insured, when of course some investigation was made as to the condition of his health at the time the assurance was effected. It was then known that he was suffering from general nervous debility, attributed to over-study at college, and it was also known that one pupil was larger than the other. The two facts were not associated by the medical adviser; but the proposer of the assurance was sent to an ophthalmic surgeon, who declared that there was no organic disease of the eye. His life was then accepted.

As regards the mental symptoms, these may show themselves by a mere failure of power, or inability to attend to business or the ordinary affairs of life, or are perhaps exhibited, in a more striking manner, by strange behaviour towards those with whom he associates, and by certain likes and dislikes in his own family. The tendency is towards dementia, seen in the memory failing and the mistakes which are made in the transactions of business. Then he will have delusions, and his mind becomes wholly deranged; but instead of growing melancholy, he becomes elated, or is, as the French say, "gay"—he has exalted ideas. These are not necessarily of the kind before mentioned, but, in cases I have seen, are rather extravagant ideas. The greatness of the man has been not merely exhibited by his boast of immense wealth, but everything around him appears to him on a larger scale, as a patient informed me that his house was a mile high, and that he had a hundred ribs, and that he could kill a thousand pigeons at a shot. They are often rather extravagant than ambitious ideas. I used, amongst my out-patients, to have a postman led in by his wife, as he could scarcely stand. When asked how he was, he smiled, said he was very well, was about to resume his occupation, saying he could walk for miles. A medical friend of my own, whom I was asked to see, told me of an operation for hernia which he had performed, and it was the quickest known—it being done in half a second. A literary friend began to be extravagant in purchasing books; he then began to talk about his intended travels over the world and informed his friends he could speak 100 languages. These were the first notable symptoms of his complaint. Strange that the patient, not feeling ill, will deceive others. Thus, not long ago, a patient staggered into my room, accompanied by his wife, and having all the well-marked symptoms of dementia paralytica; but both of them for some time warded off my questions regarding his brain or the state of his mind, since, as they said, they had come to consult me about the sudden "bilious attacks" to which he was liable. These persons, I may mention, are often subject to these so-called affections of the liver.

I not long ago went a few miles into the country to see a young man who I was told had had a sudden attack of illness, and had had several before. I found him in bed in a not completely conscious state, after having had a severe fit of vomiting. The doctor told me he had been called before to him in such attacks, and believed it was the liver. After a few days, when he had recovered, he came to my house, when it was a clear case of the disease I am describing. I then learned about his altered manner for some months previously. He is now in an asylum. It is very likely that the attacks are of an epileptiform nature, of which the vomiting is a symptom, for these patients, you ought to know, very frequently have fits of what is called cerebral congestion, or epilepsy. It has been thought that these attacks are caused by blood effusions, as the remnants of them are often found after death.

Dr Tuke, who has written on this subject, says you may divide the disease into three stages—first, where the quivering of the lips and thick articulation are observable, together with contraction of the pupils; secondly, where there is decided want of power, shown in the unsteady gait and weakness of the arms; and, thirdly, where all power of motion is gone, and there occur the epileptiform attacks. This division would have reference to the bodily symptoms, but as for the mental there would be first loss of self-control, of reasoning power, and perhaps excitement; secondly, the patient might have absolute delusions; and, thirdly, complete insensibility. Some have said that the patient passes through the three stages of alienation, dementia, and amentia.

Causes.—It would appear that in this form of insanity there is not found the same strong hereditary disposition to it as in other forms of the disease; and this quite accords with our knowledge that it may be set up by accidental causes. No doubt overwork, or wear and tear on the nervous system in a person of excitable temperament, might develop the disease; but at the same time it does seem capable of being produced by altogether accidental circumstances. There is the case of that man who lately left us. He came in saying that he had not been well since he fell off a scaffold two years before, and was brought to the hospital suffering from injury to the head. His wife said that he had not been right since, and that his capacity for business had altogether failed. He had a vacant expression, or rather a scared look, showing at once that we had before us a cerebral case; we noticed that one pupil was larger than the other ; both contracted, however, on the stimulus of light. When spoken to he answered quickly, but rather indistinctly, all his words being as it were thrown together—that is, he spoke thickly, with no clear articulation. His lips trembled when he attempted to utter a word, and his tongue when he protruded it. When asked to walk he set off with great determination, but his gait was very awkward. He could grasp you tightly, and there appeared to be no tremor of the hand ; but, when asked to write, his hand was very tremulous, as you will see by this specimen. As regards his mind, we could not say that he had very exalted ideas ; but, on the other hand, he had no approach to melancholy ; he always spoke cheerfully, said he could walk, or write, or read. He often had a book before him, but I question if he understood it. Now, if this man had had exalted ambitious ideas, not a single symptom would have been wanting to make it a case of the general paralysis of the insane ; but in spite of this absence I regard it as an example of the disease, for in some exceptional cases this exhilaration is wanting. The great point of interest is that it resulted from an injury. When you remember that you find in this disease thickening of the membranes, which are adherent to the brain, besides false membranes, which have probably arisen from extravasations of blood, the conditions are just those which might arise from severe concussion, accompanied by some injury to the cineritious structure, with perhaps effusion of blood on the surface.

A man, three years before I saw him, met with a severe railway injury ; his nervous system received a great shock, his mind became affected, he had delusions of a grandiloquent kind, was restless and excited, had tremulous tongue and tottering gait.

The moral causes have been due, in persons whom I have known, to disastrous money speculations, combined also with rather fast living.

As regards diagnosis, we must be led by a number of circumstances before we form a conclusion. In most cases there is a general paralysis, with the peculiar mental defect before mentioned, and a certain morbid condition found after death. Take these together, they characterise a disease deserving of a special name, but it does not follow that some one symptom may not be absent, and thus we must not style every case where there is bodily and mental weakness a case of " dementia paralytica," which has at the present day a very definite signification. For instance, a bodily and mental paralysis exists in mere atrophy of the brain, arising from many causes—as excessive spirit-drinking or mere old age— but in the disease of which I speak the destruction of brain has come about in a peculiar way, and the symptoms are necessarily almost characteristic. There are also cases of disease commencing in the spine, and subsequently creeping up to the head, and which, strictly speaking, are cases of general paralysis. Some of the most remarkable instances of this kind have been those where an injury has been received, causing in the first instance symptoms of paraplegia, and subsequently of cerebral palsy. I believe it is now generally admitted that patients who have had symptoms in the first place only of locomotor ataxy have ended with general paralysis of the insane. I believe I have in my recollection some knowledge of a case which terminated with all the symptoms of the general paralysis of the insane, and which resulted from a severe injury to the spine. An alienist physician could not allow, however, that the condition was the same, for it would contradict his strict rule that the brain, or at least the mind, is first affected. The distinction is one of immense importance in a legal point of view, for, although we may meet with numerous cases which being strictly interpreted would demand the appellation general paralysis, yet it is best to withhold the term, for I believe I am right in saying that in a court of justice an expert who used this expression would necessarily imply that the patient was mentally incapable at the the very first moment that any symptoms of general paralysis exhibited itself.

Many years ago I had the advantage of attending a course of lectures delivered by Dr Conolly at Hanwell, and I remember that he made the following statement, which I believe is quite correct— that general paralysis of the insane occurs mostly amongst the lower orders, and thus is seen much more frequently in public than private asylums; also that it occurs much less frequently in women, and I think he went so far as to say that he had never seen an example of the disease in a lady of the upper classes of society. I cannot but think that these facts tend to corroborate the idea that the disease is often induced by accidental causes, such as direct injuries to the head. I am told that it is now more common in women, and one reason suggested is intemperance. It is a disease said to be rarely seen in Ireland.

I believe I am right in saying that there are some in the profession who deny the broad statement that the disease is incurable, and necessarily fatal in a given period. I can venture no opinion on the subject, but there appears no *à priori* objection to the disease being recovered from in its earliest stages. I have never seen a case get well, and recovery must be extremely rare. On the contrary, the most remarkable fact in this affection is that as soon as it is diagnosed by the most trifling peculiarities, the patient may be regarded as having a mortal disease upon him.

This constitutes the remarkable nature of the disease, seeing we have no means of knowing why this fatal degeneration of the brain should be attended by such characteristic mental phenomena. If, for example, a patient becomes melancholic, we know of no reason why he should not recover his mental vigour; but let him be the subject of extravagant delusions, we know that an organic change has commenced which before long will be fatal.

We must conclude that the disease in the brain is peculiar to it, seeing there are other degenerative changes which also lead to bodily and mental infirmity or dementia paralytica. It is, therefore, wrong to use a term of this general signification for a special form of the disease. No doubt the only form seen in an asylum is of this nature, but in a general hospital like ours, we see instances of dementia paralytica resulting from disease of the cerebro-spinal centres of other kinds. The French term, "paralysie ambitieuse," is more defined, or better still, folie ambitieuse. I prefer it also for another reason; on signing a certificate, or acquainting the friends with the nature of the illness of the patient, who is raving mad and who still possesses all his ordinary strength, the use of the term paralysis appears absurd, and its justification only made explicable by the remarkable reason that he will some day become paralysed.

Other forms of general paralysis.—You will understand that degeneration of the cerebro-spinal centres implies a paralysis of body and mind. Now, in all probability, this degeneration may occur in different modes or may involve special structures, so that the symptoms in each form would differ. My complaint, therefore, against the alienists is that they employ a term of general signification to designate one form of the disease. They should define exactly what they mean by the expression, either in a clinical or pathological sense, and thus assist us in our diagnosis.

General paralysis from injury

CASE.—Thomas M—, æt. 41, a slater by trade, is totally unfit to give any account of himself, and thus the following history has been obtained from his wife. As long as she can remember he has been a very healthy man, never having an illness of any kind until eighteen months ago, at which time he was at work at the roof of a house, standing on a scaffold, when he fell backwards six or seven feet, striking his back across a beam. He got up, however, and continued his work until the evening, apparently little the worse for his accident, only complaining of pain running up his back on stooping to pick up anything. Five months after this, after working all day in the sun, he was seized with headache and vomiting, which became so bad that he had to leave off work and go home. In the night he raved so much that she sent for a medical man, who gave him some pills, which kept him in a comatose condition for three days. He recovered and was able to do some work for a few months, but was never thoroughly well again; he had pain in the head and was giddy, so that he never could again ascend a ladder. About five months before admission he became much worse, as one evening he found his right arm completely numbed and feeble, and after some hours he experienced much pain in it. Subsequently he had several such attacks, at which time he clutched hold of his right arm, saying that he felt something hard and heavy in the palm. During this time he has complained of pain at the back of the head, and has been losing his memory and had difficulty in speaking. Latterly, he has not been allowed to go out alone, as he cannot remember where he lives, and loses his way; for although he has his name and address in full in his pocket-book, he has not the sense to show it to anybody. Lately, also, he has had a very ravenous appetite, sometimes asking for dinner immediately after finishing it, apparently forgetting that he has just eaten it.

On admission, the report says, he is a fine well-developed man, having a wild and excited look. He talks to himself, has a difficulty in pronouncing some of his words, and often, being unable to answer, will start off on an entirely different subject. When asked to repeat numbers he will do so and in any order, not apparently knowing what he is saying. When asked to write his name he assents, but the letters are shaky, and in a kind of mechanical way he places dots over the Christian and surname. Sometimes he will laugh when spoken to, but generally the state of the mind displays no characteristic peculiarities; in fact, he is rather demented. When his wife and friends call he often fails to recognise them.

He remained in hospital a short time and was then removed.

General paralysis with dementia

CASE.—George B—, æt. 30. Was quite well until four months ago, when, without any assignable cause, he went out of his mind and was very violent. After three weeks the attack passed off and he resumed his work. Shortly, however, the symptoms returned, and his mind became quite lost. He would sit all day taking no notice of any one and not speaking, but he had not, as far as could be told, been the subject of any delusions.

There is no marked paralysis of any part, but there is a failure of expression, and there is general feebleness of the whole body. His gait is tottering, and his hands shake when he raises his arms. He is quiet when lying in bed, but as soon as he moves a tremulousness comes over him. When asked to write he makes the attempt, and the strokes are made by a sudden jerk of the pen. During some days he sat quiet in his chair, and beyond saying " yes " and " no " did not speak. He then had a fit in which the face became livid, the left side drawn up, and spasmodic twitchings of the limbs. After this he was delirious, especially at night, calling out, getting out of bed, so that the nurse was obliged to fasten him down. After a few days the attack passed off, and he was able to get up, and he soon after left. His bodily condition was one of universal failure of power, observed in his tottering gait, tremor of limb, hesitation in speech, and as regards his mind, it was simply in a state of vacuity.

General paralysis of the insane with no exalted ideas

CASE.—George K—, æt. 44, a baker. Was well until a year ago, when he began to get weak, and it was noticed that his voice trembled when he spoke, and that his memory was defective. He was better for a little while, but again got worse and quite incapacitated for any kind of work for four months. During this time he has been silent, has taken very little notice of anything, has never been at all violent, but was troublesome owing to the difficulty in keeping him in bed. He could walk for a short distance when supported, but if left alone he fell backwards. On one occasion when in bed his eyes became fixed and staring, and he trembled violently for about a quarter of an hour, but was not unconscious. A temperate man; had no exalted ideas during his illess.

A tall, well-made man, with vacant expression of countenance: when asked a question he appears puzzled and seems to be trying to collect his thoughts; never speaks unless spoken to; and generally answers in monosyllables. He remains in bed nearly all day, taking no heed of anything passing around, but occasionally attempts to get out of bed and cannot be kept quiet, unless his legs are tied down. When assisted he can walk for a short distance, and then in a shuffling manner on his heels. Passes his motions in bed simply from want of mental effort. The left pupil is smaller than the right. He could at one time write very fairly, but now grasps the pen feebly, and the letters he attempts to form are very shaky (specimen in report). As he was very dirty, and began to eat his fæces, he was sent away.

DELIRIUM TREMENS

Now, after speaking to you of the slow destruction of the brain-tissue by a chronic inflammatory process, let us look at the case of a more simple atrophy. In the disease we have just left—the general paralysis of the insane—there is not only loss of bodily and mental power, but the manifestations of the latter are peculiar. In simple wasting of the brain, however, there appears to be only a gradual failing of the physical and intellectual functions. See a man, for example, tottering along scarcely able to support himself, or "all of a tremble," and his mind at the same time impaired—that man, if the symptoms have existed for any period, has an atrophic brain. This may have been the result of age, and he may be approaching a second childishness or mere oblivion; but if he be a young man, it arises from disease and, in all probability, from intemperance. Certainly one of the causes most frequently instrumental in the production of an atrophy of the brain is drunkenness, or the excessive use of alcoholic drinks. Alcohol, as you know, causes a degeneration of all the tissues of the body. By its direct effect on particular organs, you may find an increase of the connective or fibrous tissues, and then a cirrhosis of the liver, a Bright's kidney, or a so-called chronic pneumonia may arise; in other instances a fatty degeneration of the tissues occurs, and thus death by diseased heart is very frequent in the drunkard. These two morbid processes are, however, often combined; but, whatever may be the exact pathological change, the result is a decay of all the tissues and organs of the body. It is remarkable how many morbid processes resemble those which naturally take place in advancing age, and thus a disease will, by its effects, add so many years to a person's life. A drunken man is literally living too fast.

There are, of course, other circumstances which tend to this general and nervous decay; but I think in most instances, as you have often heard me say, if you see a man come staggering in amongst our out-patients, the probability is that he has been taking too many so-called "strengthening things" all his life. If you find this so, and he is bodily and mentally incapacitated, that man has assuredly an atrophied brain. If you examine the body of a man who has died of delirium tremens, you find the effects of the long-continued narcotic on his organs; and, as regards his brain, that has a very striking appearance—the membranes are thickened, and the convolutions are much shrunken, so that there are deep sulci on the surface, and in these a quantity of fluid. Sometimes three or four ounces of serum are seen, taking the place of so much good brain which has disappeared. The figurative expression which Cassio makes use of when speaking to Iago contains a solid and substantial fact—"O God! that men should put an enemy in their mouths to steal away their brains!" The brain, then, of a confirmed drunkard is an organ which is shrunken, and weighs so many ounces less than it did when healthy. All its functions, also, are becoming enfeebled, and those qualifications which should have preserved to its owner the name of lord of the creation have disappeared, and he has become, in common parlance, a good-natured fool. Call such a man a brainless sot, and you have his exact condition—the sot implying a man stupid with drink, and the "brainless" imparting to the term his true pathological condition. It is such a brain as I have mentioned that you find in those who have died of delirium tremens. The earlier attacks of the drunkard are recovered from; but there comes a time when the last and fatal one arises. But then his death is really due to the degeneration of tissues which long-continued habits of intemperance have entailed. The last case I saw on our postmortem table is an illustration of what I have witnessed a score of times—a man, æt. 50, comes in with a fractured leg. He dies in four days from delirium tremens. The brain is wasted, the ventricles distended with fluid, and ependyma granular.

An illustration of intemperance is not far to find, for the newspapers of to-day contain the sad history of a gentleman whose will is in dispute in the Probate Court. Sir E. had been in the Government employ, and had been a refined, graceful, and accomplished gentleman, until he began a course of dissipation. "He then sank into a state of debasement, in which his whole nature seemed changed; he was held hardly responsible for his conduct, and imbecility seemed imminent. The graceful gentleman became a paralytic of wretched aspect and filthy habits; the gay attaché degenerated into a brainless sot, an indecent talker, a petty

pilferer; the trained and accomplished man of the world became
an object of scorn or compassion. Shunned by his equals, expelled
from clubs, unfit for active or social life, Sir E., after fifty, was a
mere decaying wreck, his body a mass of weakness and disease, his
mind, with some remains of intelligence, declining into premature
decrepitude."

Take a man who, from his addiction to alcoholic drinks, has the
weakened brain that I have described, and let any unusual stimulus
act upon it, he loses his balance, and he is thrown into the state
known as *delirium tremens*. This is my idea of the disease. I
believe some medical books state that delirium tremens is a train
of morbid phenomena produced by the slow and cumulative action
of alcohol. This, however, I cannot admit without some modification, for it might imply that delirium tremens was due directly to
stimulation of the brain. There is, no doubt, a morbid condition
set up by alcohol, but this would be *delirium è potu*. The patient
would be simply intoxicated, or poisoned by spirit, which has
"made havoc amongst those tender cells, and checked his power to
shape." I believe, however, that such an opinion as I have mentioned does largely prevail, *i. e.* that the brain is over-stimulated
by the artificial excitement, and that a sudden deprivation of this
starts the disease into life. My own strong conviction is, that
such an opinion is highly erroneous, as well as the treatment which
it necessarily involves, viz. that the stimulus must be restored,
and the excitement abated by opium. I believe firmly that such
an opinion and the conclusions which flow from it are erroneous
and mischievous.

My own idea is that the brain has been previously brought into
an impoverished state by the continued debauchery, and that, thus
weakened, it is ready to be further disturbed by any fresh excitant,
whether this be a mental trouble, an accident, or, more probably,
an extra amount of the alcoholic stimulus. I am led to this opinion
because, in the intervals of the attacks of delirium tremens the
patient presents the symptoms I have before mentioned—a mental
and physical weakness—as a result of the chronic and pernicious
effects which are being produced on the brain, and also because I
observe the circumstances which excite the attack. The cause may
be an accident or an acute disease suddenly set up, but more generally we find that the patient has been out with his friends drinking
for two or three days in succession, and if he has had any trouble
in business this would be a still further excitant cause. On the
other hand, I never hear of a man suffering from delirium tremens
from having got drunk only once, nor do I hear of it in the habitual
drunkard from his having left off the accustomed stimulus. I have

never witnessed such a case myself, and I have taken the trouble of
inquiring of those who have had an opportunity of seeing persons
habitually intemperate, and who have been placed on a bread-and-
water diet, and I am informed that such persons do not have
delirium tremens. I have myself on several occasions recommended
or insisted upon the withdrawal of ale, wine, and spirits from those
who have been endeavouring to live upon these stimulants, but
never witnessed anything like delirium tremens result. Why I
insist more especially upon the importance of having a true conception of the disease is that an erroneous opinion carries with it a
very harmful treatment. For whilst one theory suggests a continuance of the accustomed stimulus, the other demands its withdrawal, and insists upon repose for the excited and wearied brain.
Delirium tremens is not so simple a disorder as you may imagine ;
each case requires study, and its appropriate remedies.

Having given you a clue to its pathology and treatment, I have
said all that is important. As regards the symptoms, several very
graphic accounts have been written, and some by those who have
suffered all its horrors, as in the following example. Of course, in
a description of this kind, as in the analogous one of the effects of
opium eating, the account is written in calm moments, and therefore
can only be a feeble representation of the sensations experienced
during delirium.

" For three days I endured more agony than pen can describe,
even were it guided by the hand of a Dante. Who can tell the
horrors of that horrible malady, aggravated as it is by the almost
ever-abiding consciousness that it is self-sought ? Hideous faces
appeared on the walls and on the ceiling and on the floors ; foul
things crept along the bed-clothes, and glaring eyes peered into
mine. I was at one time surrounded by millions of monstrous
spiders, who crawled slowly, slowly over every limb ; whilst beaded
drops of perspiration would start to my brow, and my limbs would
shiver until the bed rattled again. Strange lights would dance
before my eyes, and then suddenly the very blackness of darkness
would appal me by its dense gloom. All at once, whilst gazing at
a frightful creation of my distempered mind, I seemed struck with
sudden blindness. I knew a candle was burning in my room, but
I could not see it, all was so pitchy dark. I lost the sense of feeling,
too, for I endeavoured to grasp my arm in one hand, but consciousness was gone. I put my hand to my side, my head, but felt
nothing, and still I knew my limbs and frame *were* there. And
then the scene would change. I was falling, falling swiftly as an
arrow far down into some terrible abyss ; and so like reality was it
that as I fell I could see the rocky sides of the horrible shaft where

mocking, gibing, mowing, fiend-like forms were perched, and I could feel the air rushing past me, making my hair stream out by the force of the unwholesome blast. Then the paroxysm sometimes ceased for a few moments, and I would sink back on my pallet drenched with perspiration, utterly exhausted, and feeling a dreadful certainty of the renewal of my torments."

A man suffering from delirium tremens has generally already shown symptoms of a weakened brain by his tremor and want of mental vigour. When you are called to see him in the attack, you usually recognise at once the nature of the case; he is probably up and dressed; he may, indeed, be attempting to conduct his usual business; in fact, he generally has a great desire to be about his accustomed avocations; he is constantly moving and wanting to go somewhere or do something, which his friends are as constantly desirous of preventing. As you speak to him he is sufficiently intelligent to know you, and sit down for a moment and converse, but it is only for a moment; his restlessness prevents him remaining quiet, and his mind then rambles on all kinds of subjects. If you place him in bed he constantly jumps up, and is suspicious of those around him, or fancies that he sees objects in the corners of the room; if the bed be surrounded by a screen then he perceives imaginary cats and dogs on the bed-clothes, and will attempt to stroke them, or sees creeping things like spiders; or he stretches out of bed to look beneath it, fearing there are persons hidden there. All this time he is constantly fidgetting his hands, and perpetually jumping up and lying down again; his pulse is quick, his tongue is furred, and he is bathed in perspiration; but the temperature is not raised.

When I say such a case is easily recognised, I mean with care, for I have constantly seen mistakes made with reference to diagnosis. I have seen general paralysis of the insane styled delirium tremens by the same physician on two different occasions. In the surgical ward of a hospital cases of injury to the head, followed by arachnitis, or effusion of blood on the surface of the brain, are extremely likely to be called by this name. Acute disease, as pneumonia, occurring in a person of intemperate habits, is sometimes masked by the delirium, and may, therefore, be mistaken. Typhus fever I have seen styled delirium tremens, and there is one form of complaint I have seen on three several occasions confounded with it; I allude to the so-called cerebral rheumatism, where the arthritic symptoms suddenly abate with the appearance of delirium. This may occur, as is usually the case, with a sudden onset of high temperature, and more rarely in connection with a pericarditis or other complication. The physical signs or history of the case ought to render it clear, although the delirium, the restlessness, and the perspiration, form a picture which is often the facsimile of delirium tremens. Authors have dwelt upon its liability to be confounded with arachnitis; such a confusion, however, could only arise from a theoretical view of what the symptoms of arachnitis ought to be. If you suppose they are like those of delirium tremens you might be apt to call the disease by the wrong name; but if you have any real knowledge of the symptoms depending on inflammation of the brain you could not possibly make any such mistake. I never, for example, saw a patient jumping about in bed with meningitis, much less have a low temperature, and be bathed in perspiration. The tendency in all forms of inflammation of the brain is towards lethargy rather than excitement.

It is important also to know what is the condition of the patient generally who suffers from delirium tremens, for on the state of the organs the prognosis mainly depends. He may have cirrhosis of the liver, or, what is not uncommon, Bright's disease of the kidney. If the patient have convulsions, you may find that the urine is albuminous, and this of course must endanger the case. Convulsions, or spasms, I might say, however, often occur independently: they may be due to the wasted brain, or irritation connected with the arachnitis or thickening generally of the membranes, or be temporary and due to the recent debauch.

Now, as regards *treatment*, the most important instruction I can give you is what *not* to do. Do not, in the first place, regard delirium tremens as a disease due to the sudden withdrawal of an accustomed stimulus, and therefore commence the treatment with the administration of wine or spirits—you will by so doing add fuel to the fire—but look upon the attack as due to excitement acting on a previously weakened brain. Then your mode of treatment is obvious and rational. In one word, get your patient *repose*. Now, I do not mean by this that your remedy is simply opium to procure sleep, but I use repose in a much larger sense. I have frequently seen a patient, probably a publican, sitting in a large room surrounded by a dozen friends, male and female, who are talking to him or holding him in restraint. He is bathed in perspiration, his pulse is very quick and feeble; the doctor says he has not slept for two or three nights, and he dare not give any more opium, as the pupils are already contracted. His fears that the man will die seem not ill-grounded, since it is impossible to suppose that any mortal man could go to sleep under such circumstances. I have ordered such a patient to be removed to a small room, put to bed, and he has gone to sleep in half an hour. Therefore I say that the most judicious treatment is required for delirium tremens; and do not go away with the notion that all you have to do is to give opium to procure sleep. I have many times seen the last sleep produced by it. And do not, as I before said, continually give stimulants to "keep up" the patient, for I have constantly seen the complaint aggravated by these means. I do not wish to reflect on the opinion of others, because I state strongly my impression that the disease is not to be cured off-hand by medicine, and that in a bad example of the complaint the symptoms will continue without abatement for three days at least. I see a great deal of this disease, and the time which I am called in is about the third day, when, all means having been found unavailing in checking the complaint, another opinion is sought; after this time the patient usually sleeps and does well. I think, then, except in very slight cases where a single dose of opium procures rest, that a certain interval must elapse before the commotion subsides; this is my experience under all modes of treatment. I have no objection to opium, but it should be administered judiciously; if you act on the principle that sleep must be procured at all hazards and as soon as possible, you will without doubt kill many of your patients. In the first instance you should place your patient in a small quiet room, and get rid of a number of officious friends.

As regards medicines, you must be guided by circumstances. I have often prescribed with advantage the well-known mixture of twenty drops of antimony wine and tincture of opium every four hours, keeping up the patient's strength by beef tea. If you can give a glass of wine or beer in the form of nourishment without its producing any injurious stimulating effects I have no objection; but, as a rule, I advocate the plan of giving none. Instead of administering antimony you might give the laudanum, or small doses of morphia with ether or ammonia, and, at the same time, support the strength of the patient. By judicious management of this kind you will find your patient recover in three or four days, unless, indeed, he be extremely diseased in consequence of his former dissipation. As regards restraint, I have often heard objections made to it, but I believe it is often necessary, and its adoption is a real kindness to the patient. He may wander about his house until he drops dead from exhaustion, when a forcible restraint in bed by a sheet across the chest might procure rest for the body, and often for the brain. Remember, then, your patient wants repose; do not be content with administering opium, and neglecting every measure which common sense would say was necessary to give every man a night's rest. As regards chloroform, you may quiet the patient by it for a time, but you do not in any way influence the disease. Other remedies are advocated. When digitalis was first proposed I gave it to a man in large doses, and, he having unfortunately died, I have never felt myself justified in administering it again.   I have seen enough of the sedative effects of the wet sheet to fully believe the statements which have been made as to its efficacy.   You strip the patient naked, roll around him a wet sheet until he looks like a mummy, then a blanket around this again.   In many cases of delirious excitement you will find that as soon as a hot vapour surrounds the patient he sinks into a quiet sleep.

PART II—THE SPINAL CORD

ANATOMY AND PHYSIOLOGY OF THE SPINAL CORD

A SECTION of the spinal cord shows a tendency to divide into two halves by a wide and distinct fissure in front, and by a less distinct though deeper fissure behind. On each side of it are the roots of the nerves, the anterior being the efferent or motor and the posterior the afferent or sensory ; at the connection of the latter there is a distinct groove, which is not the case with the former, so that the cord is naturally divided on each side into an antero-lateral and posterior portion. The section displays the grey matter in the centre composed of two crescentic portions joined by a commissural band. The two crescents terminate in anterior and posterior cornua, which are connected with the corresponding nerves. The posterior horns come near the surface and end close to the origin of the posterior roots of the nerves, whilst the anterior cornu are shorter, rounder, and more distant from the surface. The posterior roots, therefore, make a natural division into posterior and lateral columns. The posterior cornua are covered with a substance somewhat different from the grey matter, and called the gelatinous substance. Running through the transverse or commissural portion is the central canal. The mode by which the nerves are connected with the spinal cord is only in part known. There are clearly transverse and longitudinal fibres, the former passing across from front to back, and *vice versâ*, through the grey substance, and these vary much in size. The anterior roots may be seen entering the cord and spreading out, some entering the grey substance, whilst others pass upwards and downwards. The course and function of the different strands have been worked out by dissection, by experiments on animals, and by the methods of Bouchard and Waller. That of the latter consisted in tracing the course of the degeneration of nerve-fibres after they had been purposely injured.

The experiments of Brown-Séquard showed that division of the posterior columns did not destroy sensation, and he further proved that sensation was lost by a destruction of the central grey matter, proving that this important part of the cord, besides possessing its own true and independent function, transmitted sensory impressions. He further showed that the sensory fibres were continually crossing from one side to the other, so that they, like the motor fibres, decussated, but, unlike them, did not cross in one mass at the pyramids. He found, therefore, by severing one lateral half of the cord, that he produced perfect paralysis of motion on that side and loss of sensation on the other, sensation remaining perfect on the paralysed side. There have now been a sufficient number of experiments made for us on the human subject by disease and injuries to fully confirm the truth of this physiologist's investigations. We cannot, therefore, adhere to the doctrine, which is frequently taught, that the function of the grey matter is wholly dynamic or force producing, whilst the medullary portion is for conduction only. The sensory tract appears to be near the middle of the cord, and, according to Schiff, is probably made up of two distinct parts—one to convey ordinary sensation, and the other placed more posteriorly for the sense of pain. The sensory tract is generally taught by physiologists to terminate in the thalamus opticus. The motor tract is undoubtedly contained in the antero-lateral column, and probably more especially in the lateral; its fibres pass upwards and downwards, and are intimately connected with the anterior cornu of the grey matter. This intimacy is shown by disease of the cornu producing atrophy of the nerves which spring from it. The motor fibres which pass upwards pass over, for the most part, by the decussation to the other side, and so through the pons and crus cerebri to the corpus striatum; the sensory and motor nerves coming out of the pons also decussating in the same way.

Proceeding both from the corpora striata and the thalami optici are the radiating fibres which terminate in the convolutions. Some fibres, however, of the motor tract appear to continue their course on the same side, and do not decussate at the pyramids. The posterior columns, which do not appear, as was once thought, to be for the conveyance of sensory impressions, are intimately connected above with the restiform bodies and the cerebellum. From experiments on animals, as well as those made for us in the human subject, in consequence of grey degeneration of the posterior columns, it is believed that their function is devoted mainly to coordinate movements of the body, by uniting together the various centres of the grey matter from which motor impulses proceed. This is accomplished by an arrangement of short fibres, which are found continually passing from one grey centre to those above and below; by this means all parts of the cord can act in unison. This function seems intimately associated with that of the cerebellum, which stands between the cerebral and spinal systems, and whose influence is supposed to spread over the whole body, so as to keep it in perfect equilibrium. The upper part of the cord, the medulla oblongata, is much more complex than the remainder of the organ below, and no doubt has very important and special functions, especially those employed in respiration, speech, and various complex movements of the mouth and chest; the grey centres ruling over these being seen in the olivary and restiform bodies, &c. The anterior pyramids are mainly associated with the antero-lateral columns; the corpus olivare also joins with them below, and its fibres pass up to the corpora quadrigemina; its grey centre, the corpus dentatum, appears to be continuous with the anterior cornu of the cord. The restiform bodies are connected with the posterior columns and the crura cerebelli, and some of the strands, designated by the name of posterior pyramids, carry up the sensory fibres of the cord to the crus cerebri and brain. A part of the lateral columns separate or open out to form the fourth ventricle. A section of the cord will, therefore, show us the following functional divisions, as far as we at present can make use of it for clinical purposes. If the central grey matter be seriously injured the true function of the spine and excitor motor acts would be abolished, and if lesser changes in this part occur there would be an interference with the transmission of ordinary sensation. If there were an injury to the antero-lateral columns a paralysis of motion would ensue, and if lesser changes throughout its substance a disturbance of movements as seen in tremor or spasm; also if an injury or disease of the cord should involve the anterior cornu the motor root would be involved, which would produce not only a loss of function of the nerve, but an actual degeneration of its fibres along its whole course, together with a corresponding atrophy of the muscles to which the nerve was distributed. Again, if the posterior columns were diseased the function of co-ordination would be seriously interfered with. We may, therefore, associate disease of the posterior columns with *ataxia*, of the centres with *anæsthesia*, and of the antero-lateral columns with *akinesia*.

I have spoken merely of ordinary sensation and the columns which convey it, as well as of motion and the columns appropriated for the transmission of motor influences; but on analysing the various kinds of sensation, as well as the functions of the nerves, it may be well surmised that a nerve is a very complex structure, and may possibly have to transmit many other kinds of influences than those of common sensation and motion. A moment's consideration will show you that it is not an easy problem to solve, whether the impressions conveyed from the surface are but modifications of one another, or whether they are altogether different in kind. Take, for instance, the case of the hand being held to the fire, and a pleasant sense of warmth experienced, and then the actual pain produced if it be approached too near, and ask if the sense of pain is merely an excess of the stimulus which was just before producing pleasure.   It seems so, and yet we know that pain may exist when ordinary sensation is wanting; also that the sensation of warmth and touch may be quite dissociated, as witnessed in various forms of paralysis.   The latter fact was known to the elder Darwin nearly a century ago, who quotes in his 'Zoonomia' a letter from Dr K. W. Darwin of Shrewsbury, who was studying at Edinburgh, to the following effect :—" I made an experiment yesterday in our hospital which much favours your opinion, that the sensation of heat and of touch depend on different sets of nerves.   A man who had lately recovered from a fever, and who was still weak, was seized with violent cramps in his legs and feet, which were removed by opiates, except that one of his feet remained insensible.   Mr Ewart pricked him with a pin in five or six places, and the patient declared he did not feel it in the least, nor was he sensible of a very smart pinch.   I then held a red-hot poker at some distance, and brought it gradually nearer till it came within three inches, when he asserted that he felt it quite distinctly.   I suppose some violent irritation on the nerves of touch had caused the cramp, and had left them paralytic, while the nerves of heat having suffered no increased stimulus retained their irritability."   We are constantly verifying the truth of this in the wards where, in cases of paraplegia, ordinary sensation may be lost, and yet the patient still appreciate the difference between heat and cold.   This, of course, only occurs in modified forms of paralysis.   It is a question, therefore, whether one distinct nerve is not required to transmit common sensibility, another pain, and another heat.   To these Brown-Séquard would add a fourth for the sensation produced by tickling, and other physiologists a fifth for the cognizance of a muscular contraction during the passages of the electric current.   You will see that the question is much of the same kind as is asked with respect to the various effects of light on the optic nerve, or of sound on the auditory, in the production of colour and musical notes—whether there are distinct perceptive centres and corresponding nerves for their relative appreciation, or whether the different effects on the organs of sense are not due to modification of the same stimulus acting in various forms and degrees?   It is a question, therefore, whether there are distinct nerves for sensation, pain, heat, &c., or whether these, being but modifications of the same sense, are transmitted only by one, although it is possible there might be different modes of termination of the nerve plates.   If a sensory nerve is

really compound we must speak of it as made up of different
fibres, as some physiologists do—namely, one for tactile sensibility, another for painful sensibility, a third for thermic sensibility, and perhaps a fourth for electric sensibility. Those who
take this view would have no difficulty in adding a fibre for a
muscular sense, and even go to the extent of believing that there
are distinct nerves for the reflex acts. In this way each afferent
nerve would have a part ending in the grey centre of the cord, and
another proceeding to the brain, and in like manner each efferent
nerve would consist of a fibre descending from the brain, and
another which had its origin in the cord. There is no proof of any
complexity of this kind, nor is it theoretically required, for in all
probability the fibres descending from the brain above end in the
grey matter, or at all events there is no proof of any direct connection between our skin or muscles and the brain alone, except
through the grey matter of the cord. You must remember also what
you are taught as to the different degrees of susceptibility of the
skin in the various part of the body, this being most perfect at the
tip of the tongue and fingers. Temperature is best appreciated in
the lips, cheeks, and backs of the hands. The compound character of
a nerve is not thought to end here, for since the nerves appear to
regulate the size of blood-vessels and effect nutrition, so there are
vaso-motor, if not distinct trophic nerves; and again, as there are
nerves to secreting organs, many trunks must contain a nerve with
a glandular function. It is thus clearly explicable how eleven or
twelve nerves with different offices are supposed by some physiologists to enter into the composition of an ordinary trunk. The
functions of a nerve are best observed when the trunk itself is
affected; and what we want to discover is in what portions of the
nerves do they reside, and whence these functions come. The
trophic effects due to the nerves I shall speak of when I come to
nerves. That there are influences conveyed by nerves, besides their
power of regulating the supply of blood, can scarcely be doubted
when we see a neuritis of a sensory nerve followed by changes in the
skin, and of a motor nerve by wasting in the muscle. If the spinal
nerves be cut at their roots, the anterior portion wastes towards its
periphery, but not that portion which is still attached to the cord;
and not only does the trunk waste, but the muscle itself to which
the nerves are attached participates in the atrophy. This shows
that the motor nerve receives its nutritive influence from the anterior cornu of the grey matter from which it arises. If the sensory
nerve, on the other hand, be cut it retains its bulk in the part beyond
the ganglia, showing its nutritive influence is derived from them. If
any compound nerve be divided the peripheral part wastes, but not

the portion attached to the centres, showing whence it derives the power which prevents its elements undergoing disorganisation. This propagation of degenerative changes along the nerves in special directions may account for many remarkable conditions hitherto unexplained. In the cord itself the nutritive influence appears to run in the direction of the functional activity of its fibres.

Whether there be trophic nerves distinct from those we have mentioned may be a question, but since nutrition is affected manifestly through both sensory, motor, as well as the sympathetic, and these nerves receive some influence from certain ganglionic centres, there is nothing remarkable in the belief that trophic centres exist in the spinal cord. You know at the present time some physiologists maintain this.

Heat-regulating centres.—It is also thought that there are centres which regulate the heat of the body, and, indeed, it is difficult to avoid a belief in their existence, either on theoretical grounds or from experiments on animals and injuries on the human subject. It has long been observed that very remarkable alterations in temperature take place in injury to the lower cervical region—sometimes a very rapid elevation of temperature, and at other times as rapid a diminution. The reason seemed inexplicable until experiments were made on animals. If a dog had the cord severed at the seventh cervical vertebra, and artificial respiration kept up, the body soon cooled down, as Sir B. Brodie had already shown, and from this he concluded that the source of heat was not altogether chemical but was in part derived from the brain. The conclusion, however, was erroneous, for it has since been shown that if the body be clothed so as to prevent the escape of heat, it will become much hotter than natural. From this it would appear that the injury to the cord destroys the creature's power of regulating or adjusting heat; and if you think for a moment, it seems almost necessary to believe that such a power must exist in the nervous system; call to mind that on one day we may be taking much exercise, on another day seeking repose; on one day eating much and on another but little; and all this time the season is changing from summer to winter, and the converse. Therefore, as we are sometimes converting tissue largely and at another time sparingly, the amount of heat we are producing and giving off must be constantly altering. Yet, under all these changing circumstances, the temperature is the same in all of us even to the fraction of a degree. This normal temperature is the balance struck between heat produced and heat given off by the skin and blood-vessels, and that this should happen without some such regulating power can scarcely be believed. Now, these experiments on animals, made purposely, and those in the human subject, made accidentally, seem to show that this regulating force resides in a particular portion of the spinal cord ; so that when this is injured, the temperature will be constantly oscillating according to surrounding circumstances. Probably, in disease of the cord, the results of the disturbance may be seen in a modified degree, as well as in the sweating which constantly occurs.

Influence of nerve on muscle.—Another point of interest, and one respecting which there is much difference of opinion, is the influence exerted over muscle by the nervous system.　When the connection is severed between the brain and the muscular apparatus, any attempt to move the limbs is ineffectual, and the contrast between the healthy and affected side becomes apparent.　But it is not only when the will is in operation that the loss of influence is seen ; for at all times, and as soon as the paralysis has occurred, not only is the line of conduction broken, but a certain tonic effect has disappeared. This is well seen in the case of facial paralysis affecting one side, when the face falls and the opposite side is drawn up. This appearance is increased when any voluntary movement is attempted, but even at rest and in sleep the tonic contraction of the healthy side may be well contrasted with the relaxed condition of the other.　If both facial nerves are affected, it is not only when the patient "wills" to move the muscles of the face that the paralysis becomes apparent, but at all times the face assumes a blank expression from the tonicity of the muscles having disappeared.　This fact, however, is denied by some, and the appearance explained by the face remaining contracted after the last voluntary action upon it.　In a practical lecture it would be out of place to enter fully into the various suppositions and theories propounded to account for this supposed influence, but I may mention that it is one of those questions in nerve and muscle physiology which have by no means been satisfactorily solved.　Of course, one's first impression would be that a certain influence, allied, perhaps, to a galvanic force, was constantly passing down the nerve and preserving the tension of the muscle by keeping its elements together. Opposed, however, to such an idea there is the fact of the continuance of the irritability of muscle after the nerves which are distributed to it have been severed ; indeed, even more than this, for it has been shown by Brown-Séquard that in a decapitated frog the muscular power has been increased, the spinal cord, however, remaining.　Marshall Hall, however, stated that if the spinal cord were destroyed all muscular tone would be lost.　In his own words on the decapitated turtle—" The limbs or tail possessed a certain degree of firmness or *tone*, recoiled on being drawn from their position, and moved with energy on the application of a stimulus. On withdrawing the spinal marrow gently out of its canal all these phenomena ceased. The limbs were no longer obedient to stimuli, and became perfectly flaccid, having lost all their resilience. The sphincter lost its circular form and its contracted state, becoming lax, flaccid, and shapeless. The tail was flaccid and unmoved on the application of stimuli." Then, again, arguments have been taken from the independent contraction of muscle which occurs after death to show that the properties of muscle are inherent in itself. It seems strange, however, that a condition of muscle observed after death should be thought to resemble in any way the contractility which is regarded as a sure sign of life. Dr Radcliffe thinks that the active state of muscle is the relaxed one, and during rest it is kept in this condition by electricity in a statical form, and when the muscle is thrown into action a discharge of electricity takes place. He illustrates this by the case of the torpedo, where an electrical discharge accompanies the contraction of its muscles. The muscle deprived of the force inherent in its molecules contracts, but the contraction does not continue because the electricity is regained. In rigor mortis it is for ever lost. He thinks the longitudinal surface of the muscle is positively electrified and the ends negatively. The muscle is kept elongated by the opposite electricities and contraction is caused by their discharge.

A knowledge of the true relationship between muscle and nerve ought to be possessed by us before it is possible to afford an explanation of the phenomena of disordered movement, but at present we have not an accurate idea of the nature of the forces which bind these two structures together in health. Physiologists seem to hold the most misty conceptions regarding not only the nature but the seat of the motor force, although there is sufficient distinctness in their views to show that they are most opposed and conflicting. For example, we frequently read of a force generated in the brain which pervades the body by means of the nerves, and becomes altered in various ways as it is transmitted through the organs, or being discharged by the muscles produces the various movements of the body. Some physicians, as Todd, thus speak of the brain being over-charged with nerve-force in various forms of disease, rendering it liable to discharge itself as would a Leyden jar. Others, again, who would use similar language, would place the source of the nerve-force, which they suppose is in operation during muscular movement, in the central ganglia or spinal system, and if, therefore, anything like convulsion should occur from an irritation of the brain they would consider that it arose not so much from an emission of force from the brain, which then acts direct on the muscles, as it would be that the escape of this force allows the spinal system, normally under its control, to come into full play.

In opposition to these two views are those which place the forces which we see in operation during the movements of the body in the muscles themselves.   The holders of the latter opinion would not deny that a force is generated under the influence of the nerve-centres, but that we must look rather to the muscles and the nutritive changes which are going on within them for the real sources of power.   When we wish, for example, to move a limb, we do not transmit down the nerve the actual power which we see exerted on the muscles, but merely allow them to come into play by removing some restraint, just as would be done in driving a steam-engine on turning a handle and letting the properties of the steam come into action.   Many obvious facts appear to confirm the correctness of this view, such as the strength of an animal being proportionate to the amount of its muscle rather than of its brain ; also the fact, more lately arrived at, that the electrical apparatus of the torpedo is more allied to muscular than nerve structure ; and again, that an animal deprived of consciousness by loss of blood, and whose brain is therefore in abeyance, is thrown into convulsions.

This theory of the seat of motor force has always been most in unison with my own views, and has probably been confirmed by having been held by all those of my colleagues who have written on the subject.   Sir W. Gull, in the Gulstonian lectures, which he delivered in 1849, thus expressed himself :—" The muscle has its own inherent and proper power of contracting in virtue of its organisation, and nerve is the proper excitor of this power in a manner not yet explained.   All we yet know being this, that a nerve, when mechanically disturbed or affected with an infinitesimal amount of electricity, brings out the function of the muscle. We can compare the phenomena to nothing so aptly as to a spring set free by the easy motion of its stop."   Then, again, my late lamented and talented colleague, Mr Hinton, in his various writings, upheld a somewhat similar opinion.   He maintained that the forces operating in the animal body were of a mechanical kind, and were held in restraint by an antagonistic one—the vital force, the nature of which he did not endeavour to explain.   He held that the animal body, like any other machine, possessed a power which was regulated by another and distinct counter-balancing one ; the power in all machines being due to matter being restrained by the fulfilment of its natural tendencies, and that the removal of the restraining force permitting their play produced the action.   He says—" Organisation gives capacity for action only by virtue of the resistance it presents to chemical forces, these chemical forces being the true sources of functional activity." The movements of muscle, therefore, would be due to the liberation of the forces within it, and the stimulus conveyed by the nerve is merely unloosening the condition which prevented their previous operation. This view, you see, is opposed to that taught by Liebig and his followers, that muscular force is a conversion of vital or nervous force, " that the nervous force appears convertible into motion through the medium of the muscular apparatus." Hinton's view was that the nerves do not supply a force, but merely overthrow the balance. How the forces are produced is another matter; there are nutritive changes constantly going on, and these, no doubt, produce also the forces which come into play during motility. That the muscular structures are influenced by or receive some actual power from the nerve centres during their nutritive changes is in all probability true. But the influences are slow in operation, and therefore do not support the theory that movement is one of a discharge of nerve force. After the use of a muscle time is required for it to regain its lost power, as is seen in the case of the torpedo after it has given a shock ; we observe this every day after the fatigue of carrying a weight in the arm which obliges us to transfer it to the other. Again, in some very interesting experiments of Dr Poore, he showed that an electric current would help to sustain the power in a limb. After the arm has been held out straight for some time, and the moment has come when it can endure no longer, a galvanic current will remove the feeling of tiredness and allow the limb to remain extended for a considerable time longer. Hinton considered that an epileptic fit was due to an irritation disturbing the equilibrium of the nerve centre, which took off the chemical tension of the muscle and allowed its force to come into play. He says—" Muscular contraction from a stimulus is the analogue of the electrical discharge by means of a metallic contact, in which the restraint is removed, and the spontaneous contraction of the heart is parallel to the spontaneous discharge which ensues when the resistance is weak. In the heart and its ganglia the chemical and vital forces are so balanced that they assume a state of alternating activity." The same line of argument was also maintained by your late teacher, Dr Thompson Dickson, who, in his papers on epilepsy, opposed the view that the phenomena of motion were due to a discharge of nerve force. He was endeavouring to confute the idea of " discharge," the term used so much by Dr Hughlings Jackson and his followers in describing cases of epilepsy and other convulsive diseases. He denied that the brain or any other organ could do more than a certain amount of work, and, therefore, the term "surcharge" was a misnomer—that, in fact, in these diseases just named, there was every reason to believe that the brain was impoverished or anæmic, and that it was by a loss of its controlling power that the muscles came into action. His theory of epilepsy was founded on a loss of control of one hemisphere or both, which resulted in a hemiconvulsion or complete convulsion. He would have explained Ferrier's experiments of inducing a localised movement, by the stimulation exciting a discharge and so exhausting the particular convolution acted upon. Somewhat similar views are held with regard to the exhausted state of the cord in tetanus.

We must not, however, forget the clinical fact that want of tonicity of muscle generally implies deficient nervous influence, as seen in the want of expression in the faces of sufferers from general paralysis. It seems as if practically we must regard the muscles in three conditions—one of relaxation, as in paralysis or sleep ; one of extreme contraction, as when actively used ; and an intermediate state of tension during our waking hours. For example, I contract my sterno-mastoid and my head moves round ; I fall asleep and my head drops ; but in my waking hours the muscle is in neither of these extreme states, but is keeping my head steady on my shoulders.

GENERAL CHARACTERS OF SYMPTOMS CONNECTED WITH DISEASES OF THE SPINAL CORD

In coming to the diseases of the spinal cord, we should, according to the principles already laid down, meet with hemiplegia, if one motor tract—that is, half of the organ—were involved. Such an accident, however, rarely occurs ; for in actual practice it is much more common to find the cord more or less affected on both sides, or throughout its entire thickness, and thus a paraplegia is a far more likely effect of spine disease than a hemiplegia. But just as there are anatomical and functional divisions in the cord, so we find there are peculiar symptoms associated with its special lesions.

Now let us see if there are any general rules which can guide us as to the seat and nature of the different diseases which we commonly meet with. I have already informed you that clinical observations accord rather with the experiments of Brown-Séquard than with the earlier views respecting the arrangement of the fibres in the spinal cord—that is, that the motor fibres run upwards in the antero-lateral columns and cross at the pyramids, while the sensory fibres of the spinal nerves pass to the grey matter, and then cross to the other side. Thus, when the cord is divided longitudinally, sensation is altogether lost, but a lateral division destroys sensation on the opposite side. Cases of injury to the exterior of the cord, sparing the centre, exhibit a paralysis with sensation unaffected ; whilst, on the contrary, all feeling has been lost when the grey matter has been severed. Also in those rare instances where the grey centre has been so shaken by a blow that blood has been effused, sensation has been lost, whilst motor power has remained. Even the crossing of the sensitive fibres, as demonstrated by the physiologist, has been verified by some instances of remarkable accidents ; for example, there has been a case reported in one of the foreign journals, where a man received a sabre wound on the upper part of the right side of the neck, which so disabled him that his legs gave way under him, and he immediately fell. It was found, on examination the next day, that his right arm had lost its power of movement, though it still retained sensation. The right leg was a little weakened, but only as regarded motion. It was afterwards discovered that the left side of the body was insensible to touch ; pins were inserted into him without his feeling it ; and what was most remarkable, when he was pinched some obscure kind of sensation was felt on the corresponding part of the opposite side of the body.

In Abercrombie's well-known work, written many years ago, there is to be found the case of the Count de Lordat, who received an injury to the neck from being thrown off a coach. Six months afterwards he had weakness in the left arm and some difficulty in articulation, and subsequently the limb became withered and useless. After this the right arm became numb, and finally the whole body. After death the medulla oblongata and upper part of the cord were found enlarged, and the membranes thickened.

Now, if the motor tracts run towards the outside, and the sensory within, we should expect to find a lesion of motion much more frequent than one of sensation, seeing that the exterior of the cord is more liable to injury as well as disease, since the latter so often originates in the bones or the membranes. Therefore it is that when paraplegia has followed disease of the spinal column, it is motion which is first and especially involved, and the same also may be noted in chronic meningitis. In the case just now alluded to, where blood had been effused in the grey matter of the cord, sensation alone was lost. Whenever a softening of the whole of the cord occurs, as in acute inflammation, both sensation and motion would be affected. But, besides these conducting powers, the cord has inherent properties of its own ; the grey matter can be excited to action by external stimuli, and therefore, in the case where the cord is simply severed, as in an injury, although voluntary motion and sensation are lost, the portion below still retains its excito-motor power, and if the limb be touched, a movement would result. If the grey matter is destroyed throughout, then of course this power is gone. We shall find also that the grey matter exerts an influence of a nutritive kind, and therefore, if the anterior cornua are affected, a wasting of the muscles takes place. I think, therefore, we may say in general terms when disease attacks the outside of the cord there is a failure in motor power; if the grey matter is affected sensation is lost, but excito-motor power remains in the part below; and that if the grey matter be destroyed throughout, then the latter power also is lost.

We may note some still further facts; for just as we observed, in speaking of the brain, that an inflammation or irritation of the surface may set up an excited action in the organ within, so we may observe in the cord that an amount of disease on the surface not sufficient to diminish its functions or produce paralysis may yet be perpetually irritating the grey matter beneath it. Consequently, if the action of the cord appear much exalted, we may generally conclude that it is being irritated from without; and an observation of cases has constantly shown that in inflammation of the membranes involving the outer layer of the cord the excito-motor function is painfully active. In acute inflammation you may notice convulsive movements of a tetanic nature; and in chronic meningitis, where a portion of the true cord may be involved with an accompanying paralysis, constant jumping and twitching are amongst the commonest symptoms. Not long ago I had an opportunity of seeing a patient of Mr Birkett's, who, in consequence of an injury to the back, was suffering from an acute arachnitis of the spinal cord. He had no paralysis, for no portion of the medulla was affected; but he had severe pains all over him, especially in the limbs, his head was drawn back, and he was, in fact, like one suffering from tetanic opisthotonos. In the epidemic cerebro-spinal meningitis, of which you have heard so much of late, these were exactly the symptoms which were noticed. In some other cases of spinal meningitis I have observed not so much convulsion as extreme and constant restlessness. In those instances where there was much pain in the limbs it is probable that there was an implication of the nerves, for there can be no doubt that in many cases of spine disease where much pain exists the roots of the spinal nerves are involved in the inflammatory process. It may be indeed that in the cases of chronic meningitis already alluded to, where an exalted spinal function exists, the immediate cause of the excitability is the implication of the nerve roots, which react on the cord.

The true interpretation, however, of pain in spinal disease is one of the most important matters to understand. In the first place it is undoubtedly true that the spinal cord may be diseased throughout its entire thickness without the patient experiencing any pain, and without, indeed, there being the slightest sign of shrinking when the back is struck. On the other hand, the most intense spinal pains are those which are caused by disease involving the nerves only, as in aneurisms, which corrode the spine and leave the cord itself untouched. Again, it has been found that in those cases of meningitis of the cord where pains in the limbs have been a constant symptom the roots of the nerves have been involved in the process. With these facts before us we have some general rules to guide us in the interpretation of pain, although I could not positively teach you that pain necessarily implies an implication of the roots of the nerve external to the cord. It is highly probable that an implication of the nerve-fibres within the cord, to a certain extent, may produce the same result, but pain is not even then a necessary consequence. We may note also that the irritation of the nerve roots not only produces pain and excites the centres to corresponding movements, but causes tonic spasm or rigidity. Thus, when the membranes are affected in common with the roots of the nerves, a rigidity may often be observed. I have seen a man who, having fractured and dislocated his spine in the upper dorsal region, suffered intense pain and rigidity of the arms, but as soon as the bones were replaced the pain and spasm passed off. In chronic inflammation of the membranes of the cord the surface of the medulla is generally involved, and thus there may be more or less paraplegia; should the disease at one spot have involved the entire thickness of the cord, then sensation is lost, but the excito-motor function in the part below may be more than usually exalted. In our museum is the spinal cord of a man showing great thickness and ossification of the membranes, closely adherent and bound up with the structure of the cord. In this case the man lay quite paralysed, but his spinal centres were in the highest state of tension, so that it was painful to pass by his bed; the merest touch made his whole body quiver, the act of micturition threw him into convulsions, and, I believe, on one occasion a jar against his bed caused him to spring on to the floor. He was in the condition of a tetanised frog.

Let us see how these observations apply to cases before us, as, for instance, the case of a boy who was lately lying in Stephen Ward with disease of the dorsal vertebræ. His back projected outwards in consequence of an angular curvature, and thus, no doubt, a quantity of inflammatory or purulent material existed within the canal pressing on the medulla; and what might you think would be the result? You would suspect that he would be lying in bed with his legs stretched out, and quite paralysed as regards motion. The

pressure not having reached the centre of the cord, you might think
that sensation remained ; also, as a pressure of this kind on the
upper part of the spinal cord would involve the motor tract but
leave the sensory entire, and as the medulla would be healthy below,
that the true excito-motory function would remain. Thus you would
expect that, although he would be perfectly helpless to move in the
slightest degree, a tickling of the feet would cause a drawing up of
the legs and their flexion on the body. You might also have expected from the membranes being involved, and consequently the
existence of an irritation of the cord through the roots of the nerves,
that there would have been some convulsive movements of the legs
or rigidity. Now all these symptoms were actually present, and
considering what, in all probability was the nature of the case, you
will see how they accord with all well-observed physiological and
pathological facts.

We have already said that inco-ordination is connected with
disease of the posterior columns, and that the motor powers are
affected in various ways in disease of the anterior and lateral
columns.

There is another point very important to remember in connection
with diseases of the cerebro-spinal centres. Some of these are true
primary morbid states of the elements of the nervous tissue, whilst
others are altogether of a secondary or accidental character. For
example, the changes which take place slowly in the brain, as in
softening, are altogether different in kind and in pathology from
those which would arise from a tumour or from the severe laceration
caused by the bursting of a blood-vessel. So, likewise, in the cord,
the primary changes, such as we see arising from inflammation and
affecting the various strands, are altogether different in kind from
the diseases which would occur from its accidental implication, as
in caries of the vertebræ, or tumour. You see, then, some diseases
are primary, or central, whilst others are secondary, or accidental ;
and it would follow that each would require its own special treatment. It may be often difficult to diagnose between the two classes
of cases, but the following considerations will often materially assist
us. In the case of a primary affection of the cord, it is most
unlikely that the disease should attack one spot alone, or traverse
the cord in a plane so as to destroy all its strands on one level, just
as a section by a knife would have accomplished. Rather should
we see it progress slowly in the course of special fibres and in
particular tracts. It follows, therefore, that if a patient should
come into hospital with complete paraplegia, that is, with perfect
paralysis of motion and sensation below a certain part, and at
the same time the excito-motor function be perfect, just as we

see in a case of fracture of the spine, we should argue that the cord was affected at one particular spot through all its width. We conclude, therefore, that it has resulted from caries, or that there is a tumour, or that perhaps an aneurism has eaten its way into the canal. If, on the other hand, paralytic symptoms come on slowly, insidiously, and obscurely, showing that the morbid change has been progressing longitudinally in a given tract, the probabilities are in favour of the disease having been primary in the nerve tissues. I have found these considerations of great assistance in diagnosing cases in the wards. Whatever may be the cause the symptoms vary with the region of the cord affected. In disease of the lumbar portion there would be paralysis of the legs and the muscles of the abdomen. In that of the dorsal portion paralysis of the intercostal muscles, and in the lower cervical, paralysis of tho upper extremities, and above this again, where the bulb is involved, the special function of respiration, that is, paralysis of talking and swallowing, &c.

Morbid anatomy.—I may briefly say, as of the brain, we have to consider the nerve fibres, which many say have no sheath ; tho nerve cells of various forms, with the neuroglia, as already described, and the blood-vessels.

Changes of the same kind take place in the cord as in the brain, with the remarkable additional instance of that rapidly acute morbid process, in which, probably, some subtle nutritive function is destroyed, such as happens when the strands are divided for physiological purposes. Besides this, we have acute myelitis and softening, and chronic myelitis, with the various forms of sclerosis, or grey degeneration, as it has been called, spoken of as the diffused and the disseminated, or the miliary. There occur also simple atrophic changes and degenerations in the cord as in the brain. These are often associated with disease of the blood-vessels.

PARAPLEGIA

The spinal cord being thus a complex organ, you can well see what a variety of disorders may be referable to changes in its structure, and it is evident that the symptoms will vary as different constituents of the cord are involved. To connect the structural changes of particular parts with their necessary characteristic symptoms is the aim of modern investigators. This is being attended now with considerable success, and the phenomena observed from disease are found to tally with those which appear to have been expressed to the physiologist in his experiments. The most marked symptoms attendant on disease of the cord are those referable to altered motion and sensation, and more particularly the former. Thus it was that older writers were content to speak of paralysis of the legs under the name of paraplegia, as the only affection to which disease of the cord could give rise. If, however, the cord is composed of different elements, each having its own function, it is clear that lesions of certains parts will be found intimately associated with a special class of symptoms. It is our business to analyse them as far as possible, and this I shall endeavour to do for you. When, however, you have excluded all cases which deserve a special name and have a distinct pathology, there is still a large residuum to which the general term paralysis must be applied. Since a simple loss of power is, as you might expect, the commonest result of a morbid condition of the cord, so paraplegia is a useful term for including many cases which no doubt will one day submit to a further analysis.

In one case there may be disease of the spine which has secondarily attacked the cord; in another, inflammation of the membranes, which in like manner has involved the medulla; or in a third, the latter may be primarily affected. These, of course, have all different pathological conditions, although the resultant effect may be much the same in all. Thus also morbid growths, as cancer of the spine or tumour of the cord itself, may lead to its destruction, and during its progress one can scarcely adopt any other term for the symptoms than that of paraplegia.

When disease of the cord has commenced some loss of power ensues, which is soon evidenced by the difficulty in walking, and more especially in the ascent or descent of stairs. The inability to raise the legs is seen by the patient's stumbling up a step, or even over a stone in the street. The effort which a paralysed patient makes to move will generally distinguish the weakness of the legs from any rheumatic or other affection, and in a woman from hysterical paralysis. The feet flop down, and the power of pointing them in a given position has gone. You must distinguish between want of power and mere stumbling arising from a want of knowledge of the position of foot, as in anæsthesia. Then, again, one of the best evidences of diseased cord is shown in the paralysis of the bladder. First the want of ability to empty the organ without much straining, and subsequently its becoming fully charged and running over. The retention, in the first place, is probably due to mere loss of sensibility in the bladder, so that no stimulus is reflected back to the cord to excite its contraction as in health. It is observed mostly in cases where sensibility is lost and the urine has become alkaline in a few days, showing how much this organ is directly under nerve influence. There are some forms of disease of the cord, however, where paralysis of the bladder does not occur. Later in the disease there is a paralysis of the rectum. Other organs also sometimes suffer from the deficiency of nervous influence, as, for example, the digestive organs, and thus flatulence and sickness may become attendant symptoms. If the disease progresses, the patient becoms perfectly helpless and bedridden. In such a case death does not result directly from destruction of the cord, except the upper part be affected, when the chest loses its power of mobility, and life cannot hold out long; but death takes place from secondary causes. A bed sore may form, or as it is called " acute decubitus " which may be sufficiently extensive to exhaust the patient, or, what is more common, the bladder becomes inflamed, and the irritation there set up is propagated to the kidney, when a suppurative nephritis occurs, with a speedily fatal result. It may be observed that there are forms of disease of the motor columns without any paralysis of the bladder. In acute paraplegia, this bed sore, or " acute decubitus," is a very important symptom. After two or three days a black patch of dead skin may be seen on the back, which soon sloughs, leaving a large sore. In a lad in Stephen ward, where paraplegia had become complete in two days, a bed sore was already beginning to form, and the urine was alkaline. In a woman in Mary ward there were also large blebs or bullæ on the legs. All these conditions we now recognise as due to changes in nutrition owing to the implication of the nerve centres. These were mentioned by Bright in his Reports published about fifty years ago, and which I shall again refer to when I speak of nerves. Affections of the joints have occasionally been observed as a similar class of phenomena.

I might mention that the pupils of the eyes are often affected in spinal disease, being influenced, it is supposed, through the sympathetic nerves in the neck. Thus, we often see minutely contracted pupils, or an inequality in their size. This may or may not be associated with various amaurotic conditions which have altogether a different origin.

You know that in all spinal affections we look to the back, in order to discover if there be any disease in the vertebral column, and we generally percuss it. Now, as regards any value to be derived from this method, I think we must set it down as very small. We, of course, examine the spine, for by so doing we may discover a projection or a growth ; but as for informing us of the condition of the medulla within it, percussion seldom does that. Of course, should disease exist between any of the vertebræ, any violent jar on the back would be likely to produce discomfort ; but, as a rule, in slowly progressing disease of the cord, as in the majority of cases of para-

plegia which we meet with, there would be no pain produced. At
the same time a sensitiveness of the spine is very common, but this
generally implies a simple functional hyperæsthesia, so that I verily
believe that were you to test the value of this method of diagnosis by
the rule of averages, you would find pain mostly absent in organic
diseases of the cord, and present in those persons who suffered
merely from nervous excitability. But each case must be taken on
its merits. Thus in an adult man, who showed no evidence of a
nervous temperament, a permanent tenderness over one spot when
disease was otherwise indicated would be of immense importance in
the diagnosis. On the other hand, in the case of a girl who was
said to have a spinal affection owing to the existence of a variety of
nervous symptoms, the diagnosis of a purely functional disturbance
would be rather corroborated than contradicted by the presence of
tenderness over the spine. Moreover, the tenderness on pressure,
if indicative of local disease, is referable mostly to a change in the
bone; whereas, in the majority of cases of paraplegia, the disease
of the cord is quite independent of the vertebral column. I have
no means of referring to statistics on this subject, but I am strongly
impressed that the combination of disease of the bones and the cord
is far less common than disease of each separately. The two affections, I know, are often met with together, but yet, I think, more
commonly found apart. How often do mothers bring their children
amongst the out-patients, telling us that they have discovered their
backs growing out, when on examination we find an angular curvature, but attended with no symptoms of a nervous kind. Even if there
have been any, they have been due to implication of the nerves, and
not of the cord itself. There has been, perhaps, a fixed pain in the chest
for some weeks or months, and then the projection has shown itself.
If you look in our museum you will see angular and lateral curva-
tures to so great an extent that the spine is bent into a sigmoid
form or doubled forward, so as to become parallel with itself; and
yet the cord has followed all these contortions with impunity. You
may see other cases where the purulent and scrofulous matter lines
the canal, and is yet prevented from touching the cord by means
of its sheath. On the other hand, I believe the majority of cases of
disease of the medulla have begun in its substance or its membranes.
I leave out of the question whether these may not have resulted from
an external influence, such as a blow; but I intend merely to speak
of the direct propagation from diseased bone.

In some cases disease begins in the substance of the cord, and the
early symptoms are those of gradual loss of power of motion or sensation. In other cases the disease begins in the membranes, and in
these there may be additional convulsive movements, joined some-

times with extreme excitability of the cord or rigidity of the limbs. In some of these pain in the back may exist, showing that, whether arising from implication of the nerve roots as they pass from the cord or originating in the cord itself, uneasy sensations in the back may be present. The pain is often described as if a string were tied round the body, the seat of this being the course of the nerve whose root is in an inflamed or over-sensitive part of the cord where the healthy and diseased portions meet. This is very different from the pain induced by striking the spine. There are, then, included under the name of paraplegia a number of diseases dependent on different causes, and possessing a variety of symptoms.

A case which may be taken in illustration of one variety of paraplegia was that in which a blow on the spine caused an effusion of blood into the grey matter, and the primary symptoms were purely referable to sensation. A slight rupture of a blood-vessel in the grey matter appears to be always attended with serious results, and probably, from the disposition of the blood, to extend itself. If one side of the cord be more injured than the other, there might be paralysis of motion of one leg and anæsthesia of the other. This may sometimes be observed in the early stages of myelitis, proving the crossing of the sensory fibres; common sensation may be lost whilst thermæsthesia remains.

I may mention that a violent blow on the back, either from direct injury or shock, will produce a concussion which, resulting in a temporary paralysis, may be quickly recovered from, whilst, on the other hand, it may set up an acute softening, which shall be rapidly fatal; or, thirdly, what is far from uncommon, produce a chronic change, accompanied by a variety of symptoms which may endure for years, or be, in fact, permanent. I have more than once seen a man receive a severe injury to the back and be taken up paralysed, but in a few days he has perfectly recovered the use of his limbs, just as in concussion of the brain with loss of consciousness and rapid recovery; there was, in fact, a stunning of the cord. I have also seen such a case fatal, where the cord was found much ecchymosed with small effusions of blood within it, but without any injury to the bone.

Sanguineous apoplexy of the spinal cord.—The older writers spoke of this in order to account for the occurrence of sudden paraplegia, but the event must be excessively rare, as I have never seen a well-marked fatal case of it. I read, however, occasionally of instances where effusion of blood has been found in and around the cord, but here the possibility of injury has certainly not been excluded. It may be said, on the other hand, that an effusion into the cord not being speedily fatal would induce so great ulterior changes, that its existence, after some lapse of time, would not be appreciable, and therefore it might not be so uncommon as is supposed. As a matter of fact, however, when met with in cases of paraplegia of some standing, the probabilities would be much in favour of the clot of blood being the result of a recent effusion. The following, for example, is a case where blood was found in the substance of the cord, but it was difficult to say positively whether this effusion was the primary affection, or whether it was not an accident of the softening. In the second case the hæmorrhage in the membranes was clearly secondary.

CASE.—Charles U—, æt. 31, was admitted under my care into Stephen ward on Jan. 10th. Two weeks before admission he felt a twitching in the right leg from the hip to the knee ; very shortly afterwards a pain in the loins, and he began to limp. On the following day he could not rise from his bed. The left leg then became numb, and he had difficulty in passing his urine. When admitted he had complete paraplegia of motion and of sensation below the knee ; he could, however, tell the difference between heat and cold. He gradually improved until April 20th, when he could move his legs. He was then taken with shivering, vomiting, and a large sore came on his heel; all the paralytic symptoms were aggravated with cystitis, and he died May 3rd.

The cord was healthy, except at the lumbar enlargement, where a brown patch was observable on its surface; this occupied the mesian line and the substance of the cord just behind the grey commissure, and extended on either side of the posterior median fissure. This appearance of cord, beginning at upper part of lumbar portion, increased until the middle of the enlargement was reached, when a clot of blood was met with occupying especially its right side. It was not contained in any cyst wall, and appeared recent. It extended some distance down along the posterior fissure, and then, as a brown linear discoloration, quite to the filum terminale.

CASE.—A man was brought into my ward, under the care of one of my colleagues, whilst I was away, with paraplegia of two years' standing. Soon after admission he became rapidly worse and died.

On opening the spinal cord the membranes were found to be tensely bulging with coagulated blood, which was effused beneath the visceral arachnoid and quite surrounded it. It commenced about twelve inches from the upper part, and was thickest below, as if from gravitation. The clot was firm but recent. At the middle of the dorsal region the membranes were thickened and adherent; the cord was softened at this spot, and a few hæmorrhages were seen in its substance and along its fissures.

Acute myelitis.—If by this be understood a general softening of the whole cord or a large tract of it, the disease must be considered rare as an idiopathic affection. A portion of the cord may be acutely inflamed as a result of injury, or from other causes which involve the organ from without ; if so, it would present the same appearance as of inflammatory softening of the brain, being diffluent, and showing broken-up tissue and granule masses. Instead of the knife making a clean cut through the medulla, it would tear it and break up its substance. The grey matter might have lost its definition, or be altered in colour, being darker or containing small extravasations of blood ; or portions of it only might show changes, as the centre or the cornua. The general symptoms of a local myelitis would be slight pyrexia, disturbances of motion and sensation, a cord-like feeling round the body, alkalinity of urine, sore on the sacrum, and if the disease were in the upper part of the cord it would be rapidly fatal through the chest.

Chronic myelitis may show itself as a slow disintegration and softening, or as an induration or grey degeneration of special parts. Under this name is sclerosis or a hyperplasia of the connective tissue. This may affect the cord in parts, and the inflammatory product be found scattered through the cord ; or involve distinct districts, as the posterior columns, or, as in a remarkable example in our museum, the central grey matter, with a portion of the adjacent medullary substance. In some chronic degenerations we find " corpora amylacea," and in some cases the membranes and roots of the nerves, involved in the chronic inflammatory process. We shall speak of the affections of special portions and regions of the cord separately, as each is attended with its own symptoms, and is therefore clinically distinct.

The difficulty is great, however, in associating symptoms with distinct lesions, because there is no opportunity of comparing them by an actual examination of the cord at the time they occur, and the changes in the latter are so extensive at the time of death that they signify but little. There can be no doubt, however, that the parts which are affected in the early stages of disease may be looked upon as identical with those which, when injured by experiment, produce analogous phenomena, as, for example, in crossed states of sensation. Thus, one man lately was under my care with anæthesia of the right leg and hyperæsthesia of the left arm and leg, and another with numbness of the left arm and leg, with hyperæsthesia of the right half of the body ; in such cases we are justified in supposing that the cord was affected in exactly those regions which, if injured, would have been productive of similar symptoms.

Spinal meningitis.—As of the brain, so of the cord, an inflammation of the membranes often implies an inflammation of the surface of the cord itself, and of the nerves proceeding from it, the exudation being found beneath the visceral arachnoid. Owing, however, to the somewhat different anatomical arrangement of the two organs, there may exist a simple inflammation of the spinal membranes, in which the arachnoid surfaces are especially involved, and the exudation be found in the interarachnoid space. There may be recent lymph or pus, or the surfaces may be adherent, or the membranes greatly thickened and closely involving the medulla within; even sometimes they may be ossified.

The symptoms connected with meningitis vary with the amount of implication of the cord and the nerves, and of the irritation which may be set up in the cord itself. They have reference mainly to altered sensations, and there may be therefore hyperæsthesia, anæsthesia, or pain; the latter is often increased by movement. Then again, either from direct implication of the motor nerves, or from irritation of the cord through the sensory nerves, there is very often spasm of the muscles. Consequently, in acute meningitis, we may have all the symptoms of acute tetanus, and in chronic meningitis of chronic tetanus.

Acute meningitis arises generally from a direct blow or injury, although sometimes from exposure to cold. The symptoms, as before said, vary according to the number of the special parts implicated, and thus it is possible for an acute inflammation to attack the arachnoid surfaces only, and be attended by no marked phenomena. I have several times quite unexpectedly found the surface of the cord covered with lymph, where the inflammation has extended from the cranium; or, in other cases, where the spinal canal has been laid open by a bed sore. In some instances, where the visceral arachnoid is more involved, there has been pain in the back, aggravated by movement, and accompanied by spasm or stiffness of the muscles of the body and limbs, pains around the body and in the legs, and probably hyperæsthesia.

These facts show that spasmodic or tetanic symptoms may be present or not. Their presence may be due to the implication of the nerves in the effusion.

CASE.—A lad, æt. 15, was admitted under Mr. Birkett. He was playing with another lad three days before, when he received a blow on the back from his fist. He thought little of it at the time, but the pain becoming severe he applied to the hospital. After the application of leeches the pain was so much relieved that he thought of going out, but the pain returned more severely, and fever ensued. An abscess formed on the right side of the sacrum, which was opened. He continued daily to get worse, with much irritative fever and severe pain in the back. During the week preceding his death he was exceedingly restless, and often delirious, and complained of pain in all parts of the body, but particularly in the extremities. His head was drawn back, as in tetanic opisthotonos. He had no symptoms of paraplegia, and could move freely in bed. He died twenty-two days from the receipt of the injury. On *post-mortem*, an abscess was found extending over the sacrum and ilia, and penetrating the vertebral canal. The dura mater was softened at one spot, and the cauda equina was lying bathed in the pus which filled the sacral canal. The membranes of the cord were inflamed throughout, and the purulent effusion extended as high as the dorsal region. There was also exudation beneath the visceral arachnoid; the cord itself was healthy. The inflammation had reached the cranium, and there were traces of arachnitis over the whole surface of the brain.

CASE.—J. F—, æt. 12. Two weeks ago began to feel some weakness in his legs which made him stoop when he walked. He was a quick, intelligent boy, and, as a kind of amusement, he made himself some crutches, saying he must take to them. At the end of six days he was unable to walk, and had to keep his bed, and a medical man was sent for. He was then lying on his side, with his legs drawn up and complaining of pain in them. All these symptoms increased until I saw him on October 10th. He was then lying on his left side with his legs drawn up, his knees to the abdomen and heels to buttocks. On touching his legs or on attempting to move them, he screamed out with pain. The muscles were hard, as if spasmodically contracted, and the tendons were rigid. The skin also was extremely sensitive, the hyperæsthesia extending all over the limbs and reaching as high as the umbilicus. Towards the back and over the lumbar region the sensitiveness was extreme. His condition was one of spasmodic contraction of the legs, accompanied by pains. He had hyperalgesia and hyperæsthesia.

On inquiry about any injury, the lad stated that on the 24th of August he was playing with another boy, who was carrying him, when he fell off on to his back, striking himself on the kerb-stone. He felt great pain for a moment, but this passed off, and he thought no more of it. He was ordered liq. hyd. perchlorid. and potass. iodid., with ice to the spine. He remained in much the same state as above described for a few days, with the exception of having regular attacks of opisthotonos, and then gradually grew better, and on November 4th the severe symptoms had mitigated. After this the recovery was slow. On his visit to me on December 29th he was walking about, moving his legs well, but was still tender in the back, and seemed always in dread lest any one should touch him in this region. On January 29th he again came to see me, not being so well. He had almost recovered, so that he ran about and danced as before, but in the last few day his breathing had been oppressed, and he experienced darting pains in his abdomen, like knives passing into him. He had become quite changed in character—from being courageous was fearful and timid, and said he knew he should die. Over lumbar region pain was felt on pressure, and all lower part of body was hyperæsthetic. No paralysis of motion. He improved during the following month, when he was suddenly seized with symptoms like tetanus, and died. It was some time after his death that I heard of the occurrence, and that no *post-mortem* examination had been made.

Chronic meningitis.—This may occur idiopathically, or as the result of injury, either directly or in the form of shock, as in railway collisions; or it may follow disease of the vertebræ. The symptoms are mainly due to implication of the nerves and irritation of the spinal marrow. They consist of a cord-like feeling around the abdomen, aching pains in the back and legs, sometimes of a darting, lancinating character. There are spasms of the muscles, with a tendency to permanent contraction. The spasms and pain may be sufficient to confine the patient to bed, and yet the cord itself be healthy. This is shown by its special function remaining; in fact, from the irritation of the nerves the cord seems in over action; and you will observe that the most painful symptoms which the patient experiences are often due to its extreme excito-mobility. Actions, such as sneezing, yawning, or micturition, cause reflex movements of the whole body, or are sufficient to throw it into convulsions. After a time, the thickened membranes and the fluid effused beneath them may encroach on the cord, and symptoms of paraplegia ensue. In cases of chronic meningitis the progress is so slow, and the symptoms so obscure, that the case is often regarded as rheumatic. The following case was recorded by Sir W. Gull some years ago, and the specimen of thickened membranes constitutes one of the most remarkable in the museum. It will be seen that the patient's most severe symptoms were those due to the excito-motility of the cord, there being no pain nor tenderness in the back, and no hyperæsthesia.

CASE.—The patient was a man of middle age; he had led a life of excess, had had a fall on the back, and was in the habit of taking much exercise. His symptoms commenced with headache, dimness of sight, pains in the neck; he was better for a time, but about a year after found he had difficulty in stepping out; then pains in the arms came on, numbness in the legs, with increasing weakness, also very troublesome jumpings in them. As he lay in bed he could move his legs, but could not stand upon them. Had a sense of constriction around his lower ribs. The *excito-motor* actions were produced by the slightest touch, or by mere shaking of the bed, and he had constant startings when the urine dribbled through the urethra. The legs were more or less permanently flexed. He said his back was quite strong, and if struck with a hammer it would not hurt him. The urine became more ammoniacal, and the rigidity and flexion of legs increased; rest being quite prevented by continued spasms of lower extremities, which on one occasion were so violent as to jerk him off the bed on to the floor. He remained in much the same state a long time, the heels drawn up to the nates and the knees to the abdomen. Sensation was diminished, but moving the limbs gave him great suffering, and increased the contraction. On *post-mortem* examination the membranes were found excessively thickened, adherent to one another and to the cord, the posterior layer of dura mater in the dorsal region having become ossified; the cervical and lumbar portions of cord less affected. The surface of the cord had undergone slight degeneration.

CASE.—The following case I lately saw with Mr Manser, in the Tunbridge Wells Infirmary. A young man, æt. 21, was taken in three weeks before, for stiffness of his back and neck. The history was that a year before he had suppuration in the right hand and forearm, for which one finger was amputated. After the abscesses healed the arm never recovered itself; the skin became hard and brawny; subsequently ulcers appeared, and small fibroid lumps. Excepting this disease of the arm, he was a healthy man. A few days before admission the stiffness of the back came on, which gradually increased, and at the same time he got rapidly weaker and thinner.

The stiffness of neck and back increased until the time when I saw him, when he was in the following remarkable condition. He was lying on his side, with his head arched strongly backwards, and his throat thrown as prominently forward; this was from the strong spasm of the erector spinæ. The other muscles, which are usually affected in tetanus, were supple, as rectus abdominis and sterno-mastoid.

The chest moved well, and diaphragm unaffected. There was much difficulty in swallowing, from the arching of the neck. He was lifted out of bed on to his legs. He rested his weight on them, but his body was curved back like an arch. There was much wasting of the muscles. No paralysis of motion or of sensation, so that the opisthotonos constituted the only nervous symptom. The continuous galvanic current had no effect on the spasmodically contracted muscles, nor did faradization, although this caused pain. It was clear that he was suffering from a spinal irritation, or excitation of the spinal nerves. A distinguished physician had seen him, and suggested amputation of the affected arm, but it was clear that the symptoms were not of the ordinary tetanic kind, arising from an eccentric cause. During the three weeks he was under the care of Mr Manser he had every kind of treatment which could be devised. Under chloroform the spasms ceased. He died four days afterwards of inanition. The *post-mortem* examination showed much old thickening of membranes (meningitis) of upper part of cord in dorsal region.

The case is one of great interest, clinically and pathologically, the symptoms being clearly due to irritation of the spinal nerves from the meningitis, and the possibility of this being the result or a continuation of a neuritis of the arm. Although the nerves of the arm were not examined, the ulcers and brawny skin suggested a chronic inflammation of these structures.

It will be seen that in Sir W. Gull's case the cord itself was slightly involved on the surface; but it is clear, from the pain and increased excitability of the cord, that it was but little affected throughout his illness, thus proving that the disease was one mainly of meningitis. In many cases, however, the cord may be involved in connection with the membranes, and consequently a difficulty arises as to which portion of the structures to appropriate the symptoms. I will relate one or two more cases where the symptoms were mainly meningeal.

CASE.—The wife of a medical man got wet through on a steam-boat, and then rode home in an open carriage. She soon began to feel ill, with flying pains about her; these increased, and at the same time weakness came on, so that she was scarcely able to walk or move her arms. To this optic neuritis ensued, which destroyed for a time the sight of the right eye. The symptoms continued for many months, when she gradually recovered.

In the following case the progress of the disease could be followed from its commencement in the spine.

CASE.—Wm. B—, æt. 10, admitted under Dr Wilks, May 15, 1872. He stated that ten months ago he received a blow between the scapulæ, which hurt him very much, and required the advice of a doctor. He stayed at home for a week, being unable to walk on account of the pain in the back. He afterwards again went to school for eight months, but never lost the pain. Six weeks ago he began to lose power in his legs, and the weakness had gradually increased up to the present time. On admission he was seen to be a well-grown boy, and did not look ill; he could not walk without assistance, being only just able to stand alone. His only symptom was partial loss of power. Sensation unaffected. On the supposition that some inflammatory process might have been set up by the injury, he was ordered the Liq. Hyd. Perchlorid. ʒj, and Potass. Iodid. gr. iv, three times a day.

After four weeks he thought his legs were a little stronger, that he could move them better, and he was able to stand. He continued his medicine, but at the end of another month it was evident that there was no real improvement, for he was beginning to feel pain in his legs, at the same time they were less sensitive, and the muscles were beginning to be rigid. There was also some weakness of the bladder.

On July 9th he was galvanised. When a continuous current of thirty cells was applied to the spine, the arms were moved outwards, and the legs drawn up in jerks. Faradization to spine had no effect, and when applied direct to the legs the muscles did not respond.

At the beginning of September the spasmodic contraction was increasing, so that he had a sand-bag placed across his thighs to prevent them being drawn up, continuous current being used daily to the spine. He was ordered gr. ¼ of Ext. Physostigmatis three times a day, and after a week it was increased to gr. ½. By mistake he took a double dose, 1 gr. He soon became very ill; and when the house physician was called to him an hour afterwards he found him with a clear froth coming from his mouth, perspiring profusely, his face turning blue, hands cold, numb, and almost powerless. Pupils of natural size. Pulse 130. Quite sensible. He had an emetic powder, followed by warm water, and three hours after taking the pill he had quite recovered.

The above-named paralytic symptoms increased, and at the beginning of October the legs were quite rigid, so that by lifting the heel the whole body could be raised; the knees were bent with great difficulty, but, if so, the legs suddenly contracted or flew up to the body. Sensation had also become much impaired as high as the sixth rib. Almost complete loss of power of bladder. Ordered Succ. Conii, to see if it had any influence over the spasm, and he took it a few days, but with no result. It was observed that his abdomen was flaccid immediately after raising his clothes, but upon feeling it the recti became quite rigid. The same fact had been observed before in the legs, that they became much more rigid after being touched.

On November 1st limbs again tested by galvanism and faradization. The muscles responded to both, but their susceptibility was impaired, being much less than that of the arms.

The case appeared like one commencing externally, first involving the membranes and nerves, and then proceeding to the medulla.

CASE.—James M—, æt. 33, a sailor, had genital sore, but it is very questionable whether he has had constitutional syphillis. Six months ago, whilst in the Mauritius, he began to feel weak in his legs, and to walk as if he were tipsy. The weakness gradually grew worse, and the limbs were beginning to contract, when the doctor applied a red-hot iron three times on each side of the spine. No good result followed, and the contraction gradually went on, especially in the right leg. Whilst on shipboard he used to keep the leg forcibly down by a weight.

Admitted to hospital on Oct. 21. When placed in a chair he sat with his knees drawn up to his chin, his heels next to his buttocks; the legs quite rigid, so that by attempting to extend one of them his whole body would be lifted up. The right leg, which was first affected, was more rigid than the left. The knees came together, although one was a little lower than the other. No pain over the spine. No difficulty with bladder. The muscles hard, extensors somewhat wasted, tendons rigid. He was placed under chloroform, when the left leg was readily extended, but the right could not be moved from its position. Tested with galvanism. On applying the continuous current to the extensors of the left thigh, and making and breaking contact, the muscles responded, but this was more marked when faradization was used, the limb then becoming nearly straight. The muscles of the right leg acted in the same manner, but with no tendency to straighten the limb. Sensation unimpaired. Unrelieved.

CASE.—Robert K—, æt. 27, admitted October 22nd. A sailor, and has passed a great part of his life abroad. He states that he had a venereal sore about seven years ago, but had no secondary symptoms. About three years ago he was attacked with violent pains in the head and neck, the latter becoming stiff, which prevented him moving it in the least. He remained in this state for a month or six weeks, when the stiffness left his neck and extended across his shoulders and down his arms, so that he was unable to raise them or to flex his fingers. He thinks some of his finger-joints were swollen at the time. Subsequently his legs became weaker and weaker, although he was able to walk about, and he thinks his knees swelled. After some time the pains appeared to descend to his loins, leaving the upper part of the body, and at the same time as his arms improved his legs became worse. He was unable at last to walk, and forced to keep the recumbent position. His legs then began to contract and be drawn up towards his body. He began also to lose control over his rectum and bladder. For more than a year his legs have been drawn up towards his body.

On admission he presents the appearance of having been a well-made and very powerful man, his chest well developed, and viscera healthy. His legs are tightly flexed on his thighs, and his thighs on the pelvis. A considerable force is required to straighten the legs and keep them stretched out, and immediately you loose hold of them they fly up, as would a spring, into their former position, the heel striking the buttock with a sensible slap. Some little pain is experienced when practising this manœuvre. The joints are quite flexible, the resistance due to spasm of muscles. The legs are smaller than natural, but there is no active wasting of the muscles. Sensation is considerably impaired; he can feel, but cannot define very well the spot touched. The temperature is normal. The reflex action is very well marked, the legs, after being stretched out, being suddenly drawn up on pricking the feet. He thinks the arms are not much affected, not being more feeble than his long illness would necessitate. There is a swelling of the right ulna. He was ordered Iodide of Potass. in the Mist. Hyd. Perchlorid. He had been galvanized before he came in.

On November 12th he is better. Sensation more perfect. He is able, after forcing his limbs straight by pressure on the knees, to keep them straight for a short time, but any little excitement, and especially any one touching the bed, will cause them to fly up again. He was subsequently galvanized, and was ordered tonics. He again, however, returned to the mercurial mixture, and was so much better as to be able to sit up in a chair, but could never walk.

I have now a man in the hospital who has had syphilis, and three months ago began to have pains in his legs, and then some loss of sensation. He then found them getting stiff and beginning to contract, so that he could no longer walk or move. He is now in bed, with his legs drawn up; when pulled down they remain straight

for a short time, and then contract again. The excito-motor function is increased, so that the legs jump, and he is sometimes almost
thrown out of bed. I regard this case as one of meningitis, and it
may have a specific origin.

Another patient of mine has rigid legs, but they are extended out
straight; if one be lifted up higher than the other, and then let go,
it will fly across its fellow like a spring. We one day got him up
and sat him in a chair, but the legs remained straight out as
before, and when we raised him on his feet he was obliged to be
supported, for his legs were of no more use to him than those of a
stiffened corpse.

In these cases of paraplegia, where a slow contraction comes on,
there is probably a chronic meningitis of the lower part of the cord,
to which probably has succeeded a sclerosis of the antero-lateral
columns. Thus, a woman was lately in the hospital suffering from
pains in the legs, which afterwards became powerless, and then contracted. She lay in bed with the legs drawn up and closely flexed
towards the body, the heels touching the buttocks and with sometimes spasmodic jerking in them. Sensation was not impaired,
and there was no paralysis of the rectum or bladder.

In cases of hemiplegia with rigidity the pathology is more difficult to understand, because a disease of one side of the cord only,
through a considerable part of its length, must be regarded as a very
unlikely occurrence. We might, therefore, surmise that in some
cases of the kind there might be only an affection of the nerves to
account for the paralysis. For example, a woman, æt. 46, had found
that during nine months her left arm and leg had been getting stiff
and powerless, so that when admitted to the hospital she was obliged
to keep her bed. The arm was flexed and the fingers clenched; the
muscles also were wasted and rigid, so that the limb could not be
straightened. The leg, in like manner, was flexed and contracted.
Sensation was perfect.

Meningitis confined to the cervical region, forming a growth.—
In the case which was in Tunbridge Wells Infirmary, it was suggested
that the meningitis might have been due to an extension of inflammation from the nerves of the arm, and this supposition is somewhat strengthened by the circumstances of a local meningitis being
so often confined to the cervical region. So limited and circumscribed is it that it is sometimes regarded as a growth rather
than an inflammatory deposit. These cases of local affections
of the meninges present great difficulties in diagnosis during a
considerable part of their progress, since the symptoms are mainly
of a neuralgic character only. Thus, in the case of a young man
sent to Sir W. Gull, there had been long-continued pain in the right

shoulder and arm, which was called rheumatism. Afterwards there was weakness of the lower extremities, with cramps and spasmodic twitchings. These pains were the main symptoms, and finally became very distressing.

CASE.—The last case of the kind under my care occurred about a year ago. The patient, a female, came to the hospital with pains in her limbs and joints, from which she recovered after a short time. She attributed her symptoms to cold. Subsequently, she said she had a fall some years before, but felt no ill consequences from it. She had not been out of the hospital many weeks when she was again admitted. She said she had had a difficulty in walking, owing to a pain in the back, and this was much increased on exertion. There was tenderness on pressure all down the lumbar and sacral region. It pained her to bend her back, and she said she could not lie upon it. During the following two months she improved somewhat in condition. She was able to sit up, but still walked with pain and difficulty, the pain passing up to the head. During this time the diagnosis as to the character of her ailment was very doubtful. She then became worse and took to her bed ; she looked ill, complained of great pain, was delirious at night, and said she saw double. She finally got into a listless state and threw her arms about, and occasionally was very drowsy. Finally she sank into a complete lethargy, and never moved ; she died three months after admission. The *post-mortem* showed recent lymph at base of brain, and in cervical region a soft inflammatory growth limited to this part.

Sir W. Gull gives a very interesting case of inflammation of the cervical region of the cord, and where the arms were principally affected.

A woman, æt. 33, had felt weak for some time, but one night went to bed as usual, when on the following morning she found her joints painful, and was unable to move her arms. She recovered under treatment, and returned to her duties. At the end of a fortnight the muscles of the arms were becoming rigid, and she soon was quite unable to move the limbs. For four months the arms were quite useless, owing to the rigidity, but she still walked about ; she then began to lose power in her legs also. When admitted to the hospital she could move the legs slightly. The arms were extended and rigid ; she had a feeling of suffocation or constriction about the throat, had not much breath for speaking, and could not cough nor sneeze. On evacuating the bladder or rectum the whole body and extremities became extended and rigid ; she had also sudden spasmodic extension of the limbs. There was no pain in head or neck, as at first, but the limbs and joints were painful, according to the position in which they were placed. Left pupil smaller than right, and vision imperfect. A few days before death the power over the sphincters became lost, and the pain in the head most severe. The pulse was 48.

Post-mortem.—The membranes of cord thickened and completely adherent together about the origin of the third cervical nerves. Above this, the adhesion implicated the origin of the second and first cervical, and on the right side also some of the lower fibres of the origin of the pneumogastric and lingual. The roots of the whole of the cervical nerves and of the spinal accessory were matted together by old thickening. The cavity of the arachnoid was obliterated throughout the whole of the cervical region in front, and to a less extent behind. Few spots of softening in cord.

CASE.—I had a very remarkable case of cervical meningitis with Mr Lorimore, of Farnham, in which recovery took place. The patient, a middle-aged man, came to me in the summer of 1872, stating that for some months he had suffered from great pain between the shoulders and neck, extending down the arms. The pains were clearly nervous, and due to some irritation of the roots of the nerves of the lower cervical region. No disease of the bones could be discovered, nor any growths in their neighbourhood. No medicine relieved him, but the pains became agonising, and he was obliged to take to his bed. The right arm then began to waste, and the muscles to contract; at the same time the left arm became painful and weak. He also found his leg becoming weak, so that he could scarcely stand. The spine was constantly examined for evidence of abscess or growth, but none was discoverable; in the mean time he lay in bed, a great sufferer, almost paralysed, his right arm being most affected. The hypodermic injections of morphia were constantly practised, in order to give him relief. When it was attempted to move him, or draw his legs round over the edge of the bed, they would become spasmodically contracted, and at all times the reflex action was so marked that a slight touch would convulse the whole body. He afterwards lost all power over the legs. The paralysis was complete, and the bladder had to be relieved by the catheter. After lying in this apparently hopeless state for some months he began slowly to mend, the power of his limbs gradually returned, and now, after four years' illness, he is walking about as well as ever, and conducting his business; the arm is contracted and withered, but he can use it.

Meningitis of the cord, resulting from an extension of inflammation from a nerve.—Since the close intimacy has been shown between the nerve centre and the nerves themselves, in all that relates to their nutrition and their liability to contagious inflammatory processes, an explanation has been afforded of the meaning of various phenomena which were before obscure. For example, the condition of a paralysed limb as regards contraction, pain, &c., was attributed solely to the nature of the central lesion, whereas it is now conjectured that it may be partly due to subsequent changes in the nerve trunks themselves; and, on the other hand, there is every reason to believe that an affection of a nerve may propagate itself backwards until it reaches the cord, and so give rise to a fresh set of symptoms. One theory of progressive muscular atrophy is of this nature, for we occasionally see cases of local injury to a hand or foot, where an atrophic process, gradually progressing upwards, seems to show that an ascending disease of the nerves is taking place. This may even reach the cord, as in the cases Vulpian describes, where not only the nerves supplying the stump of an amputated limb are wasted, but the portion of cord to which they are attached. The cases, however, to which I wish more particularly to draw your attention are those where the character of the disease, which had propagated itself from periphery to centre, was of an inflammatory kind; where, indeed, a neuritis had ended in a meningitis. Under the head of epilepsy I shall mention a case where fits followed a severe injury to the arm, and where this mode of its production was the one which I suggested; and this, I should say, is not the only example of the kind I have seen. In the case just described, of the young man in Tunbridge Wells Infirmary, the facts were highly suggestive of the meningitis in the lower cervical region having resulted from a propagation of inflammation of the neurilemma of the brachial nerves along the cervical plexus. In the 'Guy's Hospital Reports' I related the case of a lad under the care of Dr Barlow, who had received a severe injury to the arm, which rendered it useless, and who subsequently had paralysis of the other arm, and then of the leg, until the whole body was paralysed. After death there was found a chronic inflammation of the spinal cord.

CASE.—A gentleman, 55 years of age, has been obliged to retire from his profession on account of paralysis of the right arm and leg. He drags his leg when he walks, and his arm shakes so that he can only write with difficulty. There is no paralysis of the face, and the intellect is normal. The case is regarded as one of hemiplegia by his friends, but the history is altogether peculiar. About twelve years ago he struck his right arm; an abscess formed above the elbow, which was opened, and found to reach the bone. As it healed, the soft parts became matted together, which prevented him flexing the limb well; it became at the same time weak and shaky, so that he never could use it as before. Between two and three years afterwards his leg became weaker on the same side, and at length he walked with considerable difficulty. He now drags his leg, and has little power in the arm. The face is quite unaffected.

Increased excito-mobility without permanent spasm.—In the cases already mentioned an increased excitability of the grey centres has apparently existed. I say apparently, since the cause of the exaggerated excito-motor phenomena might have been owing rather to the presence of some unnatural stimulus rather than to any change in the cord itself. This would seem to be the case where the spasm and rigidity are only paroxysmal, and induced by some outward form of stimulus. In many cases the spasm is sufficient to prevent movement, although there is no paralysis in the ordinary sense of the term. The principal circumstance observed is the extreme excito-mobility brought into play when the surface of the body is touched. In one case of the kind the spasms of the muscles were not only most painful, but what was remarkable, a spasm of the blood-vessels occurred simultaneously, judging from the pallor and coldness of the skin which always accompanied the attack.

The following is the case of a man now in the hospital:

CASE.—Peter C—, æt. 40, was taken, four months ago, with pains in his limb, and back, followed, after another two months, by numbness and stiffness, which almost prevented him walking. Whilst lying in bed the man appears well and sound; he says he can feel well, but when his feet are tested by heat and cold he hesitates as to which one is touched, and electro-mobility seems somewhat impaired. The muscles are firm and natural, and he can throw his legs about in any position. As soon as he gets up, however, and places his feet on the ground, all the muscles of the limbs become spasmodically contracted, remarkably hard, and he is fixed to the ground. It is only by the greatest effort he can move, and slowly walk. He is being galvanized with some good effect.

The following is a case, not so simple as the one I have related, but it illustrates a morbid state of the cord when thrown into action by an effort of the will, or through some reflex influence of its nerves.

CASE.—James D—, æt. 53, admitted May 3rd, 1871, employed as a dredger on the river Thames, and consequently always exposed to the weather. A year ago he first began to experience cramps in the lower extremities, and these have gradually increased until the present time, so that he has a great difficulty in straightening his legs after having been in a sitting posture. The cramps are accompanied by much pain. He has also suffered from spasm of the abdominal muscles.

He is a healthy and powerful-looking man. All his organs sound. He can walk steadily and for some distance, apparently having nothing the matter with him, but after sitting in a chair for some time, if an attempt is made to move, the most violent pains and spasms come on. He consequently prefers to lie in bed, for he then escapes these painful symptoms. Ordered Mist. Hydrarg. Perchlor. with Potass. Iodid. At the end of a fortnight he thought he was better, and therefore was ordered to get up and have the continuous galvanic current applied to the spine. After a few applications the patient expressed himself as feeling much better; his legs were more supple, he did not have so much pain, and the contractions were not so frequent. For about an hour after the galvanism had been applied he said he felt as well as ever, and his legs moved more easily. He continued improving, having no pains, except when walking or moving the limbs, until June 12th, when he complained that all his old symptoms had returned, he had shooting pains all down his legs, even when sitting still, and they awoke him at night. When walking he had more pain. Subsequently he had inflammation of the eye and sickness. When better of this he was put on Succ. Conii ʒj ter die. He improved slightly, and then relapsed. He afterwards had strychnia, and on July 14th he left, being scarcely any better than when admitted.

I have also had two cases of patients who, immediately they attempted to get out of bed and stand, were seized with violent tremors. Sir J. Paget relates the case of a gentleman who, immediately he placed his feet on the ground, was seized with numbness and coldness, so that he could not walk.

Paraplegia with Tremor

CASE.—Robert R—, æt. 25, a rope twiner. For twelve months his legs had been getting weak, and for the last three months he had been unable to work, and for two months had kept his bed. He was a well-developed man, and looked in good health, his legs firm, and muscles not at all wasted. He could scarcely move them, and was unable to rest his weight upon them. The most remarkable circumstance was the constant fibrillar tremor passing from one muscle to another. This seemed increased when any effort of the will was directed upon them. He was ordered galvanism, and an improvement at once commenced; he was soon able to stand, and then to walk, so that he left the hospital cured at the end of three months.

In this case so few objective symptoms existed, that if it had not been for the tremor I should have regarded it as one of malingering. I suppose it was a real temporary akinesia.

Cerebro-spinal meningitis.—Occasionally cases occur which are rapidly fatal, and where the appearances found after death show no other disease in the body than an acute inflammation of the membranes of the spinal cord and base of the brain. I shall presently tell you that there is an epidemic disease in which these appearances are found, and therefore it may be that the cases we now and then meet with are examples of this specific disease. Arising from accidental causes, a cerebro-spinal meningitis is not uncommon, as from caries or injury to the skull or spinal column. An injury, for example, to the head may affect the brain, and so set up an inflammation of the surface, or of the ventricles, and so pass down to the fourth ventricle, and even to the spinal cord. On the other hand, injuries and diseases of the spine may set up a spinal meningitis, which shall run upwards as far as the base of the brain. We occasionally, however, meet with cases of inflammation which are apparently idiopathic, as the following.

Cerebro-spinal Meningitis

CASE.—A young man, after one day getting cold, began to feel unwell. On the following day he was in a high state of fever, very restless, and short-breathed. In the evening he was delirious, and on the following day it was evident that he was suffering from inflammation of the brain; he had become unconscious, his eyes were fixed, and he was paralysed on the left side, whilst the right was convulsed. On the next morning, the third from the seizure, he died. The *post-mortem* showed the surface of the brain to be slightly greasy; some little exudation was observed at the sides, and on removing the organ a large quantity of lymph was seen covering the base and proceeding down the spine. The ventricles contained turbid lymph, and the walls were soft. On opening the spine and sheath the whole length of the cord was seen to be covered with lymph; the subarachnoid space was completely filled from its upper end down to the cauda equina. The cord itself was healthy. There was no disease of the bones, and no tubercle. Kidneys and other organs healthy.

CASE.—A young man, æt. 23, was admitted to the hospital, and died twelve hours after admission. No good history was attainable, but it was said that he had been ill only a very short time. The body was that of a strong, muscular man; no sign of injury. There was a recent meningitis, shown by green lymph covering the pons and neighbouring parts at base of the brain, and this inflammation extended downwards along the cord, where there was a slight effusion. All the other organs were healthy.

CASE.—A child, æt. 3, was lately under my care in the clinical ward. She was taken with shivering, vomiting, headache, and convulsions. Pupils dilated. The base of the brain was covered with lymph, and this extended down the cord as far as the lumbar region. There was no tubercle anywhere discoverable.

Tubercular cerebro-spinal meningitis.—I have already said that this is more common than was formerly supposed, for the sufficient reason that our post-mortem examinations are more perfect than they once were. We had quite lately, occurring on two successive days, the case of a young woman who died with all the usual symptoms of meningitis, where we found the inflammation extending down the cord, and the membranes granular ; and the case of a child, which also showed a similar complete tubercular meningitis throughout the brain and cord.

Epidemic cerebro-spinal meningitis, or cerebro-spinal fever.— The definition of this disease, as given in the nomenclature of the College of Physicians, is as follows :—"A malignant epidemic fever, attended by painful contraction of the muscles of the neck and retraction of the head. In certain epidemics it is frequently accompanied by a profuse purpuric eruption, and occasionally by secondary effusions into certain joints. Lesions of the brain and spinal cord and their membranes are found on dissection." Our first knowledge of this disease in Great Britain was about the year 1846, when it occurred in the neighbourhood of Dublin, and mostly amongst young boys of the poorer classes. It had appeared, however, as an epidemic in France three or four years before, and had received the attention of the Academy, being an altogether novel and undescribed disease. The patients were seized suddenly with vomiting and prostration, followed by reaction, when the characteristic symptoms began ; the muscles of the neck became rigid, and the head thrown back, as in tetanus ; there was also much fever. The average course of the disease was three or four days. The *post-mortems* showed exudation of lymph at the base of the brain, and along the whole length of the spinal cord.

Little was then heard of the disease until 1864, when it broke out in America, and received much careful attention in that country. In many of the cases an eruption occurred on the body, which suggested its resemblance to typhus. In the following year it created great alarm in Russia and the north of Europe. In our own country it has never existed as an epidemic, but the cases must be considered as partaking of the epidemic nature when several occur together in the same locality.

It must be regarded as a specific blood disease, in which the characteristic symptoms are of a nervous kind. The poisoned blood is shown by the rash, the petechiæ, the tendency to hæmorrhage ; the affection of the nervous system by the effusion of lymph beneath the arachnoid of the brain and spinal cord. In well-marked cases the symptoms are described as coming on with slight febrile disturbance, lassitude, and the other phenomena of pyrexia; these are succeeded by headache, vomiting, pains in all the limbs, and cramps. As the case becomes developed the nerve symptoms increase; the headache becomes more intense, and is attended by delirium. The pains in the muscles are very severe, more especially in those of the spine and at the back of the neck. The head is thrown backwards, owing to the muscular contraction, and this constitutes the most characteristic feature of the complaint. The other muscles, as of the face and limbs, are less affected, and some-times there is constant twitching of the limbs. Herpes, or a vesi-cular rash, is often noticed on the face, and a rose-coloured rash on the body. There may be no disturbance of the bowels or other organs, except anorexia. The patient eventually falls into a coma-tose state, with dilated pupils, and so dies. In some epidemics the malignant variety prevails, and then the patients have petechiæ all over the body, and hæmorrhages from various surfaces; these cases are soon fatal. Occasionally, an inflammation of the eye or ear has been noticed, and in some instances affection of the joints. The rash, if present, would enable us to distinguish the complaint from a simple meningitis, but the most characteristic symptoms are the intense headache and pain at the back of the neck, with retraction of the head. Pains, however, do occur in all the muscles of the body, especially in those of the abdomen.

The post-mortem examination shows more or less congestion of the brain and cord, with effusion into the subarachnoid space. This may be serous or purulent, and mixed with flakes of lymph. The brain and spinal cord exhibit no marked changes in their inte-rior, but it is by no means improbable that they are affected, and that the effusion is merely the outward manifestation of a more general inflammation of the nerve centres. The disease is question-ably contagious.

Isolated cases are constantly occurring in various parts of this country, and occasionally there is an outbreak in some particular locality, when several persons are attacked by it. Thus, in London, Dr Dowse has published several cases of cerebro-spinal meningitis, which lasted some weeks, and where, after death, lymph was found at the base of the brain extending down the cord. In these cases there was pain and weakness of the legs, dimness of vision, difficulty of breathing, and headache; also numbness and hyperæsthesia of various parts. In some cases there was no rigidity or tetanic spasms.

In Ireland, during the epidemic, a purpuric rash was present, and this is observed now whenever isolated cases occur. Two or three lately in Dublin, of a very severe form, were marked by a purple rash, vomiting, and diarrhœa, stiffness and pain in the neck, with retraction of the head. The arms were weak, stiff, and hyperæsthetic. The eyes were inflamed, and there was swelling of some of the joints.

Several cases also have lately been reported by the physicians at Birmingham, and it seems to be a question with them whether they have been treating a new disease or an old one only lately recognised. Dr Foster has seen eight cases, some proved fatal and others quite recovered. In the former the pia mater and arachnoid were found thickened and matted together by firm strings of lymph, brain injected, ventricles distended with muddy fluid, and walls coated with a half mucoid and half fibrillated material. In these cases the illness had begun suddenly, with shivering, pain in the head and spine, giddiness, vomiting, restlessness, and excitement, with stiffness of the neck and spinal column. In severe cases the retraction of the neck was an early symptom, and there was well-marked delirium; also shooting pains in the limbs, and hyperæsthesia of the general surface. The temperature was irregular, being about 102°—103°. The principal symptoms were restlessness, agitation, headache, sickness, stiffness of the head and spine. The hyperæsthesia had caused a mistaken diagnosis of hysteria. Sometimes herpes occurred on the lips, and occasionally red rashes and petechiæ. Occasionally disorders of sight and dysphagia. It was questionable if any treatment was of any avail.

Dr Warner informs me that he has seen in several cases in children and adults, about half of which were fatal. The symptoms at first were very obscure; the fever was not high, and there was not always pain. The head was generally thrown back, and with this there was pain in the back of the neck. There was not always sickness, nor pains, nor other affections, as of the eye, which I have just mentioned. These symptoms might be present or not, but the retraction of the neck was the earliest and most characteristic condition. There was no evidence of blood poisoning as shown by a rash or purpura. The post-mortem examination always showed signs of inflammation, although they might be slight. These were principally an increase of fluid in the ventricles and throughout the subarachnoid space; this was opaque, and in some cases showed flakes of lymph, which could be peeled of the cord. In several there was an herpetic eruption on the face.

Dr Russell has found in his cases, besides the inflammation of surface, an indication of the same process in the ventricles;

these were distended with an albuminous fluid, and the ependyma was thickened.

Dr Wood, who has charge of the hospital at Cape Town, informs me that this disease has existed as an epidemic at the Cape of Good Hope. He has seen it in all degrees of severity ; the common and characteristic symptom is the retraction of the neck, so that the peculiar attitude of the patient, when he walks in for advice, at once indicates the nature of the disease. In many cases, even though the patient is very ill, the other symptoms, which I have named, have been wanting, except the marked hyperæsthesia. He has never seen any rash, nor does he consider the disease to be contagious. The post-mortem appearances have resembled what has been described by others—an inflammatory effusion in the membrane of the cord, at the base of the brain, and sometimes on the hemispheres.

Tumours and new growths.—These are of the same kind, histo-logically, as are met with in the brain. They may grow in the sub-stance or in the membranes. The symptoms are due to their position, and to the amount of pressure they exert upon the cord. These symptoms, therefore, may be of the most varied kind, so as to in-clude alterations in sensation and motion, as mentioned under paraplegia. Tumours are rarely diagnosed, though there may be good reasons for sometimes suspecting their existence. In looking through the histories of these cases it will be seen that the sym-ptoms consist of vague pains, very often styled rheumatic, to which succeed actual loss of sensation and motion. In two cases of women, which came under my notice, the symptoms were at first thought to denote merely hysterical disturbance, being spasmodic contrac-tion of the legs, pain in the back, and a burning sensation around the abdomen ; these, however, were afterwards clearly seen to be due to a stretching of the cord. In other cases we have noticed jerking and twitching. In all a paralysis of motion, to a greater or less extent, ensues ; and, as regards sensation, a primary hyperæsthesia often gives way to an anæsthesia, and what is remarkable, although not rare, pains in the limbs may still continue. In these cases the growth has begun on the meninges, and has subsequently in-volved the cord ; but under the term new growth, by far the most important and common is that which begins in the vertebral column, and afterwards involves the nerves which issue from it, and finally the cord itself. This form of paraplegia arises from cancer of the vertebral cord, and is usually secondary to cancer elsewhere, as in the mammæ. Under these circumstances it may be suspected from the symptoms which accompany it, the actual existence of a growth being rarely made out. Take, for instance, the case of a woman of middle age, who has long had a scirrhous tumour in the mamma, pains begin to be felt in the leg ; these are for a long time styled rheumatic, and treated accordingly ; when, however, no relief is obtained, and special attention is at length paid to the character of the pains, they will clearly be seen to be of nervous origin ; then the name of the complaint may for a time be changed to sciatica. When, at last, both legs become involved, this diagnosis will no longer hold, and it is clear that there is actual pressure on the nerves, by the lancinating character of the pain, described as burning, or by other expressions of a like kind. Soon a weakness comes on in the legs, indicating still further the existence of pressure ; perhaps one leg has been affected before the other, and is in advance of it in relation to the progress of the disease. If, with the symptoms described, the patient should have pain around the body, and any other paraplegic symptoms, a cancer of the spine may be suspected. The diagnosis is almost certain if cancerous tumours be found elsewhere. I have within a few weeks seen three such cases, and in two of them the disease in the spine was secondary to schirrus of the mammæ. The pain around the abdomen may correspond to the seat of the disease; and, as a consequence of the nerve pressure, we may sometimes observe atrophy of the muscles and other signs of altered nutrition. In one case I observed the skin had changed in texture, and in another there was a vesicular rash. The symptoms vary, however, according to the seat of the disease and the portion of the cord implicated, and there may be added to those already mentioned, flaccidity or rigidity of the limbs, spasm, or exalted excitability.

It may be observed that the symptoms in cases of tumours are mostly those of pain in the limbs, implying, no doubt, in most cases an infiltration of the nerves in the neighbourhood of the new growth ; this shows that its seat is in the membrane, which it stretches, and so involves the nerves issuing from it. According to the observations of others it would seem that pain had been present in cases where the tumour has been within the cord itself. This has not been my experience, as it has always indicated an implication of the nerves after they had issued from the medulla.

In our museum are several cases of tumour of the spinal cord, and Sir W. Gull has given the histories of most of them in the ' Guy's Hospital Reports.'

A man, æt. 30, had a tumour, the size of a hazel nut, attached to the inner and anterior surface of the dura mater, at the commencement of the dorsal region ; this had pressed upon and softened the cord. About four months before his death he was taken with symptoms which were thought to indicate incipient phthisis—cough, shortness of breath, and wasting ; subsequently he had pain in the back and shoulder. After this pains in the joints, which were called rheumatic; then his legs became weak, and he had some trouble with his bladder, as well as also some impairment of sensation around his chest, and the arms became slightly enfeebled. The paralysis increased, urine became ammoniacal, a bed sore appeared, and he had profuse sweating. He died rather suddenly.

Case.—A woman, æt. 43, was under Dr Hughes for fibro-nucleated tumour, growing from the inner surface of the dura mater of the cord, opposite the third dorsal vertebra. About nine months before her death she began to feel pain in the shoulders, chest, and sides. She was treated ineffectually, the symptoms gradually increasing in intensity; she then began to suffer from spasmodic contractions of both lower extremities; the legs were drawn up towards the abdomen. On admission they were rigidly fixed, with heels to the nates. If extended with force they were again retracted. There was no affection of sensation, and no incontinence of urine. Even at this time, however, it was a question whether her symptoms were due to organic disease or not. She subsequently had retention of urine, and this became ammoniacal; and bed sores appeared. The pain in the back and around abdomen was violent and incessant, and she became greatly emaciated before her death.

Case.—*Cancer of the vertebræ.* A lady, æt. 50. For some months been complaining of pains in her limbs, which were called gouty and rheumatic. When I saw her she was very weak in both legs; one was almost powerless, and in this she suffered most severe pains; these were of a burning character, sometimes alternating with a sense of coldness, and the whole limb very sensitive to the touch. The left leg had similar but less severe pains. The right arm was also subject to neuralgic pains. On examination, a large tumour was found in the back, near the scapula, and another one below it.

Any disease affecting the spine may involve the cord, as in the following instances of hydatid and aneurism:

Case.—Elizabeth R., æt. 58, was admitted under my care for paraplegia. She began to suffer about ten months before with pain around the abdomen, in the back, and down the legs. These symptoms increased, until two weeks before admission, when they became quite helpless. On admission she was found to have much pain and tenderness over the lumbar spine, and a pain, like a cord, around the abdomen. She had tingling sensations in the feet, but, at the same time, the legs had almost completely lost the power of feeling, and all power over them had gone. On the dorsum of the left foot was a sore, which appeared to have arisen from pressure against the other foot; there was also a sore on the upper part of left thigh, behind, and of which the patient was quite ignorant. There was also paralysis of the bladder. She died in about a month, and there was found a mass of hydatid cysts growing in the bones of the spine, and in the canal at its lower part.

Case.—A man, æt. 30, had suffered for four weeks with pain in the back, and a feeling around his abdomen as if a cord were tied around him. The pain then went into one leg, and then the other, and subsequently he lost all power over them. He was admitted for paraplegia, the cause of which was not discovered; there was perfect immobility of the legs as well as almost complete loss of sensation; excito-motor action could only be induced by the application of cold. The bladder became paralysed; and afterwards a bed sore came, and he gradually sank. After death, an aneurism was found eroding the vertebræ and involving the spinal cord.

Syphilitic disease of the spinal cord.—The cases which I have now seen are too numerous to leave any doubt that the cord may be affected by syphilis. The first case which came under my notice was one where a gummatous nodule had formed on one of the roots of the cauda equina, and this increased upwards until the cord was involved, and a fatal paraplegia resulted. Dr Fagge had lately a case of a young woman with complete paraplegia below the neck, in which a cervical meningitis was shown by the membranes and cord being united together by syphilitic material, and it remains, therefore, only to enquire whether destructive changes may go on in the cord itself, irrespective of these membranous inflammations; whether, indeed, those changes may occur in the cord in connection with the blood-vessels, which are supposed to take place in the brain in such cases as syphilitic insanity.

The case I shall presently relate had been under treatment six months for syphilis, when he was seized with paraplegia. A similar case, which was under Dr Taylor, rapidly proved fatal, and here no gummatous nor other visible changes were discoverable. The case of gummatous deposit above mentioned occurred in 1860, and I have reported it in my first paper on visceral syphilis as follows:

CASE.—Mary W—, æt. 53, was taken into hospital on account of numbness and loss of sensation over the right hip, especially along the crest of the ilium. She had no outward sign of syphilis. The feeling of numbness continued until the leg on that side began to get weak, and subsequently the other leg became affected. Soon afterwards a complete paraplegia ensued, with all the usual consequences, as retention of urine, &c.

On *post-mortem examination*, the spinal cord was found to have in the lumbar region a hard deposit, three quarters of an inch in length. This involved the posterior roots of the nerves, to which it was closely adherent, as well as to the spinal cord. It formed a lengthened irregular mass, and in bulk was altogether about the size of a nut. It was composed of an opaque yellow substance, and resembled a similar mass found in the liver. This organ contained two or three nodules, of a tough, yellow, amorphous substance; one of these on the surface produced a cicatriform appearance. The lung contained a few hard yellow masses, corresponding to the similar deposits in the liver.

But quite lately a case has been under me, which I have no doubt can with strictness be called true syphilitic disease of the spinal cord; not one merely of gummatous deposit or syphilitic inflammation of the membranes, but one of a myelitis, in connection with a morbid state of blood-vessels, comparable in every respect with a similar condition found in the brain in syphilitic insanity.

Syphilitic Disease of the Cord

CASE.—A young man had been for some months an out-patient under Mr Davies Colley for constitutional syphilis, indicated by a rash, sore throat, &c., when one day he found his legs becoming weak, and in little more than twenty-four hours he was the subject of complete paraplegia. Four days afterwards he came into the hospital. He had perfect immobility of the legs, and complete loss of sensibility as high as the waist. He had lost control over the bladder and rectum, so that the catheter had to be used; the urine was alkaline; he had priapism, and a bed sore was rapidly forming. There was slight excito-motor action by pricking the soles of his feet, but none by heat.

The case, after this, simply progressed; the paralysis did not reach any higher than the chest, although he felt a numbness over the region of the great auricular nerve. The bed sore increased until an immense slough formed over the sacrum and trochanters; the urine became purulent, and it was believed that his kidneys had become involved; and he gradually sank, fourteen weeks after the onset of the paraplegic symptoms.

The *post-mortem* was made by Dr Fagge. The testes had undergone a fibroid degeneration, but there was no evidence of similar disease elsewhere. The liver and spleen were lardaceous to a moderate degree, and the kidneys were suppurating. On taking out the spinal cord nothing was apparent to the sight on the membranes or surface, but on feeling it a marked softened portion was found in the dorsal region. A section showed the outlines of the grey matter confused; it was quite diffluent, and presented to the microscope broken-up tissue and granule masses. The most striking change was in the blood-vessels; these had their coats very much thickened, so that when placed between glasses they were perfectly opaque, and contrasted strongly with the thin-coated, transparent vessels of the other and healthy portions of the cord.

Drunkard's or alcoholic paraplegia.—I do not know that this is deserving of a distinct name from its possessing any pathological peculiarities, but as arising in connection with a very well-marked exciting cause, it requires your especial attention ; and I refer to it the more because, as far as I am aware, authors have generally overlooked it. I have already told you how long-continued habits of intemperance in alcoholic drinks tend to the production of a fibrous or fatty degeneration of the various tissues of the body, and that, as a consequence, the membranes of the brain and spinal cord become thickened, and the organs within wasted. This, of course, would give rise to what might be called a general paralysis of body and mind. But, besides these general results, we often meet with more direct effects on the spinal cord, and to these I particularly refer. I have now seen so many cases of persons, especially " ladies," who have entirely given themselves up to the pleasures of brandy-drinking, and have become paraplegic, that I am pretty familiar with the symptoms. From what we hear from our continental neighbours, it would seem that that diabolical compound styled " absinthe" is productive of an exhaustion of nervous power in even a much more marked degree ; since the volatile oils dissolved in the alcohol give additional force to its poisonous effects. Of course, drunkards of all descriptions suffer from muscular and nervous weakness, but, as I before said, it is more especially in the legs that the effect is most striking. A loss of power is first observed, accompanied by pains in the limbs, and in some cases by anæsthesia, which seems to indicate a chronic meningitis of the spinal cord. There is at the same time necessarily some amount of feebleness of other parts of the body as well as of the mind, and thus an approach to general paralysis is produced ; but sometimes the symptoms are almost confined to the legs, and resemble in character those of the locomotor ataxy. Why the brain should be affected in one case, and the spinal cord in another, is due probably to the same idiosyncrasy which makes one man get drunk in the head, and another in the legs.

I am now visiting a young married woman who for some time past has taken to "drink." She first had engorgement of the liver, followed by an all but fatal hæmatemesis. She recovered from this, but, continuing her evil habits, she began to get feeble in mind and tremulous in her limbs. She appeared at times almost childish, and spoke thickly. She had a difficulty in rising from her chair, and then, by a great effort, staggered across the room. She is now apparently gradually recovering.

I occasionally see in this neighbourhood a publican's wife, who commenced business two years ago, previous to which time she was temperate and well. Since this, the constant presence of gin before her eyes has been too much for her, and she has drunk the burning liquid in enormous quantities. This could not continue long with impunity, and now she has been confined to her bed for six months. She is almost paralysed, having very little power to move her limbs, is not able to raise the heel from the bed, and has no power to grasp with the hand ; the muscles are flabby, and she has almost complete anæsthesia ; the mind is also somewhat enfeebled.

Now and then the arms are affected, and occasionally anæsthesia is the principal symptom. This may affect one side only.

Such cases I could multiply to almost any extent. Several I have seen end fatally, and in some a partial recovery has taken place. A most remarkable case is now under my care. A lady æt. 35, married, but without family, being left alone all day, had been secretly drinking, until at last she became perfectly paralysed in mind and body. She took to her bed and her end was shortly expected. She was sallow, wasted, quite unable to move her legs, and her arms but slightly ; she appeared to have lost feeling also in the lower extremities. She had become almost fatuous, scarcely recognised her friends, and, indeed, her mind was a blank. All alcohol was at once stopped, a nurse was procured and feeding commenced; quinine and opium were also ordered. She slowly recovered, and now, after five months, is able to walk and come to my house, having grown stout, and regained her intellect. Another woman, who could not be managed at home, recovered speedily after going to a hydropathic establishment.

I have given the name paraplegia as a general term for the complaint, but, as you might suppose, in a chronic meningomyelitis, the symptoms vary very much, according as the posterior cords of the nerves are involved. Thus, in the case just mentioned, there was anæsthesia, whilst in a young man who died lately there was hyperæsthesia. In his case also there were severe pains in the limbs. In one young woman the symptoms very much resembled those of locomotor ataxy, both in the character of the pains, which were like electric shocks, and also in the mode of progression. I might add that drunkards often suffer from pains in the limbs long before there is any sign of paralysis.

Acute ascending paralysis.—One of the most remarkable affections of the cord is that known by this name. It is of extreme interest, as it may be simply the result of a condition of cord which is physiological rather than pathological—a mere loss of function arising from some inhibitory action, which may cause a fatal ending before this state is recovered from. Whatever its nature may be, the rapidity of its extension is most remarkable, for it propagates itself from end to end like wild fire. In seeing such cases, I am reminded of a spark alighting on a piece of touch-paper, and the fire running through its length until the whole is quickly consumed. A patient, for example, may feel some numbness or loss of power in the feet, which is soon followed by an inability to move the legs ; then a paralysis of the arms and upper part of the body comes on, and in three or four days death occurs. Whether the change in the cord be due to some rapidly degenerative process in the fibrillæ, arising from the causes already mentioned, whether it be inflammatory, or whether the alteration be only that which may be called dynamic, has yet to be determined. It may be remembered how the nutrition of a nerve is dependent upon the integrity of the centre whence it arises, or on the neighbouring ganglion, and that in the case of the fibres of the medullary cord these will degenerate in a given tract according to the direction of their conducting power. It is not, therefore, difficult to understand how any cause interfering with their nutrition at one spot might be propagated along the whole length of the strands. That there are causes in operation which will arrest the function of any part of the cerebro-spinal centres without the occurrence of any organic change we every day see ; as, for example, in paraplegia which is entirely recovered from, or in hysteria where a portion of brain or cord is for a time perfectly in abeyance. I am right, I believe, in saying that both in this country and on the continent the most competent observers have failed to find any abnormal changes in the cases of which I am now speaking. I can call to mind several instances of this complaint.

CASE.—James D., æt. 47, was a blacksmith and was admitted under my care on December 9th. He said that he was in good health until December 4th, when, whilst at work, and without any assignable cause, he felt a numbness and weakness in his legs; but he managed to continue at his employment. On the following day he attempted to work, but was obliged to desist, and return home and go to bed. The weakness increased daily until the day of his admission. He was a small, dark-complexioned man, and as he lay in bed appeared to have nothing the matter with him. On examining him, however, he was found to have paralysis of motion and sensation of the lower limbs. An attempt to move them caused only slight contraction of certain muscles, and there was loss of sensation as high as the thigh.

On December 10th, the paralysis was quite complete, and on December 11th, on full examination, there was found perfect anæsthesia as high as the umbilicus, and there was not the slightest reaction to any external stimulus. This seemed to show also that the grey matter of the cord was involved. The arms felt somewhat weak and numb, and it was observed that the lower part of the chest did not fully expand. It was thus clear that the disease had already reached the cervical region. The pulse was quick and the tongue slighly furred, but the temperature of the body was not raised; in fact, there were scarcely any febrile symptoms. He was quite rational, though he answered questions slowly.

December 12th.—Respiration become more difficult, mucus collected in the bronchial tubes, and death occurred at last rather suddenly.

A post-mortem examination was made, and no disease in the body was found. The spinal cord was examined microscopically; but nothing abnormal could be discovered.

CASE.—A tradesman's wife, æt. 38, living at Streatham, was seen by me for an acute paraplegia, and I received the following history: Whilst in her shop on Monday, March 11th, she fell down on account of a weakness and numbness in her legs; on the following day she could scarcely walk; on Wednesday she was worse; and on Thursday was completely paralysed in the legs. On Saturday I saw her. She could not move her legs in the slightest degree, and had completely lost sensation as high as the breasts. There was not the slightest response to the stimulation of the feet by tickling or pricking. The respiration was becoming affected; it was quick, and she was already experiencing a suffocating feeling. Just at the line where the sensitive and anæsthetic parts met she felt a burning or tingling pain. Bladder and rectum paralysed. Pulse 110.

I saw her again on the following day. The arms were paralysed, the chest was becoming immovable, and abdomen very tympanitic. She died at night.

I made the most careful inquiries respecting a probable exciting cause for the paralysis. I was told she was a perfectly healthy woman, had no uterine or other trouble, had not been exposed to cold, and had received no injury.

CASE.—Within the last few days I have seen another equally remarkable case.

A well-grown lady, in her twentieth year, was brought home from a friend's, where she had been visiting, perfectly helpless in her limbs, owing to what had been called rheumatism. A medical man was then called in, who seeing her paralysed sought my advice. I found her paralysed as regards motion and sensation over the whole body; the chest was being expanded with difficulty, and mucus was collecting in the tubes. This she was vainly endeavouring to expectorate, and said she should be choked. I need not detail the history, but simply say that, without having had any previous spinal symptoms, she was taken six days before with weakness and numbness in the legs; this was followed by complete paralysis, which rapidly extended upwards, until the neck was reached at the time I saw her. Her brain was unaffected, and she was quite rational; you may therefore imagine the distress of her parents when I declared that she had not long to live. She died in a few hours, and I am sorry to add, a post-mortem examination was refused me.

In some rapidly fatal cases there may be a history of injury or exposure to cold, and therefore, under these circumstances, it is probable that, had the patient survived long enough, some well-marked inflammatory changes might have been found. In the case given below the patient died of one of the accidents of the disease, and thus sufficient time might not have elapsed to develop changes which could be appreciated by the eye. In another case of acute paraplegia, fatal in twelve days, of a somewhat similar kind, softening of the cord was clearly perceptible. In the cord of a paraplegic, which is apparently healthy, there may exist two very different conditions: there may be a state in which the so-called reflex paralysis has occurred, in which the cord is in no way structurally altered, and therefore may at any time recover its function; whilst there may be another state to which I refer, which shows no change, because the effects of the inflammation have not had time to display themselves. In the one case it may be supposed that, owing to a contracted state of the blood-vessels from reflex irritation, a part of the cerebro-spinal centres may lose its function, causing a paralysis, and yet this condition may endure for any time, and recovery eventually take place; and in the other case, although no change may be discovered, it may nevertheless be the antecedent of a true inflammatory process or softening, as, for example, in the case where a hemiplegia has ensued after the ligature of the carotid. Now, inasmuch as this change would assuredly have come about in time, we cannot regard any condition which is preliminary to softening, of the same nature as that which might endure for an unlimited period.

CASE.—A young man walked about in his wet clothes, and afterwards slept in them. On the following day he felt very unwell, with aching pains all over him. On the third day he was obliged to keep his bed, on account of the weakness and numbness in the legs. It was then found that there was complete paraplegia of motion and sensation as high as the pelvis; water obliged to be drawn off. These symptoms increased, and on the 10th day he was excessively ill, with febrile symptoms; abdomen tympanitic; breathing quick and interrupted. There had never been any excito-motor action. On the twelfth day he died. The *post-mortem* showed acute inflammation of the bladder and kidneys, and to this death was attributed. The spinal cord showed no appreciable disease. The question in such a case is—Was the cord only functionally affected by a reflex action, or was it in the same morbid condition which precedes all cases of softening, but not yet appreciable to the eye?

CASE.—A man, of middle age, was lately under the care of Dr. Taylor. He was seized, after getting wet, with pains and weakness in his legs. This gradually increased until the arms became weak, and finally his chest. His breathing became more shallow, until he died; he was conscious to the last. The case lasted about ten days. The cord appeared perfectly healthy by ordinary examination.

In another case, fatal in five days, the cord appeared healthy to the naked eye, but after being placed in chromic acid and sections made, it was thought that slight degenerative changes could be perceived.

In two cases where paraplegia was complete as far as the middle of the body, in five days, and fatal through the kidneys at the end of a month, the cord appeared healthy to the naked eye, but on microscopic examination showed degeneration throughout all the tracts, as if the change had been a universal one.

The last case I have seen was that of a man under the care of Mr Chaplin, of Shepherds Bush, and where the paralysis was complete in a week He began by complaining of weakness in the legs, which increased until perfect paraplegia of motion and sensation resulted. The bladder was paralysed, but the urine was not ammoniacal, nor was there a bed sore. The disease extended upwards until the arms became quite dead and useless; then the cord became fixed, so that respiration went on solely by the diaphragm. He then ceased to be able to put out his tongue or speak. Swallowing became impossible, and so he died, the brain and intellect remaining intact until the last.

As a proof how subtle may be the alterations in the cord which may give rise to paralysis, I may mention the case of a young man who was brought into the hospital after having been knocked down and received various injuries. He had a scalp wound, a fractured fibula, had several ribs broken, and was paralysed in the legs. His urine had to be drawn off, and he subsequently had cystitis; his motions also passed involuntarily, and he had a bed sore. When he died, at the end of the fourth week, no injury to the spine could be discovered, and the cord itself presented nothing abnormal. The symptoms were regarded as due to concussion.

Recoverable paraplegia.—I have seen several cases of paraplegia in which recovery having occurred I have had no knowledge of their nature. I have, therefore, put them in a series by themselves. In the cases of which I have been speaking the cord had become functionless, and death was due to an implication of the medulla oblongata, but where the inhibited condition ceased below this point, then recovery sometimes occurred. In other cases, associated with well-marked disease of the spine, a similar simple functionless state seems to have been induced. For example, in connection with caries of the vertebræ we meet with complete paraplegia, and yet this may be perfectly recovered from. These cases of recovery are constantly occurring, but I have no means of diagnosing the nature of the affection of the cord, and much less of forming a prognosis.

CASE.—*Paraplegia, recovery.*—Man, æt. 52, subject to rheumatic gout; never had syphilis. A month before admission he had an attack of gout in the feet; after two weeks was losing power of the legs and bladder. The paralysis rapidly increased. On admission he could not move his legs. There was a partial loss of feeling as high as the umbilicus, and reflex action was well marked. The urine was drawn off twice daily by the catheter and was ammoniacal. He subsequently had a feeling of tightness around the lower part of chest and abdomen. Numbness passing down arms to fingers. He gradually got worse, and about a fortnight after admission he became feverish, with quick pulse, red tongue, rigors and hiccough; a bed sore was forming, the urine ran away, and his mind was clouded. I suspected that he was suffering from suppurative nephritis, due to an extension from the bladder. He looked as if he had not many hours to live, and remained in a very precarious state for some days, when the constitutional symptoms abated, and some power in the legs returned. He after this made a rapid recovery, began to sit up in bed, gained power over the bladder, so as not to require the catheter, and soon was able to leave his bed and sit in a chair. He then asked for crutches, and began to stand alone, and soon was able to walk about the ward, when he wished to leave, which he did exactly two months after his admission, and six weeks from the time when his paraplegia was complete.

CASE.—A woman of middle age was under my care some years ago, for almost complete paraplegia. The symptoms had been coming on for a few days, and were accompanied by swelling and pain of the joints, with some febrile disturbance, so that it was conjectured that she might be suffering from a rheumatic affection of the spinal cord.

After admission she complained of great pain in the limbs, with twitching of the muscles, and feeling of constriction around the waist. A bed sore formed, the urine was passed involuntarily, and was ammoniacal, and the sphincter ani was paralysed. The slough on the back became deep, and all the other symptoms continued for three weeks, when she began to recover. She took tonics, and was galvanized, and her recovery was continued, so that she left the hospital at the end of five months quite well.

In cases of disease of the spine, where paraplegia follows and is subsequently recovered from, we usually attribute the symptoms to pressure. We do so because such a cause appears to us intelligible, and we have occasional proof that the paraplegia may be produced in so simply a mechanical way. We lately had a man in the ward with caries of the spine, followed by complete paraplegia, which was speedily fatal through bed sore and suppurative nephritis. A mass of inflammatory product was pressing on the cord, but it had not entered within the theca nor involved the substance. I have been recently seeing a gentleman who had long been suffering from pain in his back, when it was at length discovered that the bone was growing out; he then had weakness in the legs, which soon ended in a complete paralysis. A bed sore formed, and the bladder was paralysed, obliging the constant use of the catheter. He lay in bed in a completely helpless state for several weeks, and then gradually recovered, and now walks about with an angular curvature.

Reflex paralysis.—That this constitutes an established form of paralysis is by no means proved; it cannot at present be regarded otherwise than as a theoretical explanation of those cases of paralysis where no visible lesion is found. We are looking for an interpretation of the cases which resemble in every way those organic forms of disease which I have mentioned, but yet, being recoverable, cannot be considered as identical with them. I have told you that the changes which our naked eyes or microscopes detect are all of the destructive kind—broken fibres, or new products which have taken the place of the old; and we can only conjure up to our fancy some of those alterations which we please to call dynamic. In the case of the brain and spinal cord it is possible that changes may occur of a rougher and material kind, and yet be not perceptible; for we must remember that nerve-substance is complex, and that blood is a necessary ingredient for its integrity. Let the blood supply be deficient, and its function is gone. A good example of this is seen in the case where the carotid artery is tied, and a hemiplegia quickly follows; this is speedily recovered from when a fresh supply of nutrient fluid is supplied.

What change takes place in concussion of the brain and spinal cord we do not know, but it implies such a derangement of the integral portions of the nerve tissues that their function for a time ceases. And again, from recent experiment, we must take into account the fact of a nerve losing its function from some fault in its extremity. Just as a galvanic wire ceases to be in a state of electric tension when the battery stops working, so it is almost certain that portions of the cord depend for their integrity of function on distant parts. This fact may be only removing the difficulty to a distance, but it shows that a functionless part of the cord may to the eye seem quite natural. It has been thought, however, by many, that such instances of temporary and recoverable paralysis may find an explanation in a deficient blood supply, and, since it is known that the blood-vessels are regulated by the sympathetic nerves, that we have no more to do than to suppose an irritation of these nerves in order to arrive at a theory of the cause of functional paralysis. I should not, perhaps, use the word "functional," since an alteration of this kind, although a temporary one, is none the less real. A paraplegia induced by an external irritant is styled *reflex paralysis.* It is thought, I say, that the blood-vessels are thereby diminished in calibre, that the blood supply to the spinal cord is diminished in amount, and that its power departs. Since a temporary paraplegia, or one which, if fatal, has shown no organic change in the cord after death, has been associated most frequently with some urinary disturbance, the cases of the kind which I describe have been mostly styled *urinary paraplegia.* You are aware that the theory can scarcely admit of actual proof, and is rather accepted because we have no other explanation for those cases of disease where the usual morbid appearances are wanting. Dr Gull long ago expressed his doubts, and showed that a phlebitic process might be accepted as far more explicable, and with him I cannot but think that the negative evidence is important as resolving the question—take, for instance, the statement of one of my senior colleagues, who has had immense experience in all diseases pertaining to the urinary organs, and yet is unacquainted with reflex paralysis as one of the consequences of them : in fact, he does not remember a single case in connection with stone in the bladder. Paraplegia and urinary troubles so often go together that it may be difficult in any particular case to ascertain which has preceded the other ; thus, quite lately, I have been attending a lady who has had partial paraplegia and acute nephritis ; she is now recovering. What relation these two complaints have had to one another, or which is the primary one, I cannot say. Whatever explanation is given—and the theory of diminished supply of blood through nerve irritation may serve for the purpose—certain it is that a temporary paralysis may be frequently met with, not only in connection with urinary disorders, but in several other affections. Thus, I might mention the case of paraplegia succeeding to labour ; this is, perhaps, not common, but every now and then we meet with it. It has been said that a pressure on the sacral nerves would suffice for explanation, or a pressure on the psoas and iliacus muscles ; but these causes must be regarded as very doubtful, although I believe cases do sometimes occur where most excruciating pains have accompanied the passage of the head into the pelvis, and these have been followed by a temporary paralysis. Dr Fussell, in a paper on this subject in the ' St George's Hospital Reports,' states, on the authority of Mr Youatt, that cows not infrequently suffer from paraplegia after calving, that they lie down, are quite unable to move their hind legs, and apparently have no feeling in them when they are touched. In about a fortnight the cows get up well. In such cases the theory of reflex paralysis might be made to apply.

The doctrine is applicable, of course, to many other cases, for example, to the paraplegia which occasionally has been observed to follow typhoid fever, where the disease would be reflected from the ileum. Then, again, the case of diphtheritic paralysis which I have already alluded to, instead of being regarded as due to simple exhaustion of the whole nervous system, might be referred to a lowered condition of the spinal cord, owing to a deficiency of good blood supply arising from an irritation of nerves in the throat. So also the cases of acute paraplegia following direct exposure to cold might be considered due to a similar cause. In corroboration of the possibility of a deficient blood supply being all-sufficient to render the cord functionless, I might mention the case of paraplegia following obstruction of the aorta, which occurred here some years ago. That there are causes in operation sufficient to render the brain and spinal cord powerless, and yet beyond our ken, I have already referred to in speaking of concussion, but I might allude to the more striking case of sun-stroke. Here the effect on the brain is sufficient at times to paralyse its action and cause instant death, at others to produce those changes which are subsequently shown as epilepsy or mania. It is remarkable, however, as I before said, that changes in the brain which are invisible may be the precursors of ordinary inflammatory attacks—at least I judge so from the fact of there being well-authenticated cases of children having died in a day or two after exposure to the sun, and inflammatory lymph having been found at the base of the brain and in the ventricles.

The greatest difficulty which I have in my own mind in adopting any opinion in explanation of those cases of paralysis which recover, is that such cases are not uniform, but assume the character of every variety of nerve disease with which we are familiar. I can understand how a concussion of the spine can affect one region of the spine rather than another, or even by chance injure the exterior rather than the interior, or *vice versâ;* and I can also understand how in urinary paraplegia the lower part of the cord may be especially involved, and in diphtheritic paralysis the upper; but it is difficult to conceive how one particular strand or a certain functional area is to be alone affected by such causes as I name, and yet it is certain that limited portions of the cord can be temporarily deranged. Suppose, for instance, we hold clearly in our minds the seat of the changes in the cord in cases of fatal paraplegia of motion, paraplegia of sensation, locomotor ataxy, or progressive muscular atrophy, we are bound to believe that those same parts are affected whenever the same symptoms exist. We have, then, the problems before us which during the last few months have been asking for solution in Stephen ward—first the case of a man who perfectly recovered of a partial paraplegia of four years' standing; then the case of a man with symptoms of locomotor ataxy who recovered; also the case of a man with early progressive muscular atrophy who quickly got well under the application of the continuous current down the spine, and more remarkably still the case of the girl lately alluded to, who was little more than a skeleton owing to the same disease, and yet perfectly recovered under the use of faradisation. I say we want to know what must be the condition of the cord in these cases which recover. If we are led to believe that organic changes take place in certain portions of the cord in those instances which are fatal, we might, as a matter of easy explanation, suppose that the same changes had occurred in those which recover, but, being in their incipient stage, were removable by remedies. If we are not content with this explanation, and are satisfied with the doctrine of reflex paralysis, or some analogous theory, then we must regard the several portions of the cord as much more anatomically and physiologically distinct than we have otherwise done ; and it may, perhaps, be even necessary to admit this in order to explain the selection of parts for the severer organic changes.

The cases which have of late years been styled reflex paralysis have been those which have succeeded to local disease or general illness. If the paralysis has been confined to the legs, then we have had *reflex paraplegia*. This is usually described as following affections of the bladder or urethra, diseases of the uterus, and of the intestines, as dysentery. It has been thought that a sufficient number of cases of paraplegia have been collected which have immediately followed these diseases to show that a connection must in all probability exist between them. Of late, also, smallpox has been shown as tending to the production of a nervous affection. It is, however, more especially with stricture and bladder affection that the connection has been noticed, and this has given rise to the expression *urinary paraplegia*.

Diphtheritic paralysis.—This is considered by some an example of reflex paralysis, whilst by others it is regarded as a general paralysis, arising from exhaustion of the cerebro-spinal centres. Since the throat is the part most violently implicated in the original complaint, and since this also is the part which often first loses its power, it has been thought that the paralysis really has its origin or cause therein, and that the whole body is affected by a reflex action. Be this as it may, it is remarkable, as is true of the sequelæ of many other disorders, that some very severe examples of the paralysis have occurred where the diphtheritic affection was but slight. This paralysis is more deserving of the name *general* than any other which I know, for all the physical powers are affected, and sometimes the mind is enfeebled. The patient becomes utterly helpless, quite incapable of standing or moving the arms, the face loses its expression, the saliva runs from the mouth, there is thickness of articulation, and difficulty of swallowing; in fact, the patient has the appearance of an idiot, and more especially so if he be not completely paralysed and is able to walk, for then he stumbles along, and, with his head hanging forward and his vacant stare, looks like an imbecile. There is often amaurosis, deafness, and paralysis of the sphincters. The child generally recovers, unless there is that amount of paralysis of the palate which prevents him swallowing, in which case he sinks from exhaustion.

In some instances the depression is so great, and has come on so suddenly, that the term paralysis can scarcely be adopted, and this almost obliges us to remove it from such a cause as is suggested by the term reflex.

For example, a lad, æt. 15, had an apparently slight attack of diphtheria, when one day after the lapse of a fortnight he became very prostrate, the pulse sank to 50, and soon to 28, the respiration became slow, whilst the temperature was normal, and he died on the following day. A little girl in the same way had recovered from her throat affection, when she became collapsed, with slow pulse, and died in three days. Of the same kind was the case of one of our house surgeons who had a bad diphtheritic throat; after the secreted membrane was loosened and the swelling abated, he began to improve, and I hoped he was convalescent, when I was called out of bed one morning and found him collapsed, with a very feeble, irregular, and slow pulse. It seemed sometimes, indeed, as if the heart had altogether stopped, so long was the pause between the beats; there was also a great feeling of oppression of the chest. He never rallied, and died three days afterwards.

CASE.—A young man, æt. 25, the son of a medical friend, had an attack of diphtheria; this was followed by paralytic symptoms, affecting more especially the throat. From being a stout man he became thin, spoke with a nasal twang, and swallowed with difficulty. It was several months before he regained his health, and then he was constantly liable to a return of the throat weakness. About a year afterwards he had a very slight attack of modified small-pox, and he was recovering from this when he found he had difficulty of swallowing as well as of protruding the tongue. In a few hours the paralysis was almost complete; he could only speak in a whisper with his mouth, had not the slightest power of swallowing, could not cough in the least degree, and his breathing was quick and difficult. He scarcely moved his chest, and it was evident that not much air entered the lungs. A consultation took place as to the advisability of tracheotomy, but it was not done. The breathing became worse, with more lividity of the surface, and he died in about twelve hours after the accession of the symptoms. On *post-mortem* examination no morbid changes were found in any organ; the larynx, pharynx, &c., were quite healthy.

It is worthy of note that we sometimes meet with cases of general paralysis in children exactly of that kind which is seen to follow diphtheria. But lately, I saw a little girl æt. 10, who began to feel her legs weak in walking, then the arms, subsequently the whole body, exactly as is met with in diphtheria. There was no history whatever of any preceding illness. She gradually recovered. There may, then, be other causes not yet recognised, which produce similar effects on the nervous system.

Choreal paralysis.—The connection between irregular movements and debility of the muscles is well marked in the case of chorea. In this disease not only is there the perpetual movement of the limb, but it is proportionally weak, so that it often happens that the motion may cease whilst the debility remains, and the case then constitutes one of *choreal paralysis*. It is important to recognise this, because a child may come before you for the first time with a weakened limb, and without any symptoms of chorea, and yet this disease is the originator of the paralysis, and the one which requires to be treated. Not only may a limb suffer, but the whole body ; thus you may not unfrequently have a very bad case of chorea to treat, where, in the course of two or three weeks, all movements cease, but at the expiration of this time the child is quite unable to stand, or even move from the bed. A rapid recovery, however, often occurs.

Peripheral paralysis.—I ought also to allude to the doctrine of peripheral paralysis, that is, where the nerves are not injured at their source, but in the course of their distribution. One of the commonest examples of this form is the facial paralysis arising from exposure to cold, or *coup de vent*, as it is often styled. Of the same kind might be the case of a man who was paralysed in his legs from standing for some time in cold water, although here the theory of reflex action might come in. More than one case has been recorded where, after exposure, sensation and motion were lost, ending in a permanent paralysis, at last proving fatal, and where, after death, a chronic inflammation of the medulla and membranes was found.

In these several forms of paralysis in which recovery occurs it is clear that nothing in the shape of what we recognise as organic change can have taken place. They are clearly dynamic conditions, in which the organ, though structurally perfect, is yet asleep, or, at all events, is not working. I am in the habit of illustrating the two conditions by the comparison of two watches which may be lying on the table motionless. You examine their interiors and you find one irretrievably damaged in its most essential parts, whilst the other is not going for the simple reason that it is not wound up. So in the brain and cord; they are often not wound up, and are not, therefore, going, but they are ready to do so when set in operation. This is the condition supposed to exist in the hysterical paralysis, and where a strong effort of the will is often sufficient to start the machinery into motion. No doubt, a good supply of blood is necessary for the due performance of function, and has suggested the theory of a reflex paralysis of the blood-vessels. Then, again, an exhausted condition of the cord is one we can understand as sufficient to account for its temporary inactivity. A person, for example, has been walking all day without any nourishment, and arrives at his journey's end so fatigued that he can scarcely put one foot before another, his hand trembles, and his mind is incapable of any prolonged effort, the cause of the exhaustion is clear enough, but what the altered condition of the cerebro-spinal centres may be like is very difficult to conceive.

Hysterical paraplegia is the case where we believe the functions of the spinal cord are simply in abeyance or acting independently of the will. In the first place we find, as we might expect, a loss of motion, of sensation, and even of the vaso-motor function. The very completeness of the paralysis is an argument against its importance, since in actual disease particular tracts only are generally affected, productive of special symptoms. A case of this kind, therefore, must imply either nothing less than a complete destruction of of the cord or an unimportant temporary arrest of its function, since, to use the technical expressions, we have anæsthesia, akinesia, thermo-anæsthesia, and analgesia. The legs are usually stretched out, as in a corpse, the feet taking the position they naturally would in a suspended dead body, whilst the arms are drawn across the chest and tightly flexed. It may be observed, that in the hysterical form the girl remains plump, and she has no weakness of the bladder; at all events, she does not wet the bed, she merely has retention. The rectum is not paralysed, but the bowels are confined, and there is not that distension of the abdomen sometimes seen in disease of the cord.

The feet are often livid and cold; electro-mobility may be normal, but electro-sensibility is often quite gone both in the muscle and skin.

Electricity may some day teach us to distinguish between an hysterical or ideal paralysis and a real one, since the muscular tension is probably different in the two cases.

Concussion and railway spine.—Every medical man with a large practice had formerly, no doubt, seen cases where a permanent affection of the spinal cord had resulted from an injury, but it has only been in railway times that we have witnessed so many cases of

permanent disease from a shock to the cerebro-spinal system.
When one considers the velocity of a heavy train, it is clear that the
momentum with which it strikes any obstacle in its way must be
immense. If one remembers, also, that some of the component parts
of this mass of material are fragile human beings, composed of
flesh and blood, with delicate soft brains and spinal cords, no surprise can be felt in witnessing the terrible results of a train proceeding at the rate of forty miles an hour when brought to a sudden
stop. It has often been credibly stated that after an accident
persons have been taken up insensible, but without the slightest
mark or bruise upon them, and indeed I see nothing marvellous in
the proposition that the soft cerebral mass striking against the hard
skull-case should be thrown into confusion, and insensibility result.
In the same way, also, as you sometimes see persons after a blow
on the head, not sufficiently violent to produce concussion, thrown
into a state of maniacal excitement, so travellers meeting with such
a shock as I have mentioned have often been observed to jump
out of their carriage, throw their arms about, and behave in a
manner which they themselves have afterwards designated as acts
of madness. Well! our poor brains and spinal cords cannot be
thus roughly treated without the necessary consequences, and a
variety of symptoms from the day of the accident are set up, which
develop into manifold troubles, or even end in a permanent palsy
both of body and mind.

It is remarkable that in one or two instances which I have seen,
where there was direct injury to the back, and symptoms of paralysis in particular limbs immediately following, the patient eventually got well, whilst in others, where a shake to the centre had
been received, but with no apparent ill result, a slowly creeping-on
paralysis ensued. These cases have now become so common, and
have given rise to so much litigation, that they have become known
in the profession as cases of " railway spine." A man, for example,
receives a severe shock from the train having " collided" (to use the
American phrase), feels unwell for a day or two, and then believes he
has recovered from the shock. He soon, however, begins to be ill
again, is fatigued, and unable to pursue his business with the same
zest as before. In the course of some weeks the change is evident ;
he cannot walk as well as hitherto, staggers, his hand shakes when
he writes, his memory fails him, he forgets names, and blunders in his
accounts. If this be a chronic affection of the cord, slowly progressing, he becomes at last actually paraplegic, and in mental
capacity is verging towards imbecility. A number of special symptoms may result, dependent probably on particular parts involved.
In many cases there is tenderness along the course of the spine, and

there may be hyperæsthesia in the course of some of the intercostal nerves. In some there may be pain in the back or limbs. The symptoms vary, probably, as the cord alone is affected, or as the membranes are involved in the chronic inflammatory or degenerative process.

The following account of the result of a railway shaking is taken from an anonymous letter in a newspaper, and is pretty accurate in all its details :

"In the case we are supposing you are shot like a human bullet from one side of the carriage to the other—forwards, backwards, and forwards again, with a momentum in proportion to the force of the collision ; the parts generally struck being the back or front of the head, and the spine, either at the neck (the cervical vertebræ), or the lower part of the back (the lumbar vertebræ). Your head is moved as it were by its own weight, all controlling or resisting power of the muscular structure being for the moment lost. You have lights before your eyes, an odd taste in your mouth, sudden severe pain as the blows fall on you one after the other in quick succession ; and then there is a pause and you pick yourself up, or somebody else picks you up, and you find yourself feeling a little sick and giddy and a good deal bewildered. It is only that your spinal cord has received a jar, or shock, or concussion, whichever the men of science term it, the effects of which you will feel for many a long day to come. Meanwhile you 'continue your journey,' for of course you are not ' severely injured,' and besides, you have nothing much to show after all. A few days or weeks after this your hands begin to have a queer numb sensation, or they tremble after the slightest exertion. One arm, or perhaps a leg, diminishes a very little, just sufficient to be observed by actual measurement. If your head never ached before it will ache now, and pretty fre-quently too. You cannot hunt or ride as you used to do. You shrink from attempting a very small fence, and exposure to wet or cold brings on agonising rheumatic or neuralgic pains on the spot where you sustained the series of blows. You fall into the hands of the doctors, and their counter-irritating remedies seem to you nearly as bad as the disease. They tell you that yours is an obscure case—that you may be better, quite well, in fact, in a couple of years, or that general paralysis may supervene, and you may, to speak frankly, become an idiot within about the same length of time. You look careworn and older, and you feel older. All your habits undergo a sensible change, if only for a time. If formerly you were endowed with an almost demon-like activity, you are now reduced to accept idleness and rest as a boon. If you were good-looking, you lose a portion of your good looks ; if you were ill-favoured, your ugliness becomes more pronounced; and, young though you may be, you will find before long a few grey hairs if you take the trouble to search for them. Morally, the effects are not less marked. You are demoralised as regards your nerves, and your horse is aware of the fact, and presumes accordingly. Your temper is less elastic and somewhat irritable. You feel either cowed and depressed or sublimely reckless when you enter a railway carriage; and whereas before you used second or third-class carriages, if your pockets inclined to economy, you now confess that cushions and a hot-water tin are essential to your comfort. As to claiming damages in a court of law for all these minute miseries, in your shaken mental and physical condition you naturally shrink from the multiplied surgical examinations which you would entail on yourself, and from the bullying to which you would be subjected by the opposition counsel. You may not have the sensation of perfect health and physical enjoyment and happiness that was yours before, but unless you have lost the sight of one eye, have forgotten your own name, have fits, and drag at least one leg after you, you need not expect to receive either compensation or compassion. And then you lose in pounds, shillings, and pence. Your life is not so good to insure; sometimes a sound office will reject it altogether. You call in the doctor more frequently. Being less able to bear privations in respect of warmth and luxury, your habits are of necessity more expensive. You require more money, and you feel less capacity to earn it. You have, in truth, expended or been robbed of a good deal of vitality in a very short space of time. In general terms, it may be safely said that a person who has been really 'severely shaken' in a railway collision loses, on a favorable computation, at least three years out of his life."

It is very difficult to say, however, whether the nerve symptoms which come on are due to a myelitis or meningitis; and thus we class the two affections together. In the former we look for more direct results of impairment of the spinal function, and in the latter for symptoms which may lead on to irritation and implication of the spinal nerves. Thus severe pain in the back and around the body, increased on moving, and rigidity of muscles, we usually regard as indications of membranous inflammation; whereas more obvious paralytic symptoms we should put down to implication of the substance of the cord itself. The shock to the nervous system may give rise to numerous other symptoms, as disturbance of the special senses, and more especially of the eye, but this is due to some more direct connection between a portion of the spinal cord and the eye itself, as an actual optic neuritis is sometimes set up.

A blow on the back or sacrum may produce a concussion and a paraplegia, which may be completely recovered from, while a less injury may start into action an inflammatory process which may end fatally years afterwards.

Concussion we must regard as purely physical in its results, from shaking the cerebro-spinal centres. A *shock* is both physical and moral in its results. Various emotional and hysterical symptoms may occur in a person who had received no important bodily injury. Consequently the symptoms being both objective and subjective, it is very difficult to put a right value on each kind.

Concussion of Spine

CASE.—A gentleman sustained a severe shock in a railway carriage, but it was doubtful whether he received a direct blow. He did not appear to have suffered at first, but subsequently he showed want of power simultaneously both in mind and body. He was unable to undergo any mental exertion, and was obliged to desist altogether from business. It being a question as to the amount of injury he had received, the railway company did not see its way to compensate him largely, and he therefore brought an action. Whilst this was pending his condition was very remarkable, owing to his extreme sensitiveness and hyperæsthesia. Shaking hands almost threw him into convulsions, and any noise in the house went, he said, right through his back. When my colleague, Mr Cock, went to visit him he patted him on the shoulder in a good-natured way, and made the patient give a sudden leap from the chair. He could not bear the slightest touch on the back without crying out. He remained in this way some time, subsequently got a verdict and compensation amounting to some thousands of pounds, and then gradually and perfectly recovered.

Concussion of Brain and Spine

CASE.—W. H. P—, æt. 32. On January 7th was thrown out of a chaise on to the road; he was picked up insensible and brought home. He suffered for a day or two with all the ordinary symptoms of concussion of the brain, and then it was found that the spinal cord must have been also involved in the concussion. He could scarcely move the legs, sensation was imperfect, and he had pricking and numbness in the skin. The urgent symptoms passed off, and at the end of a month he was able to sit up in his chair. He could then only just walk across the room, his legs tottering under him, and as regards his mind it was still in a very torpid state; he was unable to read, but sat in his chair all day looking out at the window; often delirious at night. The only thing he complained of was a sense of constriction around his body and head. It was very clear that he had received a most severe shock to his whole nervous system, and whether it would end in a general paralysis of mind and body was impossible to say. He then went to Torquay, used salt-water baths and shampooing, and took tonics, mostly zinc, in increasing doses.

On his return at the end of two months he was comparatively well. He looked in good health, and walked pretty vigorously, although there was some hesitation in descending stairs. He also felt a constriction around his body, and could not occupy his mind as before the accident. After he had been at home some time and engaged in business he had a relapse, becoming more feeble in body and mind. This was in part attributable probably to an excess of stimulus which had been ordered him. He then went to Hastings, and again improved, and at the present time is quite well.

CASE.—A young man I have seen to-day is an example of what we are constantly meeting with, and about which such different opinions are given in courts of law as to the question of recovery. A post-office clerk was violently thrown down whilst in his van, but not struck upon any part. Now, some months afterwards, he is quite unfit for his work; he is thinner, can only walk slowly, his legs totter under him, his hands shake, his vision is impaired, his virility lost, and he hesitates in his speech. He is in that nervous state that the sound of an engine always aggravates his troubles.

As a consequence of a fall or injury there may arise a concussion of the spine, from which recovery may result, or an inflammatory process may be set up, which may be fatal in a shorter or a longer time. This may show itself as a myelitis or a meningitis, and be productive of various symptoms accordingly. In fact, any form of disease may possibly be started into action by an injury.

The following case is interesting from the suddenness of the symptoms, exemplifying, perhaps, the opinion held by some, of the ready laceration of fibres which have undergone softening. The small amount of mischief discovered in the cord would lead to the belief that if death had occurred earlier from paralysis of the chest no change whatever would have been appreciable in the medulla.

The next case is an example of the result of an injury at a later period of time.

CASE.—Joseph P—, æt. 32. He was a railway porter, and whilst engaged in pushing a railway truck along the line, he suddenly came to an ash-pit, when, for fear of falling, he made a jump into it, and ricked his back. He seemed for a moment to be powerless, but soon resumed his work. On the following day he continued also his work as usual. On the third day, whilst walking along the Borough, he suddenly fell in the street, and was unable to rise, owing to the weakness of his legs; this increased during the next two days, when he was brought to the hospital. He was then completely paraplegic, had no power over his bladder, and bed sores were already appearing. Subsequently the chest became affected, and he died in six weeks after the accident. The spinal cord appeared quite healthy to the naked eye. When examined by the microscope, some fatty granules were found in parts, but the change from the normal appeared very slight.

CASE.—Wm. A—, æt. 21. He fell on his back more than a year before his death, and then had symptoms referable to concussion of the spine. He gradually recovered, and resumed his work, when subsequently symptoms of paraplegia slowly came on. These gradually increased, so that for about four months before his death his legs were completely paralysed; then his arms became affected, and subsequently his eyesight. The intellect quite clear. After his death, when the cord was removed, it appeared at first healthy, as regards its general look and its firmness. A section, however, showed the presence of disease extending its whole length, and passing through the pons to the corpus striatum. There was

16

no disintegration or softening, but a remarkable change had occurred from the presence of a translucent albuminous material within its substance. This was for the most part situated towards the surface of the cord, so that a section showed its circumference converted into a grey translucent material. The contrast between the original white medullary matter within and the adventitious substance around it was very great. In some places the latter had penetrated more deeply, so as to involve the grey matter of the cord. The pons Varolii had on its surface two or three patches of the same material, and passing into the substance to the extent of about one eighth of an inch, and on the corpora striata and thalami optici, especially the former, there were some similar patches of translucent matter on the surface. These did not penetrate deeply, and were not observable in the interior. In this case death was due immediately to suppurative nephritis. I shall defer the subject of treatment until I have described some other forms of paralysis.

Some special results of spinal injury.—Besides these chronic effects arising from a general concussion of the spine, it is interesting to note some of the immediate results of injury. These necessarily vary with the part of the cord which is involved, and whether or not the nerves which issue from it also participate in the injury. Thus, if the latter are affected there may be great pain in the parts to which they are distributed, or the muscles may waste, or there may be various disturbances in the vaso-motor system, shown by alterations in temperature. According to the part of the spinal cord injured so may the respiratory process be interfered with, or the heart's action impeded, and, as I have before alluded to, the general temperature of the body may be remarkably disturbed.

I have notes of some cases which I have seen, and of others which have been related to me, showing these remarkable deviations in temperature, and which tend to corroborate the opinion of physiologists as to the existence in the spinal cord of a heat-regulating centre. Thus, in the case of a man who fractured the upper part of the cervical spine and injured the phrenic nerve, the temperature reached 107°. In another similar case the breathing became very slow, the pulse very slow, and the temperature rose to 107°, and after death it reached 109°. In another case of fractured sixth cervical vertebra the patient lived three days, and the temperature rose to 106°; and in another man, who lived ten days, it was 106°. The most remarkable case of high temperature on record is that described by Mr Teale, of a lady who injured her spine, and who perfectly recovered. Her temperature at one time reached 122°. In a case of fracture of lower cervical vertebræ, where patient lived five weeks, the temperature rose during the first few weeks to 103°, but subsequently it fell, and during the last three weeks was only 91°. Mr Hutchinson related the case of fracture in cervical region, where the patient lay like a corpse for five days from the depression of temperature. In other cases, too, the temperature has been lowered, but whether this is due to a different seat of the lesion is not explained. This heat centre is in the so-called cilio-spinal region, a part which has some influence over the eye through the sympathetic, as was exemplified in a case of fracture through the seventh cervical vertebra where the pupils were minutely contracted.

One cannot but contrast these cases with those where an affection of the abdomen exists, either from injury or from disease, and where the coldness of the body is so remarkable.

One might allude to another symptom of spinal injury—priapism. This seems to be due to a paralysis of the vaso-motor nerves which regulate the supply of blood to the vascular tissue, and in this way turgescence takes place. It has no reference to sensation, for it occurs when the pudic nerve is destroyed. Emissions continue, and cases are recorded where, owing to injury, a complete anæsthesia of the genital organs has existed, and yet fruitful copulation has taken place.

Fractured Spine. Permanent Paraplegia

CASE.—Bearing upon this, I may refer you to a case which was long under the care of Mr Cock. It is now seven years ago, and when he was 16 years of age, that he received the injury. Some wood fell upon him, fracturing his spine in the lower dorsal region. He was completely paralysed below the seat of injury; he remained in the hospital many months, when repair took place in the bones, but none in the cord, so that he returned home completely paralysed in the legs. He is now 23 years of age ; he lies in bed all day, and has grown since the injury; his legs not so well developed as his arms, but no especial wasting, as seen in progressive muscular atrophy. He has the legs flexed and everted, and he cannot feel below a line drawn around the body just below the ribs, although sensation reaches a little higher on one side than the other. All feeling is lost in the genital organs, and no sensation is experienced during micturition and defæcation. There is, however, a slight reflex movement passing up to the head when the bowels are moved. He often finds when he wakes that the penis is erect, and that he has had a seminal discharge, but no sensation has accompanied it. His chest had undergone a remarkable alteration in shape, having become quite flat, with a depression of the lower part of the sternum, and a bulging forward of the ribs on either side. He has often attacks of herpetic eruptions around the buttocks and backs of the thighs.

A case was lately related by Dr Muller of a woman who was stabbed in the back opposite the fourth dorsal vertebra, whereby the left half of the cord was severed, and also the right posterior roots of the nerves. The surgeon who was called in found her paralysed in the left leg, and with great pain in it, every touch being acutely felt. On the right side there was free movement, but no feeling. The upper extremities were unaffected. The left pupil was smaller than the right. An interrupted current was felt painfully

in the left leg, but there was no contraction. The right leg contracted, but there was no feeling in it.

While I am speaking of the effects of shock, I may observe that
these are by no means always due to concussion of the brain or
spinal cord, seeing that the sympathetic system must take its share
in producing some of the phenomena. These are best seen when we
are certain that neither the head nor back have received any injury.
For example, a little girl, æt. 9, fell down some steps flat on her
belly; she got up, and was put to bed; she was very faint and pale.
She did not rally, and after some hours she was thought to be
dying. On the following day she was very white, restless, with
dilated pupils, and pulse 140. There was no tenderness, fulness
of abdomen, or other evidence of any ruptured viscera. She slightly
rallied during the next two days, when the ankles were observed to
be somewhat swollen, and her urine was found to be scanty, slightly
bloody, and albuminous; this passed off in another three days. It
was then ascertained that she had passed no urine for several hours
after her fall. Nine days afterwards she was still very ill, pulse 140,
skin and conjunctiva slightly yellow. Recovery in a fortnight. It
seemed as if the shock to the nervous system had produced collapse
and disturbed the functions of all the organs of the body.

PROGRESSIVE MUSCULAR ATROPHY

The diseases of which I am now about to speak are connected
with changes in the motor columns of the cord, and for convenience sake I shall take at the same time those diseases which are
allied to them.

Commencing, in the first place, with disease of the anterior
columns of the cord, so as to include the grey matter of the anterior
cornu, the roots of the motor nerve are necessarily involved in the
morbid process; consequently, the function and nutrition of this
nerve are destroyed, together with an atrophy of the muscles to
which it is distributed. This, then, is the condition found in the
disease known as *progressive muscular atrophy*.

The disease had formerly been observed by Sir C. Bell and
others, but it is only a few years ago that Cruveilhier more particularly drew attention to it, and gave it a name. According to his
belief it is due to a degeneration of the muscular tissue. He stated
that the nerves of the limbs might sometimes be found wasted, as
well as the anterior roots proceeding from the spinal cord, but that
the cord was not itself primarily affected; for, if this was found

shrunken (as some had described) on its front aspect, this condition, together with the atrophy of the nerves, was altogether secondary.

Cruveilhier and his followers believed that the real seat of the disease was in the muscles, but later observers have proved after more accurate investigations the existence of a real and tangible lesion in the medulla itself—an opinion always held by Virchow and Gull, and more recently by Clarke and Charcot. The latest writer, however, on this disease, Friedreich, supports the original view, but I do not think myself that his facts or arguments weigh much against the observations of the recent investigators. There seems to be no theoretical objection against either statement—that the disease might in one case be primarily muscular, and in another be in the nerve centres, or even, thirdly, that it might commence in the nerves themselves. For one of the best ascertained facts in nerve pathology is that impairment of the motor nerve or its root in the cord is associated with muscular atrophy. There seems no reason, with this fact before us, to introduce another cause in the action of the vaso-motor nerve, since the one rests on a clinical and pathological basis, and the other is purely theoretical. I may here remind you of such a case as that of the biceps femoris, which would waste if the sciatic were injured, although its vaso-motor system in connection with the femoral artery remains unimpaired.

Clinically, the disease is a clear one; the muscles waste, and thus a form of paralysis is produced of a very striking kind. It appears to commence in the upper extremities, and is often confined to them, commencing in one arm or a part of the arm. Thus very frequently our patients walk into the hospital and appear to have little amiss with them until you observe their drooping shoulders, and the arms hanging at the sides as if they did not belong to them. On stripping the patient you see his remarkable condition —not a mere thinness or ordinary wasting from the absorption of fat, but a degeneration of the muscle itself; you see the acromion projecting and the deltoid flat, the trapezius wasted and the head falling forward; from this wasting of the trapezius and rhomboids the scapula is tilted up, and the inferior border raised so as to form a hollow, into which the hand can be placed. The biceps in the same manner is wasted, as well as the muscles of the forearm and hand; it is in the latter generally that you at once recognise the disease. The muscles of the thenar and hypothenar eminences have disappeared, and, owing to the shrinking of the interossei, there are deep furrows between the metacarpal bones. The hand is hollow, and the patient cannot separate his fingers. As a consequence, the fingers become drawn back until the hand puts on the appearance of the talons of a bird of prey, and thus the French have given it the name of *main en griffe*, or claw-like. The forearm in like manner has lost its roundness and has become flattened. The whole appearance of the patient is most striking ; he stands with his head bent forwards, or even in bad cases with his chin resting on his breast and his arms hanging down in front of him as if they were merely attached to him by strings or ligaments. His chest does not expand freely, and his abdomen is loose and protruding. If his legs have become affected, they have lost their roundness, and the muscles of the face may lastly have become involved, so that the patient presents an idiotic expression, and dribbles from the mouth. Finally, the chest may become more affected, then the laryngeal muscles ; the vital process of respiration is attacked, the voice is lost, mucus collects in the tubes, expectoration fails, and death ends the scene.

You will see that the muscles connected with the limbs are primarily affected, then those of the trunk, and the muscles of the special senses rarely. The disease usually commences with wasting of the small muscles of the hand, then of the forearm, and so progresses upwards, but this is not invariably the case, for in instances I shall presently mention the muscles of the forearm remained plump and firm, whilst those of the upper arm were wasted.

It is remarkable that just as in the locomotor ataxy the disease is more especially confined to the lower end of the cord, and as a consequence the legs may be solely affected, so in the progressive muscular atrophy it is the arms which are primarily and principally paralysed. In the early cases you will have carefully to test what muscles are affected, and to what degree, and you will find that the paralysis does not follow the distribution of any particular nerve. As you are treating your patient, you test the increase of power by making him raise his arm, then place it before him, see how far he can stretch it behind him, and then test the extension, flexion, and pronation of the forearm, &c. You may sometimes remark, as Cruveilhier pointed out, a remarkable tremor or quivering of the muscles, especially the trapezius when you attentively watch it ; or you may bring the movement out by gently tapping the surface. This last characteristic, however, may be met in numerous forms of disease where the muscle is wasted.

Galvanism affects the muscle according to the degree of change. In severe cases we sometimes find both the continuous and induced currents act, whilst sometimes the muscles react to faradization only. The sensibility of surface is not much affected, but sometimes slightly impaired.

The following case I shall have to refer to again, as it was not a simple case of the disease, but was combined with bulbar paralysis; a combination often seen, as the two affections are not pathologically distinct, different parts of the cord only being involved.

Case.—A man, æt. 46, has just died under my care from progressive muscular atrophy. He had all the symptoms which I have described in an excessive degree. He sat in his chair scarcely able to move from extreme muscular weakness; but not only did this apply to the limbs, but to the more important parts of the body. Thus he could only partially expand his chest, and had great difficulty in expectorating mucus from the bronchial tubes; he also had those parts affected which are involved in the labio-glosso-laryngeal paralysis. He could scarcely articulate in an intelligible manner; he had much trouble in swallowing his food, which collected in his cheeks and mouth; he could scarcely move his tongue, and hardly had any power to cough. I need not further particularise the symptoms described in cases of glosso-labio-laryngeal paralysis, but may say in a word that he had these in addition to the general muscular atrophy affecting the body. The case was therefore one of great interest; for the former class of symptoms, when standing alone, have been clearly proved to be due to disease of the medulla, and thus there existed an additional reason for supposing that the whole of the morbid phenomena might be owing to the same cause. All the severer symptoms increased, until a complete paralysis of the chest came on, when he quickly died.

The *post-mortem* examination did not reveal any evident change recognisable at first glance, but on more careful examination by removal of the arachnoid from the base of brain, medulla oblongata, and spinal cord, very marked alterations were seen to have occurred. Thus the hypoglossal nerve was very much wasted, being not more than a third of its natural size, and changed into a fine thread; in like manner the inner roots of the spinal accessory were much smaller than usual, and the same was true of all the anterior roots of the spinal nerves. This was more especially the case in the cervical region. This atrophy was not a questionable appearance, but one extremely well marked and evident to all the students. When sections of the cord itself were made it was found that this was not healthy; the anterior columns were smaller than natural; they were not however, softened, but, on the contrary, were firm, whilst the grey matter was sunken beneath them. The latter did not present a healthy appearance; its colour was not uniform, in some parts yellowish, in others there were deep red spots, with congested blood-vessels. At the lower part of the cervical region the grey matter was much larger than natural, and its colour more dark than seen in section below. The whole of the interior of the fourth ventricle presented an unusual appearance, and was evidently unhealthy, the surface having a reddish-brown aspect, differing very much from the ordinary surface.

Case.—The following is one of the most remarkable examples of recovery from a malady, apparently incurable, that I have ever witnessed, and one of the worst cases of progressive muscular atrophy that have ever been cured, for it is simply impossible that the disease could have existed in any more severe degree than was here present.

A girl, æt. 24, was sent to the hospital on July 4th, 1866, by my friend Dr Buzzard. She lived in the country, and owing to a number of circumstances connected with family affairs she began to fail in health about eighteen months before her admission. A weakness and wasting began in her arms, and then in other parts of the body, until in six months' time she was obliged to take to her bed. During the year she kept her bed she passed her motions involuntarily, and was in a perfectly helpless condition; menstruation had altogether ceased. On admission she was seen to be in the most pitiable condition that you can well imagine—she was so emaciated that she was little better than a skeleton. She lay on her back scarcely able to move or raise her arms from her side. Her fingers were contracted into the claw-like shape. The interossei seemed to have quite disappeared, so that the tips of one's fingers could be felt between the metacarpal bones. The radius and ulna showed their complete outline throughout. In the same manner the legs were wasted, and the abdomen was so flat that the spine could be clearly felt. In fact, all the muscles were so atrophied that I believed that they must have disappeared, a little fibre tissue remaining in their place. She had a slight blue line on the gums, which suggested poisoning by lead, and therefore Dr Buzzard took the trouble to visit her home, in order to see if she could have been poisoned unwittingly by this metal, but he failed altogether in proving it. She was, however, ordered some iodide of potassium and faradization to the arms and legs. The galvanism produced no effect on the extensor muscles. and only a slight one on the flexors. It was, however, rigidly followed up by Mr Branford Edwards and my other clerks, and to these gentlemen she owes her restoration to health. In two months' time it was very evident that she was better; she could move her limbs, and the muscles had grown visibly. In another month she could use a fork, and was able to write a little. In November she was able to get up and walk across the ward by means of a chair, and the catamenia had returned. She continued the faradization, and the cure progressed more quickly until January, when she left the hospital convalescent. It was three or four months after this that she called on me to show herself. I did not recognise her at first, as she was a ruddy plump girl, and said she was in good health.

The following case was long under notice, and presented certain peculiarities :

CASE.—Thomas B—, æt. 24, was under my care in the hospital on several occasions. The disease has been progressing in the slowest possible manner, beginning when he was about seven years of age, so that he has been quite unable to follow any employment, but obliged to live with his parents. The wasting commenced, according to his account, about the shoulder, and afterwards extended to the upper arm. About three years afterwards he found his legs becoming weak and wasted, which obliged him to walk on the outer side of the foot. He has sought relief at various hospitals, but without much benefit.

On admission it was seen that the progressive muscular atrophy had affected the greater part of the body, so that he could only walk with the greatest difficulty, and could scarcely raise himself from the bed. When he walked he trod on the outer side of the foot, the under part turning inwards. When he stood, his head projected forward, as well as the abdomen, the spine taking a corresponding curve, the arms meanwhile dropping at the side. When more carefully examined it was found that some of the muscles of the face were affected; he had a blank, expressionless face, although he was really intelligent; the muscles supplied by the facial being involved, as the orbicularis palpebrarum, prevented him closing his eyes tightly, and the buccinator was also somewhat wasted; the masseters good, and he breathed and swallowed without difficulty. No strabismus, but he said it existed once. Protruded the tongue straight. The sternal portion of the sterno-mastoid appeared to have almost gone; on moving the head the omo-hyoid was visible.

As regards the upper extremities, the deltoid, biceps, coraco-brachialis, and brachialis anticus, were found much wasted, and also the trapezius; the scapula could be rotated, in such a manner that its base became horizontal, and the inferior angle pointed inwards. He was quite unable to raise his arm at a right angle with the body. The pectorals much wasted, the left side of chest was much flatter than the right, and the eighth and ninth ribs projected forward. The muscles of the arm were so wasted that the arm could be easily spanned with the finger and thumb, and this smallness of the arm contrasted strangely with the size of the forearm, which appeared quite unaffected, at least as regards its size, the muscles feeling large and firm; the hands also were but little affected, which was unusual, the fingers being perfectly straight. The muscles of the thigh were wasted, as also were the peronei and those of the calf. All the functions of the body were properly performed, and the temperature was ordinarily normal. He was galvanised for a very lengthened period along the course of the spine. There was no evident result from it, but the patient always maintained that it did him good, and was anxious for its use. It was curious, too, that the muscles did not react to the continuous current, but more readily to faradization. On three different occasions, at some weeks' intervals, he was seized with severe febrile attacks, which confined him to his bed for several days. He was unable to rise, had pain in the back and loins, and felt excessively low. These attacks were probably due to some nervous disturbance in the cord, as I have seen similar ones in the course of locomotor ataxy and the infantile paralysis of children.

I have notes of several other cases of progressive muscular atrophy, each presenting its own peculiarities, but most of them began, without apparent cause, in the arms. For example, that of a man, where the wasting commenced in the right arm, subsequently went to the left arm, and then to the leg. Also in a man, lately in the hospital, where the atrophy commenced in one arm, and after several months affected the other. In his case the shoulder-blades stood out like the wings of a bird, the outer edge horizontal, and the external and inferior angle on the same level. This was due mainly to extreme atrophy of the rhomboids.

Pathology.—I have already told you that opinions differ at the present time as to the true pathology of this disease, although there is nearly perfect agreement as to the facts. The muscles are found to have undergone an atrophy, the various elements have withered, and granules have taken the place of the markings. Some of these granules are composed of fat, but the change is not due primarily to a fatty degeneration; it is rather a granular alteration, and with this is associated sometimes an excess of fibrous tissue. Cruveilhier, who first accurately described the disease, believed that its origin lay in the muscle, although he had observed a wasting of the motor nerves as they entered the spinal cord, together with a slight atrophy of the cord itself. He found this diminution of the nerves especially in the cervical region, where they were changed into lustreless strings, containing only a few nerve tubules, whilst the posterior was healthy. He believed, therefore, that in a mixed nerve any wasting within it was due to the change in the motor portion, and he was confirmed in this belief also by observing the great atrophy in one or two cases of the lingual nerve, this being reduced to a third of its natural size. The additional fact discovered since Cruveilhier's time is that not only is the muscle atrophied, together with the motor nerve which supplies it, but also the anterior cornu of the grey matter to which the nerve is attached.

The more recent investigations tend to show that the spinal cord is primarily at fault. Virchow maintained this as well as my colleague, Sir W. Gull, who had published a case where marked disease was found in the cervical region of the cord, the central canal being widely dilated in this region. Cases also have been published where it has originated from injury or disease of the spine, and I myself have seen cases where the disease broke out simultaneously in all parts of the body, together with other spinal symptoms, so that there could be no doubt as to its central origin. Friedreich is the author who has recently written on this disease, and maintains the muscular or myopathic view of its origin, and combats the central view mainly on the fact that particular muscles or portions of them may be affected by the atrophic changes, whilst the rest of the body remains whole ; but the objection is valueless if it is believed that every nerve fibre has its origin in certain cells of the cord, for then it might easily be conceived that a morbid change in the medulla affecting a few cells might influence a correspondingly small number of fibres of a muscle. His objection, too, that other spinal symptoms are wanting is met by the answer that the only portions of the cord which are affected are those which involve the nutrition of the muscles. Or if it be said that overwork of a muscle may cause its fatigue and subsequent atrophy, the same suggestion might apply to the grey centre which rules over that muscle.

The opinion of Friedreich is that the disease is primarily and essentially one of the muscles, a myopathia and not a neuropathia, the first change being in the perimysium, as a hyperplastic growth in the interstitial cellular tissue, between the primitive bundles ; then follows a swelling, increase of the muscular nuclei, a disappearance of the markings, ending in a waxy or fatty degeneration. The result is a fibrous degeneration or cirrhosis of the muscle. Any lipomatous state is merely accessory. Friedreich believes the nerve changes are secondary, beginning first in the intra-muscular nerve, and then continuing upwards as an ascending degenerative neuritis, or leading perhaps to a chronic myelitis of the cord.

Besides the two theories of a central cause and a primarily local onc, it is possible, as Jaccoud has intimated, that it sometimes might begin in the nerve. We know that injury to a nerve will cause wasting of the muscle to which it is distributed, as, for example, an injury to the circumflex a wasting of the deltoid, or injury to the hypoglossal a wasting of the tongue; and therefore it is quite possible that some general affection of the nerves themselves might occasionally be the cause of the malady. Jaccoud gives full details of a case where, from the peculiar distribution of the wasted muscles, the neurotic symptoms and absence of anything denoting a central spinal disease, he was fully convinced that the origin of the malady lay in a pretty general neuritis. Then, again, the disease might begin in the nerves and ascend to the cord. I myself had a case where a man injured his arm, and the limb subsequently became useless and wasted; subsequently the other arm became involved, and then the legs. Such a case would almost suggest an ascending neuritis along the arm to the cord. I have also a case in my note-book where a man had been standing in the water for several hours; this was followed by numbness, anæsthesia, and general weakness. He afterwards had the same feeling in the arms. The limbs then gradually grew weaker, and the muscles began to waste. At the end of four years he died, when the muscles were found to be degenerated, the nerve fibres wasted, and the neurilemma thickened. The roots of the nerves at the junction of the spine were small, and the cord itself, as seen by the naked eye, did not look healthy. At that time the microscopic method of investigation was not known. The case was regarded at the time as one of primary peripheral paralysis.

A case has been recorded by Müller, showing the intimate association between the cord, the nerve, and the muscle, where a patient had had a club foot and withered leg from infancy. The muscles of the limb had undergone fatty degeneration as well as the nerves supplying them; also the anterior roots of these nerves as they entered the cord, and degeneration of the anterior cornu of the corresponding portion of the grey matter. Vulpian has noticed that in cases of amputation of a limb in young subjects that part of the spinal cord furnishing nerves to the limb undergoes atrophy, especially in the posterior cornu and columns. My late colleague, Dr Thompson Dickson, had an opportunity of comparing the cords of a case of progressive muscular atrophy and a case of old amputation, and although in both he found changes in the tissue, they were of different kinds. In the latter there was merely an atrophy, whilst in the former new products were apparent.

Quite recently, a very perfect case of the pathology of progressive muscular atrophy has been recorded in the French journals by

Dr Frosier, where a young man died with universal atrophy of the muscles, which began in one limb, and then progressed in the usual manner, until the chest was involved. He found on examination of the spinal cord an entire absence of the large branched cells in the anterior cornu, a few atrophied ones alone remaining, or some granule cells replacing them. The disease was almost limited to the cervical region. The anterior spinal coats were atrophied, but not the posterior, and the spinal accessory and lingual were included in the atrophy. Some slight thickening and pigmentation of the pia mater were also present. The author believed the change to be a primary one in the grey matter, the nerves being secondarily affected, and subsequently the muscles. In the analogous disease, the infantile paralysis, a febrile condition precedes the visible paralysis, as if some acute mischief was in progress, and occasionally progressive muscular atrophy has a very acute history. Thus, in six weeks, a woman, I saw with Dr Taylor, of Kennington, was rendered perfectly helpless. She sat in a chair, with her head thrown forward, scarcely able to move her arms, legs straddling, pains in all her limbs, and the muscles wasted. The only conjectural cause was spirit drinking.

I should say that, having regard to all these cases, the conclusion seems to be warranted that the spinal cord may be regarded as the seat of the disease in most instances, although there seems no theoretical or clinical objection to the opinion that the disease may sometimes have its origin in the periphery. There can be no doubt that the muscles may waste under these three different conditions : primary morbid change in the grey matter of the cord, lesions of the trunks of the nerves, primary change in the muscles themselves.

In a purely local case there is somewhat more difficulty in believing the cause to be in the cord than in the limb itself, as in the following example :

CASE.—Adelaide H. M—, governess, was admitted into Clinical ward on account of her hands having been quite useless for six years. The fingers were contracted, the joints stiff, and the muscles wasted. The muscles of the forearm also very small and flabby, whilst those of the upper arm were natural. The thumbs flexed on hand, fingers rigid, and little finger firmly contracted, When endeavouring to hold a pen she soon lost control over it, from the pain and cramping in the muscles. Ordinary sensation was perfect. Good muscular sensibility shown to faradization, but slight to continuous current. In every other respect the girl was healthy. No improvement after several weeks' trial of galvanism.

I am seeing a gentleman who, three years ago, began to have wasting of muscles of thumb, and slightly of others supplied by the ulnar nerve. He used galvanism, and they have slightly grown. The disease appears to be quite local.

In the progressive muscular atrophy of which we have been speaking, there is disease of the motor nuclei of the cord, causing the arms to be chiefly affected ; if we suppose the same form of disease to occur higher up, we should have paralysis of the cranial nerves. Now, this does occur in the medulla oblongata, involving the special centres of speech and deglutition. It was first described by Trousseau under the name of "labio-glosso-laryngeal paralysis," and subsequently as " bulbar paralysis." It may not be difficult to declare that this part of the cord may suffer disease or degeneration like any other structure, but the explanation is not forthcoming why the morbid process should so accurately involve one important centre, why it should occur in the young, why sometimes come on suddenly, and why, moreover, it should be in part recovered from. Its true pathology and causes have yet to be learned, but in fatal cases there is found disease or degeneration of the motor centres in the fourth ventricle, whence important nerves arise, whose paralysis characterises the disease. It is a paralysis, as the name implies, affecting the lips, mouth, tongue, and larynx ; and therefore, as might be supposed, the functions of eating, swallowing, and talking, are much interfered with ; the nerves known as the seventh, eighth, and ninth, being in part paralysed. Whether the affection has come on suddenly, or whether it has been developed slowly, the phenomena are the same. These are so striking, that the nature of the case is soon evident. The face has lost its expression from a partial paralysis of the facial nerves, and should the sufferer attempt to speak, it is in vain, for beyond making a few unintelligible noises, his power of utterance is gone. The reason for this will be found in a weakened condition, not only of the muscles of the face and of the tongue, but of the larynx itself. The lips can be adjusted only for the formation of certain letters, as Trousseau has fully explained ; the tongue can be but slightly moved, and cannot be thrust out of the mouth ; and when the patient is asked to cough, he produces only the faintest sound in his larynx, not being able to close the organ. At the same time he eats with difficulty ; he cannot collect the food in his mouth ; he is obliged to assist with his fingers to extract it from his cheeks, and place it at the back of the tongue, when it is swallowed with difficulty. For the same reason the saliva cannot be retained, but is constantly pouring from the mouth ; the muscles of the soft palate sometimes hang down flabby, so that the posterior nares cannot be closed, and both the velum and the larynx may have lost some of their sensibility. There may be some question about this, as there is no loss of sensibility of the cutaneous surface. The appearance of such a patient is generally very striking and characteristic; he is seen holding a pocket-handkerchief to the mouth, which falls open while the lower lip hangs down; the expression is vacant, or varied only by the few grotesque movements of the face made in the endeavour to force out a word, and a slate or paper lies before him, on which he writes down all his wants. The speech, it may be remarked, is not merely thick, as in simple facial paralysis, nor is there that meaningless gabble which is heard in the aphasic patient; it is either utterly lost, or only a syllable in a nasal twang[1] can be produced at a time after violent attempts to set the muscles in motion; there is, in fact, a paralysis of all the parts employed in talking.

It may be remarked that though the capability of speech is entirely destroyed, from a paralysis of the nerves which supply the muscles, yet the trunk of the nerve need not be wholly paralysed, nor have other parts supplied by it lost the whole of their functions. Thus the face may be fallen and the mouth paralysed, so that the patient may not be able to move the mouth well, as in blowing or whistling, but he has power to close the eyes, showing that the orbicularis palpebrarum is not affected. In the same way, although the larynx is paralysed for talking, it is unimpared for breathing. This would show, Trousseau observes, that for its two separate functions, vocalisation and respiration, it must have two nerves, supplied from different sources. Now, the recurrent is almost the sole motor trunk to the muscles of the larynx, and, consequently, if it is injured or pressed upon, the organ is wholly paralysed and the patient is suffocated. It would follow, then, that this nerve is a compound one, and sends a twofold stimulation to the muscles by filaments having their sources in the centres of respiration and vocalisation. Marshall says: "When the roots of the spinal accessory are cut, the operation does not impair any of the respiratory movements, but swallowing is interfered with and the voice ceases, the animal emitting only a bubbling noise. Extirpation of one accessory nerve causes hoarseness. Thus it appears that the spinal accessory governs the momentary and voluntary opening or closure of the glottis and tension of the vocal cords necessary for the production of the voice, or for the exercise of general muscular effort, whilst the respiratory movements of the glottis are under the control of the pneumogastrics." It has long been considered that there is a region in the medulla which may be called the respiratory tract, a region to which

[1] It is curious that when any one has a cold we should say " he talks through his nose," when we mean exactly the reverse.

branches of all the nerves engaged in the respiratory process may owe their origin; in like manner, it would appear that as a large number of parts are engaged in the act of talking, so the nerves supplying them must be stimulated from a common centre, and herein lies the explanation how so complex a function should suddenly cease from lesion of one small spot. Now, the proof of this lies in the dissections of Mr Lockhart Clarke, which demonstrate the connection between the facial, vagus, hypoglossal, and laryngeal nerves. The latter are, in fact, branches of the spinal accessory which, joining the pneumogastric, are given off as the recurrent laryngeal motor nerves. The spinal accessory has two origins: the lower from rootlets arising from the antero-lateral substance of the spinal cord and lower part of the medulla, and collected into the external branch to supply the sterno-mastoid and trapezius muscles; the upper from a special nucleus behind the central canal, which, going to form the internal branch, proceeds to the vagus, and is subsequently distributed to the larynx, pharynx, and palate. If, then, the centre whence this proceeds be injured, the larynx loses that power which this nerve had previously induced, that is, there is a loss of vocalisation, whilst the respiratory power remains. Mr Lockhart Clarke has shown that there is a close anatomical connection between the nuclei of the hypoglossal, vagus, spinal accessory, facial, and trigeminal nerves. There is a column of cells forming the nuclei of these nerves, which supplies all the parts used in speaking, found on the floor of the fourth ventricle; and it is these which have undergone a change. They are continuous with the grey matter of the anterior columns. The sensory nuclei lying on their outer side escape.

It appears remarkable that a small area in the medulla oblongata, coinciding with a physiological centre as that of articulation, should be picked out to undergo a rapid or slow morbid change. So remarkable is the fact that it might be worthy, in the first place, of inquiry whether or not experience justifies us in declaring that definite parts of the cerebro-spinal centres having special functions are more prone to disease than other portions of the brain and spinal cord taken indifferently; whether, indeed, all parts are not equally liable to inflammation and degeneration, but it is only when certain physiological portions are affected that we are enabled to apply definite names, because then the seat of disease has made itself manifest by the implication of nerves whose function is known, whilst in other cases we are content to use such expressions as cerebral or spinal disease. Although I believe this to be to a certain extent true, yet I consider it is proved that those parts of the cerebro-spinal system which have definite physiological properties are more liable to disease than other spots taken indiscriminately; if so, it may show, as is most probably the case, either that the vascular supply of such physiological centres is accurately defined and circumscribed, or that a centre having a definite function, being the focus of a number of nerve filaments proceeding from it for a special purpose, must soon be involved if any of these filaments proceeding to it be primarily attacked, seeing that morbid processes choose given anatomical tracks. If, then, degenerative processes occur in connection with a morbid state of blood-vessels, and if the anatomical supply bears a relation to defined physiological areas, the explanation of such parts being selected for chronic disease is not so difficult ; and if, again, morbid processes proceed rapidly along nerve filaments, we can understand also how parts having intimate relations are concurrently affected.

It is a remarkable circumstance that there is no disease of the nervous system, as far as I am aware, which may prove fatal, and even show a well-marked lesion or degenerative change after death, but may have its counterpart in a functional and curable disorder. In hysteria it is known that every possible nervous disorder may be simulated, and amongst these I have seen a tolerably fair example of labio-glosso-laryngeal paralysis.

It is worthy of note that in the disease especially under consideration the symptoms appeared suddenly in some of the cases, in others they were of slower progress. In the former it is possible that an effusion of blood might have taken place in this specialised seat of the medulla, whilst in the latter a slow morbid change constituting the true progressive form of the disease. It is not remarkable that in some cases the motor tracts should be also involved, and therefore combined with the symptoms above mentioned, there should exist also various degrees of paraplegia or paralysis of the limbs, and, if the motor cells are involved, an atrophy also of the muscles.

I have seen several cases of this form of disease in private. In one, an old lady, lately dead, the disease had been progressing for some years ; her difficulty of swallowing had been so great that on one occasion a probang was passed down the throat, in order to see if there was any obstruction. In another case, of a woman of middle age, the attack came on suddenly, as one of ordinary hemiplegia. She rapidly recovered the use of her limbs, so as to be able to walk two or three miles daily, but she remained speechless ; she could not protrude her tongue, and could scarcely open her mouth ; she was fed with a spoon, and the saliva was constantly dribbling from her mouth. In the case of a lady, somewhat older, whom I watched for two or three years, the attack came on as a fit during dinner ; she fell off her chair, and was taken up to bed ; it was found that her senses had not left her, but she was unable to speak. In a day or two she got up, and appeared very well; but she never spoke again, and could not swallow without great difficulty. She subsequently attended to her household affairs, would play cards with the family, and walk three or four miles daily, but she was obliged to communicate all her wants by writing. Her greatest trouble, however, was the inability to hold her saliva, which was continually dribbling from her mouth. She had finally a fatal apoplectic attack, in which the effused blood ploughed up the pons Varolii; at its lower part there was an old brownish cyst.

In the case of a woman who was under my care in the hospital some years ago with this form of disease, combined with partial paraplegia, she was unfortunately allowed to feed herself, and on one occasion a large piece of meat stuck in her throat and choked her—an accident not unlikely to happen in this disorder.

CASE.—Dominick K—, æt. 31. The patient, a bricklayer's labourer, was a single man, and of temperate habits. He always enjoyed good health until the middle of July, when he went to bed well, but was unable to rise the following morning, having lost the use of his legs and arms during the night. His left side was paralysed in a greater degree than the right, and his speech and power of deglutition were also affected. He had been under medical treatment up to the date of admission, and his health had become slightly improved in consequence. He was able to walk without a stick, but with a tottering gait, though he was scarcely able to raise his feet from the floor. He could stand on the right leg without support, but not on the left, and his left knee was stiff.

Mastication and deglutition were difficult, and the tongue was only capable of very slow protrusion and retraction. The upper part of his face was unaffected; he could close his eyes firmly and quickly, but the lower part of his face was almost motionless; he could not whistle, and there was a want of expression in his countenance. His speech was thick, so that it was difficult to understand what he said. Six to eight ounces of saliva flowed daily from his mouth, showing that the amount of saliva is immensely increased. This, no doubt, is due to the implication of the chorda-tympani nerve.

His urine passed from him very slowly, and at times he had to wait a few minutes before he could micturate. His urine was not albuminous; his bowels were regular; his tongue was clean, and he was in no pain; his appetite was very good. Tactile sensibility perfect.

At the end of a month, there was little change. His vocal cords were seen, by the aid of the laryngoscope, to move freely, both during respiration and when he made an effort to utter a sound; but when he tried to cough the vocal cords scarcely moved at all, and he was quite unable to effect his purpose, a slight hacking movement of expiration being all he could accomplish. This was probably due to a loss of co-ordination, as the vocal cords could be moved during speech.

I may here remind you that the function of the olivary body is thought to co-ordinate the movements of speech, and that the corpus dentatum is continuous with the anterior cornu of grey matter, but it is not necessarily involved in the disease.

He was readmitted at the beginning of the following year, and remained in about two months, during which time his condition somewhat improved. He could walk about the ward, though dragging his legs, and had some more power in his arms. As regards his speech, he made a great contortion of his face in

17

order to produce a word, but it was more intellible than heretofore. On examination of the larynx with the speculum the right vocal cord moved slightly, but the left not at all. He seemed to have power over the soft palate to raise it.

CASE.—Mary Jane D—, æt. 51, had been in the hospital on several occasions; first in 1864, then in 1866, and again in 1867. She was a married woman with a large family. Her history as she endeavoured to relate it by monosyllables and by writing was, that on May 14th, 1864, she went to bed quite well, but awoke early in the morning, finding the right arm powerless and the right leg weak; the speech was somewhat affected, but this improved in the course of the day. In four or five months the arm recovered sufficiently to enable her to use her needle, and she remained tolerably well until the following May, when she had another attack, but on this occasion her jaw was almost fixed, so that she had great difficulty in speaking and eating. During the following six weeks a gradually increasing paralysis came over her, affecting all her limbs and her face.

When she was admitted on January 31st, 1866, she was observed to be a thin, short, old-looking woman, having an anxious expression of countenance, and not able to walk with any vigour from weakness of the legs; the arms were also somewhat weak, but more especially the right. She had almost total loss of utterance, so that on endeavouring to speak she only made some almost unintelligible noises; the voice was also weak. She had some difficulty in opening the mouth, which was drawn slightly to the left side; she also had some difficulty in closing the right eye, and the lower lid of the left eye was slightly drawn down; there was thus a more or less paralysis of all the muscles of the face; no loss of sensation. For a long time she had been unable to swallow any solid food, and had been living on liquids, always taking care to place every thing far back on her tongue. There also appeared to be some paralysis of the right pillar of the fauces, some dimness of sight and tremor of lips. She had headache, especially over the forehead. She used a slate and pencil to communicate her wants. Heart and lungs healthy. She remained in hospital until September.

She was again admitted under my care, October 30th, 1867, and remained in for six months without much alteration in her condition. She sat in her chair all day long, as she could not walk well; the arms were also weak, although she was able to write. She had completely lost the power of utterance, and was in the habit of putting down all her wants on a slate. She usually sat with a handkerchief to her mouth to catch the saliva which was constantly dribbling from it.

CASE.—Mr H—, æt. 27, a fine young man, who had lived rather freely, and might very probably have had constitutional syphilis, was seized with a fit on the night of January 5th, 1869. This appeared to be of the ordinary hemiplegic character, arising from effusion of blood. Feeling ill he attempted to get out of bed, but then he fell, the noise produced arousing those in the house, who found him on the floor and put him to bed, and when I saw him in a few hours afterwards he was paralysed on the left side, but was quite conscious; he rapidly recovered, and at the end of a month was able to walk about and return to his employment. He had never, however, completely regained the strength of the arm and leg. On August 3rd he again had a fit, but on this occasion it was of an epileptic nature, and soon afterwards he had another, and then a succession of them for a few hours. In these attacks he struggled violently, but he said never lost his consciousness; and between the paroxysms he talked quite rationally. On the following day he was better, and had no more fits, but it was observed that his speech was failing, and at the end of the week he could not utter a word. When I saw him again, and during some weeks afterwards (even to the present time), he was the subject of the complaint under consideration in its most marked form. His face had lost somewhat of its usual expression, and when he smiled the mouth on the right side was slightly drawn up, at the same time the lips were well retracted, so as to show the teeth, proving that the orbicularis oris still retained much of its power. He could also close his eyes. If asked to speak he opened his mouth and laughed, but could not utter a single word. Not only was he incapable of forming a word with his lips, but his larynx failed to produce the feeblest note. On asking him to cough it was only once that a slight gurgling was made; in all other attempts he could not produce the faintest sound. The saliva was running from his mouth, necessitating the constant use of a handkerchief. When requested to drink, he allowed a good deal of the fluid to escape from his mouth. He was said to have much difficulty in eating and swallowing, the food collecting in his cheeks, and thus it was generally placed far back on the tongue to enable him to grasp it. He could protrude his tongue a little distance from the mouth, but it was done slowly and with effort. On examining the throat, the velum was seen to hang loosely down. He had no power to raise it, and touching it with the feather of a pen did not excite it to action. He, however, said he could feel it being touched. His intellect was quite clear.

These cases shew that in the simplest form of the disease, there is a paralysis of articulation and deglutition; while the facial, lingual, and laryngeal nerves are especially affected. It will be observed, however, that the orbicularis oris is more involved than the muscles of expression, and that the tongue can be moved slightly backwards and forwards, probably from some of its muscles being supplied by the seventh nerve. If the degenerative process extends beyond the primary defined limit, then the paralysis of the muscles of the face also increases, and at the same time the limbs may become involved. The sensory nuclei and nerves appear to escape. In the worst form the paralysis may be associated with atrophy of the muscles. This is not often witnessed in the face, and the reason may be that the facial nerve has other grey centres of origin besides the one involved in this form of disease, but as regards the ninth nerve it does seem, from cases which have been recorded, that the muscles of the tongue which it supplies undergo an atrophy.

The simple form we may regard as that where expression, phonation, mastication, and deglutition, being alone involved, the disease is very localised. Should it spread to the spinal tracts, a paralysis of the limbs will ensue; if to the grey matter of the anterior columns, an atrophy also. Besides this more complex form, owing to an extension of the primary disease, which is a remarkably localised malady, we may meet with a bulbar paralysis, as a part of a more general change in the cord, due to various coarser lesions.

But lately I saw a post-mortem made of a case of Dr Pavy's, in which the pons Varolii was found softened on one side, the brain elsewhere being quite healthy. The man had first weakness of one arm, then of one leg, and afterwards of both arms and legs.

He had difficulty in articulating, in masticating and swallowing, and could scarcely protrude his tongue beyond his teeth.

In the simple form, I believe it was Lockhart Clarke who first showed the degenerative changes in the nuclei which are situated along the median line of the fourth ventricle, as in the case presently to be mentioned, where the focus of the change was at the nib of the calamus scriptorius. In more complex cases the same parts may be merely involved in larger degenerations, as in softening or sclerosis, which have commenced elsewhere in the cord.

Whether the primary change in the simple form is of an inflammatory or other nature is not very clear, for its pathology must include an explanation of the cases of recovery.

I have repeatedly said that I know of no organic nerve disease which cannot have its counterpart in a functional or curable one; that is, where a healthy nerve centre may not cease to be active or functionize, and be productive of the same symptoms as if that centre were diseased. We have no difficulty in appreciating the fact as regards the brain during sleeping and waking, when the terms dormant and active are supposed to correspond to our ideas of the different cerebral states; now, whatever these may be, we have only to carry the same idea into the spinal system to understand the nature of many forms of functional paralysis. Should any part of the spinal system sleep a paralysis would ensue, and should it be a part which rules over vital processes death would necessarily take place.

The following is a case of bulbar paralysis, in which the patient for a short time was on the brink of death, but this peril having been escaped, he rapidly recovered:

Acute Bulbar Paralysis. Cured

CASE.—Thomas H—, æt. 43, admitted into Guy's Hospital on Nov. 29th. Four days before, the attack came on as a fit, his limbs became rigid and powerless, but he did not lose his consciousness. Subsequently, his left side was spasmodically affected; he then was unable to speak, and he snored. The description of all these symptoms was very obscure and imperfect. When admitted he seemed in good condition, but his face was expressionless, the mouth open, and corners depressed. He lay on his back, and was too helpless to turn himself without assistance; pupils minutely contracted. Tongue could only be protruded as far as the teeth, could not touch the roof of the mouth, and was covered with unswallowed food. Palate hung flabbily down, and did not respond to any touch. Great difficulty in masticating and swallowing; could scarcely hold the beak of the feeder in his mouth, and then the food collected in his cheeks, and was obliged to be washed out. Saliva constantly running from the corners of the mouth. His voice was jerky and guttural; when asked to cough he could only make the faintest sound. Respiration chiefly abdominal and diaphragmatic; at each respiration a sonorous sound was made in the throat. The intellect was quite clear, and sight unaffected. His speech most indistinct, and he could pronounce no letter where movement of the tongue was required. He could only just raise his arms, but could not grasp anything firmly; he could move his legs, but was unable to stand upon them. Sensation and taste were perfect, and no pain anywhere. It was thought that he had acute bulbar paralysis, involving the lateral tracts of the cord. The fact of his chest being involved in the paralysis made the prognosis most doubtful. He remained in this precarious state for three days, when he began markedly to improve; he could move his limbs better, his breathing was more natural, and we could understand him better when he endeavoured to speak. On the next day the improvement was still marked, and he could swallow with less difficulty. On Dec. 11th he could be easily understood, and he was able to sit up in bed. On the following day he was able to stand. On the 19th he was walking about in the ward, and all the paralytic symptoms had passed off. He left on the 27th, all but well.

The following was a case where all the symptoms of bulbar paralysis were present, but no organic disease was found after death :

CASE.—A stout girl, looking well, came to the hospital on account of general weakness; she could scarcely walk or move about, she spoke slowly, and had slight strabismus. The house-physician was inclined to regard the case as one of hysteria, and being an authority on diseases of the eye saw nothing in the strabismus incompatible with this view. She remained in the same condition, appearing very lethargic in her manner, for about a month, when one day I found she could scarcely walk, and her speech was very indistinct. In two or three days these symptoms had increased, until she presented all the conditions of bulbar paralysis in a modified form. Subsequently she spoke most indistinctly, swallowed with difficulty, and was unable to cough. She, however, was able to get up, when one day after going to bed she was seized with difficulty of breathing, and quickly died. The medulla oblongata was very carefully examined, and nothing very tangible was found amiss with it.

As illustrating the effects of disease or disturbance of the medulla oblongata, I will briefly mention the case of a lady who had a concussion of the spine from a fall.

Concussion of Medulla Oblongata

CASE.—A lady fell off a pair of steps on to the back of her neck. This was followed by weakness of both arms, difficulty in deglutition, and sometimes choking. For a long time she had to be very careful how she drank, and if her head was slightly inclined to either side a small quantity of fluid would enter the larynx and cause choking. She had difficulty in speaking, she could scarcely protrude her tongue, and there was a considerable flow of saliva. It was many months before she recovered.

Progressive atrophy and bulbar paralysis combined.—It is very evident that the two diseases which I have been describing, progressive muscular atrophy, and bulbar paralysis, although clinically distinct, owing to different portions of the cord being involved, are pathologically closely allied if actually not alike ; in the one case the upper part of the cord being affected, whereby the paralysis is first seen in the upper extremities, in the other the medulla oblongata, whereby the cranial nerves are involved and the paralysis takes place in the face, tongue, and adjacent parts. But it is evident that should an extension of either disease happen, upwards or downwards, the two affections would be combined. This frequently occurs, and thus the labio-glosso-laryngeal paralysis is often seen associated with the progressive muscular atrophy. It requires merely an extension downwards from the medulla oblongata to the spinal cord for all the phenomena of the latter disease to arise.

Several cases have now been recorded where all the symptoms of bulbar paralysis were combined with those of muscular atrophy of the limbs and of the tongue. In these cases disease has been found in the lateral column and grey matter of the medulla oblongata and in the cord below.

Labio-glosso-laryngeal Paralysis, combined with Muscular Atrophy. Atrophy of the Medulla Oblongata, with Atrophy and Degeneration of the Spinal Motor Tracts and the Motor Roots of the Nerves.

(From the report of the ward clerk, Mr. Mallam.)

CASE.—William C—, æt. 46, admitted under Dr Wilks, November 9th, 1867, and died December 28th. He was a leather-dresser by trade. Five years ago he was in the hospital for rheumatic fever, since then he has enjoyed good health until June last, when he began to experience some soreness in the throat and difficulty in swallowing. Towards the latter end of September he lost partial use of his hands and legs, the left side being most affected ; but he continued at work until three weeks before admission, when he fell down and was unable to rise again. He has gradually been getting worse since.

On admission he was seen to be a short, old-looking man, with his head sunk between his shoulders, and a vacant expression of countenance. He was thin, and his muscles flabby. His eyesight had of late become much impaired ; but his pupils contracted under the influence of light, and there was no paralysis of the muscles of the eyeball. He had had pain for some time in the course of the fifth nerve, tactile sensibility was good over the face, and the muscles of mastication appeared to act well. There was a want of expression in the face, and although the mouth was not drawn to either side, the orbicularis oris had lost some of its power, as the saliva was constantly running out of the corners of the mouth ; the buccinators appeared quite useless for the purpose of mastication, and he was obliged to press the food out of his cheeks with his fingers whilst eating. He could close the eyes, and the hearing was good. The back of the throat and soft palate appeared sensitive when touched, but the contractility of the latter seemed much impaired. He had much difficulty in swallowing food, and had to wash it down with fluid. He appeared to have lost power over the tongue, being able to move it but slightly. He spoke very indistinctly and thickly, so that his words were scarcely intelligible. It was seen also that he could scarcely move his chest, and that the breathing was mostly diaphragmatic ; the chest was resonant, but on auscultation was found to be full of râles. He had great difficulty in expelling the mucus, being quite unable to cough out. As regards his limbs, there was a general deficiency in power, especially on the left side, so that he was scarcely able to support himself, and had very little use of his arms. The muscles at the same time were wasted, as was more especially apparent in the arms, the wrists dropped, the fingers were flexed, and the inter-ossei atrophied.

It will be observed that this man was partially paralysed in his limbs, and had almost lost the power of eating, swallowing, talking, or coughing, from paralysis of certain muscles above named. It will also be observed that, besides the labio-glosso-laryngeal paralysis, he had progressive muscular atrophy.

He was ordered to be galvanised with the continuous current every day for a quarter of an hour, one pole to be placed behind the mastoid process on the left side, the other lower down on the spine, and to take quinine mixture. The extensors of both hands were brought into action when either the induced or the continuous current was applied; but as regards the interossei, those of the right side were alone affected.

It appeared as if the galvanism was giving some tone to his muscles, and he expressed himself better, but at the same time it was evident that he was in constant danger of suffocation, from the accumulation of mucus in the air-passages; his slight hacking cough was constant and most distressing, and he would wake up in the night in fear of imminent choking.

He continued on with much the same symptoms, having great trouble in expectorating and difficulty in swallowing, so that he had to push his food to the back of his mouth. He then began to be troubled with various neuralgic pains in the face, in the eyes, in the throat, and along the arms.

About a month after admission he appeared to have gained some power, he walked in the ward, he could raise his right arm over his head, move his tongue better and articulate more distinctly. About this time he had a fall which hurt his head and kept him in bed; after this he became worse, very low-spirited, speech less distinct, and appetite bad. The mucus collected in his chest, and his power of expectoration became lessened; it was evident that he could not live long, and on Dec. 28, 1867, he died.

Head and spine.—There were no nodes, or other signs of disease of the cranial bones. The calvaria was removed, and the occipital part of the bony base cut out; the arches of the vertebræ were removed; and the cranial and spinal dura mater, their contents, and the cervical nerves, to the outer edges of the scaleni, were all removed together; the processes of dura mater in the sella-turcica, and the sphenoidal fissures, only being cut. The brain was tough and hard. There were no signs of formative disease; the changes required to be looked closely for. But on opening up the visceral arachnoid, there was a most obvious atrophy of the roots of the hypoglossal nerve, which had quite lost the natural white, opaque appearance of the nerves, and were little thin gelatinous threads as they crossed the corpora olivaria. In the same condition were the inner roots of the spinal accessory, and, also, very markedly, the whole of the anterior roots of the spinal nerves, especially the cervical, and least the sacral. The anterior view of the cord was remarkable; the outer aspect was flat, not round, yet it was harder than natural, so that mere flaccidness was not the cause of this; the anterior roots, also, came from a line much nearer the middle line than is natural. On section, the anterior half of the white matter was atrophied; it was white, harder than natural, and on the section it stood out, while the grey matter receded; the latter was larger than natural, it was darker, containing obvious vessels, and at the lower part of the cervical cord it was double the natural size, and showed a red colour finely mingled with yellowish white; the part, so affected, was not of great length; generally the redness and largeness of the grey part, and the thin, hard shell, or coat-like layer of white matter, made the pathological state of the spinal cord. In the medulla oblongata, as seen from the front, nothing diseased was visible, except the state of the nerve roots, as before stated. But on opening up the arachnoid over the fourth ventricle, and drawing down the medulla oblongata to look at the fourth ventricle, there was a very striking diseased appearance, without obvious derangement of anatomical position; there was a red-grey change of the calamus scriptorius, so that the nib of this was quite involved, and from the nib, upwards and outwards, for half an inch, there ran this change. The lining membrane of this ventricle and its choroid plexus were of deeper colour than usual.

CASE.—Mary Ann R—, æt. 22, married woman, and well until three months before admission, when she had some attacks of a very peculiar nature; her husband said he found her one day speechless, and her right side weak. From this she recovered, and remained well until the day before admission, when she was taken in the night with convulsive movements, and lost the power of speech. She was seen to be a healthy, well-grown woman. She could not speak, although she appeared conscious. The left arm and leg, as well as the face, appeared partially paralysed; the sensibility of this side was apparently exalted, and she was scarcely able to swallow. During a whole month she lay in bed in a lethargic state, as if asleep; when roused she opened her eyes and looked at persons intelligently, and gave signs of pain when the left arm and leg were moved. The respiration was irregular, and often interrupted by a deep sigh. After this time she began to improve, became more sensible, smiled when spoken to, and appeared to understand; she could move her left leg, but not the arm, which was now beginning to waste. Shortly after, she was got out of bed, and was able to stand. This improvement did not last long before she sank back into her old state, and during the next two months she lay quiet in bed, doing little more than vegetate. When roused or shaken she would open her eyes and smile, but never articulated a word; face without expression, but on emotion slightly drawn up on one side. She had difficulty of swallowing, great care having to be taken to prevent her choking, and the saliva was continually flowing from her mouth. On raising her right arm it would remain in any position it was placed, as in the cataleptic state. The left arm was flexed across the chest, and much wasted. The muscles responded to faradization, but gave her no pain, judging from the smile which she put on during the process of galvanism. The legs drawn up on pricking them. She remained in this state, getting gradually worse, until the breathing became very irregular and for the most part diaphragmatic, apparently like an inordinate action of this muscle when over-excited to continuous contraction. After being in this precarious state for a few days she again got better; she was roused out of her lethargy more easily, smiled when spoken to, and made strange noises when requested to speak; she also moved her paralysed arm slightly, and her other limbs more freely.

After this she gradually improved, the strength of her limbs increased, she grew stouter, looked fresh, and at the end of a month was able to get out of bed and walk about. She had now been in the hospital more than six months; walked about the ward slowly, but was very liable to fall; she could also move freely the right arm; the left was drawn across the chest and atrophied; she had little power in it. Her head fell forward and a little on one side. She appeared quite intelligent, and responded by action to everything desired, but could not utter the slightest sound. When spoken to she smiled, her mouth being drawn up on one side; when made to laugh she would get a little emotional, and during inspiration a crowing sound, evidently from partial paralytic closure of the glottis would take place. This was the only sound she ever uttered. She could not open her mouth well, nor protrude her tongue, and the latter appeared small. No loss of sensibility in any part of the face or limbs.

Dr Fagge has lately had under his care a fatal case of this kind, in which a careful post-mortem examination was made. In this case the disease began below and passed up to the medulla oblongata. The patient was a woman of 50 years of age, who began to complain two years before her death of weakness in the fingers of the left hand; this gradually progressed until the whole arm was affected, and subsequently the right arm also. At the end of about a year her speech became affected, she could not protrude her tongue, she could scarcely swallow, and the wasting of the upper part of the body and arms was extreme. After death the membranes were found thickened at the base of the brain, but the brain itself was healthy. The three divisions of the eighth pair of nerves were small, as also was the ninth. The spinal cord was diseased from an inch below the medulla oblongata, throughout the cervical and dorsal regions as far as the lumbar portion, which was healthy. The changes were pretty uniform throughout the cord; there was marked degeneration and atrophy in the lateral columns, including more or less of the grey substance. At the upper part of the cord the anterior portions were most effected, and at one spot the anterior cornu was hardened, so that it stood out when a section was made. In the other parts the diseased portions were soft. The anterior roots of the nerves all down the cord were thinner than natural. The nerves of the brachial plexus, as well as other nerves, appeared normal. The microscope showed abundance of granules throughout the cord, especially in the grey substance.

Sclerosis of cord combined with progressive muscular atrophy.— Although the tendency of morbid changes is to progress in given anatomical and physiological tracts, yet disease may be of a coarser kind, and attack several portions of the cord in succession. Consequently a combination of the affections of which I have been speaking may occur, and, amongst others, we occasionally meet with cases where a spasmodic rigidity of the muscles is combined with wasting. In these cases, in all probability, the cord has undergone sclerotic changes in the motor columns, and the anterior cornua of the grey matter have also become involved. The pains in the limbs which are met with may be due to the nerves being implicated, although there is no actual anæsthesia. The legs are at first seen to be feeble, and then gradually getting stiff; subsequently, when the patient takes to his bed, they become quite rigid, and either extended or flexed. The arms, in like manner, become rigid, and the hands flexed.

CASE.—A lady, æt. 40, after having given way for some time to intemperate habits, began to suffer with the ordinary gastro-hepatic derangement, and at the same time became, as is often the case, very enfeebled in mind and body. It was then evident that her cerebro-spinal centres were the parts more especially affected; she began to have pains in the legs, and an inability to raise them, as in walking up stairs. It was not many weeks before she took to her bed, and the paralytic symptoms grew rapidly worse. She was getting imbecile, her body wasted, scarcely able to draw up her legs, or to raise her arms from her side; when she did so, her hands fell down, as in lead palsy, being quite unable to extend them, and she had lost power over the fingers; the muscles were wasting especially those of the hand; she had great difficulty in appreciating touch, and, as regards the feet, there was almost complete anæsthesia of common sensation, but she could discern the difference between heat and cold. Much restlessness. No paralysis of the sphincters.

Case.—Mrs. L—, æt. 38, a married woman with children. During the last three years had become very intemperate, and at first suffered from the more usual gastric and hepatic derangements. About four months before I first saw her nervous symptoms set in, by some failure of mental power, with weakness and pains in the limbs, and at last, six weeks before I visited her, she took to her bed.

I found her a fair, good-looking woman, who answered my questions rationally, but apparently forgetful, as she did not know how long she had been confined to her bed. She was almost completely paralysed from wasting of the limbs and atrophy of the muscles. She could just raise her arm from the chest, but it remained flexed at the elbow, which was stiff; the hand was flexed at the wrist, as is seen in painters. The interossei were wasted, and the thenar and hypothenar eminences had almost disappeared; the muscles of the forearm also much wasted. She could not move her legs, which were very thin and flabby. There was a considerable amount of fat in the integuments. No marked loss of sensibility, although she hesitated when asked what part of her hand was touched, and yet the skin appeared irritable. Vision not affected. No paralysis of bladder or rectum; no sickness. Fidgitiness. From the uniform onset of the symptoms there could be no doubt about the central origin of this affection being in the spinal cord. I heard that she lived for some months after this, eventually getting blind and having convulsive fits.

INFANTILE PARALYSIS

The next disease of which I shall speak appears to approach very nearly in its nature to the progressive muscular atrophy. It was described by Cruveilhier under the name of essential paralysis, implying that it occurred without any known cause. It was seen to come on so suddenly, and without the association of other symptoms, that its pathology was involved in obscurity. Of late years, however, thanks to our better means of observation, a sufficient number of cases have been collected to show that the disease is in all probability spinal, and has its seat in the anterior cornua of the grey matter of the cord. What the character of the disease may be it is difficult to explain, since it is often rapidly recovered from, but in the permanent form, and where the spinal cord has been submitted to examination, a degeneration of the grey cells has been found. In some cases the children have had fits and symptoms of cerebral disturbance before the paralytic attack, so that such cases have suggested an origin of the disease in the brain.

As a rule, the children who have suffered from infantile paralysis have been from six months to two years old. The patient in many instances is seized with a febrile attack, which obliges it to be kept in bed or wrapped up in its mother's arms for a few days, when it is found to have lost the use of a leg or arm, or both legs, or sometimes of all the limbs. When the child is put to the ground it is found that one leg is quite powerless and dangles about like a dead limb; or an arm in the same way has become suddenly useless. The paralysis, indeed, is most complete. Recovery may very quickly occur, and in a few days no evidence of weakness remain; but if this favorable result does not take place the limb continues permanently paralysed, and the muscles waste; it is cold, is withered, and for the remainder of the patient's life he may have to use crutches, with the leg dangling helplessly at his side; or the leg becomes contracted, a club-foot is produced, and the aid of the surgeon is required. In very many cases a partial recovery occurs, and then the patient for the rest of his life is merely inconvenienced with having one limb rather smaller than the other. From what is known of the function of the cord an atrophy of some of the cells of the anterior cornu would account for the symptoms; for it may be remarked that there is an absence of paralysis of the bladder and rectum, and that there is no bed sore. Muscular contractility, as shown by galvanism, is much impaired or quite lost, and the same is true of reflex movements. It is remarkable how rapidly the natural contractility is lost, and yet in the analogous affection, the progressive muscular atrophy, it remains for some time. The reason is not very clear, except it be due to the rapidity of the disease. My experience of this disease has been very considerable, both at this hospital, the Infirmary for Children, as well as in private practice. It so happens that within a very recent period I have seen every variety of the complaint, and these I will briefly recount to you as examples.

CASE.—An infant, six months old, had a feverish attack, which lasted three days, and at the end of this time it could not move its legs or rest upon them when attempted to be jumped on the lap. At the end of the month, when I saw it, the legs were quite powerless.

CASE.—A child, 2½ years old, sent to me by Mr Roper, of Blackheath, was taken ill, apparently from cold, and put to bed. He always cried out when he was moved, as if he suffered pain in the loins. After a few days, when the febrile symptoms passed off, and the child was taken from his bed, it was found that he could not stand. After this a partial recovery took place in the left leg, so that when I saw the child, ten weeks afterwards, he could rest on the left leg, but had scarcely any power in the right. Sensation was unimpaired. Galvanism was ordered, but when I saw him, at the expiration of three weeks, there was no improvement, and the leg was wasting.

CASE.—A child, 2 years and 4 months old, was sent me by Dr Deeping, of Southend, with the following history. Ten months before, the child was seized with paralysis of the right arm, and, within twenty-four hours, of the other arm and both legs. He lost power over the bladder and rectum, the chest was paralysed, and respiration took place solely by the diaphragm. After a few days he began to improve, and faradization was used. Power returned in all the muscles and limbs, so that when I saw him there was no paralysis except in the right forearm and hand; the muscles were wasted, and the extensors had undergone some contraction, so that the hand was kept standing out; he had no power of grasping.

CASE.—A child, 2½ years old, was sent to me by Dr Paddon, of Putney; she had been seized a few days before with paralysis of the left arm and leg, the arm being first affected; no brain or other active nervous symptoms. The leg rapidly recovered its strength, but the arm was still weak when I saw her some weeks afterwards.

From a consideration of cases of this kind it would seem reasonable to locate the disease in a very limited spot of the cord when one limb only is affected, and in a larger area when several limbs are involved. But in the latter class of cases, as it so often happens that recovery rapidly takes place, leaving one limb only permanently paralysed, it might be conjectured that in nearly all cases the chief seat of the disease is limited to one spot, and that the wider range of the disturbance, with its temporary effects, is due merely to a slight change of a perhaps reflex nature.

The rapid restoration of the limb in many instances shows that no organic change could have occurred in the grey matter which rules over it ; the function must have been for a time in abeyance, in the same way as is constantly observed in cases of paraplegia where a perfect recovery occurs. The pyrexia preceding the attack is too often witnessed to be regarded as unimportant, and it may indeed be immediately due to a great disturbance in the nervous centres, and of which the paralysis is one of the results. It is true that in some instances no constitutional disturbance is noticed ; the child is put to bed apparently well, and in the morning is taken up with one limb paralysed. If recovery occur it is very rapid, that is, in a few days, but if not, the limb remains permanently weak and atrophied, with, perhaps, contraction of the muscles. This is no doubt associated with a degeneration of the grey matter, whence the motor nerves spring which rule over that limb.

As regards the treatment of these cases the recommendations are to exercise the limb, rub it with embrocations, and use galvanism.

In spite, however, of these measures, it is melancholy to see how little good results from them, except in very recent cases, and as the latter recover usually without any treatment whatever, it is a great question whether we possess any curative means. It is more probable that when restoration of the power quickly occurs no change has taken place in the spinal cord, but if this is delayed, that a hopeless degeneration of some of the grey cells has taken place. Nevertheless, I should recommend the above-named treatment, if it were only for the reason that the paralysis might sometimes have another cause ; as in the case of a boy where the treatment was eminently successful from the fact that the weak limb was owing to the patient having for some time sat upon it whilst sleeping. Then, again, the paralysis might sometimes have had a cerebral origin, when convulsions and other evidence of disturbance of the brain would probably have occurred. I would recommend, however, a long trial of that form of galvanism which seemed to act best on the muscles, for by its means their nutrition is preserved.

Pseudo - hypertrophic muscular paralysis, myo - sclerosis, or progressive muscular sclerosis.—I take this affection after progressive muscular atrophy and infantile paralysis, because it is apparently allied to both of them. It is a disease in which the limbs and other parts of the body grow weak, and at the same time the muscles become larger. This is notably seen in the legs, which, although growing more feeble in their movements, yet are becoming larger in girth. The typical cases are those of boys who are observed to be getting feeble and awkward in their gait, and on examination of their legs it is seen that the calves have grown to an enormous size. Duchenne, who first described the disease, gave the full-length portrait of a boy who was the victim of this disease, and all his limbs were so enormous that he resembled an infant Hercules. In the cases which have been observed in this hospital the enlargement has been almost confined to the legs, and at the same time some of the other muscles had undergone atrophy. The term pseudo-hypertrophy shows that there is no increase of muscular development. This is merely apparent, for when the muscles are examined they are found to have undergone a degeneration. Now, the fact of some other muscles being visibly wasted shows that the disease is really indicative of degeneration. In a boy lately in the hospital the legs were of enormous size, whilst the muscles of the shoulders and arms were wasted, as seen in progressive muscular atrophy. In many cases the calves only are enlarged, and these contrast most remarkably with the comparatively small thighs. The first notice of any change is the feeble and ungainly mode of walking; the boy rolls along, separates his legs, arches his back, and thrusts his belly forward.

Duchenne invented an instrument, called a harpoon, which being inserted into a muscle and a small portion withdrawn enabled him to discover what changes had taken place in it. He found that the transverse markings had disappeared, and their place was occupied by granular matter, and that amongst the fibrillæ there was a large amount of interstitial connective and fibroid tissue. In some parts there were fatty molecules. Dr Ord showed in one case that the affected leg had a temperature of one to three degrees above that of the arm. In this case he found very little change in the muscular tissue. The disease in most cases appears to be due to an interstitial overgrowth of fibrous and fatty tissue, and this, again, has its cause, according to Lockhart Clarke, in primary changes in the nerve centres. He has shown the existence of atrophy of the nerve cells, and disintegration of the grey matter in the cornua and central portion of each lateral half of the cord, as well as wasting of the anterior roots of the spinal nerves. Some observers have thought that the childish sufferers with this complaint are disposed to be weak-minded or idiotic. One must speak doubtfully at present as to its true pathology, as cases have not been sufficiently numerous for us to form an agreement upon it. Kesteven found spots of granular degeneration scattered through the white substance of the brain, ganglia, and medulla, but other investigators have failed to find any morbid changes whatever.

Granular degeneration of muscles.—I have already said that in progressive atrophy the arguments are in favour of its having a nervous origin, and in the last-mentioned diseases most authors are beginning to favour the same view. Nevertheless, it must be admitted that the muscle may undergo primary degenerative changes. Of this nature, probably, are those remarkable cases of hereditary degeneration of muscles, where several members of a family have been affected. The best-known examples of the kind are those described by Dr Meryon, where, in a family of many children, all the boys died of this disease. It was first observed when the child was in the nurse's arms, by his sitting heavily and not moving his body freely, and subsequently, at the time when he should walk, his being scarcely able to support himself. When the child grew up the muscles still remained weak, so that he could only walk with the greatest difficulty. Various remedies were then tried, but in vain. Some of the boys reached 12 or 14 years of age, and then suddenly died, and the muscles were found pale, atrophied, and the sarcous elements changed to granular and fatty matter.

A few similar cases in other families have been reported. Dr Meryon says: "In every case of muscular atrophy which I have seen or read of, in which either disease of the spinal cord or of the medulla oblongata has been detected, some symptoms of nervous disturbance have manifested themselves during life, either in pain, or in tremor, or quivering of the diseased muscles; but in every case which I have described as granular degeneration of the voluntary muscles there has been an absolute absence of any indication of nervous disturbance as there is in my present patient. I am therefore induced to continue in the belief which I have heretofore expressed of an idiopathic disease of the muscles, which is probably dependent on a defective nutrition of the sarcous elements. Every case, also, has begun in the lower extremities, and has appeared to descend in a centrifugal direction, respectless of the course and distribution of nerves."

I am not in this place speaking of diseases of muscles, for of course no one denies that the muscular fibre may undergo primary morbid changes. We see them resulting also from injury and from inflammation; and in some cases of chronic rheumatism a most remarkable atrophy of the muscular tissue sometimes occurs.

Cases have been recorded of progressive muscular atrophy having been hereditary, but they may have been of the kind just now mentioned.

LEAD PARALYSIS

I shall now briefly allude to lead paralysis, because the disease so exactly resembles the progressive muscular atrophy which I have just described that it is very often impossible to distinguish between them. If the metal has been throughly implanted in the system, a fatal result may ensue. All the tissues of the body degenerate; the skin assumes a remarkably waxen, and sometimes jaundiced appearance, the nerve centres more especially suffer, and the patient becomes at last paralysed both in body and mind, and may also be amaurotic; a true gouty condition is also sometimes manifested. A mania or dementia may result, accompanied by epileptic fits. In a less degree the effects are constantly seen, as in the dropped wrist of the painter, followed by a paralysis of the whole arm, in which the muscles waste, just as in the disease described. I have more than once seen a patient admitted and treated for progressive muscular atrophy, in whom there has been a lead line on the gums and a good history of plumbism; you may observe that accompanying the line on the gums there is often a corresponding patch on the lower lip. It might be thought that lead-palsy would elucidate the pathology of the idiopathic muscular atrophy ; but as yet it proves no more than that nutrition is affected, probably, through the influence of the nerves on the blood-vessels.   Lead, as you know, is given to arrest hæmorrhage, and acts by constringing the vessels,   You can there-fore see how its overaction or its continued action would produce an atrophy of the tissues.   Both muscle and nerve have been found to contain lead, and as the paralysis corresponds more with one nerve distribution than another—the musculo-spiral—there are good reasons for supposing the nerve is affected before the muscle. Duchenne states in his work that we have one means by which we can distinguish between lead palsy and the idiopathic atrophy.   In the latter, as I have told you, the most remarkable wasting is seen in the interossei and other muscles of the hand, so that the claw shape is produced.   In lead palsy the effect is most marked on the extensor longus digitorum, and as this muscle, when healthy and ex-cited by faradization, is stated by Duchenne to act only on the first phalanges, and has no influence on the second and third digits, it consequently follows that if this muscle is paralysed, as in plumbism, and the arm and wrist be supported on a table, the fingers can still be extended or raised, which cannot occur in the progressive mus-cular atrophy when the interossei and lumbricales are affected.

It is seen that the radial is the nerve first affected, shown by the paralysis of the extensor digitorum, extensor ulnaris, and extensor of thumb, whilst the supinator is rarely attacked ; subsequently the paralysis and wasting may spread to other muscles.   You may ob-serve that by the long-continued stretching the tendons passing over the back of the hand become thickened ; the protuberances sometimes are so great that you might suspect an enlargement of the metacarpal bones.

In course of time the atrophy may extend to other muscles, until those of a whole limb are wasted.   In two cases presently to be mentioned, where a remarkable cure was produced by galvanism, the wasting was very excessive and general.

A few years ago there was a woman in the hospital who had long worked in lead, and who had become at last completely paralysed. The limbs had gradually wasted, and became at last utterly power-less, and at the same time her mental faculties had almost gone. The post-mortem examination showed a marked wasting both of the brain and spinal cord.

More lately I have had a case of general softening of the brain in a plumber.   He had gouty deposits in various parts and excessively diseased blood-vessels. He had given up his employment two years before and had been lying in a demented state until his death. The lead was only a problematical cause, but its production of arterial disease has often been observed.

As regards the treatment of these cases, it is constitutional and local. The iodide of potassium has been found most effectual in aiding the elimination of the poison from the system, whilst as regards the local treatment, all means have been superseded by the use of electricity. We formerly bound the arm on a splint, applied blisters, and counter-irritants to the wrist, at the same time a small quantity of strychnia was often sprinkled on; and in this way when the lead was removed from the system the arm recovered itself. At a time when our only electrical apparatus was the faradaic one, we were in the habit of using it, and it is remarkable, that although no immediately obvious effect was produced by its application good results often followed. It might have acted in some way by stimulating the blood-vessels and so improving nutrition. It is very different with the current from a continuous battery ; this acts more readily on the paralysed muscles in lead disease than on healthy ones. The experiment is a very striking one, and you should see it for yourselves. On the muscles of a healthy arm you get contractility, both by faradization and the battery current, but in the case of lead palsy you have no effect from the former, whereas the muscle is excited by an amount of simple galvanism which would have no influence on the healthy arm.

Case.—Mr. S—, a gentleman of middle age, was brought to me on March 11th, 1872, by Dr Charlton, of Fareham, suffering from a most severe form of lead paralysis. His whole frame was attenuated in consequence of the atrophy which his muscular system had undergone ; his limbs were very much wasted, and he was proportionately enfeebled. He tottered when he walked, his hands shook, and were so weak that he with difficulty could raise them to his head or button his coat. His condition resembled that of a man with progressive muscular atrophy, only in this case it was induced by lead and was not idiopathic.

The history which he gave of his case was as follows : He lived in Surrey, about twenty miles from London, and had enjoyed good health until June, 1871, when his arms and hands became tremulous, so that very shortly he was obliged to use both hands to raise fluids to his mouth to prevent spilling. He was recommended a change of air, and took a trip to Scotland ; after being there a month he got considerably better and returned home. In a fortnight all the symptoms reappeared more severe than before. He went away again to Southsea, and there used salt-water baths, when he a second time rapidly improved, and at the end of a month returned home. Shortly afterwards, however, the old symptoms reappeared, when he was advised to consult a London physician. He was ordered to use galvanism in the form (he stated) of magneto-electric shocks, which did not benefit him, when his doctor, suspecting lead, had his drinking water analysed, and found it to be strongly impregnated by lead. He was then, of course, put

18

on a proper course of medicine, forbidden the use of water, and he improved. He had continued the use of the galvanism. He subsequently left London, and again went to Southsea.

When I saw him in March he had got into a stationary condition, and was in the state above described; his limbs wasted, and with little power in them. I ordered him some small doses of iodide of potassium and quinine, and wished him to use a simple galvanic current rather than electro-magnetism. Finding there would be a difficulty in making use of this at his own house, I advised him to go to Guy's Hospital every morning, and to this he readily assented.

Mr. Sandy, the electrician, tried the effects of the continuous battery current upon him, and also the induced current, with the following results. In the right arm the extensor muscles contracted well by the application of twenty cells of the Daniell's battery. The induced current was applied, as strong as the patient could bear, with scarcely any contraction. In the left arm the muscles contracted well by fifteen cells, and with precisely the same results as the right arm, by the induced or interrupted current. In the legs twenty cells caused good contraction, but scarcely any result was obtained by the interrupted current.

He continued the use of the galvanism to the limbs daily, and made visible progress.

On April 18th he had considerably more power of the limbs than he had a month previously, and on the muscles being tested it was found that the "induced" current, which had been powerless before, now excited the extensor muscles of the right arm. On application of the same strength to the left arm it extended the fingers much more than the right, but the hand was not lifted to the same extent.

The patient persisted in the treatment up to July, during the period gradually improving, and in August he had quite recovered the use of his hands and was following his usual occupation.

CASE.—Margaret C—, æt. 47, admitted February 29th, 1872. She has been married and has a large family. Two years ago her husband died, when she was obliged to work for her living. She gained employment in some lead mills, her business being to grind the white lead. For some months past she has been getting thin and feeble, her arms wasted, together with stiffness and pain in the shoulders. Has had slight colic.

On admission she is seen to be a small spare woman, anæmic and sallow, looking indeed extremely ill. She is thin, owing to a general wasting of the muscles of the whole body, more in the extremities, and especially in the arms. She is too feeble to walk, and therefore obliged to keep her bed. She can scarcely raise her arms from her side, owing to the atrophy and weakness of the muscles; the extensor muscles of forearm are extremely wasted, rendering the arm quite flat, the wrists drop without there being the slightest power to raise them. Muscles of hand soft and flabby, the right arm and hand worse than the left, so that she cannot use them for feeding herself. The blue line on the gums well marked, and a distinct blue stain along the lower lip corresponding to the stained border of the gums. Slight œdema of eyelids. Ordered ten grains of iodide of potassium three times a day. Tested by galvanism. Faradization:—As much power as the patient can bear has a very slight effect upon the extensors of the the thumb and not upon the other muscles. Continuous battery current :—Good and well-marked contraction of all the extensors by twenty Daniell's cells. The continuous current ordered. Mr. Sandy finds the more efficient method to be by placing the fingers in water containing a little salt; the negative pole is placed in the water, and the positive pole gently stroked along the extensors. This causes contraction of the muscles and elevation of the wrist; when the poles are reversed the current and the effect are less.

April 17th.—The continuous current has been used to the limb daily up to the present time, and the improvement has been marked though gradual. The blue line on the gums is much less. She is out of bed to-day for the first time. As the improvement has been going on, so the muscles have become susceptible to faradization, whereas they have required a larger amount of simple galvanism to affect them.

May 13th.—Improved considerably; walks about. Is able to feed and dress herself. Can extend the wrist, and the arms are larger in bulk. Blue line on gums and lips disappearing. On testing with faradization there is marked contraction in the extensors, the hands being well lifted; this is more so in the left arm than the right, the right being always weaker and smaller.

In this case it may be remarked, that besides a well-marked blue line along the edge of the lower gums there was a dark patch on the mucous membrane of the under lip, corresponding in position to that on the gums, but rather more defined and dotted. A question is always asked in the wards whether this mark on the lip is formed independently or follows that on the gums from contact? The latter is the probable explanation.

In these cases of dropped wrist the back of the hand is often observed to be rounded, apparently from enlargement of the metacarpal bones, but due in all probability to some thickening of the thecæ.

Plumbism treated with Electric Bath

CASE.—William J—, æt. 36, admitted under Dr Wilks July 17th, 1872, and left July 27th. He began to work at grinding lead nine months ago, and at the end of about five months commenced to feel ill, with loss of appetite, pains in his head and abdomen, and general debility. He continued at his work but daily grew worse, until a week ago, when he was obliged to desist, having pains in his limbs, sweating, inability to stand, and vomiting.

On admission he was seen to be very pale and very thin, having evidently lost a great deal of flesh. Skin hot, tongue furred, marked blue line on gums. Constipation. Recti abdominis contracted and painful.

July 20th, ordered an electric bath. This was made by Mr Sandy as follows : The bath being prepared, enough sulphuric acid was put in it to give it a slight acid taste (about ʒiv), the negative pole of the battery, attached to a large sheet of copper about two and a half feet square, was put upright in the bath and the patient placed in it so as not to touch the copper plate; the hand of the patient was held out of the water and in it he held the positive pole. Fifty and eighty cells were tried, but when the current was applied to the neck instead of the hand, the patient could not bear more than fifty cells. On making and breaking contact the patient felt a kind of thud through the whole of the body. A bath lined with glazed tiles was used.

The patient used the bath again on the 24th, and a third time on the 25th. He said he felt very cold after it. He always had his bowels relieved immediately after it. On each occasion he felt better, and on the 27th he was so much improved that he went out.

This method must be put into further practice, for it will no doubt produce good results. Dr Handfield Jones informs me that he is using it with success, and showing that its value is not due to galvanism alone, I may state that lead has been found in the bath after the water has been used a few times. The placing the limbs in water and sending the current through it seems in all cases to have advantage, as in the case of Margaret C—, just described.

MERCURIAL PARALYSIS

There are many other poisonous substances, more especially the metals, whose effects in small doses might be advantageously studied in connection with idiopathic diseases. I have spoken of lead, since the results of its action resemble so closely those of a well-known disease, and I shall now just allude to mercury, as its poisonous effects have occasionally been referred to spontaneous causes, and because, in a complete saturation of the system, the nervous centres seem to be most strikingly affected. Formerly the effects of mercurial vapour were constantly seen in looking-glass makers and water gilders, who exhibited the well-known mercurial tremor, and these same persons, if they persisted in the employment, became at last quite shattered in health. The mercurial paralysis was distinguished from paralysis agitans by the tremors occurring only when volition was exerted on the muscles, as in moving. When the limbs were rested they were quiet ; not so in paralysis agitans. Such instances are at the present time by no means numerous. I have seen, however, within the last few years, two cases showing in a much more striking manner the destructive nature of mercury, but in neither case was it due to the inhalation of the metallic fumes, as formerly witnessed.

The first case was that of a man admitted into the hospital for a form of general paralysis from which he was suffering. It was discovered that he had been in the habit of packing the skins of animals, and that these had been washed with an acid solution of mercury. After he had been thus employed for three years, he began to experience a general muscular weakness. He could scarcely walk, and, when attempting to do so, a general tremulousness took place over the whole body. When lying down, he had spasmodic movements of the chest and of the muscles of the trunk, resembling those of chorea. He gradually became more feeble, was delirious at times, and afterwards fell into a state of unconsciousness. The post-mortem examination showed no evident disease of any of the organs, but a chemical analysis by Dr Taylor proved the existence of mercury in many of the tissues of the body.

A more marked case than even this of the destructive effects of mercury on the body I had an opportunity of seeing in St Bartholomew's Hospital. A young man had been engaged in the laboratory in the preparation of mercurial methide for about three months, when he began to complain of dimness of sight, numbness of the hands, and general weakness. These symptoms increased, until at last he was obliged to be sent to bed. When I saw him he was almost completely paralysed; he was lying prostrate in bed, perfectly helpless, being scarcely able to move either his arm or legs, and there was paralysis of the bladder. He could not speak, and was quite deaf. The heart's action was quick and feeble. The mouth was not sore, but I was informed that the gums had been at one time spongy, and emitted a fœtor. He got weaker and weaker, and died in about a fortnight.

Another young man employed in the manufacture of the same article was also similarly affected, the symptoms being those of a complete paralysis of body and mind. He lost all feeling, all power of motion, became deaf, unable to speak, and quite idiotic.

But, recently, a girl was under my care who came from the same factory as the man whose case I have previously mentioned. She had not worked long in the skins when she began to lose power in the hands. On admission they were flexed, and she was quite unable to grasp any object; she said they felt "numby," but there was no marked anæsthesia. She stated that her feet felt like her hands but to a less degree.

COPPER POISONING

As I am on the subject of metallic poisoning, I might mention that attention was drawn by Dr Clapton, some years ago, to the probability of workers in copper and brass becoming affected by the metal. He found that they were subject to lassitude, giddiness, and disorders of the stomach and bowels. All their garments were of a green colour, as well as their skin, and Dr Clapton believed that this was due in part, to an exhalation from the skin itself. The subject was further investigated, but the reporters could not determine whether the green colour of the skin, gums, and some secre-cretions, was due to an impregnation of the body and tissues with copper, or whether it was owing to a mere staining from the dust which was always surrounding the workmen.

I remember but one case where I had any suspicion of the patient having suffered from the effects of the mineral. The young man, by occupation a brass finisher, having been taken with shivering headaches and other feverish symptoms, sought admission to the hospital. He appeared not to have fever, for the pulse and temperature were normal; his chief complaint being extreme lassitude and colic. When the latter was relieved it again occurred on several occasions. He was in bed a month, and had a green line on the gums, which could not be scraped off.

This is a disease where the muscles are in a state of tremor caused by some fault in the innervation. Generally speaking, trembling implies a want of nerve power, as may be experienced in one's self when the hand shakes from fatigue; indeed, it is one step towards absolute powerlessness; the shaking arm is the midway condition between a well-knit limb indicative of muscular force and one which is helplessly paralysed. The trembling limb is often seen in old age, in which case we can only attribute it to a degenerative change in the cells of the spinal cord; and therefore we should not be surprised to witness the same symptoms in any disease which impairs the function of the cord, such as chronic alcoholism. From these facts we might be inclined to infer that tremor implies a powerless condition of the nerve centres, arising either from some organic cause or some temporary exhausting one.

Unfortunately I cannot say that this is the pathological law, since the very disease of which we are about to speak often occurs in persons of middle age, and who exhibit no loss of muscular power whatever. Eventually a weakness may ensue, but the disease may last for some considerable time, without evincing any feebleness on the part of the muscles.

It is therefore necessary to speak of *active* and *passive* tremor. The tremor of old age and of exhaustion is of the passive kind, and is only witnessed when any voluntary power is exerted on the muscle, for when at rest the limbs remain motionless. The other or active form of tremor is observed whilst the arm is rested, as if the motion were due to some intermittent action of the spinal centres; being, in fact, a kind of convulsive phenomenon. If the hand is held, it still continues to shake, and, as regards its power, it may be as strong as the other. We cannot regard, therefore, such a case as showing a want or failure of supply of nerve influence, but rather as if an abundance of nerve force stored up in the centres had passed beyond the control of the higher nervous influences, and so escaped intermittingly.

Paralysis agitans generally begins slowly by a weakness in the fingers of one hand, which is often first discovered by a difficulty in holding the pen in writing; subsequently the hand is observed to shake. After a few months, when this is more fully developed, it will be seen that the hand is in constant movement; if the hand be held the motion for a moment ceases, but it soon begins again. The trembling may remain confined to the hand for several months, and then it may gradually extend to the arm; after a time, the other arm is affected, until both limbs are in constant agitation. Subsequently the malady may affect the legs, so as to cause tottering in walking, and finally, the muscles of other parts of the body. As regards the face, the muscles do not participate in the movement, but after a time there may be observed a loss of expression; this may be attributed to the slight rigidity which nearly all the muscles finally undergo. The head, too, is not affected in the same way as the limbs. In persons where the head is constantly seen shaking the tremor is generally due to age, and the paralysis is therefore of the passive or senile form. The disease makes a very slow progress, but when after some years it has affected a large part of the body the patient may be observed sitting in his chair, with a vacant expression, and his hands in his lap, in a state of constant movement. In some cases the hand is stretched out in a conical form, as if holding a pen; in other cases the fingers are bent backwards so as to form a right angle with the metacarpal joint. The joints themselves sometimes become distended, as if the subject of chronic arthritis, or as sometimes seen in various spinal affections. The speech is not characteristically altered, but it may be somewhat trembling, from the constant movement of the muscles. There is no nystagmus. When the patient rises from his seat he tumbles forward, without scarcely moving his legs from the ground, so that he looks as if every moment he would fall were he not quickly to take another step to save himself. To use the words of Trousseau, he looks as if he were always pursuing his own centre of gravity. This mode of gait is so striking that it has received the distinct name of *paralysis festinans*.

The movements cease altogether during sleep or under the action of powerful narcotics, but during waking the movements are continually going on, and quite independently of any efforts on the part of the patient, although they are aggravated by any voluntary act. We may notice that in paralysis agitans, although the limbs tremble involuntarily, the patient can arrange his movements. Now where the movements are very irregular and purposeless, to these I should rather give the name of chorea, even though the patient were advanced in years.

As regards its pathology nothing is positively known. A patient with a marked form of this disease died in the hospital of typhus, and his cord presented no marked morbid change. In old people a tremor may no doubt be reasonably attributed to senile changes in the cord, but in the middle-aged and the strong the cause is not so obvious. In a case lately observed by Dr Murchison some degenerative changes were observable in the cord. I believe there is no reason to believe that an injury might induce the disease, but several cases have been recorded where a fright or severe moral shock has preceded its development. I have myself known this as the most probable cause in two cases.

These patients do not experience pain, but they seem as if conscious of some irritable state of their nerve centres, and thus suffer from what are commonly called " fidgets." They sometimes feel an uncontrollable want to move from one position to another; they have no sooner seated themselves than they again rise and walk round their room. When tired they resume their seat, but it is not for long; they are up again, as if influenced by some indescribable desire to keep in motion. I think you may form some notion of what these feelings are like by reflecting upon the uncontrollable desire to move one's limbs or fidget them after having been long in a restrained position in a railway carriage; and some of us are very liable to them without apparent cause. I have on several occasions been consulted in the cases of patients, generally old people, in whom the "fidgets" has been the main symptom, and a most distressing complaint it is, both for the patients and those around them. I know a gentleman whom on my visit I find sitting in his chair with a slight tremor upon him, and who immediately asks, if nothing can be done for him; he gets up and wanders round the room, continually complaining; if he is pacified and made to sit down, it is only for a few moments, when he again rises and roams round his room, saying he cannot bear his feelings. This patient and others have been otherwise intelligent and evinced no delusions, although their constant restlessness has certainly resembled what is so often seen in maniacs.

The following are the two last cases recorded in my note-book of paralysis agitans :

CASE.—A man, æt 49, said that a year before, a trembling commenced in his right hand; this extended to the arm, which was then constantly moving, and was not relieved by resting it. It was quiet when he was asleep. He was a well-developed, healthy-looking man, and had no other complaint but the tremor of the arm; this was not at all wasted. It had been gradually getting worse, and continued to do so during the following year, when I lost sight of him. All the remedies he used were unavailing.

CASE.—Mr S—, æt. 59. Three years ago he commenced to have tremors in his right hand when he grasped anything, and after some months he found the hand shaking at all times. At the end of a year the other hand and arm began to be troubled in the same way, and after some months more the legs began to shake. He still continued at his business, which compelled him to travel about; he was able to walk good distances, but shuffled and shook as he went, and thus attracted a good deal of attention to himself. He had sought advice from the best men in London, had taken the usual medicines, and also used galvanism in its different forms, both to the spine and the limbs, without any good effect, the disease having gradually progressed. At the present time he is very thin; as he lies on his couch he trembles all over, the muscles of the face being also involved in the movement, and which gives a stammering character to his speech. The muscles are also getting somewhat rigid, more especially in the arms and on the right side, where the fingers are held stifly out. There is no loss of sensation. He is quite intelligent, but fears his mind may fail, as he occasionally has fits of irritability in which he can scarcely control himself.

Treatment.—Paralysis agitans is a disease generally considered incurable. It is true that patients are prescribed medicine, because they insist upon taking something, but never with any good result.

Since, however, we have been using galvanism in its simple form, we are in hopes that we have a remedy which may sometimes be useful. In the following case it appeared to be acting beneficially, and in the second, which is recorded in the 'Gazette,' the result was certainly better than under any other form of treatment which I have seen. A reasonable doubt might arise as to its efficacy founded upon the question whether they might not have been cases rather of the choreal type.

CASE.—J. B—, æt. 40, had been suffering for three years from paralysis agitans. The complaint commenced in the right hand, afterwards proceeded to left, and then to the legs, until a general tremor of the whole body took place, including the face, and affecting the speech. He had been under different kinds of treatment, but without any benefit. I wished to try the continuous galvanic current to the spine, and accordingly fifty cells (Cruickshank's) were used for ten days. After the second application the patient, who had previously had very restless nights, obtained refreshing sleep. After four or five applications he began to experience a decided benefit, saying he always felt lighter and steadier directly he had been operated upon. The duration of this improvement lengthened day by day. The patient then left for the country, and has not since been heard of.

CASE.—A man, æt. 48, came to the hospital with shaking of his arm, so that he was unable to dress himself, could not raise his hand to his head, and was unable to cut his food; his head also had some movements from side to side. The galvanic current was applied to his spine by the positive pole being placed above and the negative drawn slowly down the back. After the application he felt more comfortable, and after the galvanism had been used six times he was much improved; he could raise his hand to his head and could wash himself; at the end of two months he could dress himself, could hold the arm out without its shaking, and when he left the improvement was progressing.

In another case I tried the effects of a simple current running through the neck and body, but it produced no appreciable effect.

CASE—Jane D—, æt. 40, was a sufferer from an extreme form of paralysis agitans, all the limbs being in a constant state of tremor. She was seated in a chair, and behind her was placed a battery of thirty cells, one pole of which was attached to her neck and the other to her wrist. It was allowed to work for three hours, for five days in succession, but no effect was produced.

This form of disease has long been confounded with paralysis agitans, but it now appears to be a distinct form of affection both in its pathology and symptoms. As soon as the knife of the morbid anatomist began to be used to discover the cause of paralysis agitans a peculiar condition of the cord was found to exist in some cases which had been treated under that name. Further investigation, however, showed that these cases had their own characteristic symptoms, and had therefore been classified erroneously. The symptoms attending this disease of which I am now about to speak, somewhat resemble those of paralysis agitans in respect to the tremor, but they are unlike, inasmuch as the movements are not continuous and rapid as in this affection. They are slow and rhythmical, and do not come into play unless volition is acting upon the muscles; the disease also has a more rapid course, and terminates by spasm and contraction of the limbs. One of the best accounts of this disease is to be found in the 'Guy's Hospital Reports,' by Dr Moxon, where he has recorded all the cases which have been in the hospital since the peculiarities of the disease have been recognised. I myself had observed years ago scattered patches of deposit in the cerebro-spinal centres, but had failed to associate them with any special form of malady; subsequently Charcot described this sclerosis, disseminated through the cord, with the prevailing symptoms which accompany it.

The disease differs from paralysis agitans in being observed most frequently in young people ; it commences by a feebleness in walking, as in many forms of paraplegia, but there is no loss of power over the rectum or bladder; nor is there any loss of sensibility, as is generally met with in ataxia. After some months the feebleness and tottering gait increase, until the energies of the patient are altogether impaired and the disease is fully developed. When this has occurred, we are struck with the remarkable regular or rhythmical movements of the body and limbs. This cannot be so well recognised in the gait, but is readily seen in the action of the arms. If the arm is attempted to be held out it moves slowly and in an orderly manner, so that if the patient be told, for example, to put a spoon to his mouth, the limb will ascend in regular stages until the mouth is reached, when, if the spoon is put in, it will clatter against the teeth. The rhythmical movements, however, are best observed during the action of the muscles of speech : the words are brought out one by one, or syllable by syllable, as in a child learning to read. This interrupted or jerking mode of talking is most characteristic of the affection. If the patient be sitting up, his head also may be seen moving in a regular manner. The eyes are also constantly rolling from side to side (nystagmus). If the head is supported and the body is at rest there is no movement, as in the case of paralysis agitans, where, as I told you, movements cease only during sleep. In the sclerosis of which I am speaking, it is only when the patient rises from bed that his head and shoulders undergo an oscillatory motion.

The disease of the motor tracts often extends upwards into the cranium, and even reaches the brain proper; the muscles of the face then become affected, and the ordinary expression is lost. Tremor also may come on, and then the disease resembles somewhat that of the general paralysis of the insane. There is this difference, however, in the intellect: in sclerosis it is simply impaired without any of the positive delusions which exist in general paralysis. The patient, however, is not depressed; he is more often happy, although emotional, and is always ready to cry or laugh when spoken to; more commonly the latter.

After the continuance of these symptoms for some time, the next stage commences; the legs become stiff, and the patient is unable to walk; he takes to his bed, the legs are stretched out, and become absolutely rigid. Sometimes the legs are bent up and stiff, but more generally they are stretched out straight; there is no weakness of the bladder or rectum, as in paraplegia, nor do bed sores appear; there is no loss of sensation, and the electro-irritability of the muscles seems to remain. If the foot is struck, or firmly bent, a tremor will sometimes take place, passing through the whole limb.

In what may be called the third stage, the powerlessness is complete, and the case has become one of dementia paralytica. The disease thus lasts altogether from two to three years.

Pathology.—There is found after death a chronic myelitis or hyperplasia of the neuroglia, characterised by hard masses of connective tissue scattered through the anterior or motor columns of the cord, often reaching as high as the corpora striata, and passing into the brain itself. These are of a pinkish-grey colour, not well defined from the surrounding tissue into which they gradually pass. A section of the cord will display these grey patches of various shapes and sizes, and in sections of the cerebral hemispheres they become well marked, in contrast with the white medullary matter.

It would seem that this disease interferes with the motor function of the cord in a way by which the forces therein produced are transmitted in an irregular manner to the muscles, and so give rise to the rythmical movement, or it might be said by those who regard muscular action as the result of a number of very rapid contractions, that in sclerosis the action is slower, so that the intervals of time between them can be appreciated.

CASE.—A man, æt. 33, a veterinary surgeon, was lately under my care in Stephen ward. He said that he was well until two years before, when he had a fit and fell to the ground; however eight months previous to this he had a kick from a horse. Since this time he had been getting very feeble. When admitted he was unable to walk, but could for a moment stand, his legs all the while trembling under him. If asked to raise his arm or to take hold of any object, it would move up in jerks or rhythmically, and his speaking was exactly of the same character; his words came out separately and singly, reminding one of a child attempting to read. He had perfect control over his sphincters, and sensation was not impaired. His eyesight was misty. He had an involuntary laugh when spoken to. The muscles appeared to be more responsive to faradization than to the continuous current. He had one or two febrile attacks of a similar kind which I have noticed in ataxia and in other other chronic spinal disorders.

The following case was the first in which Dr Moxon had an opportunity of examining the brain and spinal cord:

CASE.— She was 25 years of age, and after she had been laid up for some time with febrile disturbance she found her left arm and leg weak and unsteady. This spread in the course of a few months to the other limbs and trunk, until she lost her power of walking and standing. When at rest her head and limbs were quiet, but any attempt at movement caused a jerking and tremulousness, but with a regularity of action which at once distinguished the disease from paralysis agitans. When trying to feed herself she jerked the food about, and this jerking affected the whole body. The lips and tongue showed the same jerking and unsteadiness; the speech also was very peculiar, every syllable being followed by a pause. Her mental condition was very low. After her death insular patches of grey tissue were found scattered through the cord and brain.

DISEASE OF THE POSTERIOR COLUMNS AND LOCOMOTOR ATAXY

I have already told you that the posterior columns of the cord, according to our present knowledge, are intimately related to the cerebellum, and also, by means of short fibres, connect one portion of the grey centres with another. These parts are therefore intimately associated with the cerebellum in the regulation of movement, and when diseased give rise to a condition known as "inco-ordination." The patient under these circumstances could move his limbs, but could not properly control them. He would resemble the monster which Frankenstein made, so far as the attachment and movements of muscles are concerned, which would contract when excited, but in a manner as devoid of method as in the wooden figures whose arms and legs are pulled by a string. You see that another function must be superadded to that which merely excites muscular movement—one which regulates or guides them in an orderly manner, just as a fly-wheel in a steam engine controls motions which would be otherwise unequal and irregular.

If this regulating power in the body be lost we witness the complaint known as locomotor ataxy ; and should the case eventually be fatal we find the cause of the malady situated in the posterior columns of the spinal cord.

I have already told you that pathological changes correspond very often to distinct anatomical regions, and as these changes are accompanied by special and characteristic symptoms, it tends to corroborate the opinion that certain regions have their distinct physiological functions.

In this disease, then, known as " locomotor ataxy" or " inco-ordination," we have a *chronic myelitis, grey degeneration, or sclerosis of the posterior columns of the spinal cord.* In connection with this change in the cord we also very commonly find that the posterior roots of the spinal nerves are involved in the inflammatory process ; and thus it happens that, besides the inco-ordination, we have various symptoms affecting sensation associated with it. We are indebted to Duchenne for clearly separating this form of disease from other forms of paralysis, and also, I think, for showing its true pathology ; nevertheless, the facts belonging to the disease were already known under the name of " tabes dorsalis."

I should state that Dr Todd more than twenty years ago described a form of paraplegia in which the co-ordinating power was lost, and also that my late colleague, Dr Gull, had observed these symptoms of ataxy in his paraplegic patients, although he did not separate those in whom they existed into a distinct class having a new name ; and more than this, he had associated these symptoms with a degeneration of that part of the spinal cord to which I have referred, and which is now said to be the true seat of the disease. Dr Gull, in describing such cases, says : " One patient told me he could not walk without looking down at his feet all the time, because he felt as if his legs were cut off below the knees. Another patient said he had to do so because he had no apparent weight." Again, in a case recorded in the ' Guy's Hospital Reports' for 1858, he relates how a patient, whilst in the recumbent position, could flex and extend his legs with some freedom, but the movements were sudden and vague from want of control over the action of the muscles ; the spinal centres, when stimulated by the will, seeming to shoot off their influence at once, making the feeble muscles contract to their full extent with a jerk. In other words, there was no power to regulate the muscular contraction. The movements of the fingers were also wanting in precision, and he was awkward in handling objects." Dr Gull then gives a drawing of a section of the cord, exhibiting a very defined disease of the posterior columns, and adds the following remarks :—" This brings us to the theory of the posterior columns proposed by Dr Todd, that they ' propagate the influence of that part of the encephalon which combines with the nerves of volition to regulate the locomotive powers, and serve as commissures in harmonising the actions of the several segments of the cord.' The want of power in this case to regulate the action of the muscles was very characteristic. The legs when drawn up, as they could be freely, were drawn up with a sudden jerk, and extended in the same manner. The voluntary movements of the hands were also fumbling and vague."

In the year 1851 Mr. Landry described cases of paralysis of the legs where there was no anæsthesia, but a loss of muscular sense, as he proved by placing a metal goblet in the hand, which it seemed to weigh no more than a feather.

Before this, however, in 1849, in his Gulstonian lectures, Dr Gull had described the disease, and refers to a paper by Earl, who had also been struck with the peculiarity of these forms of paralysis, having observed that the patients could not walk in a straight line, that they threw their limbs forward, and had a great difficulty in turning round. Dr Gull's description was in these words :—" There is often a numbness of common sensation, and the paraplegic affection is in part owing to the affection of sensation, so that the patient cannot direct the muscles, rather than to an actual weakness in their contractions. One patient told me he could not walk without looking at his feet, because he felt as if the legs were cut off below the knees, and another because he had no apparent weight. In walking, the patient complained of difficulty of keeping on the ground, which seems to be due to the want of sensation which is necessary for directing the movements, as well as to the sudden contraction of the muscles."

Duchenne designated the disease " a progressive abolition of co-ordination of movement and apparent paralysis, contrasting with the integrity of muscular force." He exemplified this to his class by causing the patient so affected, and who had been called paraplegic, to take a student on his back and carry him across the lecture-room ; he then made him sit in a chair, and showed that when the patient made resistance he was quite unable to bend the leg, proving that the muscular power still remained as good as ever. When, however, the patient was made to stand up and walk, he was seen to throw his legs about in a most extravagant manner, as if he had lost control over them, was constantly looking at them to see where he was placing them, and if he closed his eyes he fell down. By this method we test the ataxic condition. We place the patient in an upright position, place his feet close together, and tell him to shut his eyes. If he be a sufferer from the complaint he immediately falls. He appears to be unconscious of the position of his body, or where his movements are leading him. Having lost the proper control over them, it is necessary that he should always be looking at his feet when he walks, or otherwise he would fall. He is unable to walk in his own accustomed rooms and passages when it is dark without feeling his way; and you will thus see why, with this want of control and disorderly mode of performing movements, the term "ataxy" has been applied to the complaint.

From the want of control over the limbs the patient totters when he walks, or resembles the man alluded to by Sir C. Bell, who steps along a narrow ledge. He throws his feet out as if pawing the air, and feels almost as if he was walking in the clouds, or, as is sometimes the case when a partial anæsthesia is present, as if he was stepping on wool, or as if he had no legs and possessed no weight. The limbs are thrown out with a jerk, or as if pulled with a string, as though not altogether under the control of the will, and, when standing, the patient has a difficulty in maintaining the equilibrium, and is liable to fall. This staggering has been considered worthy of a new name—"titubation." Unlike a blind man, who holds himself backwards, he precipitates himself forward, and, unlike a drunken man, does not roll from side to side, although I know of the case of a man affected with this disease who on more than one occasion was charged with being intoxicated. His occupation was a rent collector, and he had a remarkable way of precipitating himself into people's houses, which suggested to the ignorant the notion of inebriety. The gait is rather like that of the man on a ledge, who is attempting to balance himself, and who is constantly looking at his feet to preserve them in position. The patient has a difficulty in starting, as some of you witnessed in a patient lately in the hospital: when requested to walk across the ward, he would stand motionless for some time, as if he were winding up a clock for the performance, and then off he would go in a most precipitous manner. When the patient is on the trot he often feels the same difficulty in arresting his own progress or in turning round. Thus, a patient of mine assured me that whilst at Margate he walked a long distance along the road, and then, wishing to stop in order to return, was obliged to guide himself up the bank, where he fell down. Some years ago one of our out-patients afflicted with this disease apologised for being late at the hospital, as his friends had started him out of his house in a wrong direction, and when or where he would have stopped no one knows, had he not been met by an aquaintance, who turned him round and sent him back again.   I know the case of a gentleman afflicted with this disease, and who, if he stops to look in at a shop window, is unable to start himself again, and asks some one near to give him a push.

For this want of control over the movements or want of knowledge as to the movements of the limbs the eyesight compensates, and thus the patient is continually looking at his legs.   Whilst his eyes are fixed on them he may walk for miles or stand quite steady ; but let him place his feet together and raise his head, he immediately falls.

The same facts apply to the arms.   He cannot make a straight-forward thrust, for in the endeavour his arm would strike from side to side.   A good method of testing the power of control which is left to the patient is to ask him to perform the manœuvre, suggested by Dr Hughlings-Jackson, to place the thumb to the nose and at the same time extend the fingers, or, as it is usually called, to " take a sight."   If he be troubled with ataxy, he will experience considerable difficulty in performing the operation.   Just as the patient can walk for miles if not interrupted, or make a great resistance with his leg whilst in the sitting posture, so with these tottering arms he can carry great weights.   When in bed the difference between his condition and that of an ordinary paraplegic patient is well seen ; he has no difficulty in throwing his arms and legs about, but he does so without apparently a fixed purpose, and in a most ungainly manner. In this disease, I should have said, the muscles do not waste.   Not only the limbs, but muscles of other parts of the body, may be affected, or rather the nerves supplying them, especially those of the cranium, as the third nerve.   Thus the pupils are very often contracted and unequal in size, as occurs so frequently in many other chronic disorders of the cerebro-spinal system; there may be also strabismus, and amaurosis. In some instances there has been anæsthesia of the face as well as of the extremities.   Finally, the bladder and rectum may lose their power.   Towards the end of the disease, the sexual powers, as in many other nervous affections, fail ; but at the onset it has been observed that they have been much increased—so much so that venereal excesses have been considered mainly instrumental in the production of the disease.

Then there are other symptoms which are pretty constantly present, and assist in characterising the disease.   These are more especially pains in the limbs which, at a former time, when the peculiarity of the affection was not recognised, were thought to be rheumatic, but now are known to be spinal. They are deep-seated aching pains, as if, to use the common expression of patients, they were seated in the bones. Sometimes, instead of being constant or enduring like those of rheumatism, they resemble electric shocks darting through the limbs or muscles. This comparison was volunteered by a patient lately in the hospital, who, having had the limbs galvanised, described the pains which he suffered as being exactly similar to the effects of the battery. They were not persistent, but would come on at intervals, and were described as most excruciating.

The pains are therefore of two kinds, and besides these, there is also a sense of constriction or pain around the body. The pains have given rise to the French name " inco-ordination motrice et douleurs fulgurantes."

With these pains there may be some amount of anæsthesia, for, as you know, neuralgic pains may be accompanied not only by an over-sensitiveness of the part affected, but by the opposite condition. Some writers have endeavoured to discover in this fact the cause of all the phenomena of ataxia, but this we shall speak of again presently. Duchenne, who first wrote systematically on the subject, divided the *ataxie locomotrice progressive* into three stages (see 'Archives Générales') :—1. Where there was paralysis of the nerves of the eye, shown by inequality of the pupils and amaurosis, with darting pains in the limbs. 2. Characteristic unsteadiness of gait, with diminished sensibility and pains in the legs. 3. Symptoms still further increased, and the want of co-ordination progressing upwards.

Now as regards *anæsthesia*, I believe I have seen cases where sensation has not been lost, and in others where the sense of feeling has merely been retarded. You know that time is required for a sensation to be conveyed along a nerve ; this being computed by some to be as slow as 300 feet per second. You can understand that the periphery of the nerve in the skin may be intact, and the sensorium likewise sound, and yet the conduction be impaired from disease of the nerve. Now in these cases of ataxy, and in some others, the sensory nerves are sometimes so thickened and their conductibility so lowered, that a considerable time is required for the patient to appreciate any tactile impression made on his skin. You touch or prick his legs, and no response is made until sometimes two or three seconds have elapsed ; and I have heard of a case where it was possible to count eight after the patient was touched and before he obtained a knowledge of the sensation.

As regards the *eye*, it is very remarkable that although as a rule

19

the brain is affected long after the cord, the disease progressing
slowly upwards to the fourth ventricle and to the cranial nerves,
yet that the sight is so commonly affected from the very commencement of the disease. You may notice a tendency to contraction of
the pupils and an inequality in their size; or perhaps a slight
strabismus arising from a want of accommodation or ataxia of
the ocular muscles. The vision may be also affected from an
atrophy of the optic disc, and this is said to be due, not to an inflammation or a true neuro-retinitis, as occurs in cerebral tumours,
but to a progressive atrophy. Mr Higgens, who has examined many
cases for me, says he finds that, as a rule, a white disc in cerebral diseases results from a neuritis, and in spinal cases from a
simple degeneration. In a case under my care lately, and in some
other ill-defined cerebral cases, I have observed a tendency to a
cataleptic condition of the levator palpebræ. You ask the patient
to look at you, and he will continue gazing for any length of time
without once winking. In cases of ataxia and some other forms of
spinal disease, arthritic inflammations and swellings may be met
with. Charcot has, of late, especially drawn attention to the subject, but it had been previously observed, for I well remember Dr
Addison and Dr Gull conversing about it many years ago, and
Dr Alison had distinctly described it. Several observers have also
drawn attention to a severe gastric disturbance which is sometimes
met with in locomotor ataxy.

I might here remark that as disease of certain regions of the
cord will produce definite and characteristic symptoms, such as
sclerosis of the posterior columns those of a locomotor ataxy, so it
is conceivable that a limited disease of the same kind might produce a localised ataxia, and an explanation be afforded for some of
those strange cases of movements of a single limb, which may
be really due to want of co-ordination.

I should also say that since other portions of the cord may fall
into temporary inaction, so as to produce paralysis, there is no
reason why the posterior columns should not equally become functionless and be productive of an ataxia. I throw this out as an
explanation of cases which are completely cured.

Now what is the morbid anatomy of this disease? My own experience in the pure cases of ataxy has been limited, but the statements of Gull and Duchenne have been confirmed by numerous
other observers, that a marked change has existed in the posterior
columns of the spinal cord. These have become changed into the
grey translucent substance which I before described to you. This
is composed of new connective tissue, the formation of which necessitates a certain amount of destruction of the normal tissue, so that

it is found degenerated and containing amylaceous bodies. The disease may extend through the whole length of the cord, but in the less severe cases it affects more especially its lower end, together with the cauda equina. The posterior roots of the nerves have been found involved in the disease, and sometimes the cranial. The membranes, too, may be thickened and adherent.

Charcot, to whom we are indebted for much knowledge of spinal disease, says the degeneration or sclerotic change is limited to the outer portions of the posterior columns, avoiding the tracts next to the fissure which are known as the cords of Goll. If the latter are affected, it is not a simple case but one in which the posterior columns have been involved by disease advancing from below. Cases of this kind are met with on the post-mortem table where no ataxic symptoms have existed.

Now, having described to you the disease in its pure and simple form, you must be careful not to give the designation "ataxy" to many forms of spinal affection because certain symptoms are present which may be found in ataxy. For you may easily suppose that if this disease be due immediately to the changes described, such morbid changes may not infrequently involve other portions of the cord, and so produce a complex case. Wherefore, it constantly happens that a patient comes before us complaining of weakness of the legs, at the same time informing us that he has great pains in them, and perhaps accompanied by some anæsthesia. We apply the test by requesting him to stand with his feet together, and to look upwards or shut his eyes. Immediately he does so, he falls; but the man has not ataxy, for he is scarcely able to walk across the room or move his legs from the bed; he has, in fact, true motor paraplegia. You will find constantly that patients with ordinary paralysis of motion are assisted in walking by their eyes. As for the so-called characteristic pains in the legs, these are constantly met with in paraplegia. Some of you may remember the case of an old man in one of my beds with progressive muscular atrophy, in whom the muscles were so wasted that he could scarcely move his hand to his head, and could only just stand. When on his feet, he invariably staggered and fell if he did not look downwards. If his arm was out of bed, he was quite unconscious of its position until he looked at it; and as for darting pains in his limbs, these were his most urgent symptoms. The case, indeed, was one of progressive muscular atrophy and ataxy combined, if the symptoms which he had are considered in any way characteristic of the latter disease.

The disease, then, known as locomotor ataxy is believed to be due to this grey degeneration of the posterior columns of the spinal cord, and consequently various explanations of the phenomena observed during life have been advanced. There are those who would at once place the function of co-ordination of movements in this portion of the medulla, whilst there are others who still holding to its cerebellar origin, maintain that the disease spoken of acts by severing the cerebellum from the nerves distributed to the muscles. There are some also who maintain that the affection necessitates an anæsthesia, which is always present, and in this absence of common sensation lies the explanation of all the phenomena of the disease, whilst others somewhat modify this opinion by stating that disease of the posterior roots cuts off communication from the motor spinal columns, whose office it is to co-ordinate movements.

Another theory is this, that wherever the actual seat of the disease may be, want of control means a loss of muscular sense. This theory of course necessitates the idea of the existence of such a sense. The truth of it need not oblige us to renounce the opinion that the equilibrium of the body is dependent upon the integrity of all our senses, and that they all contribute to a knowledge of our position. The action of our muscles aids us in this knowledge, as do the feeling the ground with the feet, the relation of objects around as seen by our eyes, and even probably the sense of hearing. If our muscular sense or sense of feeling be gone, we must use our eyes continually to assist us. If we close them our equilibrium is lost.

Theory of muscular sense.—The necessity for supposing the existence of a muscular sense is so great that I believe most professors of mental philosophy speak of it as an additional sense to the usual five, giving us the knowledge of resistance, force, and weight. If from disease this sense be destroyed, the patient would have no control over his muscles, and would present the symptoms of ataxia. They might be good and powerful, but he would have lost the knowledge of directing them, in a somewhat similar way as in the case mentioned by Sir C. Bell, where a woman who had lost sensation of one arm, but not the use of it, could carry her child on her arm as long as she looked at him, but directly she directed her eyes away the arm would fall. In this illustration there was simply a loss of common sensation, whilst in ataxy this is only occasionally wanting, but I mention the case to show that we must have a knowledge of the position of our limbs, as well as having possession of a power over them, in order to use them rightly. It is thought by some that this is a special sense which conveys to our sensorium a knowledge of the actions of our muscles, and that in the disease known as "ataxy" this sense is lost. The doctrine was first advanced by Sir C. Bell, who says, "When a blind man, or a man with his eyes shut, stands upright, neither leaning upon nor touching aught, by what means is it that he maintains the erect position? The symmetry of his body is not the cause: the statue of the finest proportion must be soldered to its pedestal, or the wind will cast it down. How is it, then, that a man sustains the perpendicular posture, or inclines in due degree towards the winds that blow upon him? It is obvious that he has a sense by which he knows the inclination of his body, and that he has a ready aptitude to adjust it, and to correct any deviation from the perpendicular. What sense is this? for he touches nothing and sees nothing; there is no organ of sense hitherto observed which can serve him or in any degree aid him. Is it not that sense which is exhibited so early in the infant in the fear of falling? It can only be by the adjustment of muscles that the body is firmly balanced and kept erect. There is no other source of knowledge but a sense of the degree of exertion in his muscular frame by which a man can know the position of his body and limbs whilst he has no point of vision to direct his efforts or the contact of any external body. In truth, we stand by so fine an exercise of this power, and the muscles are from habit directed with so much precision and with an effort so slight that we do not know how we stand. But if we attempt to walk on a narrow ledge or stand in a situation where we are in danger of falling, or rest on one foot, we become then subject to apprehension, the action of the muscles is, as it were, magnified and demonstrative of the degree in which they are excited. It must be a property internal to the frame by which we know this position of the members of our body, and what can this be but a consciousness of the degree of action and the adjustment of the muscles."

Such a doctrine as here laid down of the existence of a muscular sense is by no means, I believe, generally held, for, in the first place, Sir C. Bell's statements that sensitive fibres do pass into the muscles, have not been proved. Moreover, as a matter of fact, the muscular tissue is almost devoid of feeling, except when spasmodically contracting, as in cramp. Again, in ordinary contraction of a muscle, as when we grasp anything in the hand, no sensation is experienced in the forearm.

Supposing, then, that we have no knowledge of the active or passive state of a muscle, how do we account for the fact which Sir C. Bell alludes to, that a man has no difficulty in preserving his equilibrium, or that he knows the weight of any substance which he holds in his hand, or the amount of strength which he puts out when making any exertion. This is accounted for by some on the supposition that the cutaneous nerves are sufficient to acquaint us with the posture of a limb or of the whole body ; an explanation founded on a fact of which Sir C. Bell was probably ignorant— a fact which there is every reason to believe was first demonstrated in this theatre more than thirty years ago by Mr Hilton—that the same nerve which supplies a muscle also sends a sensitive branch to the skin over it, as well as to the textures which form the neighbour- ing joint.   It would follow that when the arm, for example, is flexed, a corresponding nerve to that which supplies the muscles is giving notice of the state of the skin and of the joint which is moved, and so we become possessed of a knowledge which is equivalent to that of a muscular sense.   You might infer from this, that co- existent with cutaneous sensibility would come the knowledge which we all possess, of the position and action of the various parts of our body.   To a certain extent this is true, for I have seen only this day a lady who, being paralysed many years ago, recovered her power of movement in part, though sensation is almost gone.   She is in the habit of using her hand, as, for instance, in carrying her bag along the streets, but only so long as her arm is across her chest can she retain it ; if her eyes are turned away she imme- diately lets it fall.   To strengthen this view, it is found that in cases of " ataxie locomotrice" anæsthesia does frequently exist, and there- fore there might appear to be a ready explanation of the pheno- mena ; but unfortunately in many cases perfect sensibility remains. Besides, we meet with anæsthesia constantly, as in hysterical women, where there is no ataxia, or want of co-ordination.   Experiments have been made of destroying the sensibility of the feet by freezing, and then observing whether there be any unsteadiness of gait when the person walks, but the results do not accord.   The muscular sense, according to this view, would be nothing more than the measure of pressure on the skin over the muscle.   I do not think it will explain all we understand by muscular sense.   A weight placed on the hand will give a certain amount of information, but a movement of the hand will make our knowledge more precise.   The pressure also may be made to vary by the kind of handle which the weight pos- sesses, but a difference of this kind will not deceive us.   Every one certainly is impressed with the idea that he knows what amount of muscular exertion he can put out.

You will see, then, that the question of muscular sense is not decided—its existence is not yet accepted as a fact ; but I think it must be admitted that we must possess something equivalent to it, even if the sense resides in our bones and ligaments, since we should otherwise be in the position of the statue, or the patient with ataxia. From another point of view, it seems that many physiological phenomena can scarcely be accounted for except on some such knowledge. This was hinted at by Sir C. Bell, but, as far as I know, has never been thoroughly worked out by his successors. He says: " We owe other enjoyments to the muscular sense. The divisions in music, in some degree, belong to the muscular sense. A man will put down his staff in regulated time, and the sound of his steps will fall into measure in his common walk. A boy striking the railing in mere wantonness will do it with a regular succession of blows. This disposition of the muscular frame to put itself into motion with an accordance to time is the source of much that is pleasing in music, and aids the effect of melody. There is thus established the closest connection between the enjoyments of the sense of hearing and the exercise of the muscular sense." I say this subject, as far as I know, has not been thoroughly developed, for I cannot help thinking that that knowledge which we possess of the contraction and relaxation of a muscle has more to do with many phenomena of our lives and pursuits than we are aware of. Why is it, in Sir C. Bell's illustration, that the leader of the band regulates the time with his staff? The beating time is in reality the contraction and relaxation of certain muscles. This proceeds with such regularity that there is no appeal by the ear against it. When the music teacher counts to the young player on the piano, he could do so equally well were his ears closed, the time being regulated by the contraction and relaxation of the muscles concerned in vocalisation. May not the notion of time, then, come from the muscular sense? What is meant by rhythm, by accent, and quantity in verse? What is the rising or falling of the voice but an operation brought about by muscular action? If we remember that the chest must expand and contract for purposes of life, and that we can only use for the requirements of speech as much air as can be breathed in and out in a certain time, also that during this period we contract the muscles of the larynx and again relax them, we must arrive at the conclusion that all our movements during speaking, reading, or singing, are due to alternate contractions and relaxations of muscles; that there must be a rising and falling of the voice, and what we call rhythm is dependent on simple physiological action. Accentuation of words would follow, and many other interesting points which it would be out of my province here to dwell upon. In order to prove whether this feeling of rhythm depends upon the ear or some other sense, we should naturally turn to the deaf and dumb, and inquire what knowledge of music they possess, for I think any evidence in favour of or against their possession in any degree of the perception of rhythm would be a correspondingly forcible argument as regards the existence of a muscular sense. Now, the fact of many deaf and dumb persons having written poetry or being able to versify might appear sufficient to warrant an affirmative to the question, whether they possess a sense of rhythm or of tune? having taken, however, some little trouble in order to investigate the matter, I find that most of these writers of poetry were not born without the sense of hearing, but, on the contrary, distinctly remembered the time when they had the enjoyment of this sense. Still there are those who never heard a sound in their lives and are able to read poetry with the appropriate accent and rhythm; but to this the objection might be made that the method had been learned by some special means. It is well known that the deaf and dumb are now made to speak by the method of watching the movements of the speaker's mouth and by feeling his larynx, and I am much indebted to the Rev. Mr Watson, the Principal of the Deaf and Dumb School in the Old Kent-road, for the opportunities he has given me of testing his scholars, and for giving me his own opinion on the subject I am now discussing. It is this gentleman's opinion that in speaking his pupils are conscious of the effort by a sensation in the larynx, and that a degree of cadence is necessarily produced by the alternate contraction and relaxation of the muscles, although, no doubt, the pronunciation and accentuation of the word are learned by merely tracing the movements of the speaker's lips. I might state that two of the teachers in this school were born deaf and dumb. On asking them to read aloud, they did so in a somewhat discordant tone, but with a proper accentuation of the words. A book of poetry written by a former inmate was presented to them, and they read without hesitation the following verse with the correct emphasis :

"When friends a lasting farewell take,
    There comes the ill-suppressèd sigh;
The sad fond heart is nigh to break,
    And tears quick rising floód the eye."

The author of this book of verse, Mr Simpson, believes that the deaf and dumb are as capable of writing and reading poetry as other people, but then it must be stated that it was not until he was seven years old that he totally lost his hearing, and thus the memory of words remained to him. It was so, also, with the deaf authoress, Charlotte Elizabeth, who wrote much poetry, but who did not lose her hearing until the age of twelve. Then, again, Dr Kitto—who, I believe, lost his hearing at twelve—when writing on deafness, says : "It is not wonderful that deaf mutes, and those who have become deaf in childhood, never do attempt to contend with those difficulties which seem absolutely insuperable. I am utterly ignorant of any verse written by any person under such circumstances." In answer to this I was presented with a piece of poetry composed by a late inmate who was said to have been always deaf and dumb; but of this I feel somewhat doubtful. These contradictory statements show that some further observations are required in reference to the capabilities of the deaf and dumb. At present my own opinion is in agreement with that of the Rev. Mr Watson, that they do possess a knowledge of rhythm. This gentleman informs me that they dance to music, and move their limbs in true time. If this be satisfactorily proved, it will show that we human beings are in possession of a muscular sense, or some sense equivalent to it; that is, that we are conscious of the contraction and relaxation of our muscles, or of the movements which those muscular contractions entail; and that to this sense is due not only that knowledge which is necessary to enable us to preserve our equilibrium, but that sense of tune or rhythm which is an essential part of our appreciation of music. In corroboration of this view, I am informed by a singer that when reading a piece of music he mentally "hums" the air; but if he has a sore throat, or is hoarse, he cannot appreciate its merit to the same degree as when he is well; and a lady who has become absolutely deaf reads new music with appreciation and pleasure. If this be so, it would tend to prove what I believe is correct of the deaf and dumb, that the movements of the larynx provide us with a sense of some of the qualities of music, and also that if we have a full knowledge. of all our movements, we must be in possession of what is equivalent to a muscular sense.

It might be further said that we need not be conscious, in the ordinary sense, of muscular movements whilst walking, &c., and yet the spinal cord, or the centre which rules over the muscles, may possess a knowledge of its own. By giving up the idea of consciousness, or the idea of an "ego" which must be behind all our sensations and movements in order to rule over the body, and regarding our actions rather as automatic, we at once rid ourselves of many difficulties in explanation of these phenomena. The spinal cord is sentient, but not conscious.

In the *treatment* of locomotor ataxy I apprehend that if an actual change has taken place in the cord little can be done, except, perhaps, in some cases which have a syphilitic origin, in which specific remedies are required. In most cases we cannot tell the amount of disease which may have occurred, and therefore we are bound to employ remedies. If purely functional, tonics and galvanism will be sufficient to cure. I have only seen three cases in women where ataxic symptoms existed. In all of them exhausting causes had been in operation, and in one marked sexual excesses. They all got well under the use of tonics and the continuous galvanic current to the spine.

PART III—GENERAL AND FUNCTIONAL DISEASES

EPILEPSY

WE understand, by an attack of epilepsy, the case where a person suddenly loses his voluntary power, falls into a state of insensibility, and is at the same time convulsed, these symptoms being followed by profound sleep. The two main conditions to be observed are the coma and convulsions, which the older writers used to explain by saying that there was a torpor of the brain and an excitement of the spinal marrow. The severe and characteristic symptoms which I mention as being present in well-marked cases do not necessarily exist at the commencement of the disease; but, nevertheless, we are obliged to apply the term epilepsy to the minor indications, since these may be merely the precursors of the thoroughly developed complaint. We often find that long before the patient has severe convulsions he merely "loses himself"—that is, his consciousness. This state, therefore, must be styled one of epilepsy, and it is to this that the French give the name "petit mal," in distinction to the "grand mal." But you might ask, is there any other symptom which may inaugurate the attack; may there be a convulsion occasionally occurring without any loss of consciousness, and which in time passes on to the true epilepsy? Such cases are described, as for instance, that of a young man who had attacks of convulsions of the face, but no loss of consciousness. Now, if these can be truly relied upon, we shall scarcely be able to frame a definition of the disease; for if sudden coma and convulsions are the two facts which characterise epilepsy, and yet we say in some cases the loss of consciousness may be absent, and in others the convulsion, we altogether fail in our definition. If I judged entirely by my own experience, I should say that loss of consciousness was necessary to constitute a case of true epilepsy, for I have constantly seen this symptom precede all the other phenomena of the disease, but I have never witnessed a case which eventually proved to be true epilepsy ushered in by other symptoms. The cases where convulsive movements alone occurred, were invariably due to a local disease of the brain.

The term epilepsy, then, as generally understood, is a malady characterised by convulsive attacks, in general of short duration, with sudden and complete loss of consciousness, turgescence of face, distortion of mouth and eyes, immobility of pupil, bloody froth issuing from the mouth, &c. This is the usual attack, or the " grand mal." Many of these symptoms, however, may be absent, leaving only that which I consider as essential—the loss of consciousness. If, then, we say that epilepsy is characterised by these symptoms, is the converse true ?—do these symptoms always denote epilepsy ? Certainly not, as I understand the meaning of the term, and thus we can scarcely say that an accurate definition of the word has yet been given. If we say that epilepsy is marked by certain symptoms, are we to call every case by this term where similar symptoms exist ? If, for example, a man has an injury to the head, and he occasionally falls into a state of unconsciousness and has convulsions, is he to be called an epileptic ? Is the name to be used to the case where such symptoms occur in a puerperal woman or in connection with Bright's disease ? Now, we are here much in the same position as we were in regard to apoplexy; the question is not so much one of scientific inquiry as it is one of usage. Formerly the disease had a much wider signification than at present, for the existence of certain symptoms appeared sufficient to warrant the adoption of the term, and thus you will find in the older works of medicine a list of the various causes of epilepsy, such as exostosis of the skull, tumours of the brain, syphilis, &c. ; but at the present day I believe that if the patient had such symptoms as loss of consciousness and convulsions, and it was discovered after death that he had a tumour in his brain, we should state that the latter was the disease from which he suffered, paying no regard to the symptoms which accompanied it. So also in other cases where a substantive morbid condition has been found, we allow the latter to determine the name of the disease. Now, it so happens in true epilepsy, in that disease which is protracted over so many years, with intervals of comparative health to the patient, no very definite change is found in the brain, and thus, as a matter of practice, I believe the term epilepsy is now used only in that class of cases where we consider that there is no special form of disease in any part of the body which is exciting the symptoms. I, of course, am not speaking of those slighter and chronic changes which are so frequently met with in old epileptics. On the other hand, if we consider that the symptoms arise from any definite cause within the cranium or in any other part of the body which may excite the brain, we rather use the term epileptiform. The obstetric physician has long adopted the term "eclampsia," and I see no reason why it should not be used for all cases of functional epilepsy.

You have all no doubt witnessed the *true epileptic fit*, and still retain a vivid impression of the horrible sight. The person subject to the seizure generally experiences some slight mental disturbance or premonitory sensation, styled the aura; at the same time an observer would note some pallor of the face; then perhaps a cry is uttered, and the patient loses his consciousness. Now commences a contraction of the muscles; the thumb is placed on the palm of the hand, and the fingers are clenched, while the arm often describes a rotatory movement. The sterno-cleido-mastoid muscle is violently convulsed, so that the head is turned to the opposite side, the muscles of the face are twisted, the eyes and lips distorted, and the whole aspect is hideous. The pulse is quickened, the chest is fixed, and the respiration suspended, so that the face becomes red and purple, whilst the veins of the forehead swell as if ready to burst. Froth oozes through the teeth, which are fast set, and if the tongue has been bitten a bloody saliva is projected from the mouth. There may be also an involuntary discharge of the secretions. This spasm lasts a moment, and is succeeded by another, so that the whole duration of the fit is made up of a number of alternate contractions and relaxations, which last a minute or two, when a complete resolution occurs. The patient then takes a deep sigh, his head falls powerlessly on one side, the stertor and coma pass off, and a deep sleep succeeds. He generally remains in a dull, stupid, or apathetic state for some hours, and occasionally it would seem as if the mind were quite unhinged, for a temporary mania or dementia may be the result.

We will now analyse the symptoms of this epilepsy, or *morbus comitialis*. At first, the loss of consciousness I regard as the most essential and characteristic symptom of the disease—in fact, it is sometimes the only symptom. I remember a child, some years ago in Lydia ward, who would be sitting on a chair stitching; suddenly she would fall, but before the nurse could reach her to pick her up she would have reseated herself in a chair and be again at work. This was the example of the *petit mal*. I have for some time past had under my care a young shopman in this neighbourhood who is subject to these attacks; he tells me that he often has them whilst serving a customer, but he thinks they are quite unobserved. A clergyman, a patient of mine, has suffered many years from momentary loss of consciousness. Whilst in the pulpit he will lose himself for an instant, and again go on with his sermon. A lady, whom I had long known, and who at last died in a demented state, was the subject for many years of a momentary forgetfulness, so that whilst conversing she would suddenly lose the thread of her discourse, and experience what she called a bewilderment. This was merely the precursor of a very severe form of epilepsy.

The *warning*, or *aura*, is a very striking and remarkable symptom, though by no means always present. It is the more remarkable because sometimes the seat of the sensation is in a veritable morbid state, and an irritation in this part has appeared sufficient to excite the paroxysms of the disease; whereas in the large majority of cases the aura is a truly subjective sensation. It would seem that the whole brain becomes suddenly troubled, and, as the sensorial function is departing, some curious feeling is referred outwardly to the surface, or to one of the organs of the body. The cause of the variation has not yet been made manifest.

It might be thought that if there be a true cause for irritation on some part of the body, and that a sensation be felt in that spot, the source of the fit might be found there, and that such case would not be strictly one of epilepsy, but of a convulsion arising from an eccentric cause. For instance, in a case related in which a sensation preceding the fit was referred to a painful corn on the toe, and in which a cure was effected by removal of the source of irritation, a question might arise whether such a case can be included in the definition of true epilepsy. I should be inclined, however, to apply the term to the case of a girl, a patient of mine, who referred her sensations preceding the fit to a sore spot on the face, although her father assured me that the application of laudanum to this spot was sometimes effectual in arresting the paroxysm. Sometimes the sensation is that of coldness in a part, or coldness all over the body; sometimes flashes of light are seen; sometimes the patient starts up as if he were mad; sometimes a sudden and piercing cry is uttered.

As the first cases of disease which we witness often make the strongest impression upon us, so I have a vivid remembrance of this epileptic cry. When quite a youth I was walking behind a gentleman in the street, when he suddenly gave a most painful shriek—

"Sent forth a sudden, sharp, and bitter cry,
As of a wild thing taken in the trap"—

and then rushed across the road, where he fell momentarily, as if dead; he then commenced to struggle, and I learned from the crowd around him that he was in a fit. My own impression was that he had been shot dead, but had force enough left to run a few paces; since a bird or other animal, when shot, will often exhibit phenomena very like what I witnessed in this poor gentleman. Sometimes the impressions are conveyed through the sympathetic to the viscera, and the stomach may be the organ where the aura is felt. Thus a gentleman of fifty years of age consulted me on account of attacks of vomiting. On questioning him I found that he suddenly became giddy, and was for a moment unconscious during these attacks; and also that when a youth he had fits. Or sometimes the heart is affected in the form of a violent palpitation, or as an angina; more rarely as a pain darting through the head. Thus, a young lady is seized suddenly with a violent palpitation of the heart, sometimes she loses herself, and at the same time wets her linen. Some of you also may remember the case of a little boy who experiences a sudden pain in his head, and then falls.

Some years ago a girl, nine years of age, was sent to me on the supposition that she had some disease of the brain, but the case was evidently one of epilepsy. It appeared that for some considerable period she had been suddenly seized with a pain darting through the head, and a momentary loss of consciousness. These attacks occurred once a week until lately, and now there are three or four daily. She would be sitting in a chair, feel a pain dart through her head, and then suddenly fall back insensible. The child's intellect was dull, and her whole appearance answered to that of an epileptic. This is called *epileptic neuralgia*.

A boy, æt. 15, was brought to me on account of sudden attacks of a strange feeling in the nose and forehead. Another lad has peculiar feelings in his limbs, and is momentarily aphasic. Besides the phenomena connected with common sensation and visceral disturbances, the special senses may be affected, shown by noises in the ears, flashes of light before the eyes, or objects gradually fading away and lengthening out before the sight.

One of my patients complained of a feeling of soreness all over him, another of a feeling as if the floor was sinking under him, and another as if he were rising in the air. This sensation has been dignified by the name of *levitation*. Another of my patients, whilst sitting in his office, used suddenly to exclaim, " Yes, yes, yes," was for a moment lost, and then resumed his work. Sometimes the exclamation is accompanied by some strange thought or actual delusion; at another time it is preceded by a feeling which more than one patient has styled " the horrors." Occasionally you will be consulted for these premonitory symptoms only, as in a man who came to me on account of strange feelings suddenly coming over him, when, although he would go on walking and talking as before, he would scarcely know what he was about.

These patients, with all varieties of epilepsy, do not, as a rule, complain of headache, although a pain in the head may be the principal symptom of the disease in those exceptional cases already mentioned where there is a local lesion.

As regards the convulsions, these, as a rule, are more on one side than the other, and sometimes are almost unilateral. The head is turned round, and one arm violently twisted on itself, so that sometimes dislocation takes place. The dressers must be very familiar with a woman who repeatedly comes here to have a dislocated shoulder reduced after her fit. Not only is the convulsion on one side, but the whole body is sometimes violently twisted round. There is now a boy in Stephen ward, under my care, who at the onset of the paroxysm rotates two or three times. I know a child who sometimes, instead of having a regular fit, throws out his left arm, and for a moment appears strange. The spasm of the thorax causes suspension of respiration and consequent redness of face ; the consciousness is gone, and consequently all sensation. On closely watching the patient, you observe alternate contractions and relaxations. He is in a state which was formerly called one of apoplexy, being insensible, with froth issuing from his mouth.

The pupils are dilated during the fit, and it is very remarkable how this is associated purely with the epileptic condition. You may, for example, as occurred to me not long ago, be speaking to a patient, and suddenly he falls in a fit ; you immediately raise the eyelid, and find the pupil dilated ; in a few seconds the convulsions cease, and at the same time the iris contracts to its normal size. The convulsion, however, is the most important phenomenon in epilepsy, and therefore it is that after a fit is over the circumstances attending it constitute the main features by which the fact of its occurrence is known. Thus, if the paroxysm be violent, an actual rupture of the tender capillaries in the muscles and skin may take place ; and, as Trousseau has observed, a general mottling of the skin may sometimes exist for a day or two afterwards. I remember the case of a young man whom I saw some years ago, with Dr Farr, of this neighbourhood, who, for the first time in his life, had a most violent epileptic paroxysm, and when seen shortly afterwards with an almost purpuric condition of the skin, we suspected that he had blood-poisoning, and that probably the fit was the onset of smallpox. As it eventually proved, however, the mottling was due to the violence of the spasms and the injury to the capillaries. If a person have fits in the night, his only knowledge of the occurrence may be the sight of the disturbed bed-clothes. In some instances he might discover some soreness of his limbs, or the presence of some of the evacuations. I might mention, as a circumstance worth remembering, and one no doubt intimately connected with the immediate cause of the fit, that there are persons who have been the subjects of epilepsy for many years, and never had an attack except in the night.

After the attack is over the patient awakes, looks round him like a drunken man, and mumbles some words, as if his faculties had scarcely returned. He allows himself to be put to bed, or led anywhere, without any remonstrance. Sometimes one side of the body remains weak for several hours. The mind is affected in various ways in epilepsy; it becomes, in course of time, enfeebled, and at the time of the paroxysm often much excited. At the onset of the attack the patient is sometimes furiously maniacal, so that it becomes a serious matter to know in what state the brain is, in those subject to fits. Indeed, a French physician, Dr. Falret, has come to the conclusion that no epileptic is a responsible agent. Sometimes those who have epileptic fits show a remarkable derangement for a short period before the attack, as well as afterwards. I know the case of a lad who, according to the mother's account, is quite mad about the time of the occurrence of the fits. Sometimes the maniacal state constitutes the principal feature of the attack. Thus, a lady, æt. 42, came to me, saying that for the last two years she had been the subject of hysterical attacks, but from her description, and that of her friends, they would be more rightly styled attacks of epileptic mania. She would suddenly become violently convulsed, with her head thrown back, and she would clutch and bite at any object near her, and would often scream violently. During the whole day in which the attack occurred she was quite unable to control herself. When she came to me she was perfectly calm and quiet in her manner.

The newspapers constantly record such cases as the following :—

"At Worship Street, William Bustin, æt. 27, described as a carpenter, was charged with having assaulted his wife with intent to do her some grievous bodily harm. The prisoner was further charged with attempting to commit suicide. On the 18th of December, between four and five o'clock in the morning, he suddenly jumped up in bed, and, without saying a word, struck the woman a heavy blow on the head with an earthen vessel. She endeavoured to get out of the room, but the prisoner seized her, threw her down, and beat her about the head and body. Her screams brought some lodgers in the house to her assistance, and eventually she succeeded in effecting her escape. On being taken into custody the prisoner asked if she were dead, and, being replied to in the negative, said he was sorry for it. At this time he was bleeding from a wound in the throat, apparently inflicted by himself with a pocket-knife which was found in the room. Though a severe one, it was not of a depth sufficient to sever the great arteries. The prisoner

20

was removed to the German Hospital, where he has since remained. Besides the injury in question, he had to be treated for a succession of epileptic fits. The prisoner, in answer to the magistrate, said that he knew nothing of the assault. The fit came on, and he was quite irresponsible for what he did. In early life he had had three falls, and now whenever he went to an unusual height he invariably fell. He was fully committed for trial."

I think there can scarcely be a doubt that, as a rule, the mental condition of confirmed epileptics is low. In the course of time the mind becomes impaired or obtuse, with a loss of memory. The mental impairment and epileptic attacks may stand in the relation of cause and effect, or be associated only as symptoms of an antecedent derangement. Thus, I might say that the same cause which produces epilepsy tends to produce imbecility; but I also believe that epilepsy occurring from some violent exciting cause in a previously healthy brain will, if continued, affect the integrity of the brain structure. On the other hand, there are well-marked organic changes in the brain, which lead to impairment of the mind at the same time that they induce fits. Thus, I believe, ten per cent. of all those mentally affected are at the same time epileptic. It is for this reason that a permanent paralysis may be sometimes observed in epileptics; it is, however, attributable directly to the organic change in the brain, and is not a part of the paroxysmal disorder.

The mental state may, therefore, be regarded under three distinct aspects—the temporary derangement preceding the fit, the dulness and tendency to dementia as a consequence of it, and the peculiar mental condition which often permanently belongs to the epileptic. It may be, however, that this third state is often nothing more than a prolongation of the aberration which accompanies or precedes the more regular paroxysm; or it may, indeed, take the place of the true epileptic discharge. The epilepsy is said to be masked, and the case is one of "epilepsie larvée." Very often, in young people, the mind is observed to be very active and acute, prone to be visionary and to strong religious emotions. The visions which they see probably arise at the time when the grey matter is in an extraordinary state of tension before the so-called discharge occurs. A very good example is published in 'Guy's Gazette,' by Mr. Ashby, of a young man who was brought in here from the street:

CASE.—F. G. P—, æt 34, was admitted at Guy's in August, 1874, under Dr Taylor. He was found lying in the streets, near Billingsgate, and was brought in a comatose condition to the hospital. His pupils were contracted, insensible to light, and he could not be roused. A few hours after, he was sick and more sensible, but confused. On the following day he was well enough to give an account of himself. He said that his mother, though never an inmate of an asylum, was latterly strange in her manner at times, and looked upon as not quite right. One of her brothers and one of his own brothers were also strange. He had been subject to fits all his life, but they had been more severe during the last two years. He is a tutor by profession, and has followed his employment successfully. About two years prior to one of his fits he distinctly saw a number of soldiers with rifles and commanded by officers drawn up to execute him. He could hear their voices, and distinguish the details of their dress. This continued for about an hour before he became unconscious. On five separate occasions before fits he has had the same delusions. He positively states that no effort of his mind guided by past experience can expel the delusions, and that at the time he is perfectly certain of their reality. After losing his consciousness he often walks about in an apparently rational manner before falling down and becoming convulsed. On the present occasion he was coming up the river from Boulogne, and remembered nothing after passing Gravesend, but appears to have conducted himself rationally until he fell down in the street in Billingsgate. A few months ago he went to Paris with an idea of perfecting himself in the French language, and supporting himself meanwhile by teaching English. Having the misfortune to lose what means he possessed, he wandered about the city for several days and nights, and the exhaustion which ensued seems to be the exciting cause of an attack of acute mania. He wrote an account of himself during that period, which is of interest as giving a vivid picture of the reality of the delusions of an excited brain :

" I arrived in Paris the first Sunday in June. Robbed the same evening of the greater portion of my money. Continued walking (occasionally resting on a bench) until Thursday evening, when I gave myself up to the police. On Wednesday afternoon I fancied that persons in the streets were calling after me, every one speaking English. I became very excited, and took refuge in one of the public gardens, where I lay for some considerable time under the shade of the trees. There I imagined a conversation between several persons relative to my intended destruction by some soldiers in the ensuing evening or night. I passed a most fearful night, running from street to street, being pursued by omnibuses full of armed men. Towards morning the hallucination gradually dispersed, and left me pretty quiet for an hour or two. I was not permitted to remain in peace for a longer duration than this. Passing down the side of the Seine I was met by a young soldier, who made a thrust at me with a sword-bayonet. I cleared out of his way, and continued my walk by the river. Now the most frightful scene appeared before my eyes : I am riveted to the spot, I dare not move a limb, am ordered to immediate execution by hundreds and thousands of voices. The barges on the Seine are filled with spectators and bands of musicians playing funeral marches. One vessel was decked out with all my funeral paraphernalia and the men that were to be my executioners. All the people I had ever known in my life appeared before me. I expected to fall at every moment. The bullets cracked and cracked. I could feel them hit me. At last I felt a hand upon my arm, and turning round perceived a policeman, who asked me to pick up my umbrella. I could not be persuaded for some little time to alter my position, but remained where I was till, if I rightly remember, I nearly fainted, and walked off and sat myself down on the first bench I could find. What passed for two or three hours my memory does not serve me to relate, but I found myself racing through one of the environs pursued by dozens of armed soldiers. My strength must now have entirely forsaken me. I entered a small café, and sat

myself down. There I remained some little time, till at last I was escorted by
two officers to the Commissariat of the Police. I suffered a great deal there from
the excited mob which I imagined to be on the side of the office. From this place
I was escorted to some other station of the police, and afterwards taken in some
conveyance to the Préfecture, where they locked me up in a small chamber for
about twenty-four hours, after which I went to St Anne. What I imagined in
this chamber surpasses description. I fought nearly the whole time, bathed in
perspiration, until I suppose I wore myself out and fainted. At St Anne I was
very ill, and put to bed in a straight waistcoat, secured firmly to the bed. There
were three other persons in the same room, but I could count no less than from
thirty to forty. These men, I believed, had followed me for the purpose of taking
me off to a place of execution. I remained delirious for four or five days. (I had
not tasted intoxicating drinks of any kind for several days.) At St Anne I
believed myself in my uncle's house, and he all the time was bribing them to
spare my life. I applied to the doctor for an indulgence from Rome, and this I
believed to have arrived, and I was free to return to England. I left St Anne in
pretty good health, and with five others was taken to the Asile de Ville Evrard.
It is a large place and very beautiful, capable of accommodating two hundred and
fifty men and as many women, divided and subdivided to suit the character of
the different patients. I was very tranquil all the time, and inhabited the best
quarter. I have suffered from these visitations some three or four times. I am
frequently lifted and thrown with great weight from one side of the bed to the
other. I believe firmly in the possession of evil spirits, the same that are frequently alluded to in the Gospels of our Lord. These disastrous occurrences have
almost brought me to ruin.—F. G. P., August 20, 1874."

A few days after, thinking we were rather too interested in him, the patient
took himself off suddenly.

Unconsciousness without falling.—One of the most remarkable
conditions associated with epilepsy is that where neither of the
characteristic symptoms are present—complete loss of consciousness
or convulsion. The patient is in the condition which is popularly
called "lost;" he is scarcely conscious of acts and conversation
going on around him, and yet he may continue walking in a given
direction, showing that his movements must still, in a measure, be
guided by his senses. He is in a kind of dreamland, and is indeed
much in the same state as a somnambulist. This condition, under
many varieties of form, is called the *status epilepticus*, although the
term is more usually applied to the case where the patient lies for a
lengthened period in a kind of trance or stupor, as, for example, in
the case of a man lately in the hospital, who, after a succession of fits,
lay for hours in a state of lethargy. In the milder forms it is one
of great interest from a physiological point of view, and seems to
point to the possibility of a semi-conscious state, in which the brain
is sufficiently active to control the spinal system and yet not awake
enough to excite the feeling of consciousness. In reference to the
influence of the brain on the muscles and the necessity of consciousness to preserve their tone, the condition is one full of interest.

Any circumstance which renders us for a moment unconscious causes the muscles to relax, and we fall. If we sleep standing, we fall; if sitting, our head tumbles on one side and the jaw drops; there is also a paralytic relaxation of the palate, producing the phenomenon of snoring. The explanation of this must depend on our views of muscular action. If we think the muscles are kept in a state of tension by nerve force we have only to consider that it momentarily ceases to flow for relaxation to occur; whilst, if we regard the contraction of muscle as due to a discharge of force let loose by the action of its nerve, we can still regard the phenomenon as due to the removal of the governing influence of the brain. According to one theory the brain is powerless to sustain the muscle in contraction, according to the other theory, it is powerless to let loose the latent forces in the elongated muscle. Whatever view we take of the ordinary physiological fact, we must regard the case as remarkable where the patient becomes quite insensible to things around him and yet does not fall. He may, for example, remain motionless until the fit has passed off, or if he is walking in the streets he may go on in his course, sometimes running into danger, or if, in the country, walking over a precipice. He may occasionally, like the somnambulist, avoid obstacles, as if the senses were still guiding him. It is, therefore, not quite correct to say of Lady Macbeth that though her eyes were open "the sense was shut." In the case of a patient whom I know, and in whom marked attacks of epilepsy sometimes occurred, there were occasions in which he lost himself in the most gradual manner; so that when one day he arrived at a friend's door, and was about to knock, a strange feeling came over him, which propelled him to walk on; he paced at least three times round the square before he became quite unconscious, and then he was found lying insensible and convulsed on the pavement. A girl patient, as the mother tells me, will get off her chair and describe a circle round the room before she falls. A medical man who has had a few attacks of an epileptic nature informs me that on three occasions he suffered from loss of memory which lasted two or three days. On one occasion he took his wife with him in the carriage, in order to write down the prescribed medicine immediately on his leaving the house, and to direct him where next to call. If it had not been for her assistance he would not have known in the evening a single occurrence of the day's proceedings. This continued for two days, and then passed off. I apprehend a modification of this condition must be familiar to many persons when they have been in such an abstracted mood or reverie that they have performed many acts after the manner of an automaton.[1] Workmen have been known

[1] It is probable that that state of brain which in its highest degree is associated

to suddenly lose their consciousness, and yet any peculiar movement in which they have been engaged will continue. A carpenter, for example, will stand for some time unconscious hammering or sawing in the air. I have seen a case reported lately of a sailor who would have a slight fit, and remain in any position he might at the moment be placed, as pulling a rope, &c. Another man would walk through the streets of London in a perfectly unconscious state, and on one occasion was found counting the rails around St. Paul's Church-yard. I have now a patient under my care who for two years has been subject to fits. Whilst sitting in his chair his wife sees him turn pale, and he then becomes quite unconscious; his wife leads him up to bed, or anywhere she chooses, when he goes to sleep and remains dull for two or three days. The condition of these people seems very like what is often seen from a slight concussion, where, for example, a man will be thrown from his horse on to his head, get up, remount, and ride home, but has completely lost a knowledge of every circumstance since the moment of the fall.

A lady consulted me on account of her daughter, æt. 13, first describing in a letter the following strange attacks:—"She enjoyed good health until two and a half years ago, when, whilst at breakfast, she got up from the table to put a basin on the sideboard; instead of doing so, however, she went in an opposite direction, and, after taking a turn in a circular direction, she put down the basin on another table. She was soon herself again, and answered when spoken to. About a year afterwards, whilst looking over a house, she was missing for a few minutes, and then presently came walking towards us in an unconscious state; she was very pale, and her nose like alabaster. She presently vomited, and was soon herself again. Our medical man said it was her stomach. Some months after this, whilst sitting on the beach at Eastbourne, she complained she could not see properly; she looked very strange, made a slight noise, and twitched a little. She was taken home, when she was sick."

with the epileptic paroxysm exists in every modification, not only in the disease known as epilepsy, but under all circumstances where it may be impaired by deficient circulation or other causes affecting its nutrition. A gentleman informs me that he had an attack of typhoid fever during the time of the American war. During convalescence he was made conversant with all particulars of it, and became so interested that he believed he himself was an actor in it. It was not for some time that he discovered the "muddle" he was in, and that he could not be a London merchant and an American soldier at the same time. The interesting point, however, is that the story has left the same indelible traces on his memory as the actual facts of his existence, and on reviewing the past the American war comes in its place in his personal history as strongly as any of the events which are founded on fact. He thinks when he grows old he may believe the fiction to be true.

I shall presently speak of epilepsy arising from local disease, where the phenomena are generally peculiar; but this is by no means always the case, as a local lesion may cause attacks where there is a sudden and general discharge (as it is called) as in ordinary epilepsy.

For instance, a gentleman, æt. 21, a highly intelligent and cultured man, has had three or four fits, in which he fell suddenly on the floor without any warning, and had general convulsions. He has a large scar and depression over the right eyebrow, where the bone is deficient and the brain may be seen pulsating. The injury arose from the bursting of a cannon nine years ago, causing a fracture of the skull and injury to the brain. Some bone and cerebral matter were removed at the time. He has suffered no pain since the accident, but has entirely lost his smell. His taste remains, except for flavours. The sensation of the nose is perfect, and he still sneezes. There might be a question as to the injury being the cause of the fits, as he had nervous symptoms when a child and has a brother the subject of fits. It is curious, in a case of this kind, to observe the movements of the brain; there is, first, a movement synchronous with the heart, then another with the breathing, and, besides these, various undulations of another kind may be observed, and which are supposed to be related to the varying activity of the organ, at least this is the opinion of those who have watched the exposed brain during sleeping and waking moments.

You will be prepared to learn, from what I have already said, that the epileptic attacks in different individuals do not always present the same phenomena. There is the case in which the patient experiences nothing but a simple loss of consciousness, that in which there is a slight convulsion, and, again, that which I have described as the grand attack. Whether the case of convulsion without coma can strictly be called epilepsy is questionable. Then the premonitory warning is sometimes absent and sometimes present. It may occur as a sensation styled the aura, which, beginning at the surface, creeps up to the head, when consciousness departs; or it may occur in the various other ways described. A little girl, æt. 7, is subject to fits, having one nearly every day. They are of two kinds: in one she screams, and in the other she falls convulsed. The former lasts but a short time. Then, again, the period of the day in which the fit takes place varies, sometimes occurring only after the patient has retired to rest. The intervals between the attacks also vary much, these consisting of days, weeks, or months. I am now seeing a lady who was a confirmed epileptic for nine years, during the first five of which she consulted numerous medical men, and took all the usual medicines. Receiving no benefit, she desisted from treatment for two years, when they gradually became less severe and left her. They were absent for seven years, when again, two years ago, they reappeared, and now she has a fit about once a week, and always at night time. This is a very common occurrence, and shows how states of circulation affect the brain. In some cases the fit always occurs towards the morning, as in a young lady who has had epilepsy for seven years. She experiences a strange feeling in the right arm, leg, and face, and then becomes insensible. She sleeps heavily all night, and after being called in the morning sometimes goes to sleep again. She then wakes up with this strange feeling, becomes unconscious, and again sleeps for several hours. There are also other classes of cases which entirely preclude the idea of epilepsy being an organic disease situated in one spot of the brain. For instance, I know several cases of ladies who, being subject to slight fits whilst they were child-bearing, have entirely lost them at a later period of life. Then there are those remarkable cases of epileptic fits occurring only once in a person's life. Such have been described as being occasionally fatal, but I have never seen a death under these circumstances, although I expected its occurrence in many of them. Intemperance may be the cause of temporary epileptiform attacks, excessive smoking also. A short time ago I was summoned to Mitcham to see a gentleman about forty-five years of age, who had been seized with a fit. He had driven home as usual, and appeared in perfect health until the evening, when he fell in a convulsive fit. He had several attacks, and when I saw him he was in the lethargic state which commonly succeeds the paroxysm. He had no more, and rapidly recovered. Every possible cause was gone over by the medical men and his friends, but no light was thrown upon the case. A somewhat similar case I saw a few months ago in the person of a remarkably fine young man ; the same in whom the convulsions were so strong as to rupture the capillaries and induce a purpuric appearance, which suggested the advent of an exanthem. In this case the fits were of the most violent kind, and could be attributed to no cause. Sometimes a severe shock or mental emotion will induce a fit. A young man who had been somewhat irregular in his accounts was called into the room of his superior to receive a reprimand ; he fell down on the floor in a fit as if he had been shot, and, remaining in an insensible condition, medical advice was sought. He recovered, but continued in a remarkably apathetic condition for some days. He had never had a fit previous to this occurrence, nor has he had one since.

I have now seen several cases of this kind, so that it would seem that the same causes which determine hysteria in the female will occasionally and exceptionally produce epilepsy in the male. It has always been a belief of mankind that passion will make a man convulsed and foam at the mouth. At least we have reason to think so if we find it recorded by the great observer of human nature. Thus, when Iago works Othello into such a passionate rage that he falls insensible, Cassio asks what is the matter, and Iago replies, " My lord is fallen into an epilepsy. This is his second fit; he had one yesterday. The lethargy must have its quiet course ; if not, he foams at mouth, and by and by breaks out to savage madness."

Exciting causes of this kind do not often induce a fit in a confirmed epileptic, but occasionally we are told of examples of it. Thus, an engine-fitter who has had the disease gradually increasing upon him cannot now go on into his workshop without experiencing an attack.

Diagnosis.—In a well-marked case of epilepsy the diagnosis is easy, owing to the insensibility and unconsciousness of the patient. If he show any signs of feeling or voluntary power, there is a supposition of the fits being feigned. If he fall so as to injure himself, of course there can be no doubt of the reality of the seizure ; but, on the other hand, the avoidance of danger does not mark the patient as an impostor, for sufficient warning of its approach is sometimes given to the real epileptic. The most difficult diagnosis is that between epilepsy and hysteria ; indeed, the symptoms in a particular instance may partake so much of the nature of both diseases that we are forced to employ the term hysterical epilepsy. And this compound condition is by no means confined to women, for I have seen several men suffer from this complex state. I know a gentleman who for many years has been subject to fits, and they partake quite as much of an hysterical as an epileptic nature. In a severe hysterical fit it is, however, important to be sure that the patient is no epileptic. This may be told by the want of utter insensibility, by the long continuance of the paroxysm, by the contraction of the eyelids, and the resistance used to their being raised ; the contraction of the pupil, the choking, throwing the arms about, or screaming. In fact, the hysterical patient is very noisy, whilst the other is quiet. If told that a patient has a fit, the circumstances under which it has occurred may often form a guide as to its nature ; for example, if a woman have a fit whilst alone in her house or room, and fall into danger, there can be little doubt about its reality ; but if, on the other hand, it never occurs except the patient be in company, there is a strong surmise as to its hysteric character.

I have already said that when a well-marked cause exists for the fits, as renal disease, the term epileptiform fit or eclampsia is used ;

but as regards the character of the fit itself, it cannot be told from true epilepsy, although the following circumstances may be noticed. In uræmia there is no aura, and the attacks do not come on momentarily, and there is usually a prolonged coma or lethargy following, which is not so marked in epilepsy. It is an interesting fact, however, that, the theories of uremia being so unsatisfactory, other explanations are sought for the symptoms, and that these partake very much of the nature of those theories which obtain in epilepsy. Thus, in fatal Bright's disease, where uræmic symptoms have been present, the brain is usually pale and œdematous, and this condition is considered by some sufficient to account for the phenomena. Blood effusions, however, are sometimes found.

When fits arise from organic disease of the brain there are generally some other cerebral symptoms to denote the cause, and it is remarkable that in many of those cases where there has been a local disease of the surface, giving rise to violent paroxysms of convulsions, the coma has been often absent, and thus the disease has not fallen strictly into the category of epileptic complaints. Many years ago Dr Bright was led to make the observation that in those cases where consciousness had not altogether departed during the fit there would be discovered a local disease of the brain.

There are many other conditions somewhat resembling the epileptic state. Thus, patients who have strange feelings in their heads constantly fear that they may be the subject of fits. There is such a thing, no doubt, as *epileptic vertigo,* and it may be that when a giddiness comes on in paroxysms it is a precursor of a more serious malady. There is, however, a disease to which the term simple vertigo must be applied, where it constitutes the sole malady from which the patient suffers. A French physician has found the cause in more than one case to be in the internal ear. Then there is a giddiness connected with organic disease of the brain and a rigid state of the blood-vessels, also a functional variety in connection with indigestion and its accompaniments. I have observed a very striking difference in one respect which sometimes obtains between the dyspeptic vertigo and that arising from real cerebral disease. In the latter the patient, when he feels the sensation coming over him, immediately stops if walking, or if standing lies down, whilst in the former the person feels well whilst walking or in the upright position, but immediately he stoops the giddiness comes on. The dyspeptic person lays his head on his pillow at night, and immediately the "room goes round with him;" he jumps up, and the vertigo passes off; this again and again occurs until it is only by the most cautious and gradual descent he is enabled to compose himself to sleep.

As regards the *nature or pathology of epilepsy* I can say nothing very positive, although no disease, owing to its very striking nature, has received more attention in the attempt to unravel its mysteries. Many have placed the seat of the disease in various parts of the brain, as in its centre or in the medulla oblongata, but I think every consideration of the subject must show that the brain as a whole is affected; it is true that epileptiform fits may accompany local tumours in the brain, but these must be regarded only as exciting the paroxysms. The conditions which produce fits are those which imply that the brain as a whole is involved, as, for instance, when the blood is poisoned by urea, alcohol, or other matters. Then, again, the want of blood, as in hæmorrhage, will produce an epileptiform fit. I have now had three patients with remarkably slow pulse, anæmia and epileptiform attacks. They have been told that their fits are due to their slow circulation, but inasmuch as I have never seen a pulse as slow as 28, 30, and 32, which was their number in these patients, without a nerve cause, I should hesitate in accepting this explanation. I have one patient where the slowing of the heart is clearly of nervous origin, for it is only during the epileptic attacks that his pulse is as low as 40. At other times it is normal. The decision of such a question is not always an easy one, for, if we refer the loss of consciousness to a sudden deprivation of blood in the brain, the case is similar to that of syncope. For example, a thin impoverished lady comes to me saying she is subject to fits. These consist in a sudden loss of consciousness on first rising in the morning, suggesting a simple anæmic cause for it.

I might also refer to the case of the young man I mentioned, where sudden fear was sufficient to provoke the paroxysm. Again, the fact of convulsions being excited by a distant irritant tends to corroborate the idea that the seat of epilepsy is not in one small spot in the cerebro-spinal centres. A few years ago an American visited Europe for the purpose of having his testes removed in the hopes of finding a cure for his epilepsy; the operation was performed, but with no good result. This was because he had heard it stated that epilepsy had its seat in the genital organs. Dr Marshall Hall had a notion that the disease was caused by a spasm of the glottis, and therefore that tracheotomy would prevent the occurrence of the fits. The warranty of his great name favoured the performance of the operation in a few cases, but with no success. Then, again, the post-mortem appearances of the brains of epileptics display but very slight changes, although the organ may be found wasted as a whole or the membranes thickened. If, again, we consider what is the nature of the paroxysm, we shall be led to the conviction that the fit is due to a commotion of the brain as a whole.

It is a circumstance worthy of remark that an epileptic fit can be closely imitated. An eminent French physician was thus purposely deceived by a medical student, and mistook a feigned for the real disease. The process by which this is induced is by the person making use of his voluntary powers, or putting his cerebral hemispheres into action in order to excite the ganglia below which rule over the limbs and muscles of the body. He can thus produce a paroxysm resembling that of a true fit ; nay, more, he might, if on the stage of a theatre, work himself up to such a pitch of excitement that the mind would almost lose its balance, and a corresponding exhaustion ensue. We have only, then, to suppose the cineritious surface of the brain to be unduly excited in order to conceive how it might induce a corresponding action in the ganglia below (which rule over the limbs), whilst, itself being overwrought, all mental processes would cease. As the older writers expressed it, there is coma associated with excitement of the spinal cord. Now, if the upper portion of the latter be none other than the central ganglia of the brain, we believe that the doctrine may be received as correct. These ganglia must be therefore healthy, and not structurally diseased, whilst, at the same time, the function of the surface, which is associated with the intellectual processes, is lowered in tone.

When in a case of epilepsy the mind fails until imbecility results, a structural change is often found on the surface of the brain, and it is well known that epilepsy is often an accompaniment of the general paralysis of the insane. It so happens, also, that if epilepsy, or a disease approaching to it in character, does present any positive post-mortem appearances, they are nearly always of one kind—an adhesion of a portion of dura mater to the surface of the brain, arising from injury to the skull, syphilis, or other disease. I say in cases which may be called true epilepsy, judging, not only from the symptoms, but from the general history and duration of the complaint, very little definite change is discovered in the brain ; but in cases of shorter duration and fatal, indicative of positive disease, the change is generally found to involve the surface. Why a condition which is permanent should excite occasional disturbances of the organ, does not constitute a difficulty peculiar to epilepsy ; for there is no more difficulty in supposing that the whole cineritious surface of the brain should be occasionally set in action by a local excitement in the organ itself, than it should be affected by some altogether unknown cause at a distance. Moreover, there are certain peculiarities about these fatal cases which, at the same time as they are not representatives of true epilepsy, yet tend to corroborate the idea that the first disturbing causes of epilepsy originate in the cineritious structure—for in some of those cases where consciousness is not altogether absent a local disease is found, showing that irritation of one spot is sufficient to produce the fit. Now, in these cases one side of the body is often more affected than the other, showing that instead of the disturbance being propagated throughout the surface, it is confined to one part of it, and thus the ganglion on one side immediately beneath that part is especially stimulated to action.

I think, therefore, that in cases where such local disease exists, the frequent occurrence of the fits, the consciousness sometimes remaining, and the temporary hemiplegia being more marked, all tend to show that the symptoms are produced by local irritation of the surface acting on the ganglia below. In the truer forms of epilepsy, also, I cannot but think that the morbid processes are the same, only the whole cineritious surface is disturbed at once, as well as both pairs of ganglia beneath. All the phenomena of the disease confirm this view, as well as the conditions which usually produce eclampsia, such as those resulting from blood-poisoning or mental shock. In all these cases it may be that the immediate cause for the phenomena arises from a temporary change in the calibre of the blood-vessels. The opinion of Dr Radcliffe has always been in favour of convulsive movements being due to a lowering or diminution of nervous power, rather than to an excessive action, and he would illustrate this by the epileptic seizure which is often a result of severe hæmorrhage. The difficulty in such explanation in the case of an ordinary fit is that, while one portion of the nervous centres is apparently for the time dead, another is in full activity, and yet the amount of blood in both would be the same. It might be said that the cerebral hemispheres, as already explained, have a restraining force over the ganglia below, so that immediately the function of the former is in abeyance that of the latter comes into play, but I have no facts to support such an opinion. It may be true that, thus unrestrained by the brain proper, the spinal system is more excitable on the application of stimuli, but I am not aware that it would under these circumstances spontaneously let loose its inherent forces. I would, therefore, rather believe that these ganglia, which communicate directly with the medulla, are excited by the deranged condition of the hemispheres above them, the latter undergoing such changes in their cineritious structure that a true mental alienation is produced, ending shortly in insensibility. The exact condition in which that grey matter may be is a question, but at present we may adopt in part the theory of Brown-Séquard, that loss of consciousness is due to a contraction of the vessels of the brain, brought about by irritation of the sympathetic nerve; but whether due immediately, as this physiologist thinks, to the subsequent circulation of black blood after this contraction has ceased, must at present be regarded as doubtful.

The theory of an explosion of nerve force is more or less in accord with that which was propounded many years ago by Dr Todd, and is held in the main, I believe, by Hughlings Jackson. The former physician was one of the first who investigated nervous diseases by a scientific method, and he was enabled, therefore, to throw much light on the pathology of epilepsy, so as to unravel its various phenomena of loss of consciousness, convulsion, and the maniacal state. He regarded it as a disease of the cerebrum, and, owing to the clinical fact of convulsion being caused by loss of blood, was inclined to believe that an anæmia was the precursor of a fit. He also performed experiments in order to discover the effects of irritating the convolutions. With this object, he thrust fine bradawls into the brains of rabbits, and passed through them currents from an electro-magnetic machine; in this way he obtained various movements of the limbs, face, and eyes. He therefore concluded that in epilepsy there was an undue development of nerve force, and that when this had reached a certain measure of intensity it manifested itself in an epileptic paroxysm. Just as a Leyden jar charged with electricity up to a certain state of tension gets rid of it by "a disruptive discharge," so the brain gets rid of its nerve force by discharge through the system during an epileptic convulsion. This view is essentially the same, I believe, as held by Dr H. Jackson, who speaks of the brain of epileptics being in a state of *unstable equilibrium*. The idea seems to be that there is a sudden discharge of nerve force from the brain which sets the whole muscular system in commotion, and at the same time this emptying of the brain leaves the patient for the moment unconscious, followed for some time afterwards by an enfeeblement of both mind and body. The great activity of thought also and the remarkable subjective phenomena to which some epileptics are subject support the view of the high state of tension in which the brain is supposed to be prior to the attack. I have already alluded to this theory of epilepsy in relation to the action of nerve on muscle, and stated that there are many objections to it, such as were enforced by my late colleague, Dr T. Dickson. He would not admit that an organ could perform more than a certain amount of work, and, so far from the brain evincing any such tendency to over-functionize in epilepsy, it showed an impairment of function. He thought, as did Mr Hinton, that the true motor forces lay in the muscles themselves, and that therefore in the epileptic attack, although there might be a discharge, this left the brain in a state of impairment, which produced the unconsciousness, and at the same time allowed the spinal and muscular systems to come into play, for the same reason as similar phenomena are produced if the head of an animal be cut off. Ferrier's experiments would be explained on this theory by the withdrawal of a localised force, a not altogether satisfactory argument, seeing that the effect of destruction of a convolution is not the same as that of its electrization. This accords with Dr Radcliffe's view, that muscular motion is a power in the muscle suddenly let loose, and not a nerve force finding its development or place of storage in the muscle. Whatever view be taken, the brain must be regarded as much impaired after the seizure, as the state of mind is quite sufficient to show.

Nearly all the explanations of the cause of epilepsy are quite in accordance with the fact of convulsions being associated with anæmia, and therefore we see the origin of the theory now increasing in favour, that epilepsy is due to a sudden contraction of the small blood-vessels depriving for a moment the brain of its blood. The arguments favouring this view are, that bloodletting to the verge of death in man and the lower animals will cause convulsive attacks, and, on the other hand, that extreme congestion will not; also that if a patient is seen at the very onset of the attack his face will be pale, and therefore in all probability there is an anæmia of the brain corresponding with his outward appearance. A very striking experiment in confirmation of these views has of late been made with the very powerful antispasmodic, nitrite of amyl. If, as is well known, the vapour of this be breathed, a flushing of the face, fulness of the blood-vessels, and mental confusion occur. This suggested its trial in epilepsy, and the result was very remarkable. The pallor passed from the face, and the fit was immediately arrested. Of course, it can only be tried in an asylum, where several known epileptics can be watched, as in the cases reported by Dr Crichton Browne. We might also allude to the frequent occurrence of fits in the night or early morning as corroborative proof of their association with anæmia rather than congestion, for, as Mr Durham has shown, the brain is more bloodless in the sleeping than waking state. Although all these arguments exist in favour of this view of the pathology of epilepsy, considerable difficulties still remain in explaining the cause of convulsions under apparently widely different circumstances. Either a number of conditions may by direct excitation or reflex action produce contraction of the vessels, or convulsive movements may be due to other causes. Thus, tumours of the brain, blows on the head, sudden fright, poisoning by urea and a number of other substances, will produce convulsions.

The conclusion, however, which must be arrived at is that epilepsy is not a disease in the strict sense of the word, and that the symptoms must be regarded as physiological phenomena rather than pathological. If a person have only a few fits during a long life, and in the intervals be able to follow the ordinary pursuits of business, he cannot have a diseased brain. The organ is, in fact, healthy; it is producing, as usual, its nerve force, when from some slight derangement an explosion takes place. If tumours exist, it is not to them we look for the production of the phenomena, they are only the exciting cause for the symptoms which have their origin in the healthy portions of the brain.

As regards the *treatment of epilepsy*, it must be considered entirely empirical. The term rational can scarcely be introduced even in the minor questions of diet, air, &c. I have certainly known patients reduce their amount of food and drink, especially in the article of meat, and with a corresponding diminution in the number and severity of the fits; but, on the other hand, I know other cases where a generous diet has been equally necessitated. Epilepsy is one of those cases where particular drugs may be of service, and beyond their administration little can be done. If any old woman had the possession of a herb or a salt which could antagonise the disease, her knowledge would be worth more than that of the whole College of Physicians. I am happy to say that this does not apply to many other diseases, where the knowledge you have acquired of their nature will serve you far more than all the medicines in the Pharmacopœia. The remedies, then, are empirical; those that have hitherto been most in vogue have been the metallic tonics. It is remarkable that such classes of drugs seem to have more efficiency in morbid states of the nervous system than those which have a more direct physiological effect. In the whole range of nervous affections you will find this to be the case. In my own experience the only remedies of this kind which I have seen useful have been belladonna and nux vomica —drugs having different physiological actions. I have had cases where both remedies have been apparently beneficial. The metals have been used with a certain amount of success from time immemorial, such as arsenic, silver, iron, and zinc. Some years ago I used all these remedies largely amongst the out-patients, and should certainly give the preference to zinc; I know now more than one case of epilepsy where the patient is always better on the resumption of this remedy. If you suspect any local cause in the brain, you may adopt other measures; thus, I have seen a case apparently cured by mercury. Those which were benefited by iodide of potassium had, no doubt, a syphilitic origin.

It was whilst I was examining the effects of the various remedies that I discovered the very superior value of the bromide of potassium. I was at that time trying this remedy against the iodide in bronchocele and some other disorders, and being in the habit of often using the iodide in epilepsy, I substituted the bromide for it. I was at first under the impression that it was acting as an absorbent, and was picking out for its operations those cases where the disease had a syphilitic or local origin; but when the cures came to be numerous, the explanation would not apply, and it was evident that a very valuable specific remedy had been obtained. Various writers had certainly mentioned the drug with a host of others, but only to again lay it on the shelf with them. I was not aware at that time that Sir C. Locock had recommended its use, for it does not appear that his observations had been especially brought before the Profession, much less been confirmed by others. As far as I know it was only when Sir C. Locock was President of the Royal Med. and Chir. Soc., on the occasion of Dr Sieveking reading a paper on epilepsy, that he alluded to it in the following remarks, which I quote from the 'Lancet' of May, 1857 :—"Some years since he had read in the 'British and Foreign Medical Review' an account of some experiments performed by a German on himself with bromide of potassium. The experimenter had found that when he took ten grains of the preparation three times a day for fourteen days it produced temporary impotency, the virile powers returning after leaving off the medicine. He (Dr. Locock) determined to try this remedy in cases of hysteria in young women unaccompanied by epilepsy. He had found it of the greatest service in doses of from five to ten grains three times a day. In a case of hysterical epilepsy which had occurred every month for nine years, and had resisted every kind of treatment, he had administered the bromide of potassium. He commenced this treatment about fourteen months since. For three months he gave ten grains of the potassium three times a day. He then reduced the amount, and the patient had no return since the commencement of the potassium. Out of fourteen or fifteen cases treated by this medicine only one had failed." It was in the early part of 1860 that I commenced to use it; in the following year about a dozen cases were published in the 'Medical Times and Gazette,' being the first series of cases systematically described (that I can find) in which the remedy had been found eminently successful. It was thus evident that the bromide was not simply supplanting the iodide in the cure of some special form of the complaint, but that the drug had some remarkable influence over the pure and simple form of epilepsy. This has now been confirmed by others, and even by those who had previously merely administered the bromide, as they had done many other remedies, without sufficient trial,

and had discarded it.   Of course, like every other remedy, its success has been overrated, and thus the disappointment which naturally
accompanied the reaction of opinion, more especially when it was
employed for almost every disease in the nosology.   As regards
drugs, then, I should say that the bromide and zinc are the most
important ; but you will have no lack of opportunity of trying the
effects of remedies, for epileptics often insist on being physicked
year after year when no good result is obtained by treatment.

The bromide has of late been given in very large doses as 20 or
30 grs. several times a day, and often in combination with the bromide of ammonium.   The large doses are apt to produce an acnous
eruption which may sometimes be prevented by combining arsenic
with the bromide.   It is a great question whether this remedy
which has so powerful an influence in checking the fits is really
curative ; whether indeed it has a permanent effect on the brain to
render it less unstable.   Whether or not, it is of great service in
keeping the disease in check, and so mitigating the force of the fits
which are often so detrimental to the patient.   Although this is the
rule we sometimes meet with patients who express themselves as
being greatly relieved after an explosive fit has taken place and feel
worse if it has been prevented, as if some injurious force within
them were struggling to get free ; this view is confirmed in their
minds when they find lesser fits are retarded, and then one violent
paroxysm occur apparently equal in strength to all the suppressed
ones put together.   For another reason also the prevention of the
fit has been discouraged, as for example in the case of a young lady
who had a temporary neuralgic pain in her head preceding the fit ;
after taking the bromide for some time an influence was obtained
over the disease, sufficient to prevent the full development of the
paroxysm, but not the premonitory painful sensations.   Thus the
pain which had usually preceded the fit would come on as before,
but, no insensibility following, would last about five hours before it
abated.   This was so intense that she begged that she might have the
fit, and thus work off the disease at once.   I have also known other
epileptics who say they would rather have the fits than undergo the
unpleasant sensations which from time to time seem to be substituted
for them.   I believe, however, they are exceptional cases.

I ought to mention the occasional value of counter-irritants to the
back of the neck and of setons.   I well remember two cases of men
who some years ago attended at the hospital, and whilst the seton was
open the fits were absent ; when this dried up they returned.   I have
seen other cases of the kind since this time.   I have already mentioned the case of a man whose life we saved by bleeding.   I do not
know that it is a remedy against the disease, but that it acts in the

most beneficial manner if the paroxysm is long-continued I have no doubt. In the case I referred to, the man had had a succession of fits, had swallowed nothing for some hours, and must shortly have died from congestion of the lungs, had not the lancet relieved his circulation and almost immediately restored him to consciousness. I think it very probable that in those cases in former times which were considered apoplectic, and in which recovery rapidly took place after bleeding, epilepsy was the real disease. I am convinced that I have seen several such, and therefore think they cannot be uncommon. A man, for example, is seized with a fit; you are called to him, and find him comatose, with stertor and apparent paralysis of one side; you consider it to be a case of apoplexy, and recommend bleeding; he soon afterwards recovers his consciousness, and after a few hours the weakness of the limb has passed, and the patient is comparatively well. Whatever the diagnosis, the remedy has succeeded, and thus, in a severe fit of epilepsy which becomes protracted, I have no hesitation at all in recommending you to open a vein. It might appear strange, after declaring that an epileptiform fit may be induced by loss of blood, immediately to recommend venesection as a remedy, but it does not follow that the theory and the treatment are antagonistic, for whatever may be the immediate cause of the seizure the result is a spasm of the chest, which, producing congestion of the lungs, is best relieved by liberating the blood from the overcharged venous system.

I must not omit to mention the remarkable circumstance of our capability of arresting the attack by acting on the spot whence the aura proceeds, as by tying a bandage around the limb. If the attacks were due directly to an irritation reflected from one spot, then the removal of this cause would stop the fit, as in the case of the child I mentioned, whose father assured me that the application of laudanum to a sore spot on the face would arrest or mitigate the paroxysm. But when the sensation on the surface is altogether subjective, a great difficulty in the explanation arises; unless we are content with supposing that a distinct portion of the brain must be more especially involved in order for the sensation to be felt in one part of the body rather than another, and therefore some external application to that part may cause a corresponding reflection backwards, and arrest the process that had already commenced. Or it may be that we stop the vibrations or undulations in the muscle, which are taking place in the limb.

Occasional arrest of the disease by various causes.—We must not overlook the fact, in considering the nature of epilepsy and the circumstances which may induce it, that very trivial causes may sometimes arrest it. Thus, the placing a ligature round the arm and staying the attack ought to corroborate the opinion that an epileptic seizure, frightful as it seems, is only a slight departure from a healthy natural state. Dr Buzzard says he has arrested the disease by using a blister to the part where the aura has commenced. I know a gentleman who has fits associated with an aura in the left arm; he wears a strap around it, and immediately he feels the attack approaching he pulls the strap tight, and on several occasions has warded it off. He is not anxious, however, always to do this; for, although he may arrest the violence of the convulsion he feels more affected in his mind. Patients themselves know that they sometimes have a power over the disease, and say they can prevent the paroxysms. Thus, a young man who has had a slight fit tells me that as he is walking along he feels a paroxysm approaching by a strange sensation in his arm, he then steps out quickly and manages to throw it off. A girl also tells me that when she feels a fit coming on she can sometimes prevent it by getting up from her seat and occupying herself by something in the room.

Then, again, it is worthy of notice how the epileptic state is arrested by, or is incompatible with other morbid conditions. Thus, a young man came into the hospital for epilepsy, of which he had on an average a fit every day. He caught typhus fever, and during a whole month he never had an attack. As soon as he had quite recovered they returned. A girl also came into the hospital for fits, of which she had one every day. She took erysipelas, and for a fortnight, whilst ill with this fresh complaint, never had an attack. I knew the case also of an epileptic young lady who was quite free from her complaint whilst ill with rheumatic fever. Even very slight affections may be sufficient to influence the complaint. Thus, a child who has very frequent epileptic attacks suffered from a decayed tooth, and on two occasions had alveolar abscess, accompanied by some feverishness. During the day or two she was ill no fit occurred. A case of this kind proves to us that there is nothing remarkable in the fact of a seton, whilst open, arresting attacks of epilepsy, and even without the necessity of its being placed in the neck over the medulla oblongata, the supposed seat, at one time, of the disease. As regards the apparent incompatibility of epilepsy with the febrile state, this may have reference to the condition of the blood-vessels and the circulation through the brain. I occasionally see a young lady whose epilepsy appears to have been cured by an attack of variola. She had, for a very long time, very severe fits, on an average one a fortnight, when, three years ago, she took smallpox. For a whole year afterwards she never had a fit; then one occurred, and another at the end of the second year. She has not had one since this time, and therefore only two since the attack of variola. The father was so much struck with the fact that he asked me if this might not be a clue to the cure of so frightful a malady.

Unilateral and syphilitic epilepsy.— I have already said that we designate that form of epileptiform seizure by the name of epilepsy where the disease may continue over any lengthened period with intermissions of comparative health, and where the brain is not diseased, in the ordinary acceptation of the word. In these cases the loss of consciousness is the most important characteristic; the convulsion may be of different degrees of severity, but if at all marked is pretty general. In distinction to these cases are those where loss of consciousness very often does not occur and the convulsion may be local, or always confined to one side, and sometimes associated with an actual weakness of that side. Under these circumstances we are bound to conclude that the disease is not one of simple epilepsy, but a local affection of the brain, which, by occasionally disturbing the hemisphere of that side, produces the hemispasm described. I am now speaking of cases where this unilateral convulsion is well marked, and the same side invariably affected; because, in ordinary epilepsy, although I say the convulsion is general, we often observe one side more affected than the other, but then no rule is followed; for I have noticed that where two fits have rapidly followed one another in the same patient, first one side and then the other was the more convulsed. Even in uræmic attacks, where we consider that the fit is due to a poisoned state of blood which flows equally through the brain, we constantly observe that the convulsions are much more marked on one side than the other. Then, again, the existence of an aura would almost lead us to the belief that in ordinary epilepsy the various parts of the brain are not equally disturbed; for, should every region of the body have its correspondingly associated spot in the brain, it would show that the subjective feeling must be due to an over-preponderating disturbance in some particular locality. For instance, in a mild form of epilepsy the patient may feel a little bewildered, and experience a strange sensation in the arm, which is slightly convulsed; nevertheless, we believe the brain as a whole is involved. The important clinical fact, however, remains, that should a patient have a fit without loss of consciousness we should at once suspect a local affection of the brain.

Any departure from the usual symptoms of a true epileptic attack should excite our suspicion as to its nature, and suggest some special

exciting cause for it.   In the "petit mal" or "grand mal" the loss
of consciousness exists but for a minute or so, and after the attack
the patient slowly recovers, and remains well until the next fit. But
in albuminuria, or epilepsy arising from a local cause, as syphilis, or
in renal and syphilitic eclampsia, as the fits might be called, the
paroxysms occur in rapid succession, and coma may exist in the
intervals; there may also be convulsion without loss of consciousness, or the attack may be accompanied by paralysis of one side.
These constitute a certain class of symptoms which at once suggest
to my mind a local cause, as syphilis, even before I obtain the
history, more especially if there has been a succession of fits occurring at short intervals, accompanied by a partial hemiplegia.
Under these circumstances the disease may be considered due to
a syphiloma between the membranes and the brain, and if this be
situated in one hemisphere, as is usually the case, the irritation
causes the convulsion to be unilateral or predominant on one side,
and often followed by a partial paralysis of that side.   At the same
time, as only one hemisphere is involved, the consciousness sometimes remains.   I had observed this peculiarity on several occasions
before I was aware that Dr Bright had called attention to the
circumstance in the first volume of "Guy's Reports," in reference to
a case which, in all probability, was syphilitic. The case was that of
Philip D—, admitted under Dr Bright's care July 1, 1835.   He
had syphilitic scars upon him, and was admitted for fits.   During
these attacks the right arm was convulsed, and remained afterwards
weak, whilst the man appeared to be sensible during the whole of
the paroxysm.   Dr Bright gave it as his opinion that these fits
were due to "some local disorganisation affecting the membranes
and cineritious portion of the brain on the left side, and probably
influencing the deep-seated parts about the posterior portion of the
corpus striatum;" and the post-mortem showed what was, no doubt,
syphilitic deposit on this side.   Dr Bright says, in reference to the
case, "My reason for supposing that the epileptic attacks in this
case depended rather on a local affection than on a more general
state of cerebral circulation or excitement, was *the degree of consciousness which was observed to be retained during the fits;* for
although we meet with great variety in this respect, yet in two cases
which have occurred to me the fact of the patient generally remaining conscious has been a remarkable feature, while in each, the
injury on which the fits depended was of a local rather than a constitutional or general character.   The epileptic character seemed to
point to the membranes and surface of the brain as the parts most
affected.   For of this connection I have pretty well satisfied myself
by an extensive induction of facts."

Bright, in his well-known 'Medical Reports,' had already made similar observations. He gives a case of fatal epilepsy, with drawings of the skull, which was much thickened and carious on the surface. The dura mater was adherent over the middle of the left hemisphere, and the material uniting it to the brain was of cheesy consistence.

CASE.—The patient was æt. 37, and admitted into the Clinical ward, Nov. 7, 1827, for epileptic fits. These occurred at irregular intervals, and were accompanied by paralysis of greater or less duration of the right side. On admission he complained of constant headache, and was subject to a tremor of the right leg, occurring daily and continuing for irregular periods, as an hour or longer. The tremor began in the foot, running up the leg to the thigh, and occasionally extending to the body and head, when he was deprived of the power of speech, but was aware of what was passing at the time. He also closed the lid of the left eye involuntarily and unconsciously. His right leg he dragged when walking. The fits became afterwards more frequent and intense, so that he lost his consciousness. The following plan was then tried. When the tremors were coming on a tourniquet which was held ready was applied to the lower part of the thigh, and with the effect of immediately arresting the fit. This was successful for three successive attacks, but on the next occasion he found the same sensation commencing in the arm, which, after continuing for a quarter of an hour, terminated in a fit. He then procured a ligature for the arm, and on the following day, after the aura had commenced in the leg and he had arrested it by the tourniquet, it went to the arm. This he stopped in like manner, and continued to do so on several occasions, when the method began to fail, and the fits returned as bad as ever. He left Guy's after this, and subsequently went into St. George's, where he died.

Dr Bright also gives the case of a man who fell from a cart on his head, and subsequently had fits. These began as an aura in the left leg, passing up through the body until it reached the head, when he lost recollection and fell convulsed. He also had a pain and "twitching" in the leg, and complained that the foot of that side was in a constant state of perspiration. After death the dura mater was found much thickened at the posterior part of the falx, and closely adherent to the posterior lobes, especially the left, so that when removed it tore away the cineritious substance.

He then makes the following remarks :

" As far as I have been able to infer from my own observation, I should say that the organic causes of epilepsy connected immediately with the brain are more frequently such as affect its surface than such as are deep-seated in its substance. Slow change, producing a thickened condition of the membranes, will not infrequently be found attendant upon epileptic attacks. Tumours pressing on the surface, or amalgamated with the cineritious substance, will also be found in cases of epilepsy. It is an idea entertained by Dr Foville that the cineritious is the more active part of the brain generally, with regard to all its functions, and that the medullary part is more particularly employed in the conveyance of the motions and sensations, or whatever else may be acted upon or produced in the cineritious part. And supposing for a moment this to be the case, we might expect that lesion of the cineritious substance would produce disordered action in that part, and that such action might be transferred to the distant parts of the body, producing disordered and involuntary motions; whereas if the great injury were done in the substance of the brain, the means of communication with the active part being cut off, paralysis might result, more or less mingled with convulsion, in proportion as the cineritious substance is more or less involved."

I have already told you how disease of the corpus striatum produces hemiplegia, and that an irritation of this body causes convulsions of the opposite side, so that effusion of blood in its neighbourhood, as well as tumours, abscess, or aneurism, may be the cause of hemispasm, but as a rule it is an irritation of the surface of the brain over the corpus striatum which excites its action; unless this irritation directly causes convulsive movements. This is seen in cases of injury to the surface and in effusions of blood. Now, in the absence of such evident causes and of such severe form of disease as tumours, we should rather suspect a chronic disease like syphilis, and the more so when we learn that its favourite seat is in that spot which is most likely to produce the attack. The syphilitic deposit is most frequently situated in the Sylvian fissure, in connection with the middle cerebral artery, so that we find on examination a large hard gummatous mass uniting the dura mater firmly to the brain, the surface of which is more or less involved in the disease. The convolutions affected being those around the corpus striatum (the superior termination of the motor column) would of necessity be always ready, when their equilibrium was disturbed, to set the whole motor tract in action. The result would be convulsive seizure of the opposite side of the body; and if the disturbance was confined to the one hemisphere there would be no necessary loss of consciousness. Now, in the course of time other phenomena occur in these cases, either by the direct implication of the Sylvian artery, whereby the blood supply is cut off and destructive changes ensue, or by a slower alteration in the tissue from a true syphilitic disease of the arterioles.

Under these circumstances the corpus striatum becomes involved, and a partial hemiplegia results. Remember, then, that unilateral epilepsy, especially if combined with some weakness of the side, should make us suspect a local lesion of the brain, and more especially if there is no loss of consciousness. Now, of all local affections syphilitic disease would be most likely to induce such an attack, and for the reasons before named. The further investigation of the case would in all probability confirm the diagnosis. Dr H. Jackson has made an analysis of these cases, and shown that special regions and muscles of the body are affected according as particular convolutions are involved. In one case it may be the face, in another the hand, thumb, or one particular finger, which is most convulsed. He has also observed that these convulsive twitchings may precede for a long time the regular paroxysm. This seems to support the conclusions drawn from the experiments of Ferrier on the localisation of the functions of the brain, as he produced exactly the same movements of which I have been speaking, by exciting particular convolutions on the surface of the organ. For example, if in the epileptic attacks the hand and arm were most convulsed, we should suppose that the superior frontal convolution of the opposite side was affected; if the leg were convulsed, the neighbouring convolution ; if the face and eyelids, then the middle frontal convolution ; if the mouth and tongue, the lower frontal convolution near the Sylvian fissure. Dr Jackson considers the fingers and thumb as the parts which are first affected in hemiplegic convulsion, and alludes to them as being the portions of the locomotive organs which are most highly developed or differentiated ; afterwards the face and tongue would be affected, and finally the foot.

The character and site, therefore, of the spasm would at once suggest a local lesion of the hemispheres, but that which so often leads to the suspicion of syphilis is the addition of a number of other symptoms indicating that other portions of the cerebral structure are probably implicated, as they have no connection with the presence of a circumscribed mass of disease, as a tumour. These symptoms are both of a physical and mental nature; the latter shown by a strangeness of manner, obtuseness, and other mental vagaries, the former by various local paralyses, as of the different cranial nerves. These paralyses are so varied that no small localised disease could account for them, and therefore we have recourse to the opinion that there is a large patch of morbid material on the surface, such as we meet with in syphilis. Then, again, as indicative of this, the bone may be affected, and there will be the additional symptoms of a lump on the head, with pain and tenderness.

Another fact I have especially noticed in epilepsy having a local origin, which is, that the fits occur sometimes so frequently and in such rapid succession that one paroxysm has scarcely ceased before another has begun, and so lead to death, an event very rarely seen in true epilepsy. I believe Dr Jackson has said that the epileptic cry is also wanting in these cases. Of course, in making a diagnosis of a local disease of the brain from the symptoms mentioned, we are probably assisted by the fact that the syphilitic inflammation of the membranes occupies the site of the middle meningeal artery, and therefore of the motor region of the brain. Observations are still wanting as to the character of the phenomena displayed when other portions of the brain are involved. But in all cases special nerves may be implicated which may at once suggest the nature and seat of the lesion. For example, a woman had been several times in the hospital, under the physicians and surgeons, for complaints of a syphilitic nature, which during the last year of her life had been more defined, and limited to the cranium, such as ptosis of the left eye, deafness of the right ear, some affection of the optic nerve, and afterwards of the portio dura. Subsequently pain in the head and neck came on, with dysphagia, followed by weakness of the arms, and finally the respiration was affected. After death the whole of the base of the brain was found covered with a thick layer of gummatous and connective tissue, binding together the arteries and nerves, but no distinct tumour was present.

If I am right in what I have gathered from Dr H. Jackson's writings, he does not seem to frame any broad distinction between ordinary epilepsy and that dependent on a local lesion, as he speaks of cases where there is loss of consciousness and where there is none in the same category. If the discharging lesion affects the higher centres there is loss of consciousness, but if it affects only the subordinate centres there may be none. No doubt this is true, but this variety, after all, depends upon rough pathological differences, and the clinical distinctions between them are, I think, most important. It seems to me of essential clinical importance to distinguish between an ordinary epilepsy, in which, from some unknown cause, the brain is occasionally thrown into unstable equilibrium, exciting paroxysms which may occur at long intervals and not interfere much with the welfare of the patient, and an epilepsy dependent on local causes which will before long be complicated with permanent nerve symptoms and lead soon to a fatal result. I have rarely examined the body of a person who has died of simple epilepsy, whilst my post-mortem experience has been large amongst those who have suffered from a local disease.

Since we have recognised the existence of epilepsy arising from a local affection such as is produced by a syphilitic inflammation of the membranes, we have no difficulty in discovering amongst the records of the past many similar instances, although their pathology was not then known. For example, in 'Morgagni's Morbid Anatomy,' we read of the case of a woman who had had venereal disease, and suffered from delirium and pains in her head, and in whom the membranes and brain were found adherent to one another. Also of another woman who had syphilitic tumours on the head, and the subject of epilepsy, in whom after death the bone was found destroyed at one part, and the brain covered with a membrane as thick as pasteboard, and the cortical substance beneath it as firm as the texture of liver. Morgagni attributed the scirrhous state of the brain which he found in some epileptics to syphilis. Guy's museum contains a preparation showing a portion of brain with the membranes firmly adherent. The part affected is the anterior surface of the right hemisphere; around it are several granulations which were called fungoid, but which appear to be inflammatory. It came from Elizabeth S—, æt. 50, a night nurse in Charity ward, in the year 1828. For two years she had been subject to pains in the head and to fits of epilepsy. In one of these, which occurred a few days before death, she fell into the fire; from this time she had repeated fits of tremor and loss of speech, but continued sensible. The post-mortem examination showed the surface of the right hemisphere to be firmly adherent to the dura mater. The other organs were healthy, with the exception of there being two firm tubercles in the liver (syphilitic).

As the earliest cases in practice make the greatest impression upon us, I will relate briefly one which came under my care at the time when syphilitic diseases of the brain were not generally recognised, which is nearly twenty years ago.

CASE.—Robert C—, æt. 36, admitted into Job ward under Dr Wilks, Sept. 1, 1858, reported by E. B. Truman (now of Nottingham). He was a carpenter, living at Bermondsey, but had been a soldier, and in India. He said he was invalided on account of rheumatism of the head and limbs. Two months before admission, whilst walking in the street, he had a fit which was described by a person who was near him. He did not foam at the mouth, nor bite his tongue, but he clenched his left arm, and turned almost black in the face, the left side being drawn up. Just previous to this attack he felt a tickling sensation in the foot, which passed up the leg, thigh, and body, until it reached his head, when he lost consciousness, and he also felt his heart flutter as the aura passed up. On recovering, the left arm and leg felt numb, and he could not use them perfectly. Since this attack he has had about a dozen fits. After the first two or three he bit his tongue and foamed at the mouth, but never lost his consciousness.

On admission, he complained of pains in all parts of him, and soon had fits in the manner described; the leg moved up and down, and then the arm. These members were permanently weak, or partially paralysed. His wife subsequently came to the hospital and confirmed the account of the friend in the main points. She also said that sometimes he screamed out in the fits, and was for a time like a madman. His wife came to sit up with him, thinking he would shortly die, as he had now become quite paralysed on the left side, and was scarcely conscious. Sensation was perfect. On his next visit, after a few days, Dr Wilks made a more careful examination of the case. He found the left os femoris very much enlarged, and also the right clavicle. The patient was not conscious enough to give any history of himself. He was ordered 5 grs. of iodide of potassium, and this was increased to 10 grs. in four days. He at once began to mend, the fits ceased, the limbs grew stronger, his mental powers returned, and at the end of the month he left his bed and walked about the ward. In a few more days he walked out of the hospital convalescent, with only the slightest possible dragging of the left foot.

The cases of syphilitic epilepsy are so common that any number may be found in our Records. As before said, they are noticeable by a number of special symptoms not observed in simple epilepsy. Thus, I have seen the case of a man suddenly lose the use of his right side, and then quickly recover it, which, I believe, was due to a local syphilitic disease causing a sudden discharge from the corpus striatum or its neighbourhood.

CASE.—J. P—, a young man in whom a good history of syphilis was wanting, began to feel strange in his head, desponding and unfit for business. He then had fits. After taking the iodide for some time the fits ceased. About a month after this he was taken very ill, with feverish symptoms, retching, and other cerebral troubles. He again recovered, when he found himself getting weak, staggering in his walk, talking slowly, and feeling quite confused. He again took iodide, and was better. This was some months after his first symptoms. He returned to work, but was again seized with fits, and at the same time became almost maniacal. Took his medicine again, and became calmer, but was peculiar in his manner, and said he had strange feelings come over him. Whilst talking to him he experienced an odd sensation over the forehead, and a feeling of heat; at the same time the pupil of the left eye became very much dilated. He took to the iodide again, and went to work, breaking out in pustules, probably due to the remedy. He then again became ill, with pains in all his limbs, strange sensation in his right arm, and attacks of quick and irregular breathing. At another time he had a strange sensation in the left toe, which passed up the leg to his side. He was under my care for two years, and then got well, during this time either having epileptic fits or those curious modifications of them, accompanied by mental troubles ranging from a mere bewildered feeling to actual mania.

Syphilitic Epilepsy

CASE.—James H—, æt. 37, admitted April 15th, 1871. He was a soldier, married, with three healthy children; he had had a venereal sore ten years before, but gave very little history of any constitutional disease. He was well until seven months ago, when he was suddenly seized with a fit. He first vomited, and then was convulsed. At the same time had pain in the head, which continued after the fit was over. Since this time he has had a fit about once a fortnight, and his health has suffered. He has also felt confused, and his memory has failed him.

On admission, he is ill and weak enough to keep his bed. His left arm and leg are partially paralysed, that is, he has much less power in them than in the opposite limbs. His great complaint is pain in the head, extending across the forehead and towards the right side. The right pupil is small and irregular, from old inflammation. The left tibia is enlarged and uneven on the surface, and here he experiences nocturnal pains. He states that before a fit comes on the pains in the head are worse, and that in several of the fits he has not lost his consciousness, but has tolerably distinct knowledge of what is going on around him, and has even spoken; that he knows he is convulsed, and that it is his left side principally affected. After the fit the left arm and leg remain weak.

He was ordered Liq. Hyd. Perchlorid. 3j, with Potass. Iodid. gr. x, three times a day. After a few days he began to improve, had no more fits, lost the pains in the head, and began to recover the use of his limbs. Occasionally had attacks of dizziness, but they never reached the stage of a fit. June 9th, left convalescent.

CASE.—Mr M—, æt. 25. He had syphilis, but constitutional symptoms slight, and soon passed off. About a year afterwards he began to be, what he called nervous, and subsequently had two fits of an epileptic nature. These left him very weak, so that he could scarcely walk, his gait being somewhat like that of ataxia. He felt confused and unable to follow his business, and had pains in all his limbs. He was ordered Liq. Hyd. Perchlorid. with Potass. Iodid., and gradually improved. He lost his pains, and was able to walk about. He then had simple tonics given him, and the improvement continued for six months, when severe brain symptoms suddenly came on. I found him sitting in a chair in a lethargic state. He could be roused out of this to a partial extent, and made to stand on his legs, when he managed to stagger across the room. He complained of great pain in the head, and also in the left hip and leg. He had double vision, slight ptosis of right eye, and some numbness of the face. He was ordered a scruple of the iodide every four hours, and began to improve; at the expiration of a month he was much in his usual state, being rather weak in the legs, having slight thickness of speech, and his mind scarcely capable of conducting business. He then went to Brighton, became stouter and stronger, and it was thought that he was nearly well, when he was one day suddenly seized with a fit, and soon afterwards it was observed that his skin was yellow. In a few days he was completely jaundiced. He returned home, and soon many of the old paralytic symptoms made their appearance. His left arm and leg seemed weaker than the right, and he had partial ptosis of the right eyelid; his mind was quite confused. He took the iodide again for three weeks, and all the urgent symptoms abated. This was about a year after the first brain symptoms. He had grown stout, and looked well; walked with a slight dragging of the left leg; his speech occasionally thick, and the right lid inclined to droop. He expressed himself as feeling well, and was about to return to business. He continued thus for another year, when I was informed that he had two more epileptic fits. After this he gradually recovered, and now, at the end of five years, he presents no special features by which one could recognise the source of his long illness. He is still rather feeble in mind and body, but able to transact business for a few hours daily.

Epilepsy from hereditary syphilis

CASE.—A lad, æt. 14, with well-marked syphilitic teeth and other character-istics of hereditary syphilis, was brought to the hospital suffering from the most terrible form of epilepsy. He was covered with scars from having fallen into the fire, and cut himself in various places. He had fits almost daily for last two years, in which he uttered a cry, clenched his hands, and was convulsed pretty equally on the two sides of the body. He was obliged to keep his bed, where he lay in a kind of idiotic state, scarcely knowing what was said to him, and the saliva dribbling from his mouth.

I apprehend that the iodide of potassium is quite powerless in removing those tough yellow masses of deposit which we find in the liver, in the brain, and in other organs, as the result of syphilis. It is the softer and more translucent deposit which can be absorbed by remedies, and if the whole of the adventitious material is of this nature I believe that all effects of the disease can be got rid of. But in all probability a change takes place in the syphiloma, whereby a portion becomes tough, yellow, and hard, and it is over this that medicine has no power. We find, therefore, that iodide will remove the symptoms up to a certain point, and then altogether fails; this is because all removable matter is absorbed, whilst that which cannot be touched remains.

Syphilitic Mania or Insanity.—I have spoken of mental derangement in connection with simple epilepsy, and now I should tell you that it may also exist when this disease has a local origin, and not only during the temporary occurrence of an epileptic seizure, but may continue as a permanent condition. How it is that a tumour in the substance of the brain, or lying at the base, should produce mental derangement, is not very evident, but undoubtedly such may be the case. Where, then, there is a syphilitic tumour, there are two conditions operating in favour of a general cerebral disturbance, which may produce mental derangement; the permanent presence of the new growth, and the commotion which takes place during the probable fits. Besides these causes, however, it is thought that the syphilitic process may affect the brain as a whole and so give rise to insanity which is directly of syphilitic origin. Wille, who first observed that patients labouring under syphilis might become insane, attributed the mental state to a blood condition or meningitis. Dr Batty Tuke has shown the blood-vessels throughout the brain to have undergone in such instances a truly syphilitic form of disease, a syphilitic arteritis, in which the coats have been thickened, and distinct adventitious nodules formed in their walls, sections of vessels showing rings of new material around them. The cerebral substance in the neighbourhood of these vessels had undergone degenerative changes, and the grey matter cells were altered and amyloid bodies were present. I have seen several cases where patients suffering from syphilitic disease of the brain have had maniacal attacks, and been obliged to be put under restraint, but in these persons there has been good evidence of gummatous deposit within the cranium, as they had suffered from fits, from hemiplegia, from paralysis of various cranial nerves, from aphasia, and a number of other symptoms indicative of severe lesion of the brain.

Under these circumstances we cannot be surprised that mental disturbance is superadded. What I apprehend Dr Tuke and others who erect syphilitic insanity into a species maintain, is—that the syphilitic process may primarily and purely attack the vessels and those structures which are more intimately connected with the intellectual functions, so that a manaical condition results, which is recoverable by the usual specific remedies. More lately Dr Mickle has published cases of syphilitic insanity where the arteries were found thickened with irregular nodules upon them, the cells of the grey matter were undergoing degeneration, and there were various degrees of meningitis in different parts of the cortex. These changes, however, are not very unlike what have been observed in general paralysis, and since in many of these cases of syphilitic insanity, the symptoms are very similar to those of the latter disease, it might be suggested that certain pathological changes leading to mental disturbance may be produced by syphilis, alcohol, injuries and other causes.

Other Organic Causes of Epilepsy, such as Injury.—It seems that any permanent local disease in, or on, the brain is sufficient to disturb the equilibrium of the organ, and so produce occasional fits. Syphilis is one of the most common of these, but tumours and abscesses may be the source of irritation, and probably next to syphilis, cicatrices and adhesions, resulting from former injuries, are the most usual causes for epilepsy. Several such instances may be found reported in our post-mortem records, as for example, the case of a sailor of drunken habits who was brought here in a fit, and had a succession of them until he died. On post-mortem examination there was found a firm adhesion between the dura mater and anterior lobes of the brain ; in the cellular tissue which united them was some exudation of an ochrey colour, which extended slightly into the substance of the grey matter. There could be no doubt that this condition resulted from the organisation of blood in the pia mater, which had been caused at some former period by a contrecoup, owing to a fall on the back of the head. One day a man was brought in dead, having had a fit in the streets. It was learned that he had been subject to fits for two or three years. A scar was seen on the forehead, and the brain was adherent to the membranes underneath, besides being wasted. Dr Church described the case of a girl of weak intellect who was subject to epileptic fits, and had rigidity of the right arm. The brain on the left side over the Sylvian fissure was wasted and indurated by a cicatrix-like depression. This was composed of hard connective tissue, and some neighbouring convolutions were quite destroyed, the brain otherwise was healthy. Where epileptic fits follow an injury to the head, the bone becomes thickened, and a fresh point of interest is started in the question of operation. In the syphilitic form, iodide of course is the remedy, and in those cases where an injury has been the source of the hypertrophied bone, the same remedy may sometimes be useful. In one or two cases of fits and other symptoms of cerebral disturbance, after an injury, I have seen the continued use of mercury very useful, either in the form of the perchloride or the grey powder. The effect was too marked to doubt its efficacy.

Where medicine has failed in relieving, and the bone has been found thickened or diseased, the propriety of *trephining* comes to be considered. It is now many years since this was suggested, the idea evidently being that the pressure of the bone on the brain caused the epileptic attacks. This was gained from the experience of the surgeon, who could relate wonderful recoveries after the removal of depressed bone. From what we know now, however, we can clearly see why the operation failed, as it in nowise relieved the part which was the seat of irritation. It is the inflammation of the dura mater and its adhesion to the brain which causes the fits, and no operation is ever performed with the object of tearing this membrane from the convolutions. Where any success did follow, it might be explained by giving more scope to the growth of granulations, and so relieving pressure. It has so happened, however, that in the favorable cases which have come under my own notice, the bone was undergoing necrosis, and therefore on removing this a source of irritation was got rid of, and in this way the brain was relieved. One of the earliest recorded cases at Guy's, of trephining for epilepsy, was by Mr Morgan in 1835, and the piece of bone which he removed you will find on the shelves of the museum.

CASE.—A man of middle age was thrown from a horse upon his head six years before his admission to the hospital. From this time he became an altered man; he was irritable, his memory failed him, and he had at times fits in which he was maniacal, and partly lost the use of the right arm. Over the superior posterior angle of the parietal bone, on the left side, a lump was felt at the spot where he had received the injury. He was trephined, and it is stated that his mental condition was much improved. The bone was thickened and scabrous on the inner surface.

The following is a very interesting case, and is the last which has been under my care.

CASE.—A man was brought into the hospital in a fit, and, when he recovered, stated that four months before he had received a severe blow on the head, which stunned him; he had not been well since, having more or less pain. During the last two weeks he had had severe headache, and on the day of admission two

fits. When the attacks were over, it was observed that a slight convulsive movement still continued, especially in the leg and arm. Subsequently it was found
that the left side was getting almost powerless. It was pretty clear that he was
suffering from a local irritation of the surface of the brain, where a lump existed
near the vertex on the right side; but, nevertheless, at the suggestion of a colleague, I gave him iodide of potassium. The fits continued with the hemispasm,
and therefore I met Mr Cooper Forster in consultation, with regard to the
propriety of an operation. He considered it feasible, and thereupon the patient
was removed to a surgical ward. The trephine was used, and a portion of bone was
excised, which was very thick and scabrous on the surface. The fits immediately
ceased, and the limbs grew stronger. In a few days, unfortunately, he got erysipelas, when some suppuration with sloughing of the scalp occurred. Subsequently
he was sent back to my ward with a large surface of bone exposed, which was
undergoing necrosis. He had one or two slight fits after this, and then left his
bed, being tolerably well and able to walk about. He was to come in again at a
future time to have the dead bone removed.

In the following case, where the bone had undergone necrosis, its
removal probably prevented the fits, by withdrawing a source of
irritation.

CASE.—The patient denied ever having had syphilis, and said he had never
received an injury, but several months before his admission an abscess formed
over the left side of his head. As this did not heal he went to a medical man,
who probed it, and this threw him into a fit of convulsions, and left him with
partial paralysis of the right arm and leg. When admitted, he had constant pain
in the head, and partial paralysis of the right side. There were several sinuses on
the scalp leading to necrosed bone. On the following day he had a severe epileptic
fit, in which the right arm and leg were convulsed, and also it was said the left
side of the face. Fresh fits occurred in constant succession, threatening his life,
when Mr Bryant was called to see him. He immediately trephined over the
inflamed bone, and removed a circle from an adherent and granulating dura
mater. The fits ceased immediately, and the man expressed himself as feeling
well. He then said he was never unconscious during the fits. During the next
two months he had no fit, the headache ceased, and the limbs grew strong; several
pieces of necrosed bone were removed. The fits then recurred, and soon after
he died with symptoms of pyæmia. The diseased bone was found adherent to
the dura mater, with purulent matter beneath, and on the opposite side of the
membrane a very small portion of the brain was found involved. The organs
contained syphilitic deposit.

One of the most remarkable cases illustrating the effects of trephining was lately published in the 'Journal of Mental Science.'
It is so very interesting that I will give you the outline of it.

Insanity cured by Trephining

CASE.—A man, æt. 24, had a mass of coal fall on his head, just above the left
eyelid, causing a fracture of the skull. He lay insensible for four days, and
then recovered his consciousness. Some weeks afterwards his wife and friends
observed an alteration in his habits and whole nature. He had been formerly
cheerful, merry and sociable, but was now irritable and moody to his fellow
labourers; when at home, he sat by the fire and was always cross. He afterwards

became violent, threatened his wife, and subsequently had fits. He finally became maniacal, and was taken to the Ayr district asylum, to be put under Dr Skae. After having been watched for two months, and his disease confirmed, being morose, suspicious, &c., it was determined to trephine his skull. This was done, some depressed bone was removed, and he immediately began to improve. He was pleased to see his wife, all his old affection for her revived, he became cheerful, active and industrious, and was shortly discharged cured. Four years afterwards, when seen by Dr Skae, he was perfectly well.

Epilepsy originating in an external cause.—If a convulsive attack have an eccentric cause, as often occurs in children, the term epileptiform is not the expression we use, and, as a rule, we seldom adopt it, unless the disturbance is primary and central. Notwithstanding this, we not unfrequently meet with cases where a permanently contracted or weakened limb appears to be the starting-point of the seizure, although, of course, we must admit that the brain must be thrown out of a state of equilibrium to allow of the attack. In such cases there may have been a history of a local injury and subsequent fits; an association of events which cannot but suggest their intimate relation. In saying this, I know I am disregarding the opinion of so high an authority as Dr Hughlings Jackson, having had an opportunity of consulting with him on a case in point. For example, a gentleman went to sleep with his arm hanging over a chair; he afterwards suffered from numbness; then, in a few months, spasms in the arm came on. Subsequently he was seized with a fit, in which the whole arm was convulsed. In this case Dr Jackson believed the disease to be central, and the story of pressure on the arm to have no connection with it. The whole, history, however, was so systematic that even if each fit was immediately due to a central cause I see no difficulty in supposing that the latter might have resulted from a neuritis ascending from the arm.

I have been told of the case of a man who injured his finger, and some weeks afterwards had a numbness in it, then twitching, and subsequently convulsive movements, which gradually extended up the arm. Finally, the twitching sensation, passing up the arm, reached his head, and then he became unconscious. He after this had epileptic fits, with an aura in the arm.

I remember two or three instances where, with a permanently contracted arm or leg, epilepsy has also existed, and the part convulsed has been the weakened limb. In some instances of this kind a central cause has existed to produce first the paralytic condition and afterwards the fits; but I refer now to those cases where the affection of the limb was supposed to have been due to a local disease or injury.

The following is the kind of case which suggests a local injury as productive of the fits; but whether by a simple reflex action or by a neuritis gradually extending upwards may be difficult to say.

CASE.—Alfred S—, æt. 25, a clerk, admitted into the hospital under Dr Wilks. No history of hereditary or acquired disease. Fifteen years ago he had a fall from a tree, and was much injured about the body; he had the right hip broken and left forearm. He was laid up for some months, and the left arm became stiff and weak.

He was well up to fifteen months ago, when, some days after travelling to London on a very cold day, he was seized, whilst lying in bed, with a cramping pain about the region of the seventh cervical vertebra. He attempted to get out of bed, when a pain went down his back like an electric shock. He managed to get into bed again after much trouble, and lay there until the morning, when he rose quite well. At breakfast he was stooping down, when he became faint, and fell under the fire-place. He did not lose his senses. He then finished his breakfast, and went about his work. During the day he found he was losing power in the left hand, every finger being affected, and he kept constantly dropping a small bag which he was carrying, and by the middle of the day had lost all power over his hand, and he noticed also that his left leg seemed to drag, so that he constantly stumbled, and he thinks his face was slightly drawn to one side. He was thus completely laid up for nine weeks. He gradually improved under the use of bromide of potassium, and returned to business. About five weeks after this his affected arm became suddenly drawn up, and he fell off his stool. The doctor who then saw him said he had had an epileptic fit. After this there was slight improvement in his arm and leg; he did not have another fit for four months, then one after two months, then they occurred more frequently, until lately he has had three in one day.

On admission he did not look ill, countenance bright, was able to walk about, but dragged the left leg slightly, left arm contracted and flexed, with little power of movement. The pelvis had a large bony mass, growing from the seat of injury. Soon after admission he had a fit, which was of a severe epileptic character; the convulsion was general, but the left arm the most agitated; on moving the arm the convulsions seemed to be increased. Complete coma, pupils dilated. In evening another fit. Ordered Potass. Bromid. gr. xv. He had no return for six weeks, when he had another attack, then no more for some weeks, and he left the hospital.

I have now under my care a gentleman, æt. 30, who met with a railway accident five years ago, which severely injured his pelvis and right leg. He was laid up for several months, and his life was despaired of. He had no injury to the head. About a year afterwards he had a fit, and has had three since. Before the fits come on he has a strange feeling in the right arm and leg. At all times he is subject to numbness in those limbs, and on two occasions the toe-nails have come off.

Aural or labyrinthine vertigo, Ménière's disease.—Giddiness is the feeling which we experience when objects seem moving around or away from us, or the strange and momentary sensation which obliges us to remain stationary for a time, or sit down for fear of falling. It is a symptom of many real nervous diseases, but much more commonly is symptomatic of gastro-hepatic disorder. Thus with some persons it has been a life-long trouble, and therefore of no real importance, whilst with others it may occur after an indigestible meal, to be as quickly relieved by vomiting. Giddiness does, however, constitute a real and substantive malady, and one for which the patient seeks our advice; unfortunately, too, very often after the most rigid scrutiny no cause can be found for the existence of this solitary symptom. I have alluded to it already as sometimes being the main symptom in epilepsy, and now mention it again as associated or dependent upon the disturbance of the auditory nerve. Giddiness associated with noises or singing in the ear, "tinnitus aurium," as it is called, is by no means uncommon in persons of a nervous temperament; and in hypochondriacs a clicking and buzzing in the ears constitutes often their chief complaint.

Where disturbance of hearing is associated with cerebral symptoms, there is considerable difficulty in determining whether the affection of the auditory nerve has been the seat of all the other troubles, or whether it be not like them attributable to a common origin. Thus I am often consulted by two ladies of nervous temperament who complain of swimming, noises in the head, and disturbance of the organ of hearing, in whom I regard all these symptoms as concomitant and springing from one source. I have in my note-book the cases of three old gentlemen who are troubled with giddiness, deafness, and other slight cerebral symptoms, but probably all these have a common origin in a decaying brain and senile blood-vessels. Then, again, periodic attacks of giddiness and sickness are well known to many persons under the name of bilious attacks. In advanced age, where there is a suspicion that organic changes cause both sickness and vertigo, all movement aggravates the unpleasantness of the symptoms, which pass off when the recumbent posture is resumed. The disease, however, to which I now more particularly refer, is a vertigo supposed to depend directly upon various troubles in the ear, and if accompanied by fits of a peculiar epileptiform character, upon an actual disease in the labyrinth.

Trousseau gives a very good account of the complaint, and alludes to the writings of Ménière, who was the first to recognise the fact that vertigo and other symptoms, usually referable to congestion of the brain, might be dependent on disease of the labyrinth, and more especially of the semicircular canals. The subject is also referred to by Ramskill in 'Reynolds' System of Medicine.' Toynbee wrote a paper on giddiness in connection with affections of the ear in the first volume of the 'St. George's Hospital Reports,' and there is every reason to believe that his observations are quite original; it may be said, indeed, that every one was aware that vertigo or momentary loss of consciousness might be induced by pressure upon the tympanum, but that we are indebted to Ménière for clearly raising vertigo, as dependent upon the ear, into a distinct complaint. Toynbee states that pressure upon the labyrinth produces cerebral symptoms; and that pressure upon the membrana tympani, acting through the stapes on the vestibule, will cause a sense of giddiness, an inability to walk straight, loss of distinctness of vision, and a general depression of spirits; also similar symptoms by exhaustion of the tympanum in cases of occluded Eustachian tube. Blowing forcibly through the Eustachian tube into the tympanic cavity will cause a feeling of light-headedness and swimming. It may be remembered that many persons experience a sense of giddiness, almost to falling, on violently blowing the nose. Toynbee also mentions giddiness and other cerebral symptoms resulting from pressure of wax on the membrana tympani; and, amongst other causes, artificial drum, tumours, and even syringing the ear.

In the true auditory vertigo, as described by Ménière and Toynbee, the giddiness is often the only cerebral symptom; there is no evidence of any disease of the brain; nor is the deafness found to be due to any external cause, but having come on rather suddenly, it must be considered as originating all the other nervous troubles. The vertigo may so distress the patient that, in the course of time, loss of memory, inaptitude to business, and other symptoms of a mental character may ensue. Of numerous cases which I have seen I have no subsequent history as to their termination, nor, indeed, as to the exact nature of the disease in the ear. In many the disease has began in the tympanum, and therefore if the immediate cause of the vertigo is to be found in the labyrinth, some change has been propagated thereto through the vestibule.

The peculiar interest, however, attaching to the disease is, that epileptiform attacks occur which remind us of similar symptoms produced artificially in animals when experiments are made on the internal ear, the animals turning round in different directions according as one or other of the semicircular canals is divided, implying to the physiologist that the direction of sound and corresponding movements depend on the form and position of these canals. Now, it is well known that in some kinds of fits patients have described a circle, or turned half round, or performed other movements similar to those mentioned by Flourens when the semicircular canals of animals are injured. In many of these patients there has been long-standing disease in the ear, accompanied by deafness, giddiness, occasional vomiting, and fits of a peculiar character, so that it was conjectured that the seat of the malady was in the labyrinth. It is thought that the pressure of the endo-

lymph on the auditory nerve acts as a balancing power, and as the
pressure varies with every movement of the head, so our knowledge
of position is conveyed to the sensorium. If the canals, therefore,
are destroyed, the means of judging of locality is lost. Quite
lately a discussion has taken place on the subject at one of the
medical societies, and a sufficient number of cases were quoted by
different speakers to show the actual existence of a disease in which
the principal features were deafness, associated with certain cerebral
symptoms.   Thus, Dr Duffin related the case of a man whose
hearing was impaired, but not lost, and who had attacks of vertigo,
followed by fainting and vomiting. When he fell he always came
down on one side, and this suggested an irritation of the auditory
nerve as a cause.

The ear may be affected in many ways by syphilis, so that in the
case of a young man who came to me complaining of giddiness and
deafness, with a history of this disease, a cure was effected by the
iodide of potassium.   The same occurred in the person of a friend.
He one day walked into my study, or rather staggered in, declaring
that his head was so light and he had such a giddiness that he could
not walk in the street, and had in consequence given up practice.
He rolled about when he moved, and had a confused expression ; he
said he had disease of the brain, and that it was " all up" with him. At
the same time he had become very deaf with the left ear. I remembered that he had syphilis some years before, and had suffered more
recently from a sore tongue, and I therefore suggested the iodide of
potassium in large doses.   He rapidly lost all his unpleasant symptoms, and was soon at work again.

The worst forms of the complaint under consideration are those
where no cause can be assigned for the disease of the ear.   Thus,
at the present time I am seeing a lady, æt. 35, who states that she
took cold after her last confinement, four years ago, and from that
time began to be deaf, first in one ear and then in the other, until
both were affected. She is not at this moment perfectly deaf, as
she can hear loud sounds ; but at the same time as her hearing
became affected she began to experience giddiness and noises in
the head. These she has had ever since, although varying in intensity. She has been treated for stomach and liver without any
result. She has a distressed look ; she feels worse on moving or
stooping, and when she walks often staggers like a drunken person.
Nothing is seen on examination of the ear.

A man, æt. 50, has had deafness for about twelve years, the left
ear being the worse ; and during this time has had frequent fits of
giddiness, accompanied by vomiting, and sometimes falling. The
tympanic membranes are pronounced to be healthy. He came

to me on account of an aggravation of these symptoms. On waking early in the morning he is seized with trembling, and a feeling of depression. Whilst in the streets he has fallen from giddiness, and often finds himself not walking straight but turning to the left. He has also subjective noises in the ear.

A curious circumstance is related to me by a lady, that she not only feels giddy on moving her head, but when passing impressions are made on her retina by other people moving. She has had a discharge from one ear, and is deaf on that side. She suffers from vertigo and singing in the ears. If she moves her head she feels giddy, and the same occurs when she is in church and all the people stand up.

I should say that our present knowledge obliges us to associate vertigo with deafness and noises in the head, but by no means is it proved that their prime cause is in the internal ear, and that we are at once to pronounce upon the existence of Ménière's disease.

I am now seeing a patient who has all the symptoms of Ménière's disease, as vertigo, turning on one side and sickness, but she has no apparent affection of the ear whatever.

Also another lady, who for three years has been subject to giddiness, more especially when lying down, at which times she feels as if the bed were rolling over her; she is deaf with the left ear. I may state that noises in the head or ear are common enough in persons of nervous temperament.

A servant of a friend was sent to me on account of the following symptoms. For two months she had suffered from a feeling of the bed going round whenever she lay down; this occurred again and again on every fresh attempt until the attack for that night passed off; she afterwards discovered that it was only when she lay on her left side. She was often sick at the same time. Also on stooping to clean her steps the same thing occurred. She had no deafness nor any other trouble of the ear.

CHOREA

This is a disease characterised by irregular movements or clonic spasms of the voluntary muscles, and occurring mostly in children before the age of puberty. It occurs in all degrees of severity, from a mere twitching of a particular muscle to an implication of the whole body; in the latter case there is a necessary want of power, and the patient is confined to his bed. There may be a constant motion of the whole body, not from violent spasm, as in epilepsy, but simply from irregular muscular movement, whereby the most frightful contortions and writhings are produced. Such a horrible example you have just witnessed in the case of a man to whom we administered chloroform. We were obliged to put sideboards to his bed to prevent his falling out ; he was constantly throwing his arms about, and dashing his head on the pillow. His mouth was continually being opened and shut, and he consequently ate and spoke with the greatest difficulty. He had to jerk his words out, and make the most dreadful contortions in order to swallow his food. When asleep he was quiet. His mind was clear, and therefore he was able to give a history of his case. Often, however, the mind becomes affected, and the patient is fatuous or maniacal. Thus I had a youth under my care who was quite as ill as this man, but eventually recovered. He lay in bed in constant movement, making most dreadful contortions with his face, and with an almost total inability to articulate. After sleeping and remaining quiet for a short time he would make most horrible grimaces, throw his arms and legs about in a very violent manner, and twist his body so as to turn completely round in bed. The attempts to eat were most painful to witness : he seized the spoon in his mouth as if he would swallow it, and occasionally, indeed, bit the nurse. I think this was sometimes done intentionally, as a mad person would do. He became emaciated almost to a skeleton, and all the prominent parts of the body were covered with corns and scabs. There were pustules on other parts, and he was covered with scratches of his own making. I think a severe case of chorea of this kind is as dreadful a disease as any which we are called upon to witness, and the case of this lad, which lasted three or four weeks, was certainly the worst that I have ever seen recover. For some time afterwards his mind remained weak. He had rheumatism, then slight chorea, and subsequently a fright, preceding the more severe symptoms. The mental disturbance is so great that a mistake in diagnosis is sometimes made between chorea and insanity. Dr Hills relates in our 'Gazette' that he has had four cases, three children and one old woman, sent to the asylum at Norwich as lunatics, and who were merely the subjects of chorea, the certificate showing that the patient in one case suffered from " extreme restlessness, inability to express himself, constant gesticulation, frequently breaking cups and saucers, great irritability, slovenliness, noisiness, insomnia, &c."

In less severe forms the whole body may still be affected, but in a minor degree. The child may be up and able to walk about, although in a very unsteady manner. She cannot walk in a straight line, and if she attempts to carry any weight lets it fall to the ground. This is often one of the earliest symptoms observed by the parents, who discover that the child is commencing to be very destructive with the crockeryware. These irregular movement combined with the weakness of the arms, increase until the child (generally a girl) cannot dress or feed herself. She is also constantly making grimaces, and the tongue is thrust out with a jerk and then kept in motion. One side of the body is frequently more affected than the other, and is consequently weaker. This weakness may remain after the movements have ceased, so that the patient is first brought to us with a partial hemiplegia. In some severe cases of chorea the whole body may be left in a state of extreme feebleness; in fact, the patient has for the time a kind of general paralysis. I have seen several cases of children who after chorea lay perfectly quiet in bed, with scarcely any power to move their arms, and totally unable to stand.

I believe it was Todd who first drew attention to choreal paralysis, although it has probably always been recognised. He says in his lectures, " the choreic convulsion is often succeeded by a paralytic state of the limbs previously convulsed ; the convulsive movements cease, and the limbs remain paralysed ; the paralysis is seldom complete, although considerable. When the choreic convulsion has affected one side the paralysis will likewise affect the same side ; it will be *hemiplegia*, and will resemble very closely hemiplegia from diseased brain, for which it is very apt to be mistaken by those not aware of the fact."

A very common accompaniment of the disease is a cardiac systolic bruit. This constitutes the most interesting feature in the case, and affords a basis on which can be raised several theories respecting the nature of the disease. I think I am correct in saying that the heart affection was first observed by Bright, and ever since his time the question has been asked as to its nature and connection with chorea. I believe that it might be in part solved by the simple observation of a hundred cases of chorea accompanied by systolic bruit, and the discovery whether the bruit persisted or not as the choreal movements subsided. If it accompanied the disease and departed with it, the bruit could scarcely be regarded as due to an organic change. Those who have regarded it as functional have spoken of an irregular action of the heart, but as this does not in fact occur, they must have intended to imply such an irregular action of the papillary muscles as to allow either of a temporary regurgitation of blood through the mitral orifice, or to interfere in some way with the closure of the valve, and thus produce an abnormal sound. I wish I had some more positive information about the state of the heart in those who have recovered from chorea ; but, as a post-mortem fact, I might state that I have never seen a fatal case in which there was not some evidence of a previous endocarditis ; that is, the inner surface of the mitral valve was lined by a narrow row of beadlike vegetations. Now, in some of these cases there had been no history of rheumatism, and in some no audible bruit during the life of the patient. In the most striking examples of such cases there was a very distinct history of fright, as in the case of a young girl who was residing near the scene of the Erith gunpowder explosion. She had never had a rheumatic attack, had no symptoms of heart affection, and yet after death the mitral valve exhibited vegetations. There was also the case of a little girl, æt. 7, who was admitted under my care with a most violent attack of chorea. About a fortnight before she had been taken to the Victoria Theatre, where she was much frightened by a sensation scene. All my endeavours to relieve her were fruitless, and she died two weeks after admission in a state of great emaciation. There was no history of rheumatism, and no bruit was audible. After death we found a fringe of small vegetations on the inner surface of the mitral valve. These were firm, and therefore the question naturally arose whether they were pre-existent to the chorea. I have also constantly met with other non-fatal cases where a bruit existed without any history whatever of rheumatism.

These cases would tend to show that the cardiac bruit was organic and a consequence of the chorea. A more usual opinion, however, is, that the order of sequence is rheumatism, cardiac affection, and chorea; and thus a theory is held, which was first propagated by Dr Kirkes, that embolic particles are carried from the heart to the cord, and there set up an irritation which is productive of chorea.

This theory has been accepted and strongly enforced by Dr Hughlings Jackson, who believes that the corpus striatum is the seat of the disturbance caused in the manner described. He would say the arguments used against the theory that occlusion of the blood-vessels would produce weakness of the limbs tend to support it, since the chorea is often hemiplegic, and followed by actual paralysis. Also, there is the striking fact that patients lying in bed with rheumatic endocarditis may, without any further exciting cause, be seized with chorea. No doubt cases of this kind occur, and so far seem to give the demonstration necessary to prove the correctness of the embolic theory. Thus, a lad, æt. 15, was the subject of rheumatic fever and endocarditis, and had been in bed a month when he was suddenly seized with a weakness of the right arm, so that he was quite unable to move it; the leg also was weak, and the speech hesitating. The nature of the attack could only be explained by embolism. After two days he was able to raise the arm, but in attempting to do so found he had no control to steady it, as it moved about in various directions. Two days after this, when I saw him, there were distinct choreal movements in the right arm, he had twitching in his face, his speech was hesitating, and his words came out in jerks. One could scarcely err in saying that if the first seizure was due to embolism the choreal symptoms must have depended on the same cause.

On the other hand, it is argued that if chorea be due to embolism the conditions for producing it should always be present, but there are numbers of cases of chorea coming before us without any history of rheumatism or evidence of endocarditis. It is remarkable, however, and this it is which is thought to be highly confirmative of the embolic theory, that although these conditions are absent vegetations are almost always found on the valves in fatal cases. Notwithstanding this, it is difficult to suppose that in patients who eventually recover, and the majority do recover, that any such serious affection as inflammation of the heart could have existed, or to understand when the mischief in the brain has once happened, why the patient in fatal cases should become daily more violent, and perhaps maniacal; or why, when embolism is as common in adults as in children, chorea should be almost exclusively confined to the latter.

The discovery of a bruit in chorea without the history of a rheumatic attack and vegetations on the valves after death cannot be overlooked in studying the nature of chorea, and therefore it has been surmised by the opponents of the embolic theory that the endocarditis is a consequence of the chorea, either owing to some abnormal condition of the blood or to the papillary muscles participating in the irregular action of the muscular system, and so leading to an inflammatory process. I have already said that it has been conjectured that the irregular action of the muscles might in itself be sufficient to cause the abnormal sound, whatever may be the cause or exact seat of the irritable nerve centre.

That rheumatism and chorea are closely allied, but not in the relation of cause and effect, is seen in such a case as that of the boy lately in Stephen ward. He came in with a severe attack of chorea. After remaining in bed about three weeks he began to be a little quieter, when he was seized with acute rheumatism, involving all his joints, succeeded quickly by a mitral bruit. In this case the chorea preceded the rheumatism with the cardiac disease, just as in those which result from fright. At the same time it cannot be denied that nervous symptoms much resembling those of chorea not infrequently accompany acute inflammation of the heart. Thus delirium and spasms often constitute the symptoms denoting what is called metastasis to the heart in rheumatism, and I have myself seen marked choreal movements accompany an acute pericarditis set up in the course of Bright's disease. I wish I were able to solve the question; but I by no means can yet agree with the opinion that chorea is due to a cardiac trouble when I witness the existence of this disease before the rheumatism, and also see it suddenly arise from fright, and yet followed by a cardiac bruit. It is evident that rheumatism, inflammation of the heart and chorea, are closely related, but not that they stand in the relation to one another of cause and effect. There may be some morbid condition common to all, but whether this is more immediately due to an error in the blood or nervous system is not very clear. Our late physician, Dr Addison, used to argue from many facts that rheumatism was a disease primarily of the nervous system.

It is impossible to accept the embolic or any other theory of chorea which entirely ignores two important facts : it being a result of fright, and especially a disease of childhood. The latter is a far more prominent fact than the connection of chorea with heart disease.

Dr Dickinson has lately published some cases in which he has found changes in the corpora striata and medulla. These consisted of erosions and hæmorrhages, visible to the naked eye, and some-times showing evidence of their being long-standing by the atrophy of the tissue and presence of hæmatoidin crystals. These changes occurred in connection with a dilatation of the smaller vessels in the substance of the medulla. His conclusion, therefore, as to the nature of chorea is that it is due to a hyperæmia of the vessels, arising from causes mainly of two kinds, one being rheumatic and the other acting more directly on the nervous system by mechanical or mental shock.

I should very much doubt whether chorea is due to any special disease of the spinal cord or other part of the nervous system, but is not rather, like epilepsy, due to a disturbance of the whole of the centres. That the brain is affected is shown by the occasional maniacal excitement and the more frequent tendency to imbecility. Just as in epilepsy you might imagine a sudden disruption or discharge of nervous force exciting the ganglia below, and tempo-rarily suspending the action of the cerebral hemisphere in which the explosion took place, so in chorea you might regard the irritation as more continuous, and the movements therefore constant. Hence, when any extra work is put on the cineritious matter of the hemi-spheres, as when volition comes into play, the movements are increased. The common cause of fright would also seem to show that the first shock was mental, or imposed on the cerebrum. The con-dition is one in which the nervous centres have become irritable, lost their power, and the will is incapable of directing their action. A strong voluntary effort is capable for a moment of restraining the movements, but time is necessary for the power to be regained. In

those cases where the complaint remains chronic, and more especially in those instances where the choreal movement is confined to
one part of the body, it ceases to be a disease in the ordinary acceptation of the term; the movement is simply a bad habit. For, as
the spinal cord is educated to perform ordinary routine movements,
such as take place in walking or in playing a tune, when the mind
is otherwise engaged, so the spinal cord may be badly educated, or
have become habituated to produce certain strange movements,
which require a great effort of the will or training to entirely overcome. In such chronic cases medicine is of little use.

There are a *great variety of forms* of chorea, having reference
mostly to the part of the body affected. The strangest cases are
those which occur in young women of hysterical habit, and are therefore described very frequently under the head of hysteria; these I
shall allude to again; also cases of occasional twitchings of the face
or jerkings of the leg in walking.

There is a complaint which the Italians call chorée electrique, but
I have never seen it.

It is thought by some that a stimulus applied to the skin is the
mode by which the movements occur, and that there are particular
sensitive points of the body which give rise to the paroxysm.

There is a variety of this complaint which occurs only occasionally, and is a temporary trouble, but appears to be of a true choreal
character. It occurs in persons of a nervous temperament, and
does not appear to be common, from the inquiries I have made
amongst many persons concerning it. I allude to what is popularly
called the "fidgets." When I had a patient in the hospital thus
troubled, I wished to give it the name δυσφορίη, as this is a classic
term, and used both by Hippocrates and Aretæus. These authors,
however, used it merely as a symptom of many complaints, as equivalent to "molestia" or disquiet. The attack occurs mostly after
dinner, or after lying down in bed at night. It shows itself by the
person being quite unable to keep some part of the body quiet,
more especially the legs. If in bed he is obliged to rise, and after
walking about the restlessness or fidgetiness in time passes off.

Ordinary chorea occurs mostly in boys and girls of an irritable
temperament before the age of puberty; but occasionally we meet
with it in adults. When occurring in young women it is often
associated with early pregnancy. So commonly is this the case that
I always make inquiries in this respect when I have a case of
chorea in an adult female. In hospital practice it has often been
an illegitimate pregnancy, and therefore there may have been
moral as well as physical causes which have determined the
complaint.

From direct Injury without Fright

CASE.—I have seen three cases of this in boys. One was a lad, æt. 16, who, whilst at play, struck himself against a wall, and three days afterwards had a jerking of the arm which prevented him writing : after this the leg began to twitch; then the face; and, finally, the whole body became a prey to the choreal movements. The right side was most affected.

CASE.—A boy, æt. 11, I lately saw at Holloway, with Dr Wight. He was lying in bed throwing himself about, as is seen in a severe case of chorea. The history was that, six days before, whilst away from home, he was playing with other boys, when one of them came behind him, as he was sitting in a chair, and forcibly dragged his head back. He immediately felt very strange, and found his eyesight going. He got up, but could see nothing. A medical man was sought for, who said that a little more strain on his neck would have killed him. Very shortly he began to show strange movements in his body and limbs, which increased until he had all the symptoms of St. Vitus's dance, and was brought home. He presented no difference from an ordinary case of chorea, with the exception that he had a pain down his neck and up the back of his head; no mischief could be felt in the head, which was rotated without pain. Every now and then he called out with pain, which he referred to his head, arm, and tongue. When he swallowed he had a defined pain at the back of the tongue, as if the glosso-pharyngeal nerve were irritated. He was ordered chloral and bromide, with ice to the spine. He gradually improved, and in a fortnight left his bed ; but the choreal symptoms continued for some time, also the pain in the neck and pain in swallowing. At the end of the month he walked out, but his mind was rather obtuse, and he could not fix his attention on anything long.

Treatment.—It might be thought by the inexperienced that those drugs which exert a physiological action over the nervous system would be those which would arrest the complaint known as chorea, but, as a matter of fact, this is not the case, so that I have almost given up looking for a remedy in the direction of this class of medicines. I do not despair, however, of finding some drug which might counteract that morbid condition of nervous system which is present in very bad cases, but in the absence of such remedy our ordinary curative means are of little avail, seeing that they can act only slowly and tend to produce a change long after the time at which the acute form of the disease would prove fatal. In these very severe and bad cases we can only hope to preserve the life of the patient sufficiently long for the most approved tonic remedies to act. For example, in such cases as I mentioned just now of children suffering from acute chorea induced by fright, a fatal termination may occur in a few days, and in these the direct sedatives are suggested. Morphia, as far as I have seen, is useless. I can call to mind two cases where large doses were given, but the effect was only transitory. The same with chloroform ; the vapour produces but a temporary tranquillising effect, and our experience of it is not encouraging either in chorea or in the allied disorders, tetanus and hydrophobia. I have no objection to morphia or chloral as occasional medicines to procure sleep, for in this respect they are highly beneficial, but they are not, in the true sense of the term, remedies. I have never seen strychnia of much use in the acute affection; nor even belladonna or conia. In the less severe cases it is possible that one or two of these medicines may be useful, but I feel convinced that the class of medicines of which I speak— those which have a physiological action on the nervous system— are far less efficacious than the metallic tonics. It would seem that in order to produce a cure, a bracing up or restoration of the original nerve power is necessary, and that the mere subdual of symptoms in no way tends to cure the complaint. When I say this I speak with some little hesitation of the effects of belladonna as I have seen it, apparently useful. I remember when at Paris some years ago hearing Trousseau give a lecture on this disease and warmly recommend belladonna. On another occasion he was declaring that there was no drug in the Pharmacopœia equal to strong coffee, and on a third occasion he was vaunting the new gymnasium at the Hospital for Sick Children as the best therapeutic agent he knew. I mention this to show you that there really is no specific treatment for the disease. I might say that we thought we saw some benefit in one case after the use of cannabis indica, but none whatever in four cases in which we tried the physostigma.

I believe I can tell you something very positive about the treatment of chorea, and I only wish I was enabled to make the same boast in reference to some other diseases. Many years ago, seeing that every medicine in the Pharmacopœia, as well as several others out of it, were said to be equal to the cure of chorea, I determined to watch the disease untrammelled by medicines, and I found that in many cases a speedy recovery took place without the administration of any medicine whatever. The cases which did best were the severe ones, excepting always those which were of the most violent and acute description. The first case which I watched was a little girl who had severe chorea; she was too bad to be able to stand, and was obliged to have sideboards to her bed to prevent her wriggling out of it. This child began to improve in a day or two, and went out well in a month. This is only one example of several of the same kind. I take it that the patient, being subject to constant excitement or improper treatment at her own home, has her disease there perpetuated, whereas when brought to the hospital, being under the influence of strangers who endeavour to make her suppress the movements, and with the additional advantage of good living, she begins to recover. I should say that a weakened condition of the nervous centres, and generally perhaps a malnutrition, being at the root of the malady, good nourishment and the tonic plan are necessary. After having learned the fact that the tendency of the disease is towards recovery as soon as all the circumstances which formerly surrounded the patient were removed, I soon afterwards learned that the cure is expedited by tonic medicines of the mineral kind, and this is the experience of the majority of the profession. I have put the treatment before you in this way to prevent you supposing that such remedies as iron or zinc act in any specific manner; they are useful, but operate as nervine tonics. I believe Dr Elliotson many years ago acquired great fame by his success in the treatment of chorea, his remedy, as you know, being the red oxide of iron. We still give it, and it is one of the best of remedies; our children very willingly take half-drachm doses in treacle. An almost equally favourite remedy here is the zinc—in fact, it is the medicine most commonly given, beginning with grain doses, and increasing to as large an amount as a scruple three times daily. A favourite remedy of my late colleague Dr Hughes was rhubarb steeped in port wine; the children were thus well sustained at the same time that the stomach and bowels were improved in condition. Arsenic I also give with good success.

In very chronic cases, and those where a part of the body only is affected, medicines are of little use. In some of these electricity has been sometimes curative; in some cases shower-baths have acted with the best success. One writer has advocated the use of liniments, as of chloroform, to the spine, and more lately cold by means of ice or ether spray has been suggested. Often nothing less than a thorough change of scene will suffice to arrest the habit. If this opportunity do not occur, gymnastic exercises are of use. They not only strengthen the muscles and nerves, but they break the bad habit; they convert, in fact, an irregular movement into a regular one. If the arms are constantly moving, and are then employed in grasping a beam for swinging, a new and altered condition of the whole machinery accrues, and in time the habitual irregular actions are worn out. I am sorry that we have not a gymnasium here, and therefore all I am able to do is to order my patients a skipping rope. I believe the only method by which the chorea, which at one time prevailed in religious houses, was sometimes able to be cured, was by making the ladies dance to the notes of music.

TETANUS

This appears to consist in a state of excitement of the spinal cord induced by a morbid irritation of the motor nerves. It may be set up by a wound implicating the peripheral nerves, or by a meningitis involving the roots of the spinal nerves. This overaction of the cord is doubted by Ringer, from experiments made on animals by strychnia and other drugs, in which he finds the cord is less excitable and its reflex action depressed. There is less resistance he thinks in the cord, and in this way the tetanic state is set up. Tetanus may be idiopathic, and induced either by an affection of the nerves commonly called rheumatic, or by some increased irritation of the cord through the blood, as is seen in the frog tetanised by artificial means. The tetanic affection is first observed by a spasm of the muscles of mastication, so that the patient says he cannot open his mouth freely. The medical man sees the angle of the mouth drawn up, and if the patient is suffering from injury, especially of the extremities, he immediately suspects the nature of the complaint. The trismus does not last long before the swallowing becomes difficult, the posterior muscles of the neck or the sterno-mastoid rigid, and the head thrown back. The muscles of the chest then become stiff, and the respiration often somewhat irregular, whilst the muscles of the abdomen become hard, and the diaphragm is also involved in the spasm. It will be seen, then, that the muscles of respiration and those associated with them are those which are mainly affected. The patient often has severe pain at the epigastrium, and this may constitute his most distressing symptom; it is due apparently to the spasm of the diaphragm. When the paroxysm comes on, the whole body is convulsed, the limbs become rigid, and the muscles of the back so contracted that the body is bent backwards in a curve. As it passes off the muscles of the limbs resume their flaccidity, but the recti abdominis and the muscles of the chest and neck remain rigid. Should the patient eventually get well, this rigidity may remain for weeks during convalescence, and even whilst he is walking about the abdomen may be as hard as a board, and the sterno-mastoid equally rigid. The absence of spasms in the legs, except during the paroxysms, is important to notice as, distinguishing tetanus from poisoning by strychnia. It was made an essential diagnostic feature in the case of Palmer, who was tried for killing his friend Cook, the plea for the defence being that the death was due to tetanus. In his case the limbs were permanently rigid, whilst the lockjaw was but slight.

The idiopathic form of tetanus is not generally so severe as the traumatic; it is more chronic and more amenable to cure; the paroxysms are not so prolonged nor so violent.

The cure of tetanus by remedies seems to be exceptional. Ordinary sedatives, as opium and chloroform, are valueless; they may be given in sufficiently large doses to throw the patient into a state of stupor, but they do not otherwise counteract the disease. Chloral,

however, continued in perseveringly, is reported to have been the
means of curing some patients, and indeed every remedy may boast
of its cures—such as quinine in large doses or ice to the spine.
The last-used remedy does, however, seem to have produced some
good results, and is now on its trial—the Calabar bean or physostigma venenosum. The alcoholic extract contains the active principle, and experimenters say that it depresses or annihilates the
irritability of the spinal cord, so that the reflex excitability is
actually lost; in large doses it arrests the action of the heart and
contracts the pupil. It has therefore an opposite physiological
action to atropine. I had a case of tetanus in the clinical ward
of a supposed idiopathic form, which got well under this remedy
given in very large doses. The paroxysms certainly appeared stayed
after it was taken.

The mode by which tetanus is fatal is not always very clear. It
may sometimes be only too evident, as in a case where I happened
to be an eyewitness of its ending. The patient had some food
put into his mouth by the nurse, he took a deep breath, and his
chest never again collapsed until death had released it of the spasm.
In other cases we are satisfied in thinking that the violent paroxysms
are sufficient to exhaust the system. But we occasionally see instances where the symptoms are slight, and yet the patient sinks
rapidly from nervous depression. I have seen a gentleman with
no other symptoms than trismus die in his chair, and not long ago
I saw in consultation with Mr Quain a lady patient, from whom
he had removed a very small tumour from the eyelid, and who
shortly after had trismus and difficulty of swallowing. She expressed herself as feeling extremely ill, and feared she should not
recover. She never had any spasm except in the face, and died on
the third day. This is the usual time for the termination of the
disease, but I have seen cases in the hospital fatal within twenty-
four hours, and without any very severe symptoms. The condition
which we call tetanus clearly causes some great depression of the
nervous centres.

INTERMITTENT TETANY OR TETANILLA

In considering the spasmodic affections we must not overlook
the disease known as intermittent tetany. Like many other
conditions which have only of late received a name, it has
long been recognised and described, but was waiting for a distinct appellation to bring it formally before the profession. In
the 'Guy's Hospital Reports,' Dr Moxon relates a case of
tetany, and alludes to the description of the disease and the

cases given by Trousseau. Dr Broadbent also has stated that he is quite familiar with the class of cases known by this name. It seems to occur more especially in children and in women after their confinements. In the former it might often be overlooked under the guise of convulsions, although, if carefully studied, the symptoms would seem to be peculiar, inasmuch as they are paroxysmal, and the spasms are of a tonic kind. The old term, idiopathic muscular spasm of the extremities, conveyed an idea of the affection, and was, no doubt, used to indicate what is now styled "tetany," or "tetanilla." It is a disease characterised by tonic contractions, more especially of the legs and arms, occurring at intervals. The thumbs are drawn in, and the fingers are sometimes flexed, although as often rigidly extended in the form of a cone; the foot is stretched out, and the toes flexed towards the sole. The case, therefore, is unlike one of true tetanus, where the jaws with the respiratory muscles are affected, and the extremities free, except during the paroxysmal attacks.

The following cases have been published in the ' Reports :'

CASE.—John Thomas K—, æt. 3. He had always had good health, except during the time of occurrence of infantile disorders. On March 17th, whilst having his face washed, he complained of pain in his feet, and asked to have his boots taken off. As soon as this was done his feet were found to be contracted. He was brought to the hospital in the afternoon, when three grains of grey powder were ordered. The spasm gradually passed off, and he continued better till the following Monday, when his legs were again suddenly contracted. The spasm evidently caused him a good deal of pain, as he screamed loudly; it lasted about ten minutes, and then subsided, but his feet did not regain their natural position afterwards. The spasms occurred several times during that day, and became less severe on the next. On the 23rd the child had a severe spasm, in which the hands were also affected; the fingers remained extended afterwards, and he was taken into the ward under Dr. Wilks.

On admission both legs were seen to be affected, the feet being drawn into the same position as in talipes varus, the calf of the leg being very hard; the legs were flexed on themselves, and the thighs on the abdomen. The muscles of the arm were less affected, the fingers being extended and brought together in a conical form. The child was constantly calling out in consequence of the pain caused by the cramp, and was continually moving from side to side in the bed in order to procure relief. The legs were rubbed, but with no apparent benefit, and so cold-water bandages were applied. These evidently afforded ease, as he immediately became tranquilised. He was also ordered ten grains of hydrate of chloral three times a day. On the following day, March 24th, the spasms were not so severe, and had only occurred in paroxysms three or four times in the twenty-four hours. On the 25th the spasms were gradually passing off. No effect was produced by pressing on the femoral artery. On the 27th there were no spasms, but the feet remained contracted. After this the spasms gradually wore off, and he left the hospital.

As this is a disease which subsides spontaneously, and in this very case the paroxysms had already, on more than one occasion, rapidly abated, little can be said of the merits of the remedies employed.

The next case affords a good example of this very remarkable spasmodic affection. No surprise can be felt at the serious view which the medical man took of it, nor that he was unable to declare that the symptoms were not due to a cerebro-spinal meningitis. There is no reason to suppose that this boy exaggerated his symptoms, although they are of a kind which might be readily simulated; in fact, it is remarkable how many hysteric conditions, such as contracted hand, resemble those of tetany.

CASE.—Master M—, æt. 16, whilst at school was subject to severe attacks of spasms or cramps all over the body. The doctor who was called in was much alarmed at his condition, fearing he had inflammation of the spinal cord. The lad wrote out a history of his own case as follows, dated March 31st:—"As far as I can remember, I have been subject to cramp for about two years, having it when I first went to school at Clapham, but mamma thinks I had it before that. I have it more in winter than summer. This year the cramp has visited me more severely than ever. I have had it all over me, at all times of the day. In washing my feet, although hot water was used, I always had the cramp in my legs, sometimes very bad. When washing my chest I had the cramps about the chest, shoulders, and neck. I have often had cramps in the veins of my wrist, and often in my neck and chin when gaping, and also in my tongue, so that I could not speak plainly for some time. After skating I always had cramp severely. I often had to stretch out my leg in school or in play-time, and if it came on in my chest or side I had to get up. If I kept my hands and arms quiet for any time numbness would come on, and if I kept quite still they would get quite warm. If I let this go on for long my arms and legs would become perfectly powerless. After trying to move the limbs for a little time the numbness would gradually pass off, and then I felt cold. I often felt as if pins and needles were pricking me after the numbness had gone off. I always had numbness in chapel, unless I used to move my arms frequently, in which case it did not get bad, but by moving the fingers went off very soon. One day, having kept my arms in one position for some time, prayers being over, I tried to move my hands and arms from the ledge for the books. After a little I dragged my arms to the edge of the ledge, and they then fell like heavy weights on my knees. One day I was obliged to get a boy in the washing-room to put my socks on me, because I could not bend my leg without getting cramp. I often kept my leg in one position, generally straight out, for some time when I had the cramp. At times I did not see perfectly with my eyes. After sitting down for a short time latterly, when I got up I could not walk very straight. On February 28th I felt very queer all day; at night I took two pills. Next day I got up as usual, but, as on the morning before, my legs were stiff, and in walking I hardly lifted them from the ground, and I felt worse than the day before. I went down to breakfast, but could not eat. Drank a little coffee, which, after leaving the room, I vomited, and felt so queer that the master came, who sent for a doctor and ordered me to bed. The doctor gave me some medicine to take four times a day, which took away the feverishness and kept me in a perspiration. On the third day I got up, but did not go to school for a week. After this illness I had no more cramp, and I think it was all the medicine which did it. I felt weak, and in walking my feet sometimes bent outwards. After this illness I could not do my work well. One day, after lying down, I got up and found my arms and hands powerless through numbness, and could not move them for about a minute and a half afterwards. Had not numbness nearly so much since my illness. All that I feel now is that I want keeping up and strengthening. As I think this is all that is necessary to be told about myself, I will come to a close."

The doctor who was called in during this attack was Mr Tayloe, who found him suffering from acute spasm of the limbs, with numbness of the tongue and imperfect utterance. He seemed slightly paralysed all over the body, he could not stand, his hands were almost useless, with the fingers flexed on the palm, and sensation seemed everywhere blunted. His speech was slow and indistinct. Mr Tayloe gave him twenty grains of bromide of potassium with five grains of iodide three times a day for two weeks, when the symptoms disappeared, leaving a suspicious debility in the limbs.

On March 31st, when I saw him, he appeared quite well, and was a strong, robust-looking lad. He was not nervous or hysterical looking, and there was nothing about him indicating any malpractice, or that he had in any way fabricated his symptoms.

(Reported by Mr C. Knott)

CASE.—William H—, æt. 18, a draper, admitted under Dr Wilks, March 22nd, 1872, and left April 25th. About last Christmas he suffered from severe neuralgic pains in the head, and on one occasion his hands became clenched for about half an hour. He remained well until the day previous to admission, when he suffered much from headache, but this passed off in the evening. In the night he was attacked again by severe pain in the head and legs. Afterwards his arms became affected. He put them in hot water with slight relief. Getting worse, he was sent to the hospital on the supposition that he had rheumatic fever.

On admission he was seen to be a healthy-looking and well-grown lad ; he complained of severe pains in all his limbs, more especially in the arms. These were placed across his chest and spasmodically contracted. The muscles were rigid, and the thumbs drawn to the palms of the hands. There was a severe aching pain down the arms, increased on movement. Sometimes there were paroxysms of cramps, when he called out with pain.

On examination of the body the abdominal muscles were observed also to be slightly rigid, and it seemed that he scarcely used his abdominal muscles or diaphragm during respiration. Legs unaffected. There was considerable pyrexia, the temp. being $100 \cdot 9°$, resp. 24, pulse 94. On seeing him some time afterwards it was found that the spasm persisted, although there were paroxysmal attacks, when there was violent cramp attended by much pain. The case was regarded as one of idiopathic spasm or tetanilla, although with the presence of febrile symptoms there was a question whether a meningitis ought not rather to have been diagnosed, and, if so, that portion of the cervical cord should have been affected which included the origin of the phrenic nerves. Ordered fifteen grains of chloral three times a day. In the evening he was asleep and quiet. Temp $101 \cdot 6°$.

23rd.—Much the same. Temp. 100°, and somewhat higher at night. Herpes coming out on the face.

24th.—Tetanic symptoms passing off. Two symmetrical patches of herpes on both sides of the mouth, in course of superior and inferior maxillary nerves.

25th.—In the morning pretty well; in evening return of paroxysm.

27th.—Continued improvement. Pains passing off. Pins and needles in arms and hands. Perspires very much. Feverish symptoms abating.

April 2nd.—Slight cramps in the legs, lasting about half an hour. Left his bed. When walking complained of stiffness in his knees, and this continued for some time. Left on 25th, able to walk well.

Hydrophobia.—This disease in England usually arises from the bite of an infected dog some time before its onset, the average period being two or three weeks ; sometimes even the incubation is longer, and it is said may last for years. The symptoms of the disease are first manifested by the patient beginning to feel feverish, to have headache, to become agitated or restless, with loss of appetite, and should the remembrance of the bite come before him all this nervous distress is apt to be intensified ; these early symptoms, indeed, so much resemble what fear alone will produce that a diffi-culty is often created in the determination of the real nature of the case. After these premonitory symptoms a choking in the throat comes on, and this again is so constant a concomitant of violent emotion that a difficulty in diagnosis may even still exist. If, how-ever, you have ever seen a case of hydrophobia you will gene-rally recognise at once the features of the genuine disease. You will be struck with the expression of the patient indicative of dread or horror ; he will be sitting up in bed or in his chair with a frightened look, as if he dreaded your approach ; he may, indeed, have the distinctive features of the disease upon him, so that if you feel his pulse you may set him into a convulsion, or by breathing near him produce a paroxysm of sobbing, choking, or spasm. If a glass of water be offered him, he will become violently excited, throw up his arms, or put his hands to his throat. If prevailed upon to take it he may seize the glass and try and gulp down a draught, but it will probably bring on a most frightful paroxysm. Pouring water from one vessel to another will also excite it. Sometimes the patient becomes maniacal and attempts to bite the attendants around him. All this time there is a difficulty or oppres-sion of breathing and mucus collects in the fauces. The patient dies from asphyxia or exhaustion in two or three days.

The disease is comparatively rare, so that I have only seen three cases during thirty-five years' experience at Guy's Hospital, and in one of the largest hospitals in London there had been no case re-corded until quite lately, so that one of the surgeons, now deceased, doubted the existence of such a special disease as hydrophobia, and regarded it merely as a kind of tetanoid hysteria. He was wrong, inasmuch as he was arguing in the absence of knowledge, but it is very remarkable how the symptoms due to hysteria or nervous emotion do resemble those of hydrophobia. So much is this the case that I have known the instance of a gentleman who died with all the symptoms of this disease shortly after the bite of a dog, but who was so excited and apprehensive from the occurrence that the medical men regarded the case for some time as one merely of hysteria.

Only this week I have been told by Dr Fancourt Barnes of a case of a man, who came to St Thomas's Hospital with all the simulated symptoms of hydrophobia. The patient was a strong, powerful looking butcher, who came to the hospital one evening in a great fright and barking like a dog. He had just been bitten through the lip, causing a large wound also in the cheek. He went into the surgery and walked up and down holding out his arm in a theatrical manner, and was evidently in a state of extreme terror and excitement. On being questioned about himself he at first made only inarticulate sounds, and afterwards explained his case. On being offered water he took a little and sipped it with reluctance. Whilst dressing the wound he struggled most violently, and kept using incoherent expressions. He then left, throwing his arms about and barking like a dog. On the next day he came again to the hospital, was then tranquil, but he said he had pains all over him and felt excessively frightened. He was seen a few days afterwards, when he was well and calm enough to go and ask for a summons against the owner of the dog. The dog was proved not to be mad but was ferocious. There is a popular notion that a person bitten by a mad dog barks, and therefore this may put you on your guard against a fictitious case. It is remarkable that the attack of an animal does produce an extreme nervous depression or collapse, so that persons die from the effects of it without sufficient injury to a vital part to account for their death. I have known two instances where the shock has not been recovered from for many weeks. Livingstone described his powerlessness when attacked by a lion, and the death of sheep by "worrying" of dogs is probably through shock.

A man, æt. 43, came to me exceedingly nervous and in a constant tremor. I suspected alcoholism, but he declared he was a temperate man. He said he had pains darting through his body, especially if anything touched his hands or feet. If he took up a pin or needle a sensation would dart through his brain. This had been going on for three weeks, and he said it was in consequence of a dog having bitten him in the calf of the leg. This extreme nervous sensitiveness was still present a fortnight afterwards, and then I lost sight of him.

CASE.—Geo. R., æt. 13, admitted under Dr Rees, on Monday Jan. 16th. On Dec. 18th, twenty nine days before, he was bitten by a dog, on returning from school, on the upper lip. The dog was under treatment for rabies, but had got loose. It subsequently bit a girl and was then killed. The boy was taken to a surgeon's, and within a quarter of an hour the edges of the wound were pared and then brought together by pins. In nine days these were removed, when the wound was healed. On the days before he had two attacks of rigors. The boy was subsequently in his usual health, except that the mother thinks sharper and quicker than before.

On January 12th, *Thursday*, he complained of headache and lassitude; afterwards of stiffness of face, commencing on right side and then passing to left. On *following day* more stiffness and headache. On *Saturday* had lost his appetite, went out shopping with his mother, and was very restless all night. On *Sunday* he felt difficulty in swallowing and had a spasm of the throat when trying to drink some tea. Tried to drink several times since but could not. On *Monday* at noon he was admitted. He had an anxious or frightened expression; his intellect was clear and he answered questions intelligently, though unwillingly, as talking produced spasms of the muscles of the neck. When placed in bed he had a convulsive attack, which seemed to be due to the draught of air made by the blanket when thrown over him. A scar was observed on the lip, but it was quite sound and no pain was felt in it. The breathing was irregular and sighing; pulse irregular, 92—98. When a glass of wine was brought to him he declared that he could not take it, but when pressed to do so he raised the cup with a determined air, and threw a little into his mouth. It brought on a most terrible convulsion and violent spasm of the muscles of the neck. He then threw himself back in the bed exhausted and panting. After two hours he was asked to try again, when the very thought almost brought on a convulsion; he, however, very bravely by great exertion got a little wine in his mouth, when immediately a spasmodic attack came on. His pulse varied in a few minutes from 88 to 102. At six o'clock he tried again a spoonful of wine, but convulsions followed as before. At eight o'clock he started up in bed with a feeling of choking and pain at epigastrium, and called out for water and a spoon. This he thrust into his mouth with a determined effort, but immediately spat the water out, saying he could not swallow it. At ten o'clock it was thought advisable to try chloroform, but the inhalation brought on a spasm, and he threw himself out of bed. At midnight he lay trembling all over, and asked the gentlemen near his bed to breathe away from him as his throat was stuffed up.

On Tuesday morning at three he had been having constantly recurring spasms; his head was thrown back, mouth open, eyeballs protruded, and occasionally crying out, throwing his arms about or beating his chest.

At five o'clock he was rolling about in bed, in constant agitation. The slightest touch threw him into convulsions, and on one occasion he jumped out of bed, crawled on the floor, and got under the bed. He then became wildly delirious, spitting and retching mucus tinged with blood, and at six o'clock he was so violent as to be obliged to be restrained in bed, screaming, shouting, and spitting. At seven o'clock he was less violent, became weaker, and his movements were more like those of chorea. At eight he was much exhausted and muscles getting flaccid; could then swallow a little; the pulse then faltered and his extremities got cold, and at half-past eight on Tuesday morning, the day after admission, he died.

The *post-mortem* examination shewed nothing more than redness of the fauces and back of the tongue. The brain and spinal cord seemed quite healthy.

CASE.—Amelia A—, æt. 8, admitted under Dr Taylor, on Sept. 14th. On May 1st, she was bitten on the cheek by a dog, which had also attacked other people, and was consequently killed. She was immediately brought to the hospital, where the wound was thoroughly cauterized with nitrate of silver, and subsequently with ointments. She left quite well on May 29th. She did not seem at all alarmed about the bite and remained well. A short time before her second admission she complained of the face aching and a pustule appeared, with some inflammatory redness. The day before admission the child could not eat her dinner and had great difficulty in swallowing some tea. When her mother attempted to wash her she appeared terrified, and had some short respiratory movements.

On admission she had a wild-looking, flushed face ; cicatrix, of pink color and shining ; she was very irritable, tossing herself about and screaming, and said she had pain at the epigastrium, which came on at intervals. On the near approach of anybody she started and respiration became hurried and gasping ; when offered milk she shrank from it in terror and with a spasmodic inspiratory action. She was ordered injection of $\frac{1}{4}$ gr. of morphia. She tried to eat, but it gave her much pain, and when attempting to drink was seized with the same spasmodic movements. More intense spasms were induced by a breath of air blowing upon her. During the day she allowed the sister and nurse to touch her, but was terrified at the approach of any one else. Trembling of the whole body was produced when she took a porringer in her hands, and on raising it to her lips her head was jerked away and the vessel dropped. In the evening she was extremely restless, first sitting up and then lying down, closing the eyes for sleep and again starting up, groaning or shrieking, complaining of thirst, but if milk was brought her hiding her face in her pillow. She also complained of the draughts occasioned by the movements of the bed-clothes. A quarter of a grain of morphia was injected. She passed a sleepless night and took neither food nor drink.

On the following day she was much in the same state, but would get up; she was extremely restless and excited, clutching at her mother, stamping her feet, groaning and crying, wiping the viscid saliva from her mouth, and begging those in the room not to blow upon her ; she had less spasm than on the previous day, but more terror and restlessness.

On Sept. 16th, at one o'clock in morning, she was still standing and crouching on the bed; at four o'clock made several unsuccessful attempts to drink; at nine o'clock she lay down much exhausted. She then began to wander, and became semi-unconscious with constant chorea-like movements of the limbs and spasmodic respiration. She died at six, seventy-six hours after the first decided symptom.

The *post-mortem* examination was very carefully made by Dr Goodhart, and no trace of any morbid appearance could be found in the brain, cord, or any other part of the body.

HYSTERIA

This term is used very widely and vaguely by some, and in a very restricted manner by others. A too narrow definition would make it inapplicable to many cases which must be called by the name, and a too wide one would not only be meaningless, but harmful, by including many instances of nervous disorders having a totally different character. It can only be by a discussion of various examples of the affection that we can obtain a tolerably correct idea of what we intend by the term. When we use the expression hysteria, we mean a nervous disorder, in which the nervous system is deranged without the existence of any organic disease. It is a disorder occurring in those who possess a more than usually impressionable constitution, and in those also in whom there is not that equilibrium between the nervous and other parts of the organisation which we find in the most perfect frames. When I say the nervous system, I speak of it as a whole ; for not only may there be perversions of the functions of the body, but the whole mental and moral capacities are often changed. There are some persons, with a not very susceptible nervous system, who, from mere want of force of character, are a constant prey to every unpleasant circumstance which may operate upon them, whilst there are others of a far more highly organised constitution who, by superior mental vigour, are able to withstand the effect of impressions which would otherwise cause an intolerable disturbance. Therefore the condition of system which tends to produce what we call hysteria is common to the human race. I look upon it as merely the extreme development of a disturbance to which nearly all men and women are liable. There are few men who will not own to their "good days" and their "bad days," referring to those times when the mastery of their will is so great that no obstacles in their path can fail to be overcome by it, and then to those occasions when they are borne down or almost driven to despair by the most trifling misfortunes. Their will is then less powerful, and you will find that a want of will is one of the most marked features in hysteria. An endless variety of definitions have been attempted according to the opinions held by physicians as to its nature, and therefore I shall not add to their number. You will find, however, that an expression denoting that the higher nerve centres are in abeyance, leaving the spinal system uncontrolled, will cover as many cases as any other definition which you can frame.

It follows from what I have said that I do not consider the disease peculiar to women, and therefore, of necessity, uterine ; but owing to woman's organisation, the complaint is far more common in the female sex. A current opinion once existed that hysteria was due immediately to a disordered condition of the uterus—an idea dating back to the time of Aristotle, who spoke of the womb travelling through the body, and so giving rise to the hundred ailments to which hysterical women are liable. Some modern writers have not yet surrendered the opinion, and have even gone so far as to declare that the hysterical fit is none other than a counterfeit of the sexual act. Knowing how intimately a woman's constitution is bound up with all that relates to the more special functions of her sex, and how, therefore, of necessity all violent impressions made on her nervous system would in very many cases have reference to these peculiarities, we can feel no surprise that uterine troubles and hysteria are often intimately associated. Some authors, therefore, place the seat of the disease in the ovaries, and declare that the diagnosis of hysteria consists in a tenderness of the lower region of the abdomen over these organs. I cannot admit this, although it is true that a tenderness over one or the other ovary is not at all uncommon in nervous and hysterical women. Then, again, ulceration or other disorder of the womb is sufficient in some patients to cause a disturbance of the whole nervous system. But after allowing this, we must renounce the doctrine of hysteria as a uterine disorder, and still maintain that it is owing to constitutional peculiarities, and sometimes to circumstances of an altogether temporary or accidental kind.

In the lectures lately published by Sir J. Paget I see that he holds this view. He says : " In the defective ovarian and uterine functions of certain patients some see the centre and chief substance of the whole disease ; a very mischievous fallacy. Of course the sexual organs appear generally in fault to those who are rarely consulted for the diseases of any other part ; but in general practice they are, in a large majority of cases, as healthy as any other parts are, or not more disturbed. The close and multiform relation of the sexual organs with the mind and with all parts of the nervous system are enough to make the disorders of these organs dominant in a disorderly nervous constitution, but their relation to 'hysteria' or to neuro-mimesis, though more intense, is only the same in kind as that of an injured joint or an irritable stomach. All in their degrees may be disturbers of a too perturbable nervous system ; and equally on any one of them the turbulence of a nervous centre may be directed with undivided force."

Look around amongst your own friends and acquaintances, and consider their different organisations. Two patients have the same complaint or meet with a similar accident. The one will aid the doctor in every possible way to promote his recovery ; he would, if need be, rouse himself from his bed, and say, as Ligarius did to Brutus—

> " By all the gods that Romans bow before,
> I here discard my sickness ;"

whilst the other will nurse his complaint, and even speak of it when every trace of it has gone. Such a reflection as this will give you some insight into nervous affections. Indeed, the great occupation of medical men throughout the day in the investigation of their cases is to put the true interpretation on the symptom of pain— whether it refer to an organic disease, or whether it be a mere functional disorder, and, if the latter, whether it be of the kind denominated hysterical. I do not mean that such an analysis is to prove the presence of hysteria or not, and that all pains which have no organic seat are hysterical, or that in hysteria no real pains exist, for this is a mistake too often made, and one which I cannot too strongly warn you against.

The so-called fit of hysterics is due to a violent perturbation of the whole nervous system; an emotion upsets the equilibrium, and we witness the phenomena of laughing, crying, choking, &c. This may be easily induced in a highly impressionable person, whilst a more powerful cause is necessary in the stronger-nerved and stronger-minded. If a man have undergone great bodily fatigue, and his mind have been at the same time harassed, so that exhaustion of his nervous system results, he may be thrown into a state very like that of an hysterical woman. I have more than once seen a man under these circumstances give way to his feelings and play the part of a woman. It was reported that a well-known member of Parliament of strong and sturdy frame became hysterical as he stood over the grave of his friend Cobden. Some months ago I received an urgent message to visit a gentleman a short distance from town; when I arrived at his house he was sitting in his parlour and not looking ill. I expressed some little vexation at being summoned so hastily. He said he was now much better, and commenced explaining to me the reason of the summons, when he began to cry; presently the crying reached the stage of sobbing; this became louder and louder and more violent until it changed into a laugh, which he was totally unable to suppress, and I became a witness of the most marked attack of hysterics that I had ever seen in either sex. He presently fell back in his chair quite exhausted. He was a man thirty years of age, with a large black beard, and had as manly an appearance as you would wish to see. His wife then told me that he had been speculating, that he was a ruined man, and would have to leave his house and family. He had returned home that evening shortly before I was sent for, and the thought of the prospect before him was more than he could bear, and hence the cause of the attack. Whilst she was relating this he grew calm, and then commenced to talk to me, saying how foolish he was, but could not refrain from referring to the circumstances of his misfortune. He had not proceeded far when he was again overcome; another laugh commenced, and then he broke out into such a loud and involuntary fit of laughter that the noise could be heard throughout the whole house. It only ended with his utter exhaustion, when I left the place. I saw him a few days afterwards, and he was pretty well. This gentleman simply had an hysterical attack from a violent emotion; but sometimes we meet with hysterics in men in the more chronic and ordinary form. About three years ago I had a young man in the hospital for several months, in whom existed all the symptoms of hysteria in woman—such as headache, pleurodynia, palpitation, choking sensation in the throat, and, on several occasions, fits, for which the dresser was called to him; these fits were exactly of the hysteric kind. I mention these cases to prove to you, as they do to me, that hysteria is certainly not necessarily a uterine disorder.

I might state that my great authority on this subject has always been the celebrated Sydenham, for I consider that his epistle on hysteria contains more correct knowledge on this disease than most of the treatises which have been subsequently written. This acute observer believed that it might exist in both sexes. The following is an extract from his writings:

"Very few women, which sex is the half of grown people, are quite free from every assault of the disease, excepting those who are accustomed to labour and live hardly; yea, many men that live sedentary lives and are wont to study hard are afflicted with the same disease. It must be confessed that women are much more inclined to this disease than men, not because the womb is more faulty than any other region of the body, but for reasons to be shown hereafter. The origin and antecedent cause of this ataxy is a weak constitution of the said spirits, whether it be natural or adventitious, for which reason they are easily dissipated upon any occasion, and their system soon broke. Wherefore this disease seizes many more women than men, because kind nature has bestowed on them a more delicate and fine habit of body, having designed them only for an easy life, and to perform the tender offices of love; but she gave to men robust bodies, that they might be able to delve and manure the earth, to kill wild beasts for food, and the like."

Hysteria, then, occurs more frequently in women, but is not dependent necessarily upon any uterine disorder. In three well-known cases of my own, where the patients were bed-ridden for years, and subsequently recovered the use of their limbs, there was not the slightest irregularity of the uterine functions. When, however, you remember that all mankind is destined for some work or employment, and that women are debarred from performing the tasks which Sydenham prescribes for men, whilst there may be no opportunity of their undertaking the offices which more especially belong to them, such as the rearing of children, domestic avocations, and the like, then you will comprehend that nature having no outlet for the superfluous energies, the whole system becomes disordered, and those hysteric symptoms ensue, which we may regard

as the exponents of a wish unexpressed or a want unfulfilled.
In this case every organ of the body may suffer, and, amongst the
rest, the uterus. The latter may erroneously be seized upon as
the seat of trouble, and, being assiduously treated by the medical
man, the nervous disorder become more deeply rooted, and the
real cause being overlooked, a subjective ailment be converted into
a real one. I have seen so many instances of this that I can speak
very confidently as to its truth.

If the nervous system is unduly excited, a perturbation of the
whole system must ensue; now the superfluous forces thus produced would be got rid of by a person of energy in some occupation, and thus we find that where such nervous forces are operating to the discomfort of the patient a want of mental vigour is
often the cause. Thus the hysterical condition not only varies with
the degree of susceptibility of the nervous system, but is intimately
connected with the powers of mind. Indeed, if you placed the
strong-minded and the weak-minded at the ends of the scale, you
would be separating at the same time in great part the hysterical
from the converse. At the one extremity there would be the person
who would fall a prey to every untoward circumstance, and be the
victim of any one who chose to play upon his fancy; at the other
end would be an Alexander the Great or Napoleon I, who would
subject all mankind to their rule. Look at a case practically;
a young woman has a pain in her leg, she wishes to be up and
doing, get well as soon as she can, and forget her ailment.
Another one of a different temperament is pleased with the sympathy of kind friends, dwells upon her trouble, talks about it, so
that she probably exaggerates her suffering, and does not admit
that she is recovering. Such an example is daily occurring to us
in those cases where we say, " the patient does not make the best
of it." A third woman would not only exaggerate her troubles
but would fail to admit that she was well when the complaint had
entirely left her. Now, if a woman states she has a pain when it
does not exist, we are in possession of a case which exhibits one
form of hysteria, and then we arrive at a still further stage by supposing the case of a patient who might have been suffering from a
visible malady, such as a swelling in the leg, and who had actually
recovered, and yet perpetuated the condition by means of a ligature. We should then have a fictitious disease, and witness one
variety of hysteria which is not uncommon. From the simple
exaggeration of a symptom to the artificial production of disease
there is but one degree.

You see how the desire for sympathy, or that feeling which many
possess of taking a prominent place in the hearts of kind friends,

will prompt many a woman to pretend to be ill when she has no ailment whatever, and, in a further stage of this morbid state, to actually manufacture a disease. We thus see not only every real disorder simulated, but other various remarkable conditions produced. If the pretended complaint resemble a real one, we have often a considerable difficulty in distinguishing the genuine from the counterfeit, but in many cases the mere oddity of the disorder serves to mark it. The strangest vagaries of human nature which we perhaps ever witness are those which occur in young females in the early stages of womanhood; the whole nervous system, including the mental and moral nature, becomes so perverted that no circumstance of the most extraordinary kind may not then happen. The girl may not only present in her physical nature all the strangest maladies that can be conceived, but there may occur such aberrations of the mental and moral feelings that every one except the medical attendant would attribute her acts to wickedness rather than madness. Under such circumstances the behaviour is like that of one " possessed of a devil," for the acts are not those of an ordinary criminal who has an object in his wicked deeds, but are often purposeless, or for the simple love of mischief. Thus I have heard medical men generally unravel those marvellous ghost stories which we are constantly reading in our newspapers by the discovery of a young girl in connection with them. When you see a paragraph headed " extraordinary occurrence," and you read how every night loud rapping is heard in some part of the house, or how the rooms are being constantly set on fire, or how all the sheets in the house are devoured by rats, you may be quite sure that there is a young girl on the premises.

The following are examples of what we are constantly reading in newspapers. You will notice in both cases the prisoners declared they had no motive for their acts.

ATTEMPTED MURDER.—At Shipton-on-Stour, Warwickshire, on Wednesday, Mary Robinson, a domestic servant, was charged with setting fire to the house of Mr W. Harris, of Tysoe. From the evidence of Mrs Harris it appeared that the house had been on fire no fewer than ten times in six weeks, and attempts had been made to burn the baby and three children to death. The first fire occurred on the 15th of January, when a bed and an arm-chair were nearly destroyed. On the 5th of February the cradle, in which was a baby three weeks old, was found to be on fire, but before the child was rescued one of its ears was severely burnt. The damask drapery of one of the bedrooms was discovered on fire on the following day, and on the 22nd of February Mrs Harris left the child in the cradle for a few minutes, and on her return the cradle was on fire, and a piece of live coal was found up the sleeve of the child's nightdress. The child's arm was severely burnt. On the following day a bed in which there were three children was found in flames. The children were with difficulty rescued, and although a medical man was called in, one of them did not regain consciousness for some time. On the 26th of February a bed in which the baby was sleeping was found to be on fire, although its parents had left the room only a few minutes before. On this occasion the baby was nearly suffocated. The same day there were three more fires in the house, and on the baby being picked up out of the cradle a piece of burning coal was found in the cradle blanket. The inmates of the house were unable to discover the culprit, and could in no way account for the fire-raising. They put the matter in the hands of the police, to whom the prisoner confessed her guilt, but she could give no motive for the crimes. She has been in Mr Harris's service for fifteen months. She was committed to take her trial at the next Warwick Assizes on three separate charges of attempted murder.

CHARGE OF ARSON.—At Rochester, yesterday, Jane Ashmore, a domestic servant, was charged, on remand, with having set fire to her master's premises on three separate occasions. Mr Beveridge, a builder, of New Brompton, said that the prisoner was a general servant, and had been in his employ since she was thirteen years of age. On the night of the 27th ult., whilst sitting in his sitting-room, he noticed smoke coming through the flooring, and on rushing downstairs he found the place on fire. The flames were subdued, but not before considerable damage had been done. On the following evening the place was again found to be on fire in the same place, and the building was further damaged. The next night, about half-past nine o'clock, the witness was alarmed by a third fire, and some minutes elapsed before the flames could be got under. The prisoner afterwards confessed to having caused the fires, but refused to say why she had done so. She was committed for trial at the Maidstone Assizes.

When a few years ago the whole country was shocked by the news of the murder of a little boy in the middle of the night whilst surrounded by members of his own family, the event was enveloped in the darkest mystery, seeing that the crime was of so extraordinary a character, and was wanting in all those objects for its commission which are usual in similar deeds. No adult, especially no man in his senses, commits a crime except to attain some end ; and therefore the very purposelessness of the act (except, perhaps, for revenge) convinced me that it was perpetrated by a young woman. I felt quite sure in my own mind as to the real criminal, who even afterwards, on her own confession, was considered by many incapable of such a deed. The public press then learned what medical men had long known as to the extraordinary vagaries which may occur in the female sex at a particular period of life ; and although it is not pleasant to refer to a crime almost forgotten, yet, as it points a moral, I will read how a daily paper commented upon the case, and afforded an explanation of the dreadful occurrence to its readers :—" Hard physiologists and shrewd observers give an answer that will shock the tender mind. From twelve or fourteen to eighteen or twenty is that period of life to which the tide of natural affection runs the lowest, leaving the body and intellect unfettered and unweakened in the work of development, and leaving the heart itself open for the strong passions and over-whelming preferences that will then seize it. Youth, it must be confessed, does not feel much, and, sad to say, it is the softer sex especially which is said to go through a period of almost utter heartlessness. Girls, it is said, are harder and more selfish till the master passion takes them. In the want of active employment there is that peculiar brooding, imaginative, inventive tendency found in many young girls. In these cases the dream seems to grow and become an inner life unchecked by social feeling and by outward occupation, till a mere idea equally causeless and wicked fills the soul and masters the very act." I mention this case not because the young lady was hysterical, but because the causes which prompted to the deed are the same which lead to the commoner though less frightful vagaries; indeed, in some instances, the feigning to be ill is combined with actual wrongdoing, as seen in a case which was published some years ago under the name of the "Female Jesuit."

I have already said that it is difficult to give a definition of hysteria, much less to state strictly its pathology. I think, how-ever, as bearing upon the proper method of cure, we may comment upon the fact, of which there can be little doubt, that the nervous centres are constantly producing forces which are correlated to the other forces in nature, and that as in one case there may be an ab-sence of sufficient energy generated, so in another there may be an excess requiring an outlet. Thus, the assertion that work is a neces-sity of man's nature, and that every being should have an object to fulfil, is merely stating a physiological doctrine. If the brain centres be compared to so many galvanic batteries always at work, we can understand how, with half a million of women in the country unmated, a large amount of superfluous force is either running to waste or doing mischief either to the producers of it or to others. If the energies are not used for the more direct purposes to which they are intended, they may find a very appropriate outlet in good actions towards the poor and helpless, or even in assisting the parish clergyman in his duties, no matter whether the aid afforded be of a substantial or a frivolous kind. Better than doing nothing and becoming a prey to one's own feelings, is riding, walking, or per-forming the routine of fashionable life. Let none of these measures be adopted, the fire produced within will gradually consume the vitals, and the force thus generated, if not escaping, will disturb the whole organisation of the body. Examples of this you may see in our wards amongst hysterical women—one with a pain in the epigastrium, another with a palpitation, a third with constant sick-ness, and on seeking to ascertain what organ is diseased, you find none; the machinery is good, but it is working irregularly; it is an engine with the flywheel gone, or one deficiently supplied with steam, or perhaps over-abundantly supplied, and, having no work to keep it in regular action, is thrown into disorder. Those persons who are fulfilling their legitimate objects in life are like so many locomotives drawing their trains hither and thither with a regular and fixed purpose; those, however, whose existence is one of idleness, are not like the disused engines, but rather remind one of a number of locomotives running here and there without guidance, without a destination, injuring all with whom they come in contact, but above all themselves.

If women are not fulfilling the objects which are more especially allotted to them, they should have a pursuit, and granting this you can understand the case of a young lady who, although long bedridden, and beset with so many ailments and doctors that her life was despaired of, yet speedily recovered when, on the marriage of her elder sister, she became manageress of her father's house. The menial duties imposed upon the inmates of monasteries and nunneries are the means by which the " nerves" are kept under, but even then human nature will sometimes exhibit its mastery. Without offering an opinion upon the merits of mortifying the flesh as these people do, we cannot forget they are still human beings, and therefore you need not be surprised to hear that I have lately seen both a priest and a nun whose bodily troubles would I believe all have been dissipated if they had led the life of other mortals. There cannot be a doubt that fasting or living low for several weeks renders the body especially liable to disturbing influences and the mind ready to receive any extraordinary impressions. In spite of all measures taken to preserve the nervous system from excitement, and to employ all the energies, it is a fact that some of the most extraordinary nervous complaints to which humankind is liable have broken out in religious houses. It was in them that St Vitus's dance and St John's dance, &c., spread until these disorders became epidemic. This reminds me how contagious are complaints of this kind, and how women can exercise a self-control or not, just as the fashion sways them. This was remarked on in a magazine I was lately reading with respect to fainting. " Ladies do not faint nowadays—at least but rarely. If one can trust a certain mass of evidence, oral and written, syncope at the end of the last century, and up to the thirty-fifth year of this, was a habit with ladies. A story without a swoon was impossible until lately. Let us thank heaven comfortably that our mothers, wives and daughters have given up the evil habit of becoming cataleptic at the occurrence of anything in the least surprising."

As in hysteria the whole nervous system is deranged, so every part of the body may suffer, and the function of every organ be disturbed, as well as the nerves themselves disordered, in all possible manners. Let us examine some of these irregularities.

First, the nervous system proper may suffer, and the motor portion be depressed or excited. Thus, paralysis is a very common hysterical symptom, affecting more especially the lower limbs. A leg cannot be moved, or both legs are the subject of paraplegia. As, in such cases, the cause is want of nervous energy, so you will easily understand that rousing the will is often sufficient to put fresh vigour into the system and dismiss the complaint. A sudden alarm has often cured the patient who has been considered hopelessly paralysed, and this gives us an insight into the correct treatment to be pursued. I have already on more than one occasion shown you the importance of the moral treatment of hysteria. A young lady has a complaint of an imaginary kind, you visit her daily, and treat it as if it were a reality; the consequence is that it is perpetuated, and you have assisted towards the result, but if you understand the real seat of the complaint, and attack that, you will cure your patient. Having had under my care cases of paraplegia of years' duration, most assiduously treated by medicine, and at length cured here by moral means, I cannot speak too highly of the method. These cases are seldom difficult to diagnose, since in a real paraplegia the patient grows thin, bed sores appear, paralysis of the bladder and rectum may be present, and the patient feels ill; whilst in the case of hysterical paraplegia the patient remains plump, there is no trouble with the bladder, or if any, it is retention of urine; the abdomen is tympanitic, and the bowels confined. The physiognomy of the patient and her surroundings sufficiently indicate the nature of the case. She has taken to her bed as if for the remainder of her days, and all is arranged accordingly; the stitching, the embroidery, the religious books are placed within easy reach, and she generally receives more sympathy from the clergyman and the lady visitors than do cases of real illness. The fact is that there are no painful and loathsome circumstances attending the case, and, from the conversation of the patient and industry with her hands, it is regarded as an " interesting" one.

This is the class of cases of which we see so many examples in the hospital, and where we are so successful in the cure. Some most remarkable examples we have lately had, but they are too numerous to mention. Several cases I have published in the "Reports" of girls who after having been bedridden for years, became perfectly well in a very few weeks. The treatment is by the moral method, one which can scarcely be adopted in private practice, since, in a word, it is to excite to action the dormant will. Sir J. Paget puts the case well when he describes the hysterical state as one where the patient says "I cannot," and where the friends say "You will not," but the doctor says "She cannot will."

When we have the opportunity of comparing the melancholy results observed in the home of the patient compared with those which are gained in public institutions, we cannot but think of some of the disadvantages of the rich. In private practice we see, for example, a young lady lying in bed, receiving advice from one doctor after another, who style the case one of hysteria, but are able to do very little towards promoting the recovery of the patient. It is next to impossible to get the friends to acquiesce in the plan recommended, for the pursuasions of the patient too often overcome the advice of the medical man, as I have but just witnessed in the case of a young lady, where after spending hours in dictating the adoption of a rational treatment, I find all my labours frustrated by a spine doctor, who has condemned her to lie for another three months on a horrible apparatus of his own contriving. Should, however, the friends or parents follow the advice of their medical man, and breaking up the charmed circle in which the patient is imprisoned, allow their daughter to be removed to a lodging to be under the guidance of the doctor and nurse alone, this change is not so advantageous in my opinion as that of the hospital. The latter has advantages which no private dwelling can afford. The patient is in a ward with other people, whom she perceives have real diseases, some growing better and others worse; she finds also the physician adopts a uniform plan of kindness to all, doing his best to cure and relieve, and she herself is put on a perfect equality with the rest; she sees, too, the nurses performing their tasks in a uniform and business-like way, having very little time or ability to speak a sympathising word to any one, and with still less inclination to heed fanciful complaints, ready indeed rather to exhibit their indignation at the display of any imaginary troubles. There is, in fact, no one in the ward who is ready to play the part which is necessary to perpetuate an ideal malady; everything is real around the patient, and thus the whole pervading influence of the place is sometimes in itself sufficient to cause her to forget her self-created troubles, and at once to participate or even assist in the good work which is going on around her. I know it has been said that the placing an hysterical patient with other invalids is injudicious, but I have not found this to be the case, and it is certainly to be preferred to keeping her in her solitary room at home.

It sometimes surprises me that medical men, seeing all this, declare their utter helplessness when standing by the bedside of an hysterical patient. They will confess that all means have been tried in vain, that there is no real disease to cure, that it is an imaginary or nervous disorder, where nothing can be done, when all the while it is their own presence in the case which constitutes the very root and foundation of the malady. Let us take the case of a girl who keeps to her bed with an ideal paralysis of the legs, or some similar disorder. She sinks into a morbid state, puts on a second nature, and becomes the centre of a world of her own creating ; she is the interesting invalid, and receives in consequence the sympathies of inquiring friends, the care of nurses, the consolation of the clergyman (for she is usually outwardly pious), and, above all, the daily visit of the medical practitioner, who prescribes appropriate physic. This is her perverted life and the scene she enacts. Now and then the physician is called in, who gives his opinion that a great deal of the malady is due to hysteria, orders some iron and quinine, and perhaps galvanism, and so the play goes on. The medical man declares that he has tried every means and failed. Should he not see that the whole affair is a drama of the patient's own creation, and she the central figure of the piece ? She is to be ill, she is to have her doctor, and enjoy in her morbid way all the interesting surroundings of the invalid. Is he not aware that to cure her he must break into the charmed circle, and to spoil the play he must get rid of some of the performers ? And cannot he perceive that, even if he has no influence over others, he might withdraw himself ? Here is a young lady who says " I will be ill, and have a doctor to attend me." How can she accomplish this if the latter declines to obey her behests, or, if he accepts the post, how can he, in the name of common-sense, say he cannot break her of her fancy whilst he is a party to it ? If he sees clearly the truth of what I have been saying, his duty is, as professional adviser to the family of the patient, to retire, and use his influence to prevent the calling in of another medical man. I have myself seen, in several instances, where such advice has been given, and the parents have said to their child " We will have no more doctors," that recovery has at once ensued. In one of the worst cases of hysteria I ever saw, where after a young lady had been bedridden for three years, during which time she must have swallowed hogsheads of physic, and had her body covered with leeches, blisters without number, besides being well rubbed with tartar emetic ointment, the medical attendant suddenly died, when the father declared that his daughter was ruining him, and that he would have no more doctors ; from that time she began to recover, and may now be seen walking about quite well. Of course, if the medical man be wise and judicious, he may adopt various plans to break up the scheme of the young lady who has become the presiding genius, not only of the household, but of the whole family circle for miles around. My complaint is against the conduct of a medical man who pays a daily visit to his patient, sends her physic to be taken every four hours, besides sleeping draughts, prescribes a very particular regimen, consisting of all kinds of delicacies, commencing with rum and milk in the morning, and then says he cannot cure her of an imaginary complaint. Next to giving her physic, when he knows there is nothing the matter with her, the worst thing is to diet her, for there is nothing so harmful in perpetuating a nervous malady as this. If he would one day say, " No more physic, and as for diet eat and drink what you would like," he would be administering a moral stimulus more efficacious than all the iron and quinine she had ever swallowed.

A young lady keeps her bed for two or three years for an affection of the hip, and is seen by all the leading men in London. One day the clergyman walks in, prays over her, and she gets up and walks. The case is reported in all the religious journals as a miracle, whereupon the doctors all join in declaring that the case was one of hysteria, and that there was nothing the matter with her. Then, I would ask, why was that girl subjected to local treatment and to the infliction of physic every day for years ? Why did not the doctors do what the parson did? Of course the utmost acumen is required in order to make the diagnosis, for it is as cruel to call every female disorder hysterical as it is baneful to treat every malady as real. It is the doctor's daily labour to unravel the meaning of pain, whether it has a real seat or whether it is subjective. No rules for diagnosis can be laid down ; every case must stand on its merits. I have given you my experience towards the value of moral treatment in genuine cases of hysteria, and of the harm often done by other means. I speak without hesitation in this matter, for some of the most remarkable recoveries that ever could have occurred to any medical man have taken place in my ward, in cases which have been hopelessly despaired of when the usual routine was being pursued.

Dr Savage informs me of a case which illustrates in a striking manner the importance of forming a correct diagnosis and of the applying the appropriate treatment. A young lady, soon after the age of puberty, became nervous and irritable, with dyspeptic symptoms. Very soon she began to vomit after meals, and finally took to her bed. Being very feeble, it was thought that she would sink, especially as all treatment had been unavailing. She had seen the most celebrated doctors and been to various water establishments. Dr Savage then saw her, and found she was passionately fond of music. He conversed with her on the subject, and obtained her promise to endeavour on another occasion to rise from her bed and touch the piano. This she did ; she was satisfied with her success, repeated the experiment, and very soon was about again, with the loss of all the symptoms for which medicine had been prescribed in vain.

Then, besides loss of motion, there is perversion of motion, and we witness sometimes, associated with hysteria, some of the strangest movements conceivable. These are not of that irregular kind which we witness in chorea, but are usually of a rhythmical character; thus, instead of the body or arms being constantly writhed about in various directions, they are more slowly or regularly swayed in a given manner. I allude to these movements here because they often occur in hysterical subjects, but they may be met with in all classes of persons, and at all ages. I refer to them again under the name "rhythmical spasms." For example, there was a girl in the clinical ward, two years ago, who sat in a chair, and was constantly bending or bowing forward, as if saluting all those present, and several months elapsed before she got better. In this case, as in all others, the greatest discomfort was produced by the use of any forcible means to restrain the movements; the cause lies in the centres within, and no approach to a cure is produced by attacking the effect. In several other cases the arm is in constant and regular motion, as if acted on by clockwork. This form has received the name of malleation. I remember a case of Dr Barlow's, where the woman had constant quick breathing, and, what is remarkable, every inspiration occurred with a beat of the heart. This continued for weeks. I have quite lately been visiting a child who has died with this form of hysteria. After having various strange symptoms for some months she took to sitting at the side of the bed, and having some person or object before her which she could continually keep thumping with her fists or head all day long. Any restraint only added to the irritation. Chloroform, opium, conium, and other remedies in large doses, produced only a temporary effect, and she at last died utterly exhausted and wasted almost to a skeleton. There was no disease found in the brain. These movements are highly contagious, and sometimes, like the dances we read of in the middle ages, pervade a whole school.

We meet, too, with permanent spasm as an hysteric symptom, seen more especially in the hand, which is firmly clenched, the tendons becoming rigid and the muscles contracted when the hand is forcibly opened. Again, the whole body may be affected after the manner of tetanus. This is more often seen in an acute attack of hysteria, but the lockjaw may remain as a very troublesome and constant symptom. I draw your attention to this fact, for it requires often all our acumen to distinguish a real disease from an hysteric one. Then, also, you may have that remarkable condition known as catalepsy. This in its purity is not very common, although I have seen two cases of it in the hospital. One of my patients would sink into a kind of swoon or deep sleep, during which condition she would stand perfectly still in the middle of the ward, or, if in bed, would remain in any position in which you chose to place her. Minor degrees of the cataleptic state are frequently met with, and not uncommonly in the epileptic of both sexes, especially after the occurrence of a fit. During the drowsy stage which follows, you will frequently find that the patient's limbs will remain in any posture in which you place them. You will observe, in fact, that the whole nervous system is deranged in hysteria. You will have evidence of irritation of the cerebro-spinal system in the movements I have mentioned, and in the strange mental vagaries; then also of the deadening of the centres, as seen in paraplegia and in the disposition to lethargy. We are sometimes called in to a person lying perfectly insensible, and apparently as if near her end; it is, however, but a mere phase of hysteria. An extreme form of this condition, when continued, is usually styled trance. The whole nervous system may be so lowered in tone that the person lies helpless and insensible, but the functions of life go slowly on. This state may last for a great length of time.

With regard to the nerves of sensation, it may be said that these are in some way invariably altered in hysteria. More commonly there is hyperæsthesia of some of the senses. The patient cannot bear the light, or the least sound troubles her, but more usually it is common sensation which is affected. Thus, sometimes no part of the body can be touched without the patient shrinking—I mean the body proper, as the chest and abdomen. Often it is some particular spot, the more usual parts being those which are tender in many persons whose "nerves are low," as the middle dorsal or third lumbar vertebra, the vertex, and the left side. You will find many nervous persons flinch when you touch them in these places. Moreover, there may be some particular spot to which the whole attention of the patient is directed until that place is believed by her to be the seat of actual disease; I allude to the hysterical breast of Astley Cooper and the hysterical joint of Brodie. It is not always that the patient complains of pain, but an exquisite tenderness when the part is touched.

Then, again, there is the opposite condition of anæsthesia, where, owing generally to some violent commotion of the nervous system, the sensorium is thrown into a lethargic state, and the senses are sealed. I was once called to a girl who had received a great fright, followed by an hysteric attack and a subsequent state of lethargy. During this time she appeared to have lost altogether the sense of touch. The absence of sense of pain whilst that of touch remains I have already referred to; it is very commonly met with in hysteric women. This is called analgesia without anæsthesia. A girl may be stitching, and therefore feel the needle between her fingers, but you may run the needle into her skin without her knowing it.

Now, besides the hyperæsthesia or over-sensitiveness, our hysteric patients complain of and suffer pain. You must not think, because your patient is hysterical, that she does not feel pain, for assuredly the suffering of many hysteric women is real. There has been, and still is, much controversy as to the seat and cause of these pains. Are they merely subjective, and due, as is hyperæsthesia, to a morbid sensibility of the sensorium, or have they a local seat? and, if the latter, are they situated in the nerve and neuralgic, or in the muscle and myalgic? We have had writers who have contended strongly for one view or the other, but in all probability both are correct. I think, however, as I have already observed, we are much indebted to those physicians who have directed attention to the frequent existence of myalgic or myosalgic pains. Thus, the pains in the side and in the head so frequently met with are said to be muscular, and more especially the pains which occur in the chest or abdomen. I have now under my care in Mary ward two good cases of the kind. The one, that of a young girl, who lies in bed or sits in a chair, leaning forward, complaining of great pain at the epigastrium. She cannot bear it touched, and says she feels as if a load were oppressing her, which will presently suffocate her. She is sometimes so bad that her mother thinks she will die, but notwithstanding she is well developed, stout, and has apparently no real disease upon her. The other case is that of a woman well known to all students on account of the trouble which she has imposed upon them. Before she came under me she was in charge of one of my colleagues for several months. Her complaint is a most excruciating pain at the left side of the abdomen, which draws her double, and hitherto has not succumbed to the medicines which have been taken in vast quantities. The only relief she has obtained has been from the subcutaneous injections of morphia, which have now been practised for many months. The woman is in good condition, and does not look as if she suffered from any organic disease. On examination of the abdomen, the left side is full, rather rigid, and highly sensitive when touched. She complains, when the paroxysm of pain is on, of a most distressing bearing down and irritability of the bladder. She stated that she had passed blood in her water, and therefore the case was treated as one of calculus of the kidney and ureter. At the present time opinions are divided between this diagnosis and one of hysteria, where the pain would be attributable to a spasm of the abdominal muscles. My own opinion inclines to the latter view, perhaps being somewhat prejudiced by the fact that the great master Sydenham takes such a case as illustrative of one of the forms of hysteria, and which I will read:—"When this disease seizes one of the kidneys, it plainly represents, by the pain it causes there, a fit of the stone, and not only by that sort of pain and by the place it rages in, but also by violent vomitings which accompany it, and also for that the pain sometimes extends itself through the passage of the ureter; so that it is very hard to know whether these symptoms proceed from the stone or from some hysterick diseases, unless perchance some unlucky accident disturbing the woman's mind a little before she was taken ill, or the vomiting up of green matter, shows that the symptoms rather proceed from an hysterick disease than from the stone. Neither is the bladder free from this false symptom, for it not only produces pain there, but it also stops the urine just as if there were a stone, whereas there is none. But this last kind seizing the bladder, happens very seldom, but that which resembles the stone in the kidneys is not so rare."

Besides hysterical pains, we may have disturbances of all the various organs of the body. Palpitation of the heart is very frequent, or the breathing may be affected, and we have a kind of nervous asthma. The larynx may be implicated, so that there is a want of power to articulate, and neurophonia becomes one of the commonest symptoms of the hysteric condition. At another time the larynx is over-sensitive, and we have that troublesome and most annoying symptom, the hysteric cough. This has been considered by some, however, to be due to a kind of chorea or spasm of the diaphragm. You may recognise it by its loud hollow or barking character, want of expectoration and any evidence of disease in the chest, or, to give the description in the words of Sydenham, which is both precise and accurate— "Sometimes it (the hysteric disease) seizes the lungs, and the patient coughs almost without intermission, but expectorates nothing; and though this sort of cough does not shake the breast so violently as that which is convulsive, yet the explosions are much more frequent."

In other cases we have spasm of the diaphragm or hysteric hiccup. I have seen two such cases lately which lasted a long time, but the complaint at last wore itself out. Hysterical women also sometimes become temporarily blind.

Sickness is one of the most troublesome and obstinate of all hysteric disorders, because the organ having got into the bad habit of discharging its contents upwards can with difficulty be broken of it. It is remarkable that in these cases of daily vomiting the characteristic of the hysteric condition, the plumpness or absence of emaciation, still persists. A furred tongue may have no special signification, as it is often seen in hysteria. One mode by which we diagnose these cases as hysterical is that no medicine is of any avail; in real disease, even in such organic maladies as cancer, our ordinary remedies afford relief, but here the cure must be attempted through the nervous system. I believe the best method is to starve the patient for a while, or to use injections, so as to preserve the stomach in absolute quiet for some days, and then to commence with the smallest quantities of food. Sometimes a loathing of food is the only symptom, and patients die of starvation. This I shall presently allude to again.

The bowels, moreover, are, to use Sydenham's expression, seized upon by hysteria. Thus, prolonged and obstinate constipation is a not uncommon phase of the disease. This and minor maladies are only to be cured by the medical man having his patient well in hand, and by letting her know that he is quite aware of the unimportance of her complaint. The regular plan is for such a patient, like others of the kind, to be taken from doctor to doctor, who write the usual prescriptions, with the usual result. A good example of the influence which can be produced on hysterical patients by physicians and attendants is seen in some of the cures which occur under the direction of one of our "sisters," who introduces herself to her patients with "No nerves in Esther Ward."

As regards the nervous influence on the kidneys in the production of a large amount of water, the fact is one of importance in a diagnostic point of view. Sydenham says:—"Among all the symptoms which accompany the disease this is the most proper and almost inseparable—viz. a urine as clear as rock-water, and this hysterick women evacuate plentifully, which, I find by diligent inquiry, is in almost all the pathognomonick sign of this disease, which we call hysterick in women and hypochondriack in men; and I have sometimes observed in men that presently after making water of a citron colour (yea, almost the next moment), being suddenly seized with some violent perturbation of the mind, they presently void water as clear as crystal, and in great quantity. Three years ago a nobleman sent for me who seemed to be suffering from an hypochondriack colick. Visiting him one day, I looked upon his urine, which was of a citron colour. He was then merry and cheerful, and said he had a craving appetite; but one coming in at that very moment vexed him so much that suddenly being taken ill he called for a chamber-pot, which he almost filled with urine as clear as crystal."

Indeed, if I were to detail all the disturbances to which the body is liable in hysteria, I might occupy you for a month, and, to quote

Sydenham once more, " Nor is this disease only frequent, but so strangely various, that it resembles almost all diseases poor mortals are inclinable to. For, in whatever part it seats itself, it presently produces such symptoms as belong to it, and unless the physician is very skilful he will be mistaken, and think those symptoms come from some essential distemper of this or that part, and not from any hysterick disease."

The only objection I can make to Sydenham's idea of hysteria is that he appears to connect it with hypochondriasis. This may be owing to the long and supposed necessary usage of the term for an affection of women, whilst a corresponding disease of the male he would call hypochondriasis. I should, however, discriminate between the terms and the corresponding maladies, making each sex liable to either, whilst admitting at the same time the greater liability of women to hysteria, and men to hypochondriasis.

The hysterical condition is shown in so many different ways, that where the phenomena are very remarkable, we are obliged to designate it by different names ; for example, the hysterical fit is sometimes so violent and continued, that it approaches epilepsy, and the term *hystero-epilepsy* has to be employed. I have already said choreal movements are associated with hysteria and sometimes mental conditions which are no less than maniacal. I shall presently further allude to catalepsy and some other affections.

I have now a man in the hospital who has fits of a remarkable kind, reminding one of the so-called cases of "convulsionnaires." He has what is sometimes met with in women—a mixture of epilepsy, hysteria and mania combined. The paroxysms, which last for an hour, consist of a succession of fits, in which he struggles violently, and gnashes his teeth ; then is for a moment quiet, asks for water, or simply lies with a vacant stare or talks incoherently. The man, a sailor, was admitted with feverish symptoms, and having a large spleen was thought probably to have ague. There was considerable tenderness over the organ, so that pressing over the spleen would induce an attack of convulsions.

Catalepsy and somnambulism.—I have already spoken of catalepsy as being associated with hysteria, and I might also say that it is met with in a modified degree in many forms of nervous disease. In its extreme and simple variety it is rare. I have told you that when consciousness is lost the muscles relax, and the person, if previously standing, will fall, and I have also said that there are modified forms of epilepsy in which, although consciousness has departed, yet some influence of the higher centres remains, sufficient to preserve the tension of the muscles and direct their movement. A similar state is also seen in somnambulism, where, although consciousness is lost the impressions on the senses are sufficiently strong to guide the spinal centres, which are still awake. Although the two conditions I name are met with under different circumstances, the phenomena observed in both are much alike. Also in some peculiar forms of epilepsy of the somnambulist type the muscles will sometimes still continue to move after the manner in which they are started, as where a workman continues to use a tool after consciousness is lost. All these states, therefore, seem to show the existence of some influence of the brain over the cord independent of consciousness. In the cataleptic, consciousness is dormant, the patient apparently does not feel, and will not move when requested to do so, as if all perception and voluntary power were gone; but if a limb be placed in any position there it will remain. If the patient be able to walk he may be led into the middle of the room, and being placed in any posture, he will there stay for any time until released. If the limb be placed in a restrained position it will be there kept for a much longer period than could be accomplished by any voluntary effort. Demented cataleptics will often walk rapidly without any purpose, just as a somnambulist; and many instances of this are seen in lunatic asylums. I was shown a man in Morningside whom they could mould in any position. Whilst in bed on his back they could arrange his arms and legs in any posture, and there his limbs would remain. Dr Savage has a case in Bethlem of a young man who will keep his arm stretched out for two hours, and stand on one leg for a great length of time. If made to follow another patient he will continue to do so until he is stopped. It would seem that the spinal system is at work without consciousness, and also that it is operated on through the senses, just as in a person in a reverie, who will walk along an accustomed street avoiding all obstacles, and yet be so pre-occupied with his own thoughts that he may afterwards have no recollection of the route which he has taken.

Somnambulists will go through various performances and yet have no recollection of what has passed during the state of "noctambulation." If they avoid obstacles in their walk their senses must be acted upon, and through them, by a reflex process, their lower nerve centres also, and in this way their body is guided without consciousness or any activity of the brain. Dr Hammond had the opportunity of closely watching the case of a lady somnambulist, and published the case in the American journals. After getting out of bed, she lighted a candle and walked down stairs, passing him and other persons on the way without noticing them. Her eyes were open, and they did not wink when the hand was waved before them; her face was pale; she did not seem to feel when touched, nor did she move when smelling salts were placed under her nose. She then sat herself down in a chair and wept. On tickling her feet they were withdrawn, but it did not rouse her. Dr Hammond then took her head and violently shook it; she awoke, looked round her, and went off into hysterics. On recovering she had no recollection of anything that had passed, and had not even had a dream.

These pure cases are rare, but the cataleptic and trance-like state is often witnessed as part of hysteria. I was once called into the country to see a young lady who was said to be in a trance. After severe hysterical attacks arising from the shock caused by the sudden death of her father she remained in bed perfectly quiet, never moving nor eating. She was lying in this absolutely quiet state, like a dead person, or rather like one in a deep sleep. On attempting to make her drink she retained the fluid in her mouth some time, then made a convulsive effort to swallow, and threw up her hands. On placing her arms in any position there they would remain. She had lost all sense of feeling when touched. She recovered in a few days, when I was again sent for, on account of her having lock-jaw. This was again overcome by a little management.

Trance.—One phase of hysteria is that where the body remains perfectly motionless, and the mind apparently quite inert, so that the patient lies like an animal hybernating for days together without eating or drinking, and is apparently insensible to all objects around her. In some cases there may be moments of semi-consciousness, in which the patient takes a little food, but only in the minutest quantity. It is remarkable how small an amount of nutriment the body can subsist on during this absolute period of repose, especially when the patient is in a warm comfortable bed and all means are used to prevent the loss of heat from the surface of the body. From what I have said, you know the minds of these people are dull and perverse; there is a moral obliquity about most hysterical patients, and therefore you can quite imagine how a girl in this trance-like condition, exciting the wonder of her neighbours, ventures to add to their astonishment by taking her very frugal repast in secret, and thus making them believe that she is absolutely living on nothing. If the girl be poor the marvel excites the active and benevolent sympathy of her friends, money is deposited in the cottage coffers, and thus the strongest inducement is held out to the parents to join in the deception, and a case like that of the Welsh fasting girl is easily manufactured. Of course no one who has received a scientific education could for a moment suppose that animal life and temperature could be preserved without an adequate supply of fuel in the shape of food by the mouth to keep the bodily furnace burning. It is worthy of remark in this case that death occurring so soon after the deprivation of all sustenance, it must be attributed rather to the want of fluid than of solid food, and therefore probably to a poisoning by the animal secreta.

Ecstasy.—This may be considered a form of hysteria in which, the mind being self-concentrated, the body becomes almost dead to the outward world. In extreme forms it would pass into trance, in which the body is perfectly motionless and the mind itself probably inert, or if at all active it is in the same condition in which we dream, so that when recovered from, the patient will speak of past occurrences as of a dream-like nature. In ecstasy the mind is most intensely fixed on some subject, generally of a religious kind, and therefore the disease is rarely met with in England amongst our Protestant and leucophlegmatic people, but more frequently amongst the Continental and Catholic nations of the south. In these countries the frame of mind consisting in self-contemplation and examination is encouraged, and therefore if women of an excitable nature be shut up in a religious house, and all the influence of society cut off, the condition I am speaking of is quite ready for development. It is usually met with in young girls who, fixing their mind upon some religious idea, have become so absorbed in its contemplation that it has penetrated into their very nature and they are dead to the outward world. Since religious fervour, intensity of love and animal passion are so interwoven, it is very difficult to make an exact analysis of the mental condition of an ecstatic. It would, no doubt, be differently construed by the priest or the doctor, according to his proclivities and professional training. The most remarkable case of this kind described of late years is that of Louise Lateau, who, every Friday morning contemplating the vision of our Saviour in his agony, becomes herself the crucified one, and blood issues from her hands, her feet and her side, until she manifests in her own body all the stigmata of the Cross. Some medical men have asserted that she is performing no trick, but that in actual reality the blood oozes from her skin; the explanation they afford is that the power of her imagination is great enough, when exerted through her nervous system and concentrated on specialised parts of the body, to throw these into disorder and alter their nutrition in such a way as to produce all the phenomena described. It occurs through expectant attention.

A variety of nervous affection having a religious cause is found in the chorea or dancing mania which prevailed on several occasions as an epidemic throughout Europe. A modification of it may be seen during our own religious revivals. There are, however, other morbid

conditions induced by the same causes, which we are not unfrequently called upon to treat. We see, for example, young persons,
especially girls, naturally of a sensitive nature, who being debarred from all amusement, from its supposed immoral tendency,
find their only change or recreation in the church or chapel on
Sunday. Here may be represented to them the wickedness
and miseries of this world, which is only a place of probation
for another and better; they are encouraged to practise self-ex-
amination, and become accordingly a mass of self-consciousness,
always engaged in introspection, and a continual prey to their
own feelings. The morbid state, reacting on the body, at last
necessitates the assistance of the medical man, who has to explain
the true physiological principles of life, and that active work is the
best rule to follow for health. He sometimes has to preach a
sermon himself, and explain that the best Christian course for the
young patient to pursue is to do good rather than talk about it.

Hysterical anorexia.—It is a remarkable circumstance that in
hysterical vomiting the patient does not grow thin, showing that
the small portion of food which is retained is assimilated by the
healthy organs. There is, however, another variety of hysterical
stomach in which all desire for food is lost, so that the patient,
eating scarcely anything, becomes at last starved. This form of
nervous malady has lately received especial notice by Lasegue, in
the 'Archives Générales,' under the name of *hysterical anorexia*, or
inanition ; and by Sir W. Gull in the 'Transactions of the Clinical
Society' under the name of *apepsia hysterica.* I have witnessed
three good examples of it. One was that of a girl, who was the
thinnest person I have ever seen. She was little more than a skeleton,
and solely because she would not eat. These patients declare
that they do not care for food, and so they take less and less until
all appetite has gone, and then, indeed, a loathing may come on.
The causes are of a nervous kind, either physical or mental. In
one case the act was one of deliberate conviction that she could live
on next to nothing. The treatment must be for the most part of
the moral kind; a coaxing or scolding, according to the patient's
temperament, and tonic medicines, not forgetting arsenic, which in
some forms of dyspeptic anorexia acts in a most beneficial and
remarkable manner.

CASE.—A young lady was first seen by me three years before her death. She
then complained of pain in the abdomen, constipation, and various other symptoms
which I considered to be functional. She ate but little, and was sick. All remedies failed to relieve her, and she was taken to see several medical men, all of
whom regarded the cases as nervous. At the end of the year she began to keep
her room, and said she could eat nothing. What little she did take made her sick,
and her bowels were obstinately constipated, and she grew necessarily very thin.

All the symptoms became still more aggravated, and she would pass several weeks without a motion. Enemata were then used, and scybala removed. She lay on her back, eating nothing, with the exception of nibbling a biscuit, and drinking wine by drops. The wonder was how life could continue on so scanty a diet. Putting together the constipation, sickness, and inability to eat, a question arose whether the case was purely nervous, or not due to some partial obstruction of the intestine. Accordingly, a surgeon was called in, who seriously proposed a speculative operation of gastrotomy. Nothing, however, was done. She lay in bed, nibbling the biscuit as usual, and her bowels were never again opened naturally, but every few weeks some scybala were removed from the rectum. She died at last rather unexpectedly. After much trouble I obtained permission to make a post-mortem examination. On a most careful search through the body not a particle of disease of any kind was found. The intestine was healthy throughout; and indeed from the necropsy alone it would have been difficult to have discovered the cause of death.

CASE.—A lady, æt. 35, whom I visited on account of a gradually increasing illness, the nature of which had baffled the acumen of her medical man, was lying on her sofa in the most extreme stage of emaciation. Her face was quite hollow, so that all the outlines of the skull were plainly marked, every trace of fat had disappeared from beneath the skin, and the muscles were reduced to the smallest possible size. She could only just stand, or with assistance was able to walk into her bedroom. Her limbs appeared to be little more than the bones covered with the skin, and altogether she presented an extreme form of what is sometimes met with in such malignant diseases as cancer of the stomach. It appeared that she had led a very active life amongst the poor, but in a very exclusive religious sect; she had become a teetotaller for moral reasons, and soon afterwards thought she could diminish her diet, so as to provide for others. She accordingly took less and less, until all care for food and appetite was gone, and in this way gradually wasted away and became feebler and feebler. She was quite calm when I talked to her, but showed great perversion of mind, and I subsequently heard that some of her friends had not hesitated to call her "mad." She assented to eat more, but she was too far gone for hope, and she died in a few days.

Nervous and hysterical conditions in boys.—The same class of symptoms which are seen in hysterical girls may also be met with in young boys, and more usually at the time or approach of puberty. These cases occasionally resemble in every feature the extreme forms of hysteria in women, and the patients describe a variety of painful symptoms exhibiting a most depraved condition of the nervous system. Some years ago a boy was in the Clinical ward, who said he was paralysed and unable to move. At times he lay with his eyes shut, declaring that he could neither see nor hear. After having tried all other means, it was proposed on one of these occasions to apply the ferrum candens to his spine, and the method of using the red-hot iron by rubbing it up and down his back was fully explained to the class. After we had left the bed, and had reached the other end of the ward, a commotion amongst the nurses made us all turn round, when we saw the blind, deaf, and paralysed boy making his escape at the door.

Various emotional conditions also are often met with in boys, promoted by circumstances already named to you.

CASE.—I was asked lately to see a boy, æt. 10 years, who was said by the parents to have croup, and by the medical man laryngismus stridulus, on account of the paroxysmal attacks of suffocation in the throat. The case was considered so bad that the parents thought he had not long to live. When I saw him I found a healthy-looking boy sitting on his mother's lap, who was petting him and lamenting over his approaching end. She never ceased caressing him and calling him a good boy. On examining and talking to him I came to the conclusion that he was a very bad boy, being peevish, ill-tempered, and self-willed. He said he did not want to be cured, but wished to die, and expressed himself in so remarkable a manner for one so young that it was clear his whole nature was thoroughly morbid; and then I suspected that his symptoms were purely nervous and emotional. Presently he had an attack of the so-called croup, which was nothing more than an hysterical barking noise. I then told him he was a naughty boy to say he wanted to die, and to cause so much distress to his mother; that he would soon get well, and that I should soon call again and find him cured. After I had left the room he resented my conduct towards him, and said he would kill any more doctors who came near him. The parents' eyes being thus opened, a new course which was laid down was pursued, and in a few days this dying boy was quite well.

The parents had themselves to thank for the affair. He was an excitable emotional boy, and had been very badly educated; he had been practised for a religious object in the habit of self-reflection, and had been much to the services of Messrs. Sankey and Moody, who had a tent close by his residence. The boy had only too rigidly accepted his teaching, that it was better to die than to live, and in this way the imposture was brought about. When I relate the case, I may add that it is not a solitary instance of the kind which I have seen.

I had another lad once under my care, who caused much anxiety to his friends on account of a paroxysmal cough accompanied by a choking. He would suddenly call out, and then begin barking like a dog. Boys will even concoct maladies, like hysterical girls, as in a case lately related in the journals of a boy who passed "worms." These turned out to be portions of arteries which he had procured from a sheep.

A very remarkable case of an extreme excitability of the nervous system in a very clever boy, brought about by a distinct cause, I saw at Egham with Mr Roberts.

CASE.—A very precocious clever boy, æt. 12, had been in a state of maniacal excitement for some months when I saw him. After lying quiet for a few minutes he would throw himself across the bed, scream out, and declare that he had pains all over him, put himself into dreadful contortions, one of which was hanging out of bed with his head near the floor. As soon as we left the room he screamed out for his parents and had another paroxysm. He was a very intelligent lad, described his symptoms and spoke in a very old-fashioned way for a child. One of the most remarkable things in the case was the opinion which had been formed by some most distinguished medical men that he was the subject of spinal meningitis. Mr Roberts, who had charge of the case, suspected the existence of bad practices. We agreed that the complaint was functional, and aggravated by the presence of sympathising parents and friends. We obtained the assent of the parents to allow the doctor to take him to his own house. Here he was well watched, the symptoms rapidly abated, and he was restored well to his parents in a fortnight.

Hyperæsthesia is met with in boys as in hysterical women, and I have known the slightest touch throw them almost into convulsions, like exciting a tetanised frog. This has more usually occurred when the neighbourhood of the neck has been touched. One of the most remarkable cases recorded of an influence exerted over the cerebro-spinal centres by touching the periphery of a nerve was that of a lad under Mr Holden in St Bartholomew's Hospital.

CASE.—A lad, æt. 12, had a fatty tumour on the right side of the neck. If this swelling were touched gently by the finger, or even by a feather, he instantly lost consciousness, and was thrown into a most violent tetanic spasm. The body got quite stiff, the pupils dilated, and he was quite unconscious, having no feeling whatever in any way he was tested. After about a minute he drew a deep breath, and recovered. As often as the lump was touched the same phenomena occurred, and even whilst he was asleep in bed. Touching not only the lump but the region around and over the spine would produce the same result. By continuing pressure on the lump Mr Holden kept him in an unconscious state for twenty minutes. If he were raised in bed or taken out, he became quite rigid. Besides these paroxysms he had other attacks of an hysterical kind, such as barking and crowing. Generally about once a day, when not touched, he would go off in a paroxysm, calling out, " Oh, my bump !" sometimes jumping out of bed and throwing himself about like a maniac, or biting at the attendants like a dog, He was a clever precocious boy, but his mental state was very strange. Mr Holden subsequently removed the tumour, and all his nervous symptoms ceased.

I have seen nervous anorexia also in boys. A lad, æt. 16, grew so ill and emaciated that malignant disease was suspected. The case appeared simply nervous, and he subsequently perfectly recovered. Vagaries in respiration are not at all uncommon in hysterical women, and the same may occur in boys.

Nervous dyspnœa in a boy

CASE.—Robert W—, æt. 16, admitted March, 1871. About three months ago he began to feel sharp pains in the back and abdomen, which were relieved by rest. He often also vomited his meals. Some years ago he had shingles. The pains, accompanied by difficulty of breathing, still continuing, he came to the hospital.

On admission, he was seen to be a boy of delicate constitution, fair complexion, with great tenderness over epigastric region, and along the spine to the loins; also a tightness in the chest. He said he could not breathe, the respiration being remarkably quick, and amounting to 84 per minute. The heart and lungs, on examination, were found to be healthy. Was ordered gr. j of valerianate of zinc three times a day. He improved, and at the end of a week his breathing was 40 and pulse 72. He afterwards had a return of the symptoms, and his respirations increased to 60. He continued better and worse, sometimes the respiration decreasing to 40, and then rising again to 60, whilst the pulse was 70. After a trial of the zinc he took quinine and the bromide of potassium, besides having blisters to the spine. He generally kept his bed, for he said he was invariably worse after being up, the breathing being quicker and the pains in the chest increased.

As no actual disease was apparent, he was ordered to get up and to take Mist. Ferri co. He seemed better for a few days, and was then sent to the Convalescent Hospital, after having been in the hospital six weeks.

ON THE CONNECTION BETWEEN THE CEREBRO-SPINAL SYSTEM AND THE SYMPATHETIC

I wish now to speak of the intimate relation existing between the spinal and the sympathetic system. I have explained how the one regulates our voluntary movements, and the other the action of the viscera; but besides these, their special functions, many spinal nerves have a direct distribution to the organs of the body, and exert an influence over them; indeed, late observations have tended to show that in some cases the sympathetic and spinal nerves possess their own special attributes when sent to the same organ. Suffice it for my present purpose to present to you the intimate relations between the two systems, whereby impulses from the cerebral, emotional, or spinal centres may react on the whole bodily machinery, and, on the contrary, disturbances within the viscera make themselves manifest by unpleasant impressions on the sensorium.

I have already alluded to the ordinary excito-motor results obtained by touching the body, remarking how a violent blow on the spine will act on the heart, by partially paralysing, I suppose, the sympathetic ganglia, and allowing the pneumogastric to come into play; but to show the connection between the cerebro-spinal and sympathetic systems I need only refer you to the effects on the secretory organs. It is worthy of remark that there is scarcely an emotion to which the animal frame is liable but what is evidenced by an effect on the secretions. Sorrow will produce tears. Pent-up grief may influence the liver, and the sufferer becomes jaundiced. Fear may arrest secretion of saliva, as is seen in some of the ordeals of savage nations; whilst the mere thought of a luxurious repast will bring the water to the mouth. Fear, while drying one end of the alimentary canal, may cause undue secretion in the other; a fact which might be illustrated by numerous examples, scientific, pathetic, or humorous. The trail which some nasty animals leave behind them when hunted is well known, especially, I believe, in America. Another effect is seen on the skin, when the hair stands on end and the terror-stricken man is bathed in a cold perspiration. Mental emotion is seen on the kidneys, where a large quantity of pale urine is secreted; and not only on the secretions but on the irritability of the organs themselves. There is many a person who has irritable bowels or bladder solely by giving these organs thought. Only lately I saw a young man who told me he never could sit out a piece at a theatre or a sermon at church if by chance he once thought of making water. I do not know that the music had any influence, although we do read in the 'Merchant of Venice' that "some men there are, when the bagpipe plays in the nose, cannot contain their urine." You see, then, that even in the case of a supposed bladder affection you cannot overlook the nervous supply of the organ.

I will remind you again of what I said in my first lecture, that these nerve forces, once produced by emotion or passion, must have their way of exit somewhere. If acting either on the lachrymal gland or on the alimentary canal, they must be discharged into the pocket-handkerchief or into the water-closet, otherwise they will react injuriously on the system of the patient. The last illustration, although truly scientific, is also a piece of popular pathology; for a woman once told me that the cause of her son's illness was being frightened by a large dog at a house where he went on an errand, and that if the master of the house had let her son go to the water-closet, as he desired, she believed he would not have been ill. *She* said the fright struck inwards. We might adopt a different phraseology, but not a better explanation. We hear of the hair turning suddenly grey from fright, as in the case of Marie Antoinette, whose hair turned white in a single night. I would rather take an example from a man than a woman, for I myself have on more than one occasion had a visit from a lady with jet-black hair, but on the morrow, when seeing her in bed, I found that this had changed to grey. We also hear sometimes of fear turning the whole mass of blood. I believe this is literally correct. I have now seen so many cases of anæmia, some of them fatal, occurring after a severe shock to the nervous system, that I have no doubt of the fact. How this occurs I cannot tell, until the physiologists inform us in what part of the body the blood is manufactured.

These illustrations are merely to show you that, although the organic machinery is kept at work by the sympathetic system of nerves, yet that, these being associated with the spinal, they are influenced by causes which violently disturb the cerebro-spinal centres. Now let us consider, on the other hand, how, under similar exceptional and perturbing influences, we become sensible of changes going on within us. In a perfect state of health we probably ought to be in no way cognisant of the machinery which is at work within us, whether this is making new blood, or pumping it through the system. Certain it is that persons in the enjoyment of rude health feel a simple pleasure in existence: they know not they have a brain, or a heart, or a stomach. Alas! how few of us can say that! We students who consume the midnight oil know what it is to have a throbbing brain, a palpitating heart, or a flatulent stomach. I suppose there is scarcely any one, even in the best of health, who has not experienced some internal sensations: he has had a sinking before dinner, and perhaps a rising afterwards. Now, it has long been a question, with certain of the metaphysical writers as well as physiologists, to what sense these feelings are to be referred. Are they a part of common sensation, or are they to be regarded in the light of a sixth sense? Some have had no hesitation in teaching that certain sensations which we have within us, dependent on the operations in the alimentary canal, are not referable to common sensation, but must be included in a new sense, in the same way as the muscular sense (supposing there be one), must be regarded as a seventh. At the present time we have no other words to express our internal feelings than those derived from common sensation, and thus, when our patients complain of pricking, burning, swelling, &c., we must take the terms for what they are worth. You will soon observe that patients tell you of swellings within them, which you, innocently translating into tumours, examine in the expectation of discovering some positive adventitious growth; you discover nothing; but the term is the nearest to something definite in their minds which they can adopt. If it be said that all our expressions for common sensation have reference to touch, it is certain that many of the feelings which we experience within us cannot come under that category; therefore the belly ache is worthy of being exalted into a sixth sense.

This subject of pain has never, as far as I know, been scientifically considered as a whole, but is one of great practical importance. The sensory nerves are, as you are aware, distributed all over the skin, and thus we become sensitive creatures, and acquire a perception of all around us. The internal organs have no need of such nerves; and therefore you find that the brain may be softened, or the lungs be riddled from end to end, and yet no pain exist. The liver may be full of abscesses, and the patient perfectly unconscious of it; or advanced Bright's disease may be known only to the medical man. This is a matter of every-day experience, and therefore if we could absolutely state that the external shell of the body is sensitive while the interior is not, we should have an important fact to help us clinically. I myself believe this, in a general sense, to be true, and of great assistance in diagnosis. For instance, pain in the head generally indicates an affection of the sensitive nerves in the skull or membranes, but not disease of the brain itself. The pain in pleurisy means that the chest walls are involved, there being none in the simple pleurisy accompanying a pneumonia. In peritonitis, pain implies an implication of the abdominal walls, for the inflammation of the coverings of the organs or deeper-seated parts does not necessarily produce pain. In the case of the hollow organs severe pain may occur, but certainly not, as in the common sensation of the skin, by substances passing over the mucous membrane; for not only may large bodies, as stones, pass along them without producing any sensation, but ulceration may exist in the stomach and throughout the intestines without the production of any pain whatever. When these organs spasmodically contract, or are unduly stretched, as in colic of the intestinal canal, gall-duct, or ureter, then alone an agonising pain is produced. I repeat it would be an important and interesting study to ascertain what amount of sensation exists within the body, and under what exact circumstances pain is produced.

The uses of pain are obvious in expressing to us the abnormal condition of some part of the body, but it must be remembered, at the same time, that the sensitiveness of persons varies immensely. Thus, I have seen a man endeavour to hide the fact that he had received a severe blow on the head by saying he had had a slight knock, when at the time he was suffering from fractured skull. Another person, especially a woman, may grow eloquent over her sufferings, when you can find nothing but a quick-beating heart or flatulent abdomen. This shows that in judging of pain we have not only to regard the spot where the pain is said to have its seat, but we must not neglect the condition of the nerve centre on which that impression is made. We must remember the law of our natures, whereby all sensations are referred outwardly, and recollect the fact that when a nerve is irritated in any part of its trunk the sensation is referred to its extremity. The commonest example is that where, a leg being amputated in any portion of its length, an irritation of the nerves of the stump produces a sensation referred by the recipient to the foot or some part below, where the branches of the irritated nerve had at one time been distributed. A pain thus produced is a real objective sensation, but by what name shall we designate that sensation caused by an irritation made at the centre itself or by no irritation at all, but due merely to an over-sentient or morbid condition of the ganglionic centre? It does seem possible for a person to have such an impressionable sensorium, either as a natural or morbid condition, that he would, were that centre sensitive, have pain in it; but not being sensitive (the brain, spinal cord and ganglia possessing no feeling), the impressions are referred

to some external part of the body. A person, therefore, who possessed this morbid nerve centre would feel a pain in some part of
the body to which is distributed a nerve which has its origin in
that centre. He would be suffering a pain having no outward cause,
and that man I call a *hypochondriac*. You see the pain is real, for,
if not, the patient would be shamming ; but the hypochondriac does
not sham, thé pain is a reality within him.

The patient necessarily refers the cause of the pain to the spot
where it is felt ; it is a law of his nature so to do, as he can know
nothing of the centre, within which lies the sensorium or perceptive
organ ; but for us medical men our daily and hourly occupation is
the interpretation of pain ; and to discover whether that which the
patient regards as outward and local is really objective or only subjective. If the latter, we consider that the sensory centres are morbidly acute, or at least that they permanently retain that actively
receptive state which is only a temporary one during the existence
of actual pain. When, for example, from an irritation of the skin,
at the point of the finger, a pain is felt, we know that the recipient
grey matter in some centre is for the moment disturbed, since if
we cut the nerve leading to it from the finger no sensation would be
experienced. Now, if we suppose this temporarily altered condition
to continue, and that the centre does not return to its quiescent
state, the pain would also remain. In this way the abnormal conditions, with the resulting sensation, would become habitual.

It has always been assumed that a somewhat analogous alteration
may occur in the motor apparatus, and therefore, if this be true, there
seems no good reason why the same theory should not apply to the
sensory. For example, if a person voluntarily throws himself into
a state resembling tetanus, epilepsy, or chorea, he is for the time inducing a special action in certain cells which rule over the motor nerves
and muscles employed in the movements ; and the same occurs if
he merely moves one arm or leg. Now, we cannot but think that
whilst these muscles are still in action there must be some corresponding disturbance in the motor centres which rule over them ;
so that in the case of chorea we must look upon the complaint not
as a real disease but only as an abnormal persistence of a state
which in its temporary form is purely physiological. If, then, a
temporarily disturbed motor centre does not resume its normal
quiet, and we have in consequence a continuous movement, so in
the same way it can be supposed that a temporary disturbed sensitive centre not falling back into quiescence would produce what is
equivalent to a continuous pain. In such cases, where pain is constantly felt, as in the hypochondriac, the health is too good, and the
bodily functions too soundly maintained, for us to suppose that

any real morbid condition of the nerve centres can exist. It is therefore in one sense true, as the doctor assures the patient that there is nothing the matter with him.

If you take this idea in the larger sense, and say that the whole sensorium is so affected or so morbidly sensitive that impressions existing there have no reality, then the brain is diseased—the person is mad. He is conjuring up images without any corresponding picture on the retina, or hearing sounds without the membrana tympani having responded to a single vibration—the man, I say, is mad. "He hears a voice you cannot hear; he sees a hand you cannot see, &c.;" and so I take it that hypochondriasis is a species of madness, or a little madness, for some nerve-centre must be wrong when it is sensible of impressions which have never been made upon it. If this be so, say you, how difficult to ascertain the real state of a patient when we have nothing but pain to guide us to the truth. It is most difficult. It is the great difficulty which the medical man is attempting to surmount all day long. You may infer as a rule that when you see a patient very impressionable, the real cause of suffering is less than it would be in another. If tears are shed, showing the patient is very emotional, the cause may be most trivial. A short time ago I had two women patients side by side in the ward. One had most violent attacks of neuralgia coming on in paroxysms. She suffered an agony, but never exclaimed, and never shed a tear. Her neighbour, an anæmic woman much out of condition, in a highly nervous state at the climacteric period of life, could never relate her sufferings without weeping; but, as far as I could discover, she had no bodily disorder. You may now understand what I meant to imply when, showing you that old man in Stephen ward who has had an excruciating pain at the pit of the stomach for six years, I said he had a mad semilunar ganglion. The inexperienced generally argue in a manner the very reverse of this, and thus you will find that nervous women, and the most demonstrative, and the most emotional, are those who will receive the sympathy of the benevolent, while the real sufferer will keep his troubles to himself, and be uncared for.

If, then, a person have a feeling of discomfort in any part of the body, or a pain, and there be no cause for it at the spot, in the nerve, or in any other part whence the irritation might be reflected, the nerve-centre is at fault. In an extreme case he would be said to have delusions—be mad. We can form some conception of this by remembering what condition our brains are in during sleep when the dreaming thoughts are regarded as realities.

"They are the children of an idle brain,
Begot of nothing but vain phantasy."

A man in his sleep may start up fancying that he has seen some one, or be aroused by a loud knock at the door when all has been still. Such a condition during waking is madness; a modification of it is hypochondriasis.

No cases require our more serious attention than these, especially in women. Where a real pain exist, and is due to a definite cause, it must be treated; but, as is often the case when the sensation is only within, what mischief must accrue from keeping the attention of the patient fixed on an imaginary evil instead of attempting to divert her from herself. The evil of treating imaginary disorders is exceedingly great, especially in the more secret troubles of women.

Then, again, the relation of the cerebro-spinal system to the sympathetic is one whose importance cannot be overrated. The impressions made upon us from without we feel as mental processes, and therefore we are obliged to use such terms as the influence of the body on the mind and the mind on the body, but the very instances which we use in illustration are sufficient to show that we are not speaking of two distinct natures, for the forces which affect the mind are material, and the effects of mental processes on the body are no less material; indeed, the mental processes themselves are material, that is, they belong and are the usual phenomena attendant upon a particular kind of matter. John Hunter said it was possible to concentrate one's thought on a part until it became inflamed; and I have read how a lady, shortly after being shocked by seeing a child crush his foot, became herself lame, and found, on returning home, that she had an inflammation of the ancle-joint. I have already said that there are some who regard the case of the ecstatic Louise Lateau as genuine and consider that the strong mental effort which the girl is able to exert over herself is sufficient to produce the bleeding stigmata on her hands and feet. This may be less surprising if we for a moment consider that a real and positive force is constantly being produced in the brain and spinal cord. Think of the weight of these organs, amounting to several pounds, and the enormously disproportionate amount of blood which they receive. They have no ducts, like many other organs, whereby we can become cognisant in part of the work done, but they have nerves proceeding from them, and these are distributed to all parts of the body. Must it not be true, then, that they convey forces which are generated in the centres? Probably much of this force is actually used in the performance of work in the various organs; and even more than this, for it was the opinion of Sir B. Brodie, gained from his experiments, that animal heat may be in part correlated to nerve force. A large part of this force is no doubt used in the production of muscular power, even though it be true that at the moment of the contraction of muscles we do little more than unloose the latent force already existent in them. In health we have this under our control, whilst in such affections as chorea and paralysis agitans the nerve force is being emitted in jets, or is dribbling away, whilst some think in epilepsy it rushes out in an explosion.

Müller speaks of the nervous principle contained in the centres as being in a state of tension, and always ready to act; and he says that the slightest change in their condition excites a discharge of nervous influence, as is manifested in laughing, sneezing, &c. Thus every mental impulse to motion disturbs the balance of this tension, and causes a discharge of nervous force in a determinate direction. He also compares the nervous system to a musical organ with its bellows charged with air, which is ready to pass through any given pipe, according to the particular key that may be touched. Using this illustration, we may imagine the air either rushing out of the organ with a scream, roaring out through the larger tubes, or diffusing itself melodiously through a series of small musical pipes. In a similar way the superfluous nerve force is thought by some to display its operations according to the sex, age, and temperament of the patient. For example, the same cause which may produce hysterics in a mother might induce chorea in her child, the one disease being almost peculiar to the adult period of life, and the other to childhood. The same fright which may excite so great an amount of nerve force in the mother as to cause the explosion known as hysterics may operate on the child in a slower manner, and give rise to the less violent action known as chorea.

The explosion of nerve force by an hysterical attack acts as a kind of safety valve, protecting the internal machinery from danger, and although all persons are not alike impressionable there is scarcely an individual who may not find the value of this provision when he is acted on by an overpowering stimulus. Relief is, there-fore, often obtainable to an over-excited nervous system by laughing or crying. For, as Byron observes, the power which women possess, compared with men, of pouring their troubles into their pocket-handkerchiefs, is no doubt often very beneficial to them, so far as their health is concerned. A woman who is excited, if she do not have a good cry, often allows the redundant nerve force to escape through that unruly member, the tongue, and thus an extreme volubility of utterance perhaps eases her from further unpleasant-ness. Of course, the talk which flows from her lips is altogether different from the result of an intellectual process; and therefore it is still true now as in ancient times that "anger is a short madness." In other cases the superfluous force escapes by the limbs. Thus, an angry person slams the door or destroys her own property.*　A man of better sense, when vexed, takes a walk, and in this way gets rid of his extra nerve force, or, if the irritation and its results are more chronic, sits down, takes up his pen, and eases his mind by "publishing the whole correspondence." This is by no means incompatible with the view before taken, that the emotional system is excited at the expense of the intellectual, and that many of the phenomena mentioned show a weakened condition of the higher centres. A strong brain could suppress the emotion which a weak one could not.

Should this tension of the nerves not be lessened in any of these various ways, it is very likely to react on the bodily functions, and may thus produce injuries as serious as would be the pent-up steam in a boiler, even amounting to the destruction of the individual who is the subject of it. It is well known that pent-up grief induces manifest evils, and so popular a piece of pathology is this that tales of fiction and books of poetry contain numberless instances of it. Indeed, we must turn to the writings of the poets rather than to those on medicine to find its illustrations. A very perfect story of the kind is to be found in the two following verses from Tennyson's " Princess :"

> " Home they brought her warrior dead,
> 　She nor swoon'd nor uttered cry ;
> All her maidens watching said,
> 　She must weep or she will *die*.
>
> Rose a nurse of ninety years,
> 　And set his child upon her knee ;
> Like summer tempest came her tears—
> 　' O my child ! I *live* for thee.' "

Such a case is no poetical fiction, for I verily believe that a flood of tears may be the safeguard against a serious illness. This is popular pathology, I know, but also scientific. It comes home to us very positively, does this mental emotion, when we see it induce an indigestion, a jaundice, or a fatal anæmia.

Our own Shakspeare might be quoted copiously in illustration of the sentiment contained in the well-known lines where unavowed love may be the cause of chlorosis.

* The following appears in the police reports :—"A young man, a gardener, came before Mr Paget, the magistrate, to ask his advice. He had been married five months to a woman who turned out to be a great shrew, and who, acting under the advice of her mother, had broken up her home three times. On Tuesday afternoon he was engaged in a gardening operation which he could not complete by dinner time. There was delay of an hour, and when he did come home he was soundly rated by his wife, who commenced breaking the plates and dishes, destroyed the furniture, and made a general wreck of the place. She then attacked him," &c.

> " She never told her love,
> But let concealment, like a worm i' the bud,
> Feed on her damask cheek. She pined in thought,
> And with a green and yellow melancholy
> She sat like Patience on a monument,
> Smiling at grief."

Or, in the following lines, illustrating the relief of giving vent to sorrow, where Malcolm says to Macduff, when he hears of the slaughter of his wife and children :

> " What, man! ne'er pull your hat upon your brows,
> Give sorrow words ; the grief that does not speak
> Whispers the o'erfraught heart, and bids it break."

I believe myself that idiopathic fatal anæmia, or "pernicious anæmia," as it has lately been styled, may result from a shock to the nervous system. The late Sir H. Marsh related the case of a young lady who accidentally poisoned her father by giving him laudanum instead of a black draught. The occurrence so preyed on her mind that she took to her bed, became anæmic, and before many months had elapsed died, without any apparent organic disease.

The most remarkable case which I ever witnessed of a person dying from grief was one which I had an opportunity of seeing in consultation with Mr Brown, of Lewisham.

Mental shock.   Death in five weeks

CASE.—Two young ladies residing with their widowed mother were most devotedly attached to one another. The younger died rather suddenly of disease of the heart. The elder one was for the moment like one thunder-struck. At first she could not realise the calamity by which she was afflicted, but she soon saw the event in all its terrible reality. She never shed a tear. She declared that, her only object of affection being gone, she would go seek her sister in another world. She then arranged the whole funeral ceremony for her sister, and chose the grave in a neighbouring cemetery. Almost immediately after returning home she began to suffer from palpitations, sickness, and pain over the region of the heart, as her sister had done. She would eat nothing, and declared that she had her sister's complaint, and should shortly follow her. There was no reason to suppose that any disease existed ; in fact, the disturbance was clearly functional, and, as she herself declared, was produced simply by emotion. She was a well-grown, healthy-looking girl, and I had no fear that her illness was due to anything more than temporary excitement. However, I failed to gain the co-operation of the friends to have her removed from the scene of her trouble, for they not only sympathised with the girl, but agreed that her case was the exact counterpart of her sister. I, on the contrary, regarded her symptoms simply as the result of good acting. In spite of all the influence that could be used she would not be comforted. She refused food, and rejected what was given to her ; and at last, much to my horror and surprise, she died in a kind of hysterical convulsion, exactly five weeks after her sister, and was laid in the same grave.

In a case like this, one cannot overlook the possible similarity

of organisation which in the case of twins is most remarkable.
I attended a young man who died of pleurisy after a week's illness.
This so affected his twin brother that he said he should soon follow
him. He was then seized with pleurisy, like his brother, and died also
in a week, the two cases resembling one another in every particular.
We have all heard of the celebrated comic actor who, consulting
a physician for depression of spirits, was advised by the doctor (who
was ignorant of the name of his patient) to go and see Grimaldi
play; upon which the patient retorted, "Alas! I am that unhappy
man!" In such a case it is not improbable that although the actor
displayed the utmost gravity, his brain must have been in sympathetic relation with his audience, and thus its excitation from not
finding the ordinary channels of outlet, have produced an injurious
reaction on his whole nervous system. Had he laughed with his
audience, much of the humour might have been wanting, but he
would perhaps have suffered less.

These are neither metaphysical abstractions nor poet's sentiments.
The poet, if he truly describes nature, and more especially human
nature, cannot portray anything different from what the physiologist sees, since they are both looking on the same body and at
the same phenomena.

One of the most important considerations resulting from the idea
that the human body contains a living force, or rather that it itself
is an active machine, is this—that it must be employed. If it is
not used for good purposes it will be for bad, and "the devil still
will find some work for idle hands to do." The evil is most marked
in young ladies who are brought up to no occupation, and remain
unmarried. Better than doing nothing are the frivolities and
amusements of a London season, when "idly busy rolls their world
away." But the young lady who has not these opportunities, or
whose conscientious scruples forbid her to indulge in them, and who
even has not the parish church or school to patronise, falls into a
listless wayward hysterical condition, becomes a prey to her own
feelings, and is consumed by her own pent-up fires. When you
meet with these cases, may you be wise enough to adopt a rational
and moral treatment, rather than aggravate them by an injudicious
sympathy and the administration of worse than useless drugs.

HYPOCHONDRIASIS

I have told you already that the hysterical patient is often
weak-minded as well as being over-sensitive, and that a strong-
minded person would therefore be able to subdue all hysterical
promptings. The hypochondriac is by no means necessarily
deficient in either mental or physical vigour, and yet may

fall a prey to his own feelings, and be at times the most wretched of mortals. Many men of historical renown who have performed deeds famous in the annals of their country, have yet suffered from this nervous ailment. I have already alluded to the hypochondriacal patient, and what a plague he is to the doctors, how he presents them with a sheet of paper completely filled with a description of his symptoms, how he reads medical works, first imagining he has this disease and then that, until a moment's reflection sometimes tells him that many of these maladies are incompatible, and so his disease works its own cure. The hypochondriacal patient really suffers from dyspeptic symptoms, as he complains of flatulence, eructations, and palpitations, and thus, I have no doubt, he feels really ill. He looks upon his body as if it were a piece of machinery that wants his hourly superintendence, lest it should go wrong. He watches every symptom, and comments upon it with the utmost vivacity. He visits in turn every doctor in London, and carries about a heap of prescriptions, enough for a hundred people. He never looks away from himself; his mind is always turned inwards on his own feelings. He reads medical books, and finds his own case exactly delineated within. He feels his own pulse, and carefully examines his tongue every morning, and to keep it clean buys a tongue-scraper. He will tell you about his urine and motions, even to the most disgusting minutiæ. Even during his dinner he is thinking of his bowels, and is taking such articles of diet as will in some way affect his "secretions." Such patients exist in very large numbers. They are often deranged in health, and that their whole nervous system is out of gear is clear from the dyspepsia, flatulence, palpitation, &c. The proper treatment can rarely be adopted, because the patient is his own master, and no sufficient control can be exerted over him. He calls upon a medical man, and, where the latter takes a right view of the case, he would inform the patient that very little really ailed him, and that by swallowing physic for these different troubles he was only perpetuating them. The doctor may give the advice, but cannot enforce it; consequently the patient may go to some one less conscientious, or at last fall into the hands of quacks, who then keep a tight hold of their victim—unless, indeed, he be advised by some enthusiastic lady, to try homœopathy, when there may come about a very harmless termination to his difficulties; for then he may consult his book all day long, and amuse himself by calx No. 5 every morning, or tincture of sulphur to the millionth dilution every night. There is a gentleman in this neighbourhood, of independent means and therefore with nothing to occupy his time but his own feelings. He looks upon his body like a piece of machinery, which must be constantly kept oiled, and he has a file of prescriptions containing various remedies. One he takes for flatulence, another for heartburn, one when his tongue is furred, and another when his eye is yellow; medicine for his bowels when he goes to the water-closet once a day, and another when he goes twice, and becomes really eloquent over the colour of his motions, or the "secretions" as he politely calls them. He meets me in the street; I am obliged to feel his pulse, unless I can rapidly pass behind him whilst he is standing before the jeweller's shop, apparently looking at the goods, but really gazing at his own tongue.

When a person is affected by a slight trouble so as to prevent him occupying himself in business, his complaint certainly borders very closely on monomania, and thus those who like distinctive names have called it *nosomania* or *nosophobia*. There can be no doubt that in a natural state a man should be occupied, his mind should be away from himself, and all knowledge of his feelings is morbid. No wonder that a certain nation believed that the happiness of heaven consisted in activity without consciousness. In an unhealthy state, man becomes sensible of the working of the machinery within him, and in the extreme form known as hypochondriasis we have (to use an expression which I have somewhere seen) " a meditation of man on his own health." The feeling of illness has no direct relation to disease; it may exist, as I have said, without any real disease, and on the other hand fatal maladies may progress and the patient declare that he is not really ill. The amount of depression which the patient experiences is not the measure of his illness.

Ordinary hypochondriasis is a complaint very commonly met with amongst our private patients, but it would only be seen within the wards of the hospital when existing in an excessive degree—if then, indeed, it could strictly be styled by the name. I do not allude to those cases where the patient has a variety of maladies, but where there is a complaint of a never-ceasing torment in some part of the body, mostly in the abdomen towards its upper part. You have seen three patients of mine suffering in this manner. One man has a constant pain at the epigastrium, which is often so bad that he calls up his friends and says he shall not survive the night. This has been for eight years. Another man has a constant burning at the epigastrium, and declares he has got a worm inside of him. The third patient is a woman, who has a fixed pain in the left side of the abdomen. She has had it for years, but is fat and looks well. I have just seen a private patient, a woman, who has such indescribable sensations over the region of the cæcum that she constantly awakes her husband in the night to send for the doctor. She has had some of the best opinions in London, and no one can discover any disease;

in fact, there is none, but she has what I called a " mad cæcum."
It would seem that a certain class of persons, especially those who
are inclined towards insanity, have a number of morbid sensations,
but how these are to be accounted for is not very clear, whether the
sensorium—that is, the brain—or whether some intermediate
structures, such as the sympathetic ganglia, are at fault; nevertheless you must try and distinguish them from feigned pains—
these patients do not sham, as the pain is real to them. If a person
have an impression, it is the same to him whether it is a real or a
false image, but the medical man would say the impression arising
from a reality was natural, whilst the other was morbid, and would
style it a delusion. So, if a patient have a pain, and we can discover an objective cause for it, we call it a reality; if none exist, we
style it a delusion, although it is quite the same to the patient. At
least, I suppose it is all the same to him whether he has a pain or
thinks he has pain.

When we say that the sympathetic system is at fault in the
hypochondriac we are stating more than an hypothesis, for there
are very evident manifestations of its being disturbed, and probably
weakened. We may first observe the flatulence and dyspepsia, and
then the palpitations and throbbings from which hypochondriacs
sometimes suffer. I am now seeing a gentleman who, besides his
ordinary distressing feelings, suffers from attacks of fulness of the
head, flushing of the face, palpitation of the heart, and a sensation
of pulsation through every part of the frame. A lady patient, too,
who has always been ailing, has suffered all her life from strange
feelings in every part of the body; sometimes there are pains, at
other times a sensation of coldness at the back of the neck, extending
to the throat and various parts of the face; or sometimes the sensation
is that of a swelling in different portions of the body. One of my
patients, a man of 60, who has various nervous ailments, is also
hyperæsthetic all over his body, and in his symptoms very much resembles an hysterical girl. That there is a kind of paralytic condition
of the blood-vessels is known by the pulsations and throbbings of the
aorta, as can easily be detected by placing the hand over the abdomen.

I have already said that Sydenham was apt to regard hypochondriasis of men as the same affection as hysteria in women, but you
will now see the great distinction between them. Hypochondriacs
are people of middle age, dyspeptic, ill-tempered, careful of their
health, and always going to the doctor to be cured; whereas hysterical women, though young, gay, in good health, and cheerful,
like to be considered ill and crave for sympathy. Of course, great
discernment is required in making the diagnosis between these
affections as well as between a real disease and an imaginary one.

A mistake in either direction may be attended by the most disastrous consequences.

I can scarcely tell you how to *treat* these people. Occupation and diversion for the mind are no doubt the most essential elements in any treatment, but they are just those which you cannot enforce. The worst part of the therapeutical system is, that the patient will not take your advice, and by prescribing medicine for him you are assisting in perpetuating his illness. You might think that a patient who was always ailing and got no relief from treatment would not trouble medical men any further, but it is very remarkable that it is that very patient who takes our physic. He will sit down and tell you of the number of medical men he has seen, and show you a bundle of prescriptions, declaring that they have done him no good, and yet he will ask for another. On asking one of my lady patients why she continued to come to me though I and others had never done her any good, she exclaimed, "I come to you because I never feel any the worse for what you prescribe for me." You effect no purpose by offending such patients, because they go elsewhere for sympathy, and this is what they want. I have observed, however, where a medical man, in regular attendance, has had sufficient courage to give his patients a good scolding they have always been better for it. If the patient has been a woman, and has not tamely submitted to this correction, but denounced the hardheartedness of those around her, and subsequently had a "good cry," she will be observed to be considerably better for some days after. The hypochondriac wants sympathy, and he generally gets it, both disinterestedly from his friends, who believe in his maladies, and from the doctor, who is paid for treating them as realities. The prospect of cure is thus very remote. With reference to judicious treatment, I will again quote Sydenham : "One of our reverend bishops, famous for prudence and learning, studied too hard a long while, and fell at length into a hypochondriacal disease, which afflicting him a long time, vitiated all the ferments of the body and wholly subverted the concretions. He had passed through long steel courses more than once, and had tried almost all sorts of mineral waters, with often-repeated purges and anti-scorbutics of all kinds, and a great many testaceous powders, which are reckoned proper to sweeten the blood, and so, being in a manner worn out, partly by the disease and partly by physick, at length he consulted me. I presently considered that there was no more room for medicine, and I advised him to ride horseback. Had he not been a judicious man, he would not have been persuaded to try such a kind of exercise. I entreated him to persist in it daily, going further and further, till at last he went so many miles without regard to meat and drink or weather, like a traveller. He continued this method until he rode many miles a day, and at length not only recovered, but also gained a strong and brisk habit of body."

Another piece of good advice for a hypochondriac is that of Abernethy : "Live on a pound a week, and earn it."

There is another class of patients who are equally troublesome, but differ from the true hypochondriacs in not suffering from depression of spirits. Nevertheless they are solely occupied in a study of themselves and the preservation of their health. They, therefore, consider it quite correct to occupy the time of the busiest and most celebrated physicians in every capital of Europe, to ascertain what their exact regimen should be, what German watering-place they should go to in the summer, and what spot in the Mediterranean for the winter. If they take a house in England, the nature of the soil is of vital importance to their health, they talk a good deal about chalk and clay, and they are anxious to select a convenient spot for the ready call of the medical man. Whilst they are thus striving to live for ever, their health is frequently suffering from idleness and the imbibition of too much wine.

We define the true hypochondriac as the man whose mind is in possession of some imaginary evil, from which he cannot be diverted. A case of this kind is rare compared with the host of those who suffer from nervous depression arising from anxiety or irregularities of life, although probably the pathology of the two conditions is the same, and merely varies in degree. Patients come for advice, declaring that they are unfitted for the duties of life, that they wake in the morning depressed in mind and tired in body, that they have no appetite, and suffer from a number of morbid sensations, as pain or heat at the top of the head, pains or tingling in the limbs, suggestive to them of paralysis ; they have palpitation, and think that they have heart disease, though their muscular frame is good, and they can walk without fatigue. If they are possessed of one idea it is usually of a sexual character, and we are obliged to hear a number of particulars which are supposed to imply impotence. It is very evident in many cases that the so-called spermatorrhœa is a fiction of the mind, since no reasoning can disabuse the patient of his delusion. One young man was so convinced that his seed came away with his water that I was obliged to examine it microscopically in his presence, and on declaring that not a single spermatozoon was to be seen, he was quite prepared to refute me by saying that they were probably in the form of cells containing undeveloped germs, as Lallemand had described. Another kind of patient believes he has syphilis, and presents himself every now and then with a pimple on his face, being assured that it is a manifestation of the disease. The syphilophobist cannot.

however, be convinced ; for example, on telling a patient the other day that he could not have venereal disease without a sore, he at once declared that he had had a gonorrhœa. On further investigation, finding that he had had no discharge at all, he simply replied that it was the dry form.

No doubt there are plenty of persons suffering from over-work or fatigue, who really have an exhausted state of the nervous system, and are consequently martyrs to various pains and morbid sensations. They have headache, pain in the back, sleepless nights, disturbed digestion, and feel unequal to their daily task. A holiday usually sets those persons right. A judicious method is to get such a person out of himself, and by inducing him to occupy himself with new surroundings, and make use of other faculties of his brain, a cure will soon result.

If you want a good description of nervous ailments you will find them nowhere so well described as in the quack advertisements. The charlatan, whose only object is to trade on the unfortunate sufferer, details to him every morbid feeling which he is likely to have, having familiarised himself with all the commonest symptoms of the hypochondriac. Here is a bill which was handed to me to-day in the street :—" Professor L— is the most successful practitioner in England for the cure of nervous debility, loss of energy and vital power, mental and physical depression, indecision, impaired sight and memory, indigestion, loss of appetite, palpitation of the heart, dizziness, noises in the head, prostration, lassitude, pains in the back and limbs, timidity, disinclination for business, groundless fears, local weakness, impediment to marriage, impurities in the system, blotches on the skin, &c. His medicines enrich and purify the blood, &c." The quack has made himself acquainted with all the troubles which a nervous patient can suffer from, he dwells on the importance of them, causes them in this way to be more indelibly impressed on the victim's mind, and does not let him escape until he has reaped a golden harvest.

Patients with supposed impotence require moral treatment like hysterical women, for it is usually found that the malady is as often mental as physical. A gentleman who had been under several quacks in London subsequently consulted M. Lasegue, in Paris, who wrote an opinion of his case. This is conveyed in those concise and appropriate terms which are so remarkable in French writings, that I will read it to you : " J'incline à croire qu'il a exercé sur ses fonctions genitaux une surveillance inquiète, et qu'il est devenu peu à peu hypochondriaque limité, a partir de ce moment l'acte genital est devenu une occasion d'apprehension de crainte d'insucces plutôt qu'un plaisir desirable. Le malade a ainsi perdu l'appetit sexual comme une homme préoccupé de l'imminence d'un vomissement perd l'appetit gastrique. Cette influence plus morale que physique me semble jouir ici le principal rôle."

Do not let us, though with an honest purpose, be the instruments of perpetuating the evils, to cure which our advice is sought. This is not so likely to happen in the case of men who are occupied in business as in women who have no occupation but housekeeping, which can be dismissed at once if illness need it. I am sorry to say that I have seen numerous women sacrificed to medical treatment. A lady, for example, of middle age, gets out of health, and in consequence has bad nights and suffers from want of appetite and great debility. Advice is sought, and it ought to be given with the object of restoring the patient to health. Such a remote object may perhaps be in view, but in the mean time a plan is adopted which renders life completely artificial, and is so injurious that it must before long bring the poor woman to her end. A tonic or stimulant medicine is ordered because the patient is low, more wine because she cannot eat, chloral or opium at night because she cannot sleep; in a short time, from continued use, the effect of these drugs is lost; they are, therefore, increased in amount, until chloral, morphia, and alcohol are given all day long, and death soon closes the scene.

One great reason why harm is done to nervous patients is to be found in the habitual use of drugs to promote sleep. It therefore becomes a most important consideration for the medical man to discover the cause of wakefulness and the best mode to overcome it. There can be little doubt that the experiments performed some years ago by Mr Durham really exhibit the state of brain during the periods of waking and sleeping, or, in other words, during the times of activity and repose. They correspond to and confirm the general physiological law of *ubi stimulus ibi fluxus*—that where there is increased function there is a greater distribution of blood. If, then, during sleep the brain is quiet and comparatively bloodless, we can understand how all mental activity or worry, by keeping the thoughts in a constant strain, should prevent the occurrence of the needful rest; also how exhaustion and other causes which deprive the brain of its stimulus should tend to promote it; now, these conditions have a close relation to the question of digestion and eating late at night. According to the physiological dictum just mentioned, the stomach is more vascular during the digestive process, and in all probability there is less blood elsewhere. Any occupation of the brain, therefore, during dinner diverts the blood from its proper channel, and indigestion is the consequence, the natural tendency during a meal or after it being towards inaction of the brain, and even towards sleep. There is, therefore, no more important question to discuss than that of meal time in the case of sleepless patients, and if the fears of supper could be got over there is many a person who would pass a good night if he had only something in his stomach: a very common distribution of meals amongst the middle classes, however, being a dinner in the middle of the day and tea with bread and butter at six o'clock; so that for twelve hours of the day there is little or nothing in the stomach. As regards drugs, of course we are bound to give them, and it would be a most interesting thing to know how they act—whether by some direct quality of their own, or only by affecting the vessels in such a way as to induce a condition like that which exists in natural sleep.

Occasionally patients come to us where wakefulness or the opposite condition, sleepiness, constitutes their principal complaint. In these cases, sometimes after the most rigid investigation, the cause remains undiscovered. For example, a governess tells me she is quite incapacitated from earning her living owing to the tendency to sleep; as soon as she sits down she falls into a slumber. A young man also comes to me for the same complaint, and his case is interesting as showing how the tendency is always greatest when the brain is probably least supplied with blood. He is constantly sleepy, but more especially before dinner, and whilst standing or talking he will sometimes fall into a doze. His circulation is very feeble, and his fingers often become dead. You may often observe amongst persons who have difficult digestion cold extremities, or even dead fingers, together with sleepiness.

During the war in West Africa we heard of a remarkable complaint existing there, in which the patients (natives) had a great tendency to sleep. It was said that the sufferers had enlargement of the cervical glands, and that if these were removed the complaint was cured, from which it was conjectured that these glands in some way affected the circulation of the brain. Much controversy took place about the authentic nature of these statements, but Mr Gaskoin informed us that Portuguese writers had long before mentioned a complaint of coma and lethargy as prevailing amongst the islands of the African coast. Wakefulness, as a purely simple trouble, does not often present itself to our notice, but occasionally I have had patients come and consult me for this alone. One gentleman has had recourse to various methods in order to gain the wished-for blessing, and at one time was travelling backwards and forwards between Dover and Calais with some beneficial result.

Melancholia and sympathetic mania.—I have spoken of hypochondriasis as the case where, from an extreme sensibility of the nervous centres, the thoughts are directed to the existence of some imaginary complaint, and where the patient either from a morbid activity or sluggishness of his sympathetic ganglia may suffer as much as if he had a real neuralgia or paralysis. The nature of the malady and its functional character is shown by the good effect of diverting the mind and by the success attending moral treatment; and that the pain has been altogether subjective and indicative of a morbid state of mind has been often shown by the subsequent absolute mental derangement of the patient, whilst the local trouble has at the same time ceased. We have, nevertheless, to notice how a trifling ailment may react on a sensitive centre, and set up a true maniacal condition. It will sometimes require all our acumen to discover what relation the mental and local affections bear to one another; which the cause and which the effect. In some cases a supposed local affection turns out to be a purely imaginary trouble; whilst, on the other hand, a truly maniacal state is solely produced by a localised malady.

In the following cases the supposed local disease had probably no existence:

CASE.—A gentleman of middle age began to complain of an agonising pain and strange sensations in the neighbourhood of the rectum and bladder, and which, he said, destroyed the whole comfort of his life. He consulted numerous medical men, and amongst others the most eminent who had made diseases of the bladder and rectum their speciality. They all prescribed, but without avail. He became irritable, did not sleep, and declared life was a burden to him. This went on for two years, when he became maniacal, and was obliged to be sent to an asylum. His mental symptoms during the several months he was there were of the worst description; he was dirty, had extraordinary delusions, and he wasted away; but he made no further complaint of his urinary trouble. On my visiting him one day, and asking him about it, he said there was nothing the matter with him, and never had been. At the end of a year he had recovered, is now quite well, and makes no mention of his old complaint.

CASE.—A lady complained of irritable bladder, shown by a constant feeling of distress and desire to make water. She consulted the most eminent surgeons and obstetric physicians, who prescribed medicines internally, and all sorts of local remedies, including washing out the bladder with opiates. This went on for a long time without any result, when she passed into a state of melancholia. She immediately ceased to complain of the vesical symptoms. She could pass several hours of the day and night without any desire to micturate; in fact, every symptom of irritability had disappeared. Thus it seemed to depend wholly upon the condition of her nerve centres.

When patients come to us as strangers, of whom we know nothing, it is very difficult to form a correct estimation of their case. For example, a gentleman lately came to see me, stating that he was the subject of pain and uneasiness all over the right side of his chest; and that the pain sometimes went over his face, so that the whole of the right side of his body felt different from the left. He had been a sufferer from this for many years, and had been obliged to place the desk in his counting-house in such a position that no one should approach him on his right side. He had not any hyperæsthesia, but was always conscious of his side, and dreaded it being touched. No one had done him any good, and he feared that the case was incurable, as he had long ago been told that his uneasi-ness was due to the lung growing to the side. The patient looked well, but gave me a good neurotic history of his family. I believe the unfortunate explanation of his symptoms had so rooted itself on his mind that it was impossible for him to get rid of them, and that the whole of his troubles were subjective. This is the case where the theory of habit is applicable. We say that an abnormal move-ment, such as the choreal spasm of a muscle, is due to habit; and in the same way an abnormal and constant pain may also be due to habit.

I was told lately of a case where a man had for several years a most severe neuralgia in his head. During the paroxysms it was so severe that it quite incapacitated him from doing anything. Injec-tions of morphia were used, and these quickly gave relief. It was afterwards found that pure water did equally well. After his death a careful examination was made, but failed to reveal anything abnormal in his brain or head.

Cases of this kind, where the head or chest are the affected parts, are not so common as those in which morbid sensations are referred to the abdomen, and where there is often a delusion that there is something alive in the bowels. As a rule, it may be said that when patients come before us saying they have a tapeworm, because of some strange feeling in the abdomen, they very rarely have one. If, owing to their importunity, an appropriate medicine for worms should be given, and none be discharged, the delusion still remains.

It has been suggested whether, as the brain has different func-tions, special portions of it may not be affected in different cases, through some reflex action or more direct influence of the circulation, and so the mental or other nerve symptoms differ accordingly. For example, since the anterior and posterior portions of the brain are sup-plied by the carotid or vertebral arteries respectively, so the cerebral symptoms might depend upon causes affecting these vessels sepa-rately. Bright had already published cases which he thought sup-ported these views, but having reference rather to the more positive conditions of pain in the head, sleepiness, &c., than to the results of modern researches, which might lead us to think that the mental condition would be characterised by a greater intellectual disturb-ance in the one case and sensational disturbance in the other.

I may add that, just as neuralgic pains are often periodic in character, so also are the subjective and imaginary ones. Patients, for example, who are undoubtedly hypochondriacal, and have no tangible cause for their suffering, will look for its approach at a given time every day or night.

Sympathetic mania.—In contrast to the cases we have been discussing, we meet with persons whose minds have lost their balance from being constantly directed to a part of the body where actual disease exists, and with others who fall into a maniacal state from the sympathy of similar disease, but without their being conscious of its existence. A case of the first kind was related to me by the late Sir C. Hood, of a woman whose attention was uninterruptedly fixed on a small subcutaneous tumour on the abdomen. In order to satisfy her it was removed, when her mind became calmed, and she left the asylum cured. Of the second kind was a case reported by Dr Savage of a woman who died in Bethlem of cancer of the stomach and neighbouring glands. She was admitted because she was untidy, dirty in her habits, had delusions, and refused her food; when this was forcibly given her it was rejected. Should mental disturbance be associated with phthisis, it is still, as in other cases, of the melancholic type; a remarkable circumstance, as Dr Savage remarks, since the usual mental condition accompanying consumption is of the cheerful and hopeful kind. It is very probable that minor forms of melancholia and hypochondriasis may often have a real seat in the abdomen, as for example, in the case of an old lady who was so great a trouble to the doctors that they regarded her as a confirmed hypochondriac; she always had a most dismal countenance, and a long string of troubles to relate to them, but no disease of any kind was discoverable after repeated examinations. Having died of chest disease, I took the opportunity of procuring a post-mortem examination, in order to discover, if possible, why she always had such disagreeable feelings in her abdomen. I found a portion of omentum adherent to a femoral sac, and this dragged down the colon to the pelvis. I conjecture that this might, through the sympathetic system, have been the cause of her abdominal anguish and continued low spirits. But lately I have seen an extreme form of hypochondriasis in a woman with a large umbilical hernia, and also the case of a genuinely maniacal patient who had a cæcal abscess. She had eaten nothing for six weeks, as she said nothing could pass the diseased bowel. I ordered her to be fed, but I learned afterwards that she died of starvation.

A man was lately in Stephen ward with marked hypochondriasis. He said he was very well until a year before, when a plank fell on his abdomen, and he was brought to the hospital with such severe symptoms, that it was thought he had ruptured his intestines. He, however, recovered, but ever since has had all the symptoms of the most confirmed hypochondriac, complaining of a sensation at the pit of his stomach, of palpitation, of low spirits, and feelings of great depression at times coming over him. It is well known that writers on insanity have constantly noticed the fact of misplacement of the colon having been met with in persons who have died of mania or melancholia. This was found to be the case in a woman who died of mania in this hospital.

The sympathy between the mental condition and abdominal organs is a fact which may be any day proved by observing the gloomy and desponding disposition of the dyspeptic, and contrasting it with the cheerful nature of the phthisical patient. The cause no doubt is to be sought in the anatomical arrangements, whereby the sympathetic nerves exert their influence over the vascular system, so that the difference between the collapse arising from abdominal injuries and the depression with the low spirits of dyspepsia or colitis is merely one of degree. From the beginning of history low and disturbed mental states have been observed to be associated with affections of the abdominal organs, as well as an every-day experience that liveliness takes the place of apathy in children after a good purge.

Mania after acute disease.—This subject was first brought distinctly under the notice of the profession by Griesinger, and it is remarkable that it had not more forcibly attracted the attention of its members before this, since, in my experience, cases of the kind are by no means uncommon. Patients, for example, who have had an acute disease, and when the height of it has passed or during convalescence, become acutely maniacal. The brain symptoms form no part of the delirium which may have been present during the pyrexia, and therefore are not due to unnatural heat; the mania must be owing either to an impoverishment of the brain, or some morbid state of blood resulting from the previous fever. I have seen it after typhoid, scarlatina, pleurisy, and some other acute diseases. In all my cases the patients recovered except in that of a lady, where it followed peritonitis; she sank into a state of permanent dementia, with occasional maniacal attacks. These cases may have the same immediate cause as those of puerperal mania, although in them we must not disregard the frequent albuminuria, and the direct nervous influences which are at work. In the pregnant state there is often seen a tendency to nervous excitement, even to the production of chorea and similar affections. That nervous shock may be sufficient to affect the mind is proved by a fatal case of mania which I witnessed after removal of the breast, and the case of a man I have lately heard of who had a similar attack after the operation of lithotomy.

PART IV—NERVES

THE EFFECTS OF INJURY AND INFLAMMATION

HAVING described the various diseases of the brain and spinal cord, we come to the nerves. Now, you can imagine that, as each compound nerve has numerous origins for its different fibres, we must already, in treating of diseases of the cerebro-spinal centres, have spoken of cases where one or more functions of this compound nerve have been arrested or disturbed. In cases, for example, of spinal disease, there may have been loss of power of a limb, loss of sensibility, increased spasmodic action, or a total destruction of every function of the nerve.

It may be as well, however, to again look at these disturbed functions where the nerve itself is involved, and by so doing we shall find that the symptoms vary somewhat from those where the central origins are affected, this difference arising from our own want of anatomical knowledge as to the exact relation between them. We will first ask what would be the consequences of a complete paralysis of a compound nerve arising from its division? There would be loss of power, loss of sensibility, and changes in nutrition. The latter are amongst the most interesting of the results which would follow the lesion of a nerve, and the question is now asked, why do they occur? Are there special trophic nerves which rule over nutrition, having their own centres in the spinal cord, or is nutrition dependent merely on vaso-motor nerves which regulate the blood supply to the part? and again, are these nerves none other than the sympathetic originating in their own ganglia? The question is at present a difficult one to answer, since, although the effects of nerve lesions are very obvious, the necessity for nerve influence on growth is not so clear. For example, the lower animal life proceeds without a nervous system, and in the human body tumours increase rapidly without any exertion of nerve power. Besides, a considerable amount of disease of the cerebro-spinal centres can exist without any apparent effect on nutrition; although this, of course, may arise from the circumstance that the especial centres which rule over growth and decay are not interfered with.

Lesions of the nerves do, however, indisputably show the influence which these structures have on nutrition, although from clinical experience we find the motor and sensory are far more potent in this respect than the vaso-motor ; one of the most striking facts in pathology being the power which the motor nerve has over the muscle, just as the grey centre from which it springs has over the nerve itself.   Thus, wasting of the muscle following an injury of the nerve which supplies it, is one of the most evident and important facts observed in practice, and this is quite independent of vaso-motor influence, as is exemplified in the oft-quoted instance of division of the sciatic nerve.   If this be done the muscles supplied by it will waste, although its vaso-motor nerves accompanying the blood-vessels are entire, as they arise from the crural and lumbar plexus.

Then, again, the influence of the sensory nerve is seen in the ulceration of the cornea after division of the fifth nerve, and also in the eruptions which accompany the various neuralgias.

Another very striking fact is the influence exerted over the various secretions through the nerves, as the salivary, lachrymal, mammary, &c.

If a *compound nerve be cut through*, various changes take place in the parts to which it is distributed ; the skin becomes dry and atrophied, and, as Paget mentions, the fingers become glossy and painful to the touch, the nails curved and wrinkled, and occasionally vesicles appear, but this arises generally from irritation of the nerve, and not from its division.   A supposed natural example of this atrophic change is to be seen in lepra anæsthetica, or joint-evil, as the wasting away of the fingers is called ; the skin loses its sensitiveness, and becomes brown and atrophied; sometimes the bones also.   The disease is supposed to be due to a fibrous degeneration of the nerve.   Where, as in infantile paralysis, the limb wastes, the cause is often put down to disuse, but in all probability a failure of nerve influence is the main cause, and if so, it shows that every tissue, even the bone, is controlled in its growth by the power of the nerves.   I have seen a case where the skin of the whole body became remarkably thickened, and, being associated with cerebral disease, I attributed it to a nerve origin.   It is well-known, also, that some persons grow fat and that others waste in different nervous disorders, and that the secretions become fetid.

If a *nerve be irritated or inflamed*, as in the neuritis from injury, more marked and active changes occur; the skin may ulcerate and become very painful, or vesicles and blebs appear on it.   The temperature is probably lowered, and the secretion from the skin may become excessive or fetid, and the skin sometimes of a reddish colour.   You will see that worse effects arise from nerve irritation

or neuritis than from nerve paralysis or nerve division. A continuance of the irritation will produce intense pain in the limb, with
spasm or flexion and great constitutional fever and irritability.
Several remarkable instances are mentioned by Mr Hilton in his
work on ' Rest,' where irritable ulcers were cured by dividing the
nerve proceeding to them. These cases show also how the healing
process can take place without nerve influence.

I have already alluded to the nutritive changes occurring in
paraplegia, as bed sore and vesication. It is interesting for us to
know that Bright observed this tendency to sloughing and production of vesication in cases of spinal disease, and suggested that
nerve influence might be very important in the nutrition of the
body. He mentions the case of a man who had a fracture through
the first dorsal vertebra, where "two or three large bullæ, the size of
pigeons' eggs, or larger, full of clear yellow serum, appeared on the
ankles and feet, and where any particular pressure had occurred."
Also another case, where vesicles formed on the inside of the knees.
He then gives the case of a woman who died of acute paraplegia,
with slough on the back, cystitis, and where " several large oval
vesicles, filled with clear serum, appeared on the feet and legs." After
speaking of the paralysis of the bladder, Bright remarks, " Another
curious circumstance connected with the paralysis of the lower
extremities is illustrated by this case — the tendency which is
observed in such affections to the formation of vesications, or bullæ,
which frequently make their appearance in a night on some part, as
the knee, the ankle, or the instep, where accidental pressure or irritation has taken place. They contain a limpid fluid, which, after a
few days, becomes opaque. It has sometimes struck me that this
connection between interrupted nervous action and the formation
of bullæ might be found hereafter to throw light on the nature of
that most singular disease, herpes zoster, which, from the peculiar
pain with which it is accompanied, as well as from its strict confinement to one side of the body, seems to be connected with some
peculiar condition, perhaps the distension of the sentient nerves."

The experiments, therefore, made for us in the wards are of this
kind : that injury to a motor nerve produces atrophy of the muscle ;
that irritation of a sensitive nerve produces inflammation of the
skin, with eruptions; and that division or disease of a whole nerve
produces atrophic changes in the skin, as I have described. What
amount of participation the vaso-motor nerves have in the result is
not determined.

In Bernard's well-known experiment of dividing the sympathetic
nerve in the rabbit's neck no further changes ensued than dilatation of the vessels and rise of temperature. I believe no immediate

alterations in nutrition were observed, but that after a time the ear grew larger.

Allied to these facts is the state of blushing, which is evidently a sudden dilatation of the blood-vessels arising under nervous influence; also, forms of erythema and other skin affections having a nervous origin, as well as partial sweatings corresponding to the distribution of nerves. Thus, there is now in the hospital a man with some of the symptoms of exophthalmic goitre, who is constantly observed to have a perspiration on one side of his face. Cases of partial *hyperidrosis* are occasionally seen. I used to know a lad who perspired on one side of his face, and I have now a patient who is distressed by the upper part of his shirt always being wet, and this occurs mostly in cold weather. Then we know that certain drugs acting immediately on the nerves will cause contraction or dilatation of the vessels, notably digitalis and nitrite of amyl. We see also in cases where there is a tendency to general paresis, as in hysterical paralysis of a limb, not only loss of power and sensation, but the limb congested. A division of a nerve produces, I might have said, in the first place, a kind of chilblain and there is a very interesting fact, too, in connection with ataxia and other forms of spinal disease, viz.: the occurrence of an inflammation of the joints. A case is mentioned by Sir James Paget, of a gentleman whose feet whenever he commenced to walk grew cold, white, and dead. He could only attribute this to a reflex action on the vaso-motor nerves. I might allude to a somewhat similar reflex condition observed in the white and senseless fingers of young dyspeptics, when their fingers grow dead, as they say. I am now seeing a woman with paraplegia, who has occasional attacks of cramps in the muscles of the legs, which at the same time become white and cold. Numerous other illustrations I might give which clearly point to a nervous system ruling over the supply of blood. I lately found a lady of highly nervous temperament sitting by the open window on a very cold winter day complaining of the excessive heat, her face flushed, breathing very quick, and pulse very rapid. The case reminded me of what one sees after the inhalation of nitrite of amyl, where a warmth and a glow are produced all over the body, from an evident dilatation of the blood-vessels. This is frequently witnessed in exophthalmic goitre, which is supposed to be due to paralysis of the vaso-motor nerves at their central origin, and where, besides the well-known characteristic symptoms, the patient is often breaking out into violent heats and perspiration. The flushes of heat at the climacteric period are also well known. The palms of the hands may be noticed of a deep red in many nervous disorders, or the joints inflamed.

It is remarkable that, although from physiological considerations we are constantly speaking of the influence of the vaso-motor nerves on nutrition, our facts at present show no more than altered states of vascularity when these structures are interfered with; whilst, on the other hand, the most evident changes in the tissues occurring in connection with disease of the nerves are seen when the motor and sensory nerves are implicated. I might allude to the fact mentioned by Dr Owen Rees as perhaps bearing upon the chemical element in nutritive processes, and that is, that the nerves of the cerebro-spinal system end in an acid fluid, whilst those of the sympathetic end in an alkaline one.

I have already told you that if a nerve is cut it gradually degenerates, as if it received a nutritive influence from its centre. If it be simply injured it inflames like other structures, becomes pulpy and infiltrated with exudative products. In chronic inflammation it becomes indurated from infiltration of connective tissue, and at last undergoes a cirrhosis, in which little is left of the original nerve fibre. Neuritis resulting from an injury causes paralysis of the muscles and some loss of sensation, but during the active inflammation there is often great pain and burning in the part, accompanied by ulceration or vesication, which may pass on eventually to atrophy and contraction of the limb; and, as before mentioned, many of these active and painful conditions may be removed by division of the nerve.

It is curious that in nerve injuries, where a complete motor paralysis has resulted, sensation should still remain, as if the least possible remnant of a nerve were sufficient to carry impressions to the sensorium, but that much more is required to preserve the muscles in their proper tone. The one function is passive, and the other active. It may also be remarked that after division of a nerve sensation may often return, as if sensory impressions had found a path by other channels.

The pain in neuritis is often very severe, and if the nerve of a limb be affected it may be referred to the periphery, where the distribution takes place, just as is observed in the case of a stump after amputation. Not only are these subjective feelings referred to a distance when the sensory nerve is touched, but what is remarkable, as pointed out by Mr Weir Mitchell, if an electric current be passed through the stump of a limb the patient becomes conscious of certain contractions of the fingers and toes, and painfully so if the force of the galvanism be increased.

Having now seen the results of marked injuries to the nerves, we can take up the several cases where we find them affected as a result of various morbid conditions. We may first remember the case of altered sensations, including hyperæsthesia, anæsthesia, and analgesia, which we have already considered in speaking of various paralytic states. We can then pass on to painful conditions of nerves known as neuralgia; afterwards to affections of the motor nerves, as witnessed in local paralyses and spasm.

NEURALGIA

We understand by neuralgia some affection of a nerve, whereby it becomes the seat of excessive pain. We do not use the term where there is a local lesion affecting the peripheral terminations of the nerve, nor apply it to those cases where there is manifest disease implicating its trunk; we limit the use of the expression to pain in the nerve itself. If in any case of pain in the course of a nerve we are led to exclude every tangible cause which may produce it, we place the pain in the nerve itself, and style it neuralgia. In a case of pain in the legs, due to pressure by a growth in the abdomen, I do not think we should with propriety call it one of neuralgia. It seems as if any nerve may be the subject of some morbid state causing it to be highly sensitive, and so we can have neuralgia of every part of the body. What the pathology of this condition is we do not know. Dr Anstie thought the nerve was inflamed, or sometimes wasted, but he offered no evidence in proof of it. I believe Dr Radcliffe states that neuralgia is associated generally with anæmia, and that inflammatory and hyperæmic conditions are not accompanied by pain.

Under the name of neuralgia we have all those sensations which I have mentioned in connection with irritation or inflammation of a nerve; there is not only pain but a sense of burning, coldness, and every feeling with which you must be familiar when you strike the "funny bone."

I have already alluded to the nutritive changes in connection with injuries and irritation of nerves, and therefore it is that we so frequently meet with them in connection with neuralgia. These are mostly seen in the form of vesications of the skin, as bullæ, or herpes. In the case of intercostal neuralgia the eruption of *shingles* is often regarded as the essential disease, and the pain as a nerve symptom accompanying it, but the opposite is the truth, for the pain may exist long before the appearance of the eruption, and may continue for a long time after it has gone. You may some-

this organ and involve its structures in inflammation. Mr Hutchinson has pointed out how the corymbiform or cluster-like character of the eruption corresponds to the branching periphery of the nerve. In my own experience the herpetic rash accompanies only simple neuralgia, and very rarely the cases of nerve pain due to organic disease. Therefore I have been able to give a favorable prognosis, and have argued that the pain has been merely functional in every case, where I have heard that an herpetic eruption has at any time existed. It may be that in organic disease the function of the nerve is interfered with, whereas in cases in which herpes exists there is merely irritation of the nerve. Probably every nerve in the body may be the subject of neuralgia, and not only the ordinary sensory nerves but even the sympathetic, owing to their close connection with the spinal. For example, the viscera cannot be called sensitive organs, since the food will pass along the alimentary canal without there being any consciousness of it, but at the same time under morbid conditions they may become the subject of most exquisite pain, as in gastrodynia or enteralgia. The solid organs also, as the heart and liver, may apparently also be the subjects of pain.

You must remember that the symptoms attendant upon irritation of a nerve may be of the most varied character, passing on from pain to spasm and paralysis. And then, again, under the name of pain, there are sensations which can scarcely be put into words, such as feelings of burning and coldness. You can imagine that there would be all varieties of strange sensations when a sensory nerve was irritated, in the same way as there may be all stages of colour-blindness or deafness, leading up to a complete loss of vision or hearing when the optic or auditory nerves are affected. Thus patients constantly complain of a feeling of coldness in a leg or an arm, which I regard as a variety of neuralgia, as it corresponds to it in all other particulars.

The causes for neuralgia are diverse, and yet in individual cases rarely discovered. Direct cold to a part can no doubt produce a nerve pain in it. Malarious influences are amongst the most marked and certain causes of neuralgia, as well as all circumstances which tend to impoverishment of the system, or produce cachexia. An exhausted state of the nerve centres renders them very susceptible, and at the same time, probably, starts the nerve itself into morbid activity.

Neuralgia of the fifth nerve.—This is one of the most common and painful forms of neuralgia, and is often styled tic douloureux. It may affect the whole nerve, but more usually a single branch or even one filament only. The pain may be limited to the course of one nerve, or it may by sympathy involve other nerves, and thus the pain be radiated from the original seat as a centre. It may be pretty constant, or may come on in paroxysms, and then is often of the most excruciating character; it is sometimes of a burning nature, or as if a red-hot iron were being run into the face, the pain being felt all over the forehead, eye, cheek, or jaw; it may be accompanied by lachrymation or by constant reflex spasmodic movement of the face. An herpetic eruption may accompany the attack, and if the tic be paroxysmal there may be a corresponding tendency to a vesicular rash. Neuralgia of the *first division*, or ophthalmic branch of the fifth, is often seen as a temporary complaint, more especially in the hemicrania of migraine. In the permanent neuralgia or tic it is the supra-orbital nerve which is mostly affected, and it is this which is met with generally in the malarious kind known as brow ague. An herpetic eruption attending neuralgia of the first division is very important, on account of the nerve supply to the eye, through the ciliary nerves of the nasal branch. Where the eye is not affected, this branch has probably escaped. This may set up an inflammation which in most cases is superficial, but may proceed to the iris and cornea, producing destruction of the eye. This is the reason why the patient so often presents himself first to the eye surgeon, who recognises the disease as *herpes* or *zona ophthalmica*. I have had several such cases under my care. At the present time there is a man in Stephen ward who was sent to us by the ophthalmic surgeon; his temples are covered with scabs of the herpetic eruption; his eye has recovered, but he is now suffering from severe neuralgia of the first and second divisions of the fifth nerve. I am constantly seeing a lady who has been for two years the subject of tic of these two upper divisions; her face is red and rough from the constant occurrence also of vesication, although this rarely reaches a true herpes. Another and less common form of neuralgia affects the eye alone, and is known as *ciliary neuralgia*. The patient is seized with paroxysms of violent pain in the globe of the eye and eyebrow; it is accompanied by lachrymation, is very sensitive, and becomes highly vascular. After a time some opacity or ulceration of the cornea may ensue. The attacks come on like migraine, without much warning, but induced often by disturbing nerve causes, in neurotic or gouty subjects.

Neuralgia of the *second division* sometimes accompanies that of the first, or it may exist independently. Some of the worst cases which I have ever seen have been those of tic of this nerve. We had in the hospital at the same time two men who might be seen sitting up in bed violently rubbing their faces as paroxysms of pain came on.

Neuralgia of the *third division* we often meet with alone, and it also constitutes a very frightful form of the complaint. Occasion-

ally, as in the other case, it may be relieved by friction, but generally the skin is very sensitive, and the pain is much aggravated
by movement. This depends very much upon which branch is
most affected, whether, for example, the mental or lingual; but we
find all movements of the lower jaw, as in eating, bring on a
paroxysm of pain. Probably the same would occur in the tic of the
two upper divisions of the fifth, if the parts with which they are
connected admitted of movement. It is curious, however, to observe
how, in some cases of tic, rubbing the part will relieve, whilst in
some other cases the merest touch will start the paroxysm.

Neuralgia of the tongue alone I have met with, but it is comparatively rare. I have seen also a sore mouth with tic, which
probably had an herpetic origin.

There has been no disease in which so many remedies have been
tried, and in all probability every one of them has at times
been successful. In choosing a medicine we are guided in part by
the speedy action of some drugs over others, and these we naturally are inclined to give first. Now, amongst the most efficacious
of these is the chloride of ammonium, an old and approved remedy
for neuralgia. I generally give fifteen grains three or four times a
day, and a few doses suffice if the remedy be likely to act beneficially.
In other cases quinine at once cuts short the disease, administered in
large doses to begin with, followed by lesser ones. It is thought to
be peculiarly efficacious in the supra-orbital form when paroxysmal.
If clearly of miasmatic origin the quinine will undoubtedly cure;
but even if it be not of this nature it will often be successful, for the
periodic nature of neuralgia by no means proves its malarial nature.
In other cases iron will cure. The old sesquioxide had long a cele-
brity against nerve pains. Lately we have had a new medicine—gelseminum, and of this I have given the tincture in twenty-drop doses
three times a day with good effect, but it is a remedy still on its trial.
In other cases chloral is very serviceable, and some give the preference to croton chloral. If all the more powerful remedies fail, I
usually give arsenic, one of the most efficient drugs we possess, but
time is required for its beneficial action. Then there are various applications of the irritant and soothing kinds, such as blisters, mustard,
&c., or ointments of aconite and belladonna. If these fail we have
recourse to injections of morphia. It is very remarkable that in
cases of periodic neuralgia, plain water may sometimes be substituted for the morphia with equally good effect, but whether this is
due to any mechanical pressure of the fluid, to the irritation caused
by the puncture, or to the effects on the imagination, is not very
clear. Galvanism is sometimes a very potent remedy, and ought
always to be tried, the continuous current being better than faradiza-

tion. In a case just now alluded to of tic accompanied by herpes it slowly effected a cure. I will give one example more at length.

CASE.—Sophia C—, æt. 60, came into the hospital, having suffered from neuralgia of the left side of the head for five months. The pain was in the temple and forehead extending down the nose, involving nearly the whole of the first division. At first the pains were slight and transient, but they gradually became worse and more frequent until they were most excruciating in character, and were brought on by slight causes, such as blowing the nose, talking, or eating. She had lost all her teeth on the affected side, and in consequence of this and the pain caused by mastication, she was obliged to cut her food very fine. She was ordered three drops of the tincture of aconite three times a day, and this was increased until, a fortnight after admission, it reached seven drops, She was no better, and the pains, which came in paroxysms, were of the most excruciating character. She was then ordered a dram of chloride of ammonium every four hours. This she took, and subsequently iron and quinine, without effect. Galvanism was now ordered, one pole of the battery being placed over the forehead and the other stroked down towards the temple. She began slowly to improve, so that at the end of a week she was much better, and at the end of a fortnight she had scarcely any pain. It was continued, however, for another week, when she left the hospital quite well.

Occipital neuralgia, that is, a pain running along the course of the great occipital nerve, is not uncommon. In any of those conditions where headache is liable to occur, and more especially the neuralgic, this form may be met with. Sometimes the back of the head is the exclusive seat of headache, but nevertheless it may only be of a neuralgic kind, as is shown by its evanescent character and its curability by appropriate remedies.

Cervical neuralgia is a form occasionally met with where the pain is distributed over the neck and shoulders. If extending down the arm it may be called a *cervico-brachial neuralgia.*

Phrenic neuralgia is supposed to account for a pain sometimes met with, running from the neck and shoulder and across the chest, in the position of the diaphragm.

Intercostal neuralgia is one of the most common and painful forms, and, being very frequently associated with herpes, is known as "shingles." The pain may precede the eruption for a short time, and when this has disappeared may continue for many weeks or months afterwards. The nerve affected is generally one of the middle dorsal, although it may be one of the upper or lower dorsal. The complaint occurs mostly at the middle or later periods of life, and by its severity and persistence makes the patient really ill. Sometimes the skin is excessively tender if touched. When the eruption is developed a little starch powder is usually applied, and remedies are given for the neuralgia. Local irritants, which again bring out a rash, often give relief, but this is only temporary. As the pain is likely to be persistent, I usually give arsenic, and this, if continued, will often cure. I sometimes give it with quinine, and in other cases with the bromide of potassium. I have tried galvanism in several cases, but without much success.

Pleurodynia is an intercostal neuralgia, but the term has reference to those cases where the pain is fixed at one spot. It is very commonly met with in women.

Brachialgia is not so common a form of neuralgia as that of the leg, but I have several times met with it. I once saw a gentleman suffering from agonizing pain in his arms, as if the whole brachial plexus on both sides was affected. He got well, but I never ascertained the cause of the attack. In one case the pain involved the *musculo-spiral nerve*, and there was great tenderness on the outer part of the arm, with paroxysms of intense pain on the back of the hand. In some cases the *ulnar nerve* is the one affected, and in one instance of a lady the pain alternated with a feeling of coldness along the outer side of the forearm. I may here remind you of the pains in the arms associated with angina pectoris, and those met with in real spine disease, besides others which are of a neuralgic character associated with gout. We are constantly consulted by patients on account of pains and numbness in their arms and legs, which they fear are the onset of paralysis. This has often, too, been suggested by their medical man, who knows that pains of this kind do sometimes have a central origin, and who consequently warns his patients of the possible consequences. They are, however, more frequently of a neuralgic character, having their origin in morbid blood conditions. Let me ask you to remember neuralgia of certain joints, which are mostly met with in young women, under the name of *hysterical joints*.

Sciatica is one of the commonest of all forms of neuralgia. It is a pain commencing at the point of exit of the nerve, and extending all down the back part of the leg. Sometimes the pain takes the course of the tibial nerve, and at other times of the peroneal. In severe cases there may be tenderness over the course of the nerve, and the pain is usually aggravated by movement. The thigh may be slightly bent in walking, and the muscles spasmodically contracted. This causes a difficulty sometimes in diagnosing it from hip-joint disease. In the latter the surgeon believes that pressure against the sole of the foot will cause pain, which would not be the case in simple sciatica. Sometimes even slight wasting of the limb will occur; this probably is of nerve origin. It is a disease much more common in men than in women. We may with advantage give as remedies all those before mentioned. In chronic cases I have used with some benefit the Tr. Actææ racemosæ, in half-dram doses, with a few grains of iodide of potassium. Sometimes galvanism affords a speedy cure, and, when other remedies have failed, we may have recourse to the daily injection of morphia. Hot salt-water baths and shampooing are also excellent remedies in very chronic cases. I have had a patient in whom the pain ran down the thigh in the course of the obturator nerve, and no disease being discernible to cause it, we gave it the name of *obturator neuralgia*. When pains take the course of other nerves around the back, hips, and genital organs we suppose the existence of *neuralgia of the gluteal and lesser sciatic nerves*.

You may also remember a pain which we call *coccydynia*, met with in women, in whom an operation by cutting down on the bone has been performed. If the pain run down the front of the thigh and leg we may have a true *crural neuralgia*, an example of which we lately had in the hospital.

CASE.—John N—, æt. 41, blacksmith. Well until ten days ago, when he began to feel a stiffness in the left hip, as well as pains and weakness in the leg. These symptoms increased so that he was obliged to apply for relief. He was a strong, healthy-looking man, and nothing on examination could be discovered. He had pain in the left hip, extending along the front and inner side of thigh to below the knee. There was also sense of numbness. The pain was increased on flexing the thigh. He was ordered to keep in bed, and take iodide of potassium, with Tr. Actææ racemosæ. He gradually improved, and at the end of a fortnight could walk about the ward, and in nearly a month from the time of his admission he left the hospital well.

Quite lately a new operative measure has been performed for the cure of neuralgia, called *stretching the nerve*, and more especially in those cases where a neuritis has been thought to exist, as in contraction of the limbs, or in painful stumps. It is mentioned by Billroth, although I think he was not the proposer of the operation; it was first attempted in this country by Mr Callender. In the German case the median nerve was isolated, and the surgeon pulled forcibly upon it, and the patient was cured. I suppose the *modus operandi* is by the breaking down of adhesions, and so altering the relation of the nerve fibres. A man was lately in this hospital who had had the operation performed at Chicago. After an injury to the back the legs became drawn up and rigid. Both sciatic nerves were cut down upon and stretched. He was relieved for a time, but the contraction again returned. Mr Durham has lately performed the same operation on the infra-orbital nerve with some success.

Internal and visceral neuralgias.—In the diseases of the several organs, which we have to treat, neuralgia may be the principal symptom; as, for example, in the *mastodynia* or irritable breast of young women, which is sometimes associated with small hard tumours.

Angina pectoris, or cardiac neuralgia, is a hyperæsthesia of the heart. It is usually associated with degeneration of the muscular tissue and disease of the blood-vessels. The nerve fibrils are no doubt involved in the morbid process. The connection of the cardiac plexus with the cervical nerves directly through the cervical ganglia explains the pain which is often felt down the arm. Why this should often be more on the left than the right side is not very evident. It is a complaint which is necessarily incurable, but showing its neuralgic nature, I have seen more than one case in which great temporary benefit was obtained by arsenic. It is also sometimes relieved by inhalation of nitrite of amyl.

Gastralgia is a complaint we shall be constantly called upon to treat. It seems to be a pure neuralgia, but is not so evident where its exact seat may be; or whether the sympathetic or pneumogastric is most instrumental in the production of the pain. That the latter is capable of causing it, is seen by cases of painful chronic ulcer examined by Dr Habershon, in which he found fibres of this nerve exposed on the walls of the ulcer. Addison's attention was long ago drawn to the fact of this disease occurring in young impoverished girls, who exhibited other evidences of neurosis in the frequent pleurodynia and other signs of what is called spinal irritation.

Enteralgia is an abdominal pain having its seat in the intestines, where some irritant cause for it may exist; or it may be due to a true hyperæsthesia of the nerves of the hypogastric plexus. Neuralgia of the *rectum* is often associated with hæmorrhoids or ulcer, but it may exist alone as a most painful disorder. Neuralgia of the testes, or *spermatic neuralgia*, is occasionally met with. I know the case of a gentleman who for many years has had attacks of most excruciating pain in the testes. Nothing tangible has appeared, showing the purely nervous nature of the attacks. Whether other solid organs suffer from neuralgia I cannot say; that is, whether there be, for example, a *hepatalgia* or *nephralgia*. It is certain that pains may exist over the liver and the region of the gall ducts without any evidence of disease, and also that patients may suffer with pain passing down from the kidney to the bladder, exactly as if they had stone, and yet that these organs may contain none. I have seen several such cases in women.

In nervous and hysterical women it is remarkable what a great variety of pains may exist, especially in the abdomen. It was the opinion of Briquet that these were situated in the muscles, and were therefore *myalgic*. There is no difficulty, however, in regarding them as purely neuralgic, when we consider the relation of the nerves of the solar plexus and mesentery to the neighbouring parts, and more especially when we observe the disturbance of the abdominal organs, as shown by dyspepsia, flatulence, and general depression. In lead poisoning, where we know the nervous system is affected, there is no reason why the sympathetic should not also be involved, and in this way enteralgia is explicable in those cases where colic is absent. Where, however, as in true colic, the pain seems to be due to excessively powerful contraction of the bowel, and yet we know that ordinary peristalsis takes place without the production of any sensation, we are forced to believe that ordinary impressions are carried no further back than the next ganglion which takes cognizance of them, but that in the extreme contraction of colic the impressions proceed further on to the spinal system, and so to the sensorium, until we become conscious of them. In all visceral neuralgias we should have to acknowledge the same law, viz. that the sympathetic or other nerves being stimulated beyond their ordinary and due measure force their impressions beyond the sympathetic ganglia into the spinal system, and so on to the brain.

Cephalalgia, or headache.—Of all the lesser ailments to which mankind is liable none is more common than headache. It may have its origin in organic disease within the cranium, or it may merely arise from sympathy with some other organ. In various morbid conditions of the blood, headache is a very common symptom, not only in those dependent on a specific cause, as the contagious fevers, but where the blood is contaminated with some extraneous matter, as alcohol, urea, or opium. It is, therefore, merely a symptom, and does not require our direct attention. It is only when headache becomes the prominent trouble or special ailment from which the patient suffers that we have to regard it and treat it separately. It would be a question of much interest and importance to decide, where is the seat of the pain. In cases of organic disease we find very often that tumours, softenings, and abscesses have existed in the brain substance without the production of any pain; whilst, on the other hand, disease of the membranes has given rise to the greatest suffering. It is difficult, however, to argue from these facts as to the seat of the pain in ordinary headache. A certain neurotic disposition is necessary for its production, since many persons declare that they have never had anything like a headache, and cannot conceive the nature of the pain.

Looking upon it clinically, you must remember the general and very severe headache accompanying morbid states of the blood in various fevers; then the pain associated with organic disease within the cranium. If the bone and membranes be the seat of it the pain is constant and localised; if the disease, as a tumour, be within the brain substance, then, if there be pain at all, it is very severe, and generally paroxysmal. Headache arising from gastro-hepatic disorder is generally characterised by a pain and feeling of weight across the forehead; whilst that of anæmia and exhaustion is altogether of a different character. In this case the pain and other sensations are felt at the top of the head, and it is in this region that the patient often feels a burning or sensation of creeping, intermixed with actual pain.

Almost the only case where we are consulted for headache alone is the case of nervous headache, and this requires, therefore, our special consideration.

Hemicrania, or sick headache.—This I shall dwell upon for a short time, because it is so common and peculiar in its nature that it deserves a place by itself, and because, as far as I know, a good description of it has not yet found its way into medical literature. The reason of this omission probably is that medical men have cared little to make such a study of a mere functional disorder as they would of a more marked or tangible malady, and perhaps also because, personally, they have had no conception of the symptoms attending it. I once knew a very eminent medical man of the melancholic temperament who told me he had no conception of what was meant by the term headache. I am sorry to say that this is not my case, and therefore I take this opportunity to draw your attention to a malady the particulars of which are, for the most part, gathered from personal experience. I am alluding to the affection popularly known as sick-headache, or technically as hemicrania or migraine. Like many other complaints, it is hereditary, and in a most marked degree. Thus, it is a complaint met with in members of particular families, and transmitted from father to son, whilst there are other families of different temperaments in which a headache is unknown. All the members belonging to a particular family may suffer, both male and female, and therefore the complaint is not to be considered as identical with the clavus hystericus, although the latter, I have no doubt, owns the same immediate cause. This might be included with hemicrania under the term nervous headache; but if, because styled hysteric, it be regarded as a trifling disorder, there can be no doubt that the true hemicrania is a reality of the gravest kind, unfitting its victim, while it lasts, for all the avocations of life.

Being so frequently associated with stomach disturbance, it is often styled sick headache; head and stomach disorders standing in the relation of cause and effect, though constantly changing places. Remembering, however, that the gastric disturbance is often greatest when the primary cause is in the brain, the term hemicrania is not an unfit one; for, as a rule, although the pain may reach far over the head, it is most usually fixed to one spot, or is more concentrated on one side than the other. It may commence as a dull pain over the forehead, then, as it increases, pass down to one eye, and so to the temple, where it remains fixed. Exceptionally, the pain is at the top or back of the head. The pain is sometimes so violent as to deserve the name of neuralgia, but generally it is somewhat duller and of a most sickening character. Its great peculiarity is the throbbing that occurs with each beat of the heart, aggravated by every movement of the body, and more especially of the head itself. The movements required for washing and dressing on rising can scarcely be endured. The sufferer walks slowly, since everything which tends to make his arteries beat a degree more violently adds to his misery; in his head he perpetually hears or feels "throb," "throb," "throb," and his only relief is to support the head against a pillow or rest it on the hand, and to avoid all possible excitement. His whole attention is distracted by the painful throbbing, and he becomes utterly incapacitated for business; every movement, every word spoken aggravates the pain. His only desire is to be let alone and be unspoken to. During this time he looks exceedingly ill, very pale, with a dark margin round the eyes, and the pupils contracted; he may have also indistinct vision, or hemiopia; there is a general feeling of chilliness over the whole body, excepting the head; the pulse at the wrist is feeble, whilst that in the head is strong. The anorexia is complete; the loathing of food being so great that it is often impossible to swallow a single mouthful, and sometimes there is actual vomiting. In a bad attack the stomach generally refuses food for twenty-four hours. There may have been no error of diet to account for the attack, nor any constipation of the bowels, as is often thought; although a disturbance of these parts is often one of the symptoms of the complaint. It is remarkable, however, that anorexia is not always present. Some persons, although suffering from headache, eat as usual, others vary in this respect, according to the severity of the attack; and, what is very remarkable, the sympathetic affection of the stomach is occasionally related to the seat of the pain. Thus, I have a patient who, if she awake with a pain over her right temple, is a sufferer the whole of the day, can eat nothing, and would be made worse by stimulants; whereas if the pain were seated over the left temple she would be relieved by a stimulant, and, as a rule, be better after dinner. She is always made worse by lying down, and this peculiarity in her own case she has observed as distinguishing it from a headache arising from a bilious attack. This case is not singular, for many patients tell me how the general symptoms vary with the seat of the pain. Another lady is always sick when the pain is over the right eye, but not over the left. She is always sleepy, but is not better for sleep, as twenty-four hours are necessary for the cure. The duration of a bad attack is generally several hours. If the person awake with it, the headache persists during the day, and it is only after another night's rest that he rises free.

If it should come on during the day it gradually increases in force, and then the night brings little comfort, for the throbbing, aching head entirely precludes sleep; indeed, the recumbent posture becomes impossible, and a sitting or upright position is the only one in which the patient can endure his suffering. I know several persons who cannot lie down, and whose only comparative comfort is found by spending their night in an easy chair. This is my own case during a bad attack. A young lady whom I occasionally see has attacks of migraine, with throbbing over the temple; as lying down increases the pain, she is compelled to sit up until the attack is over. All remedies have failed to relieve her. When the attack is less severe, there is, fortunately for the sufferer, a strong disposition to slumber; he lays his head against any surface, and readily sleeps. I have observed myself that if ever, during the evening, I feel drowsy, and, on lying down, very quickly fall asleep, on the following morning I rise with a headache. How far the sleepiness induces the subsequent attack, or how far it is a mere symptom of the approaching disorder, I am uncertain, but I am inclined to think the latter.

Now, as to the cause of this misery, I have already said that whilst the body is cold the head is hot, and that whilst the radial artery is small the carotid is full; in fact, if the term determination of the blood to the head is applicable to any malady, it is assuredly to this. This irregularity in the circulation due to nervous influence has created much interest of late years, as I have already told you in describing various diseases. It has been clearly shown that the blood-vessels are regulated in calibre by the sympathetic nerves, and that the supply of blood is immediately under nervous control. Now, in this complaint of which I am speaking, the carotid on one side with its branches throbs inordinately, is apparently very full, and is sending too much blood to the brain and its coverings. This I knew when quite a boy, for when leaning my head on my hand I distinctly felt the increased size of the throbbing temporal artery on the side of the pain, which would be sometimes on the right and sometimes on the left side. I remember mentioning the circumstance to more than one medical man, and they received the statement with incredulity. I knew it, however, to be a fact, nevertheless, and am sorry to say I have been too fully aware of it up to the present day. The fact is, that in this distended throbbing carotid and its branches lies the source of the trouble. The vaso-motor nerve on one side seems for the time paralysed, the vessels of the head dilate, more blood is sent to it, hence the increased heat, throbbing, and pain which the patient has to suffer until the tone of the nerve is restored. There can be little doubt of the truth of this condition when the head is felt to be hot and throbbing, whilst the feet and hands are cold, coupled with the fact of the aggravation of the symptoms induced by the recumbent posture. I may even mention another fact in proof of this increased arterial flow of blood. It is well known that activity of organic function is in proportion to vascularity. It is remarkable in the form of headache of which I am now speaking how quickened the mental powers often are. In spite of the distressing pain there are none of the low spirits evinced in a much less painful abdominal affection, and the mind, as far as is possible, may be occupied with subjects requiring much vigour of thought. Indeed, some patients have told me that before the onset of an attack their minds have been in a state of the utmost tension, and that they have been the subjects of an intensity of thought which seemed to make them all but ready for the solution of the most subtle and difficult problems. I may remark that, although I have no doubt of the fact of the throbbing vessels and increased flow of blood, the explanation perhaps is not to be found in a vaso-motor paralysis, but rather, since we have been enabled to measure better the pressure of blood, in an increased tension of the vessels. This would, according to some theories, imply that there is a decreased flow of blood to the brain, a supposition supported, perhaps, by the pallor of the face, contraction of the pupil, and the evidence of general nerve disturbance, at the onset. It would be very important, therefore, to ascertain whether in migraine there be more or less tension on the vessels, since the appropriate remedies might then be suggested. If the altered tension be due, as it seems to be, to a nerve cause, it is also important to discover why the nerves are thus acted upon. Since a stomach derangement usually accompanies the hemicrania, it is very frequently thought that the source of the trouble is always gastric, and that medicine of a particular kind will relieve. That this is partially true is no doubt correct, but just as frequently the complaint arises from a direct influence on the nervous system. Besides, if it does arise from the stomach, the cause is not the same as that which operates injuriously in the mass of people from over-indulgence in eating and drinking, where a more general headache is the consequence, but the gastric disturbance is dependent upon a number of trivial circumstances of which the sufferer himself alone has any knowledge. It may be said, no doubt, with truth, that gastric derangement is a very common exciting cause in those who are subject to the complaint, but very frequently no cause for the attack is apparent, and certainly none attributable to the stomach. When the cause is evident, it is very often one which has acted directly on some portion of the nervous system, and to the

non-susceptible would scarcely be credited with so powerful an operation. Thus all worry, excitement, or overwork will readily produce
a headache; walking in the sun is a very sure method of inducing
an attack; strong impressions on the olfactory nerve, as the smell of
paint, and in some persons the odour of spring flowers; also impressions on the retina, as long use of the microscope, or a protracted
visit to a picture gallery. An atmosphere overcharged with carbonic
acid is one of the most fruitful sources of headache, as that of a
crowded assembly-room, and what would affect myself at once
and in the most intense degree would be the presence of uncon-
sumed carbon from candles or lamps. Loud noises in the ear will
also cause a headache. One patient says that listening attentively
will also bring it on; in fact, it would seem that a strong impression
made upon any part of the nervous system is sufficient to induce
an attack. Probably derangements of any organ might cause it,
not only of the stomach, as we constantly see, but of the uterus in
women, more especially at the catamenial periods. Some of the
most violent attacks which we witness are in women at these periods.
There are those who are doomed every month to an illness of a few
days, with intense headache, prostration, and sickness; if there be
much uterine pain, the case is styled one of dysmenorrhœa, and the
cause attributed to the uterus, which may or may not be true.
Those, of either sex, who are thus liable to violent headache,
are proscribed many of the pleasures of life, since irregularities of
any kind are so apt to lead to their wonted complaint. Under the
most favouring circumstances, however, it is my experience that
they can never escape an occasional attack. There is no doubt that
persons subject to gout are liable to headache, and that the same
causes which are favorable to the production of one may induce the
other.

The immediate seat of headache is not known; various opinions
have been given. Many have denied that it is in the brain itself,
seeing that the organ may be diseased in various ways without pain
being present. Some have considered that the pain resides in the
dura mater, and originates in the branches of the fifth nerve, which
are distributed to this membrane. Briquet, who has in his investigation of various hysterical conditions shown that many local
pains are in the muscle, or myalgic, considers that headache is of
the same kind, the seat of it being in the temporal and occipito-
frontalis muscles. I suppose that one's own feelings ought not to
influence the judgment, otherwise it would be thought that the
pain is situated in the very depths of the brain itself. I once had
an opportunity of testing the power which the individual has in
discovering the seat of pain. Having scalded my head with steam

issuing from a pipe to vaporize a sick room, I endeavoured to analyse the character of the pain which followed, but was unable to discover how it differed in kind from the pain of ordinary headache.

As such trivial causes are sufficient to induce an attack of this hemicrania or migraine, it might be supposed that some equally slight circumstance might be sufficient to counteract or cure it. I should think it probable that such is the case, although, after long searching for the remedy, I have not yet discovered it. Certainly the ordinary aperient doses which the medical man so commonly prescribes for a headache are useless ; besides, the attack may have spontaneously subsided before there could be any expectation of a result from the medicine. The act of vomiting, however does in some cases afford very speedy relief. This is not by getting rid of any crudities, for the stomach may be empty, and therefore the effect must have been through the nervous system. As regards stimulants, as a rule they cannot be prescribed ; they often aggravate the complaint to an intense degree, although I have found that in some milder cases a little brandy-and-water or a glass of champagne has, after a short period, been apparently beneficial ; sometimes a cigar. Of all remedies, perhaps tea is the best, but I am not quite certain what amount of benefit is to be ascribed to the tea and what amount to the hot water. Tea, of course, is well known to have a direct and marked action on the nervous system, and thus it might appear absurd to raise a doubt as to its efficacy did not I know more than one person who obtains more relief for a headache by sipping very hot water than by any medicine which has ever been prescribed.

Those who have any knowledge of the perpetual and horrible throbbing in the brow or temple, also know that nature prompts us to seek relief by pressure on the aching part. The leaning the head against the hand or any other object is in obedience to what instinct dictates. On lying down, if the attack be not too severe to allow it, the aching brow is always pressed against the pillow. By more direct and intentional pressure a more marked relief is obtained. Thus, a pressure on the carotid in the neck will produce a suspension of the throbbing and the pain, but the effect is only for a time, as the blood apparently soon finds its way to the head by other channels. Although the use of pressure may at the present time have a show of reason, it has no doubt always been adopted at the dictate of nature. It is probable that medical authors may allude to the method, but we need only go to our own Shakespeare, who appeared to be possessed of universal knowledge, to lead us to the belief that it must always have been in common use. Thus, in the scene between Hubert and Arthur, in ' King John,' when the latter is petitioning for the preservation of his eyes :—

" When your head did but ache,
I knit my handkerchief about your brows."

Also, as you know, in ' Othello ' the main feature of the play lies in the loss of a handkerchief, which Desdemona produced for the object I have been mentioning.

Desdemona.   Why do you speak so faintly ?
          Are you not well ?
Othello.   I have a pain upon my forehead here.
Desdemona. Faith, that's with watching ; 'twill away again :
          Let me but bind it hard, within this hour
          It will be well.

Then, again, besides pressure, the application of cold gives relief, as a wet cloth bound round the temples. I have already alluded to the effects of cold and heat upon the nerves, and the resultant influence on the blood-vessels ; thus, cold is said to depress the action of the nerve centres or ganglia, and heat to excite it ; consequently the former would be used when we wished to remove nerve stimulation and cause a greater flow of blood, whilst heat would be used for a contrary purpose, as to check hæmorrhage—at least, I believe it is said, that heat to the spine will repress hæmorrhage from the uterus, whilst cold will produce warmth in the extremities. However this may be, and on the supposition that the theory is true, there must be immense difficulties in the way of making the application to the appropriate part ; so we must be content with the fact that it is cold, and not heat, which affords most relief in headache. It may be that the cold acts directly on the vessels to constringe them, and thus causes the diminution and lessened blood-supply. The object required is to lessen the force of the vessels, for it is certain that whilst the pulse at the wrist is low, and the whole body inclined to be cold, the head is hot and throbbing. As the cause is nervous, our agencies should be directed to the fountain-head, and thus it is by no means improbable that something may be discovered which may have the power of affecting the sympathetic, and consequently curing the malady. For this purpose various remedies have suggested themselves, and theoretically galvanism to the neck or head would be beneficial. I have only used it a few times, and with some success.

As regards remedies, when I first gave a lecture on this subject I said that I knew of none, and even now, in many cases, every known drug fails to have any effect. The fact that I can at the present time declare that means have been discovered which can ward off or even cure a sick headache, speaks volumes in favour of an advance in therapeutic knowledge. As these remedies act specially on the nervous centres, it is probable that they relieve by influencing the tension of the blood-vessels, especially as we know that simpler agents apparently operate in this way. For example, relief is obtained by the application of cold to the head, by a warm bath, or by placing the feet and hands in hot water. Amongst the simpler remedies in the more constant form of headache, bromide takes a good place ; either in continued doses or in one large dose. There are persons in whom a scruple dose will produce sleep or quiet, and a speedy relief to their headache. In similar cases where headache in a lesser form occurs every day or two, the persistent use of cannabis indica acts very well. I am in the practice of giving this once or twice daily for some weeks, with the best effect. Opium or morphia injection is sometimes the only remedy which has any power over the neuralgia, but unfortunately the unpleasant effects often forbid its use. In very bad attacks, however, the relief obtained is speedy and marked. Aconite, I have been informed, when given in a full dose, is sometimes followed by a very rapid relief to the pain, but I have no experience of it. In the same way nitrite of amyl has been recommended, and since we know its effects in dilating the capillaries in the head and brain, I should think that there can be no doubt that some striking effect must follow its use either for good or for evil. Then comes the last new remedy, guarana ; this I had been acquainted with for some years, and even taken it myself before I made its value known to the profession. It had been quite useless in my own case, but considerable success with others made it clear that a really efficacious remedy was in our hands. My first method was to give about a scruple of the powder when the attack was coming on, and this would often avert it ; to be followed by another powder if the first was not successful. In some cases where the headache has been very frequent, I have given smaller doses three times a day for several days together. We have now a liquid extract of guarana, which is a pleasanter preparation. Although the remedy altogether fails in some cases, there are others where the effect is so marked and speedy that no doubt can be entertained as to the power of this drug over migraine. I have received communications from a great many persons bearing witness to its efficacy. Amongst my notes I find one patient, a farmer's wife, who says she could never go out to tea, to market, or for a drive, without returning with a headache accompanied with much sickness, lasting always twelve hours. She carried guarana about with her, and took half dram doses in tea at any time when she was squeamish or had a headache, and was always relieved within an hour. After taking six powders she had headaches much less frequently, and after another half dozen could do anything or go anywhere without any headache coming on. I could quote a large number of cases in the fashion of the quack advertisements. Thus, a young lady writes, " I have found much benefit from the guarana powders you recommended me; they almost invariably cure my headache." A gentleman writes, " The headache remedy has been a wonderful success with me, and I am now almost a stranger to the pain." A young lady says, " I write to let you know the result of the guarana powders which you recommended to me. I think I may say it is decidedly satisfactory, as last Wednesday I woke up with all the symptoms of one of my bad headaches, and I took one of the powders immediately before breakfast. By midday I felt better, and by three o'clock my head was quite well. I felt all the more satisfied, because I know that otherwise by that time I should have been fit for nothing. It is the first thing I have tried with any success."

Various local applications will sometimes cure neuralgia—when this is probably not of the migraine type—as galvanism, blisters, and the vapour of bisulphide of carbon, applied by means of a bottle containing a small quantity of the liquid placed on wool.

LOCAL PARALYSES

Paralysis, besides arising from disease in the spinal centres or at the origin of the nerve, may occur from some affection of the nerve itself in the course of its trunk or at its periphery. The mere fact of the palsy being localized is sufficient very often to show its external character, for the tendency would be for it to extend or spread should it have a central origin. Moreover, it is not yet decided whether every spot in the cord does rule exclusively over a few muscles or muscular fibrillæ, and therefore, if diseased, necessitate a corresponding paralysis. Our knowledge seems rather to indicate that the roots or origins of the nerves are spread through so large a field of surface, and are so interwoven with others that disease of any one spot cannot produce a complete paralysis of a muscle, and therefore whenever this is met with, it would imply some involvement of the nerve trunk itself. An example of this is seen in the case of paralysis of the seventh nerve.

The effects of galvanism also are very different in the two cases of central and local paralysis. Marshall Hall had already distinguished between cerebral and spinal paralysis, with regard to their electrical susceptibilities; in the one case referring to such a paralysis as hemiplegia, where the connection between the brain proper and the spinal cord is severed, and in the other case to a paralysis arising in the cord itself. Of late we have further found that the two forms of galvanism act differently, according as the paralysis is local or central. For example, when a nerve has been injured from cold, pressure, or other cause, so as to produce paralysis, its trunk loses its susceptibilities to the electric current; and as regards the muscles which it supplies, these begin to waste, and very soon lose their excitability to the induced current or faradization, whilst they become more susceptible to the primary or battery current. We shall see this in the case of ordinary facial paralysis and other forms of local palsy. By thus discovering that a particular form of paralysis is more susceptible to one kind of galvanism than to another, we are led to know which will be the one to effect a cure. We use electricity also not only to test the electro-mobility of a muscle, but also its electro-sensibility, since the latter varies very much in different forms of nerve disease.

We will first take paralyses of the cranial nerves, and then pass on to some local paralyses of the extremities and muscles of the trunk.

First, or olfactory nerves.—As regards paralysis of these nerves, if they be affected alone, the cause is found in disease of the cranium, and then there is total loss of smell. For the complete integrity of smell the mucous membrane of the nose must be sensitive, and its secretion healthy. It may therefore be impaired, as well as that of taste, in paralysis of the fifth nerve; but, showing the two senses are absolutely distinct, we often meet with cases where common sensation is perfect, whilst the sense of smell is lost. There is now a patient under my care where this is the case; and I often see a young man who, when a child, had a fracture of the frontal bone and injury to the brain, followed by a total loss of smell, but ordinary sensation is perfect. He has also partial loss of taste, but it must be remembered that this is a very complex sense, dependent upon odour and common sensation as well as upon its own special power. Dr Ogle has related several cases of *onosmia* resulting from injuries to the head, which involved the olfactory bulbs, and in this way the experiment is made for us by which we can analyse the sense of taste. In such cases it is common for patients to say they have lost the sense of taste, but it will be found to apply only to those substances which have flavours, since acids, sweets, salines, and bitters, can still be appreciated, which shows that the true sense of taste can apply only to these qualities.

Second, or optic nerves.—These are constantly affected in connection with brain disease, as I have frequently mentioned; optic neuritis being very common, but the mode in which it comes about, from the presence of such a disease as a tumour in the brain or cerebellum, is not very clear.

Pupils.—Little, too, can be said positively as to the interpretation of alteration in the size of the pupils. Their movements depending upon different causes, you may see how their condition may be frequently deranged. You know that the circular fibres of the iris being supplied by the third nerve, a paralysis of this nerve will necessitate a dilatation of the pupil; and you know how late experiments have shown that the radiating fibres are supplied with motor filaments, which run into the cervical sympathetic; and thus how pressure on this nerve, as is seen in cases of tumours of the neck, will cause contraction of the pupil. It has been found that irritation of the upper part of the spinal cord and upper thoracic nerves will influence the pupils through these same nerves, and also any affection of the first division of the fifth pair through the same channel. The exact nature of the nervous filaments which supply the radiating fibres of the iris has not yet been satisfactorily determined. It is clear that the ophthalmic nerve of the fifth has a power over the pupil, but this may be derived from the third or the sympathetic, seeing that the ciliary ganglion has sensory nerves from the ophthalmic, motor from the third, and sympathetic from the plexus. Through these it is connected with the so-called cilio-spinal centre, which is situated between the sixth cervical and second dorsal vertebra. Apart from these very definite and well-ascertained causes which influence the pupils, it is observed that they become altered in size under a great variety of circumstances. In the general paralysis of the insane, and in locomotor ataxy, the pupils are, as a rule, diminished and unequal in size; and the same condition may be observed in some other chronic diseases affecting the cerebro-spinal system. In various acute diseases of the brain the pupils are altered, but I am not aware that any definite lesion is necessarily associated with the inflammation. In cases of pressure on the brain from effused blood, the pupil is often larger on one side than the other; and I have seen a case of chronic hydrocephalus in which the pupil of the side on which the patient lay was always more dilated than the other. But although a difference in the size of the pupils will do little more than denote cerebral trouble, an extreme condition of dilatation or contraction always indicates a very serious affection of the brain. For instance, in ventricular effusion the pupils become, as a rule, widely dilated. This fact you should well bear in mind, because, on visiting a child suffering from head disease, you may be informed by the mother that it is in a "nice sleep," but when you raise the eyelids you find the pupils widely dilated. You know that in sleep the pupils are contracted, and so remarkable is this circumstance that, if you place the child before the window, their size remains unaltered, but immediately you awake him the pupils expand, although the light is still shining upon them. At the same time remember that a contraction of the pupil is an equally unfavorable sign in many cerebral disorders, and more especially in apoplexy. In those fatal cases where blood has burst into the ventricles, or diffused itself over the base of the brain, the pupils are often found minutely contracted,

just as they are in cases of effusion into the pons. In all cases contracted pupils denote a grave disturbance of the sympathetic nerve,
either in the cervical ganglion, in the course of its fibres, or in the
cilio-spinal region of the cord.   Thus it is that in many chronic
spinal affections myosis is seen.   It is thought by some that the
state of the pupils will inform us of the nature of the cerebral
lesion.   Certainly a well-marked change in the pupils will always
indicate something serious, but a natural condition may exist in
most important diseases Thus, I do not know that you can tell
concussion of the brain or drunkenness by the pupils. In uræmic
coma I have observed the condition of the pupils over and over again,
and should say that they are not affected.   I have repeatedly seen
cases brought into the accident ward where injuries to the brain had
occurred with various amounts of effusion of blood, and where the
pupils, although at first contracted, afterwards became dilated, showing that the slightest change will affect them whilst the main lesion
remains unaltered.   You must also not only observe the size of the
pupil, but whether the iris can still be stimulated by the influence of
light upon the retina.   In recent injury or acute disease, if there be
immobility of the pupil when the light of a candle is thrown upon
the eye, it generally is an evidence of some affection of the brain in
which the optic tracts are implicated.   You must also remember
the sympathy which exists between the two eyes, and how the
perfect susceptibility of the one will influence the other, hence
the necessity of examining each eye separately.   As regards the
state of the pupils after death, I think no conclusion can be formed
from their size, as they alter when life departs.   We want more facts
with reference to their post-mortem state in poisoning by opium.

   Paralysis of the third nerve.—There is paralysis of the muscles of
the eye, excepting the superior oblique and internal rectus, as well as
the levator palpebrarum and sphincters of the iris ; there is generally
dilatation of the pupils but this is by no means constant.   In a case
of paralysis of the lower branch only of the third nerve there was
no ptosis, but the pupil was dilated.   If the pupil is dilated to the
utmost, atropine has no further effect upon it.   It is curious to observe how readily this nerve becomes paralysed, either in association
with disease of the brain or standing alone without any other symptom.
The paralysis may come on suddenly, without any apparent cause,
and in a slight degree is often seen in drunkenness, hepatic disorders,
migraine, and other temporary disturbances of health.   In connection with syphilis, mydriasis may be seen with or without any ocular
paralysis, as if a portion of the nerve, as the ciliary branch alone,
could be affected, and cause the dilatation.   In sleep you know that
the eye is turned upwards and inwards, and the pupil is contracted.

Some physiologists have traced the roots of the third nerve to the corpora quadrigemina, which may indeed be also the origin of the optic, a fact of importance in relation to the connection between the muscles of the eyeballs and the retina. In tumours at the base of the brain and the cerebellum we often meet with complete blindness, dilatation of the pupils, and nystagmus. It is really very difficult to know what importance to attach to strabismus apart from other symptoms. In a case of severe illness, where cerebral mischief had been for a long time doubtful, its occurrence would be of serious moment, whilst under other circumstances it might be of no importance. The cases of squint, for example, so often operated on by the ophthalmic surgeon, may have occurred without any marked illness. I know now of the case of a child who has intermittent attacks of strabismus ; every few days it is observed that her eyes converge, and after a few hours become straight again, but otherwise she appears to be in perfect health. What appears more alarming, is the occurrence of a sudden complete paralysis of the third nerve with ptosis. I have known this to happen without any other symptoms whatever, and where effusion of blood in the cavernous sinus, or some similar diagnosis, has been made, although founded on pure conjecture. I need not trouble you with the mode of testing which eye is affected in diplopia, as this is taught by the ophthalmic surgeon. I might allude to a remarkable case now in the clinical ward, of a man who has double paralysis of the third nerve, and sees every object as three ; that is, he sees double with the left eye. The explanation is difficult, as he has not astigmatism.

Paralysis of the fourth nerve.—The pupil is slightly raised above the lower lid. There is no power of rotating the eye, and it moves with the head. If an object is looked at on a plane, the image approaches when the eye is directed upwards, and recedes when directed downwards. The simple test is this, if the person affected look at an object (say a pen) below the horizontal median line he has diplopia, and that eye is the one affected which forms an image inclined towards the other.

Paralysis of the fifth nerve.—There will be loss of sensation of half of the face, from the middle line running backwards to the front of the ear ; the posterior half of the ear and lower lobe being supplied by the auricularis magnus. There is also loss of sensibility in the eye, the nose, the front part of the mouth, the gums, and the tongue ; so that, on eating, the tongue may be lacerated, and the patient not know it. The taste is unaffected, and this confirms Reid's views, that the sensory nerve to the tongue is not that of taste. The smell and hearing are not necessarily affected, although they are sometimes impaired, owing to the mucous secretions having become altered. In a case related by Althaus of double paralysis of the fifth there was no lachrymal secretion and no salivary secretion, but there was an abnormal discharge of mucus from the eyes, nose, and mouth, which formed a thick frothy layer on these parts.

In most cases of paralysis of the fifth nerve the cornea becomes ulcerated, and this fact has given rise to much discussion as to the influence of the nerve on nutrition. Those who have disputed this inference have offered in its place the explanation that irritating substances having got into the eye, which has lost all knowledge of their presence, an inflammation has been set up, and this is proved, they say, by the fact that if the eyelid be kept closed by strapping, the ulcers will heal. Experiments on animals, however, seem to show that covering the eye makes no difference in the result, as corneitis always takes place, as well as ulceration of the mucous membrane of the nose. It is thought that the trophic nerves proceed from the Gasserian ganglion, and the ulceration depends entirely upon whether the disease of the nerve is on its distal or cerebral side. But here, again, experimenters do not seem to confirm the opinion, that the nutritive changes depend upon injury to this ganglion. The pupil, also, is often contracted, as is seen in paralysis of the sympathetic of the neck; but as the ciliary nerves from the otic ganglion are in connection not only with the third but with the ophthalmic and sympathetic, it may be accounted for in this way. In connection with the inflammation of the cornea you ought not to forget the herpetic eruption in the neighbourhood of the eye, as well as the conjunctivitis often associated with neuralgia of the ophthalmic nerve.

Then, besides the loss of sensation, there is paralysis of the masticatory muscles, the pterygoids, temporal, masseter, anterior belly of digastricus, mylo-hyoid, and tensor palati, so that the patient cannot chew on the affected side, and if the hand is placed on the face no muscular contraction is felt. It may happen occasionally that each branch is separately paralysed. If it be the *first* division, there would be anæsthesia of the upper eyelid, conjunctiva, ala of nose, and the pupil would be contracted; if the *second* division, the nose, cheek, and upper jaw would be anæsthetic; and if the *third* division, the cheek and front of tongue would be anæsthetic, and the muscles of mastication paralysed.

It is interesting to observe the experience of a good clinical physician before the introduction of any physiological knowledge into the wards to bias him; thus Abercrombie says, in respect to paralysis of this nerve, " A remarkable circumstance connected with the affections of the fifth nerve is the tendency to inflammation and sloughing in parts which have lost their sensibility, particularly in the eye. A very instructive case occurred to my friend Dr Alison. The patient had loss of common sensation of the left side of the face, the left nostril, and left side of the tongue, with insensibility of the ball of the eye. There were frequent attacks of inflammation of the left eye, with dimness of the cornea, which were relieved from time to time, but at length the cornea sloughed out, and the contents of the eye were discharged. The muscles of the left side of the jaw were paralytic, and felt quite flaccid when the patient chewed or clenched the jaws, but the motion of the muscles of the cheek was unimpaired. The nerve was found, after death, to be almost destroyed near the ganglion."

Paralysis of the sixth nerve. — This nerve has its nucleus adjoining that of the seventh, and is occasionally paralysed with it.

Paralysis of the seventh or facial nerve.—This may arise from disease at its origin, at its course through the temporal bone, or at its periphery. The first form existing as an independent disease is almost unknown, for central paralysis occurs generally with other nerve lesions, as in hemiplegia. The second form is more common, and may remain as a permanent lesion, owing to a destruction of the nerve trunk. The third or peripheral variety generally arises from cold, and is the commonest form met with. You can understand that, the nerve having various connections, the symptoms would differ in these three forms, and how a peripheral paralysis would vary from one arising from disease of the trunk whilst it lies in the Fallopian canal.

In *peripheral paralysis of the facial* the expression of the face is very striking, for, owing to the loss of muscular tension on one side, it falls, whilst the opposite side is drawn up. This distortion is much increased in smiling or talking, or whenever the influence of the will is exerted on the muscles. Thus, the patient cannot wrinkle his forehead, cannot close his eyes from paralysis of the obicularis palpebrarum and corrugator supercilii, and when he attempts to do so the eyeball is seen to roll upwards and generally inwards. When he sleeps the eye remains open, this is caused mainly by the drooping of the lower eyelid, as the upper is counteracted by the levator palpebræ. Horner's muscle being also paralysed, the tears run over. If the patient take a deep breath the nostril is observed not to expand on the paralysed side, but falls in, and sometimes the tip of the nose is turned to the opposite side. Owing to paralysis of the lip, the angle of the mouth falls, and the opposite one is drawn up. The patient cannot speak as he should naturally, and is unable to whistle, the mouth is screwed on one side, and he cannot draw back his lips to display his gums ; the saliva also runs from that side of the mouth, the buccinator is paralysed, and the cheeks puff out.   He can chew well, though food gets in the cheek.

In *paralysis of the trunk* it was long ago observed by Romberg that the velum palati is sometimes paralysed; it cannot be lifted on one side, and the uvula is turned towards the weak side, the voice at the same time becoming nasal. He believed, with these symptoms, the nerve was diseased in its course through the passage, and that the large superficial petrosal nerve had become involved, this nerve being attached to the seventh in the Fallopian canal, and communicating with Meckel's ganglion.   In this way the levator palati and azygos would obtain their nerve supply.   Romberg and others have also thought that the tongue was sometimes paralysed, but I think that this is very doubtful.   It was accounted for on the supposition that the chorda tympani nerve supplied the styloglossus and some fibres of the lingualis, a belief not now entertained. This is one of those cases in which Waller experimented according to his method.   He showed that if a nerve were divided its branches wasted or underwent fatty degeneration in two or three weeks. Therefore, in reference to any supply to the tongue by the facial, he cut this nerve, and a fortnight afterwards examined the tongue and found no change in its nervous fibrillæ.   He concluded, therefore, that the chorda tympani did not supply the tongue.   This nerve is now known to have another function.   It passes to the mucous membrane of the tongue, and to the submaxillary gland, and is known to be intimately associated with the *sense of taste.* Physiological observers had long discerned the function of this nerve, and have now pretty well agreed that it is the nerve of taste. Numerous cases of disease have occurred to corroborate the conclusions drawn from their experiments, where disease of the facial has affected the sense of taste, whilst it has not been destroyed in disease of the fifth.   Cases under my own care have sufficiently established this, especially one where it was evident from other symptoms that the disease of the seventh nerve was in the trunk.   In this case, when the tongue was protruded, common sensation was found to be perfect, but if any salt or sweet substance was placed on one side of the tongue, towards the tip, the patient could not taste it. In an instance related by Dr Noyes, of New York, where there was disease of the temporal bone, any pressure on the ear would cause a peculiar sensation at the side of the tongue, and a flow of saliva.   Subsequently, from an extension of disease, a complete paralysis of the facial ensued, together with dryness of the tongue and loss of taste. Other observers have also shown how the chorda tympani sends fibres to the papillæ of the tongue and submaxillary gland; so that if it be involved in disease there ensues dryness of its surface, loss of taste, and diminished secretion of the submaxillary gland. If it be the nerve of taste, as now appears to be conclusively shown, another question arises how it becomes so. It has been conjectured that this property is derived from the superior maxillary through the petrosal and Meckel's ganglion. An objection to this is to be found in the case published by Dr Heslop, where although a tumour destroyed the petrosal nerve in common with the fifth, no loss of taste ensued. Others have thought that the chorda tympani is really derived from the facial, but has its special function obtained from the root called the nerve of Wrisberg. You may remember, therefore, that there is only loss of taste in those cases of paralysis where the lesion is above the spot where the chorda tympani is given off. In testing this sense you should not forget that sweets and salines are appreciated only by the tip of the tongue, while bitters are recognised by the posterior part which is supplied by the glosso-pharyngeal. Much care also is required in judging as to what is meant by taste, since many so-called tastes are merely smells appreciated by the olfactory organs. This fact may be proved by closing the nostrils when the taste disappears; also it may be nothing more than the sense of touch expressed by such terms as cool, hot, burning, &c.

The seventh nerve supplies also the platysma, the muscles of the auricle, and the stapedius. *Deafness*, therefore, may be a symptom of disease of this nerve and it may occur under three different conditions. The paralysis may have a central origin, and the portio molls being likewise affected, a perfect deafness would result. It may also arise from destruction of the organ of hearing in the petrous bone, and thus have the same cause as the paralysis of the seventh; or the deafness may arise simply from paralysis of the small muscles of the ear. Occasionally, noises in the ears have been observed, and the explanation has been a paralysis of the stapedius and over-action of the tensor tympani.

I might say that the movement of the eyeball upwards when an attempt is made to close the eye is not quite satisfactorily explained. It seems to take this position in sleep, for if the eye be opened gently it will be seen to be turned upwards and inwards. This contraction of the superior and internal rectus and inferior oblique muscles is generally due to an active reflex state, for we observe the upper movement of the eye in coughing, sneezing, and some kinds of convulsions. The difficulty lies in explaining why it takes place during the quietude of sleep when all other muscles are relaxed.

When a facial paralysis has occurred for some time the same changes may take place as in other muscles ; that is they undergo contraction and the face becomes drawn up on the affected side. I have seen a mistake made in the diagnosis owing to this. I have more than once heard the contraction attributed to the protracted and indiscriminate use of galvanism, but there is no reason to suppose this. Inasmuch as in every chronic case such treatment would have been adopted, the inference has been very naturally drawn. A double facial paralysis is sometimes seen, and is recognised by the face being fallen and expressionless. It is rarely observed except in connection with other lesions such as bulbar disease.

Treatment.—This depends upon the seat and cause of the disease. When central, the paralysis is only one feature of the case. Also when there is disease of or injury to the temporal bone, the treatment is directed to that. It is only in the peripheral cases that special treatment is required, and here galvanism appears the most effectual. It acts merely by stimulating the muscles to increased activity, and both forms of galvanism are sometimes used to advantage. It is, however, the continuous current which has so marked an effect, both physiologically and therapeutically, in contrast to faradization, and indeed, facial peripheral paralysis affords a good example of the extreme susceptibility of the muscles in local paralysis to this form of galvanism. Thus, if both kinds are tried on a healthy face, a reaction naturally takes place, but if one side be paralysed, faradization will produce but little excitation of the muscles, whilst a continuous current which is so weak as to be scarcely able to rouse the muscles on the healthy side, will produce an instant contraction on the paralysed side. If the nerve be permanently damaged, the patient for the rest of his life has his face drawn up, and for this unfortunate the French have invented a new name,"zygomanique," applicable to Victor Hugo's "l'homme qui rit."

The eighth pair.—Paralysis of the *eighth pair* I have already alluded to under the name of labio-glosso-laryngeal paralysis—a form of paralysis in which the organs of speech are affected owing to an implication of the seventh, eighth, and ninth nerves ; the portions of these nerves which act harmoniously for the purposes of vocalisation have their centres in close proximity, and consequently disease covering a very small spot in the medulla oblongata is sufficient to produce a well-marked paralysis. The fact of the larynx being also partially paralysed is explained by the observation of Lockhart Clarke, that the internal root of the spinal accessory nerve, which mainly forms the motor supply to the larynx, proceeds from the same spot as the lingual. The recurrent nerve, or motor nerve to the larynx, has, however, other filaments than those derived from the spinal accessory—viz. those from the lingual, facial, and cervical nerves; when, therefore, the grey centre of the spinal accessory is involved, the nerve is only partially affected. When the nerve is pressed on in its entirety, a complete paralysis occurs, and if both recurrent nerves are compressed, then the larynx is paralysed and suffocation ensues; just as in Reid's experiments of dividing the recurrent nerve when the dilating muscles of the larynx were paralysed, and the aperture no longer opened during inspiration, the abductors, and more especially the crico-arytenoidei, being rendered inactive. Under these circumstances tracheotomy has, on more than one occasion, saved the life of the patient. This double paralysis is not common, but paralysis of one side is often met with where the left recurrent, as it passes under the arch of the aorta, is compressed by an aneurism. Such specimens you will see in our museum. A living example of this I believe I have now in my ward. When the larynx is examined by the laryngoscope, the vocal cord on one side is seen to be immovable, and the effect, which you hear, is what you might expect—a stridulous voice and a peculiar brassy cough.

Dr G. Johnson has observed cases where pressure on one nerve has been sufficient to cause paralysis of the larynx, and I think I have seen the same. The explanation has been sought for in a supposed reflex inhibitory action on the centres whence these nerves proceed, or on the supposition that one nerve alone is the active agent in stimulating the larynx, and that its fellow ganglion does no more than follow its impulse. A disease therefore of one side would be sufficient to cause paralysis, a theory in vogue to explain many of the phenomena of aphasia, &c.

I was shown one day at the Veterinary College an exact counterpart of a larynx in our museum, where the muscles on one side were wasted owing to pressure on the inferior laryngeal nerve. It came from a horse who was said to be a "roarer," and I was informed that this atrophy of muscles on one side was the pathology of "roaring." I could learn nothing about aneurisms as a cause.

As regards the trunk of the pneumogastric, I need not enter here upon the effects of its lesions. You will find them dilated upon in your works on physiology. The interest for us clinically is, that they are not unfrequently involved in cancerous disease which attacks the œsophagus, and, as a consequence, a low form of pneumonia with sloughing of the tissue, results; and here the same question occurs which we have already had before us—whether this effect is due simply to a nutritive influence being removed, or whether it does not arise from altogether secondary causes, as in the cases of the bed sore and cornea already mentioned—that is, that the nerve power being lessened, the efforts of expectoration become more difficult, mucus collects in the tubes, and thus the further inflammatory processes are set up.

Diseases implicating this nerve are well worth studying, owing to its numerous relations and the consequent disturbance of more than one function. Thus, knowing its distribution, you might suspect that an irritation of the stomach would produce cough ; or that disease of the lungs would cause sickness. I have seen more than one case of phthisis treated for a long time as a stomach affection, and in one patient the vomiting was so obstinate that a scirrhous pylorus was actually diagnosed. I apprehend that it is through implication of this nerve in disease of the base of the brain that many well-marked symptoms are observed. In arachnitis, for instance, the influence on respiration is seen in the irregular breathing or sighing ; on the stomach in vomiting, which is so constant a symptom ; while the influence on the heart is apparent in the diminution of its pulsations, producing the slow labouring pulse of cerebral disease which is so well known.

The influence of the pneumogastric on the heart has been made the subject of such laborious investigations that I cannot but allude to them, especially as they have given rise to the opinion that this nerve acts as a regulator to the heart, and even to the suggestion that many organs of the body are supplied by similar inhibitory or restraint nerves. You know that while there are certain influences which excite the various organs to increased action, there are others which tend equally to arrest action—or, at least, so it would appear. If the organic machinery is kept in operation by certain nervous forces, originating in ganglia and distributed amongst its different parts, you can conceive how a diminution of these forces will arrest the action, and with this explanation we have hitherto been content. But experimenters of late have made it appear that there is some direct and active power at work which can restrain the different movements of the animal mechanism. Thus an injury to the spine will often cause a diminution of the number of the heart-beats. This might be explained by the paralysing of the motive nerves of the heart ; but the same result is said to arise from an irritation of the pneumogastric nerve in its course. It has been conjectured, therefore, that the heart, with other organs, has two sets of nerves— one to excite it to action, and another to control or arrest it. Certain it is that many organs have more than one supply, not only that proceeding from the sympathetic and the spinal system, as seen in the intestine or uterus, but a distinct supply from an independent source ; the one is the motor, and the other the regulator, or the inhibitory nerve. Thus, for example, if you take the heart, this is supplied from the ganglionic or sympathetic system, which, as you know, is associated with certain spinal nerves coming off from the lower cervical and upper dorsal portion ; it is also supplied by the pneumogastric. Now, it has been shown that if the sympathetic nerve be galvanised the heart's action is much increased ; and, in like manner, if the pneumogastric be galvanised, the heart's action is retarded, or, as the engine driver would say, " slowed." It is thought, therefore, that the sympathetic stimulates the heart to increased action, and the other tends to retard or regulate it. Now, if these facts be true, they are of importance in practice, and, as Dr Handfield Jones has well shown, appear to be to a certain extent corroborated by clinical observation. We find, for example, that irritation of the sympathetic produces palpitation, and it is thought that the remarkable disease, exophthalmic bronchocele, where the eye protrudes, the thyroid enlarges, and the heart beats violently, is due to implication of the sympathetic. At the same time, if true, division of the sympathetic would cause the heart to gradually cease to beat. The phenomenon of its arrested function is probably seen in those recorded cases of sudden death from a blow on the epigastrium, as also in those cases which, no doubt, some of you must have witnessed, where an injury to the spinal cord has brought down the action of the heart. Such an example I saw not many weeks ago, where, from a fracture of the dorsal spine, the heart's beats were reduced to 40. The cardiac plexus was there being paralysed through the spinal nerves. But this, it is said, is not the whole explanation ; it is allowing the pneumogastric, whose influence is to arrest the action of the heart, to come into play. It is further said that if the pneumogastric, or nerve which " slows" the heart, be divided, the organ commences to beat violently, not being overruled, until at last it runs wild, and its action ceases from sheer exhaustion. If the divided pneumogastric were galvanised, this would, on the contrary, gradually " slow" it until it stopped. In oposition to this, I have seen it lately stated that Zermack has been able in his own person, by pressure on the pneumogastric nerve at the border of the sternomastoid, to produce a decided diminution in the frequency of his pulse. Kuss also relates how a medical student pressed on his pneumogastric, and his heart ceased to beat. He became unconscious, and some time elapsed before he completely recovered. I need scarcely say that this inhibitory action of nerves is by no means satisfactorily proved, for it has been shown that a slight stimulus on the same nerve will excite it to moderate action, whereas an increased stimulus will arrest it. You see, however, into what a large sphere of disease a discussion of such nervous influence might lead us ; and if these observations be true, they are surely of practical importance to us. If we have a notion why the heart's action is retarded in one case or accelerated in another, we may be on the road towards appropriate remedies, besides its affording suggestions of the *modus operandi* of medicines.

Suppose the inhibitory action of the pneumogastric be true (but which is by no means proved), it is argued that as this nerve supplies stomach and heart it can be seen how dyspepsia and flatulence give rise to cardiac disturbance, and how also in heart disease gastric disorders are amongst the most distressing symptoms. Suppose, then, that we have a case of dyspepsia which we believe to be due to a want of tone in the stomach arising from exhaustion of the nervous system, or due directly to a deficiency of good nerve force supplied by the pneumogastric, we might expect that the influence of this nerve, being also removed from the heart, the sympathetic would come into play, and a palpitation result. This is certainly what we see, and I might also inform you, as a matter of experience, that such palpitation would not be cured in the same manner as that dependent on organic disease. It would be relieved by a tonic, as iron, or, as a temporary remedy, by a glass of brandy-and-water. Digitalis would have no effect in arresting the irregular or quick action, whereas this would be the drug in organic disease, where, again, alcohol would often aggravate the excitement.

This apparent divergence from my subject arises from my being obliged to dwell upon the functions of the pneumogastric and the symptoms of its paralysis. The result on the lungs is one well known to you, and much dwelt upon by physiologists who follow the experiments of John Reid ; but these other results, as seen daily by us, on the functions of the stomach and heart, are, I think, of more practical value, and come into my province as a teacher of clinical medicine. As I have alluded to this regulating action of the nerves, or the inhibitory influence, as it has been called, I may state that the same doctrines have been attempted to be applied to other organs.

As regards respiration, the two kinds of nerves, it is stated, are not so well marked. A stimulus applied to the pneumogastric certainly does not arrest the breathing in the same way as it stops the heart's action and causes vomiting ; but it is said that galvanising the superior laryngeal will arrest the respiratory process.

As regards the intestines, it is said, in like manner, that the splanchnic nerves are the inhibitory or restraint nerves. The doctrine is carried still further, for since the excito-motor functions of the spinal cord do not come into play until the communication with the brain is severed, it is thought that there are special restraint nerves under the dominion of the will. Remember, I do not want to teach this dogmatically, for there are some who I know would feel satisfied with the older theories of exhaustion; that is, that an organ is stimulated by the nerves which supply it, and that its opposite condition is simply one of exhaustion. Of late, however, the theory of inhibition by distinct restraint nerves or Hemmungsnerven has been in vogue in some German medical schools. The theory will certainly account for some of the abnormal conditions of stomach and heart which I have mentioned, and it is for this reason that I have thought it right to bring the subject before you.

Paralysis of the ninth or sublingual nerve.—This is partially paralysed in hemiplegia, in bulbar paralysis, &c., but it may be the only nerve affected from various local causes. Under these circumstances the tongue becomes visibly smaller on the paralysed side, the muscles undergoing atrophy. This leads to some difficulty in speaking and swallowing, but actual loss of the tongue does not render a man dumb, for only lately there was a man in the hospital from whom the tongue had been excised, who could speak fairly well.

Sir James Paget has described a remarkable case where there was atrophy on one side of the tongue in connection with disease at the base of the cranium, which involved the ninth nerve. An abscess formed over the seat of the disease, and then some bone was removed, the nerve was released, the power in the tongue returned, and it gradually increased until it had reached its original normal size. The furring of the tongue on one side has been associated with irritation of the sensory nerve of the fifth.

As we pass down the cord we find other nerves affected. Many of these instances I have already mentioned in connection with disease of the cord, but occasionally the nerves are the subjects of independent lesions.

The *phrenic nerve* may occasionally be affected. If the nerve be implicated in disease a paralysis of the diaphragm results. About two years ago a man came to the hospital, who said he had had increasing shortness of breath for twelve months. On examining his chest the diaphragm was seen to be perfectly useless, the abdomen sank in during each inspiration, and the lower part of the chest dilated in a most remarkable manner. He attributed it to a blow in the neck from a man's fist. The lungs became congested, and gradually indurated, and in a few weeks he died. You, of course, must be careful to distinguish such a case from disease of the diaphragm itself, where, for example, it becomes useless from fatty degeneration of the muscular substance.

The *long thoracic nerve* may be the nerve especially involved, and then we have *paralysis of the serratus magnus*. This would no doubt occur to you as the probable cause whenever you saw this muscle inactive, but in many of these local paralyses it is the muscle itself which is at fault. Indeed it would be rare for the thoracic nerve to be injured, but over-straining the muscle is common enough. We therefore find the affection in sawyers and in men who use their arms in a similar manner to them. The diagnosis is easy. The serratus, more usually the right, is seen to be wasted, and its action entirely lost. The patient is unable to lift his shoulder or raise his arm above the horizontal line. When the arm hangs down the scapula sticks out, and seems quite loosely attached to the back, the inner angle being nearer the spine and projecting in such a manner as almost to enable us to pass the hand under the bladebone. From the serratus being paralysed the other antagonistic muscles pull it out of place. A man lately in the hospital was a shipwright; he had found the arm getting weak for two months. On examination, the serratus was found flabby and wasted; he could not raise his arm above the horizontal line. The scapula seemed to be hanging loose on his back, with its under edge projecting outward, and in order to replace it in its proper position it had to be pulled forward and outward and slightly upward. The muscle did not react to either form of galvanism, and but little good was done by their repeated application.

I may remind you of what I have before said, that the use of galvanism is still speculative and experimental. We do not know always which form will be useful, nor the direction in which it should be applied; nor again does it follow that a tired muscle always requires galvanism to stimulate it afresh.

The *circumflex nerve* may be paralysed, in which case the deltoid and teres minor become powerless. It occurs for the most part from injuries and diseases of the joint, especially chronic rheumatic arthritis, where considerable wasting of the muscles takes place. The arm lies at the side motionless, and cannot be raised or stretched out. This is the case where the different effects of the continuous and interrupted current are well seen. A man was in the hospital lately with wasting of the deltoid and an inability to raise the arm. Faradization was used without any effect, but immediately a simple galvanic current was used, by placing one pole on the top of the shoulder and the other at the insertion of the deltoid, and contact was broken, the muscle responded and the arm rose from the side A continuation of this plan completely cured him. I have repeatedly referred to the influence which the nerves have on the integrity of the muscles which they supply, and therefore in any injury to the circumflex nerve wasting of the deltoid might be conjectured; but the wasting of the muscles is often equally marked in inflammation, of the joints, and therefore we might infer that the nerves were here involved. In explanation of this, Sir James Paget suggests that there may be such a thing as reflex atrophy, because he has observed that the wasting is proportionate to the pain, as if it were due to the disturbance of some nutritive nervous centre irritated by the painful state of the sensitive nerve fibres. In cases, therefore, where the arm is wasted, weak, painful, and not very movable, the diagnosis between local paralysis and commencing chronic rheumatic arthritis is often difficult.

Pressure on the *axillary plexus* may cause paralysis of the whole arm, as I have witnessed from the growth of tumours, from injuries, and dislocations. We were therefore justified in supposing that in a case where pain and wasting of the arm had existed for some time, and was subsequently recovered from, that a neuritis of the plexus had occurred.

Paralysis of the musculo-spiral nerve I have frequently seen. It occurs most commonly from the patient sleeping on his arm whilst resting on the back of a chair, or, as happened in two patients who were lately in the clinical ward, from heavily sleeping on the ground whilst they lay drunk with their arms under them. When such persons wake they find their limbs quite helpless, and often numb; the muscles at the back of the arm, as the triceps, extensors of forearm, and supinator, are flaccid; the hand falls, and the fingers and thumb become flexed on themselves. There is generally a partial anæsthesia of the back of the arm and the hand, as far as the second phalanx. If the pressure has occurred, as is very often the case, between the biceps and brachialis anticus, the triceps escapes, and it is only the muscles of the forearm which are paralysed. When speaking of lead poisoning, I said it had generally been observed that the supinator longus escaped, and this, therefore, would be one of the diagnostic signs between it and the case where the trunk of the nerve was compressed. The electro-motility would not be impaired.

If the *median nerve* is paralysed, it is generally from injury. The flexors and pronators are powerless, and flexion of the second phalanx cannot take place; the thumb is extended and turned in; there is also loss of sensibility in the palm of the hand, and also on the last phalanx on the dorsal side, owing to the nerve curling around from the front. After a time a wasting of the muscles may ensue. If the nerve be the subject of inflammation or irritation, then, in addition, certain changes on the skin may be observed, such as vesication, ulceration, shrivelling, and roughness of the cuticle, with loss of the nails. There would be also pain along the course of the nerve, and all the other sensations belonging to this painful affection would be present.

The *ulnar nerve* is not infrequently affected, either from injury or growths near the olecranon. Its exposure at this part may perhaps account for the frequency of its paralysis. There would be loss of power in the flexor carpi ulnaris and flexor digitorum, as well as in the muscles of the thenar and hypothenar eminences and the interossei. This is seen in the increased difficulty of flexing the wrist and adducting the hand, also of flexing the fingers, especially the little one ; also the thumb cannot be adducted, and the difficulty of separating the fingers from one another is very distinctive. After a time all the affected muscles waste, the thenar and hypothenar eminences are flattened, and the spaces between the metacarpal bones become hollowed out. Sensation is also diminished.

I have said that in these local paralyses the continuous current usually effects a cure, but this is not always the case, as in the following example, where faradization was more efficacious.

Paralysis of the Right Arm

CASE.—George J—, æt. 30, admitted January 12th, 1870. Six days ago, whilst attempting to rise from bed, his right arm fell powerless at his side, and he has had no use in it since. He is found on admission to have loss of sensation and motion of the right arm ; slight sensation above the right elbow, increasing gradually upwards. He has slight power in the biceps to flex the elbow. Can neither feel a touch, nor the prick of a pin, nor the sensation of heat and cold. There is thus akinesia, anæsthesia, analgesia, and thermic anæsthesia. There is electric sensibility to faradization, but the continuous current produces scarcely any excitability.

Faradization was ordered him, and he gradually improved. Left February 10th ; arm much stronger.

Local paralyses of the leg.—We occasionally meet with these without any apparent cause, and we either suppose the trunks of the nerves are the subject of some morbid affection, or that the paralysis is of a peripheral nature. In the cases of neuralgia of the *sciatics, gluteal, obturator, and crural,* already mentioned, some loss of function accompanies the pain, and if the former symptom prevailed we might sometimes use the term paralysis as an alternative name for the same case. The paralysis is indicated by the want of power and the wasting of special muscles, such as the glutei. That the atrophy would be the effect of disease of the nerves is seen in those cases where growths have existed in the spine and involved these structures. As regards the leg both the tibial and peroneal nerves may be affected, causing wasting of the muscles. In a case where enlargement of the head of the fibula pressed on the peroneal nerve, the muscles were flaccid, the foot could not be flexed, and walking was very difficult.

I will mention two cases to show that local paralyses of limbs are by no means uncommon, although their causes are very often involved in obscurity.

Paralysis of Leg

CASE.—George W—, æt. 36, admitted into Stephen Ward, June 19th, for weakness of the leg, and left July 23rd. This man was the subject of a remarkable enlargement of the veins, on the surface of the abdomen, indicating some obstruction in the vena cava. He had observed this fourteen years, but it had given him no inconvenience nor had it interfered with his employment.

Patient stated that in March last he was seized with very acute pains through the left hip and groin, which gradually spread down the leg; and these pains were worse at night. Went to Swansea Hospital, where the knee became contracted and he took to crutches. He was then sent up to Guy's Hospital. He was put to bed, being quite unable to walk, on account of pains and weakness in the left leg. No local cause was discoverable to account for the symptoms; the leg was somewhat drawn up, it was perceptibly wasted, being smaller than the other, and sensation slightly impaired. On testing the limb the muscles were found to respond to both the faradaic and galvanic currents. He was then ordered the continuous current to be applied daily to the front and back of the thigh. After the first application he expressed himself as having much relief from the pain, and in a few days it had altogether left him. At the same time the strength returned in the muscles, so that in a few days more he could walk. The current was still applied with a daily improvement in the strength of his leg, so that on July 10th he was walking about, and on the 21st he sufficiently recovered to be able to leave the hospital convalescent and nearly well. Patient took no medicine.

CASE.—William N—, æt. 50, admitted January 4th, 1871. Two months ago he began to feel difficulty in walking, on account of pain in the right thigh and buttock; subsequently weakness of the limb came on. On admission he is seen to be a healthy man, with the exception of his present complaint. He lies in bed, and is unable to raise the right leg. Tactile sensibility is considerably diminished, points being only recognised as separate when ten inches apart. No tremblings of the muscles. On his chest are two scars, as of diseased bone, and he has also had iritis. He owns to having had venereal disease some years ago. With these hints he was put on the Mist. Hyd. Perchlorid. He gradually improved in strength, so that on the 31st he walked down the ward, and on February 9th was discharged convalescent.

CONTRACTION AND SPASMODIC AFFECTIONS OF MUSCLES

These diseases, in my experience, constitute the most difficult to explain or to treat in the whole domain of clinical medicine. The physiologist has not yet determined the true relation of muscle to nerve, for we have, on the one hand, evidence of the power of contractility in the muscle when separated from the nerve,[1] and, on the

---

[1] The contractility of muscle by direct excitation is constantly observed in the living subject when percussion is made over it, as in the pectoralis or trapezius. This was pointed out many years ago by Dr Stokes, and is more frequently met

other hand, there is the fact of the muscle losing its tone in para-lysis, as seen in the muscles of the face when the portio dura is affected, and in the experiments of Marshall Hall already referred to. If the dropping of the muscles of the face argues want of nerve influence, and the falling of the head in sleep the same fact, how are we to regard the case where a muscle is kept in undue con-traction? An over nerve stimulation would be the most reasonable answer, for if the muscle relaxes when we become unconscious in sleep, we seem bound to believe that we are sending a stronger force through the nerve when by a voluntary effort we cause it violently to contract. And yet I have already told you that an exactly opposite theory is held by some, viz. that the contraction is the passive state of the muscle and its elongation is caused by an active process under the influence of the nerve; the effort of the will merely allowing the inherent powers of the muscle to come into play when it immediately contracts. This might be an explanation of the rigor mortis and also of the contraction of an hysterical limb when the patient has lost all power over it. We know, however, that during sleep a relaxation has occurred, and therefore we fall back into our old difficulty of hypothetically regarding two opposite states of muscles as evincing passivity. The fact is, as I said before, that the physiological problem of the relation of muscle to nerve has yet to be determined. When this has been done the physician will have no difficulty in making it applicable to epilepsy, chorea, tetanus, and all spasmodic and convulsive diseases. A solution of the problem is the great desideratum in nerve pathology.

Cases of spasmodic contraction of the limb I have already spoken of under paraplegia, and they are mostly found in connection with meningitis where the nerves are involved.

Local spasms.—I use the term for those cases where a few muscles only are affected and there is no evidence of a central cause of irri-tation. Spasms are generally divided into tonic and clonic, according as the contraction is persistent or intermittent; as, for example, in tetanus and chorea. One of the best known forms of spasm of muscle is that temporary condition known as cramp, which, it may be interesting to note, occurs so often during sleep. This, although local, has usually a general cause for it in the blood, as witnessed in cholera The forms of spasmodic affection to which I wish more particularly to draw your attention are chronic and persistent; they

with in phthisis, owing to the greater liability of consumptive persons to be per-cussed. It is not peculiar, however, to this disease, for it is observed in the wasting of cancer and other disorders. An interesting physiological fact may be noticed under these circumstances, that the contraction does not radiate from the point struck, as a centre, but along a line in the anatomical course of the fibres.

are, of all nerve disorders which we are called upon to treat, the most difficult to cure. We may still regard them as persistent or chronic, even if they occur in paroxysms, and the muscle has intervals of healthy rest.

Beginning with the head, we may meet with cases of spasm of the ocular muscles, causing squint; this may be of a temporary kind, as in the case of the child I mentioned, who was the subject of attacks of strabismus, which lasted a few hours, and then passed off.

Trismus is not a common affection, but when it occurs is most persistent in its nature. I was consulted some time ago about a woman who, nine months before, was seized with a pain in her face, and an inability to open her mouth. From this time the spasm continued, varying in intensity, and becoming worse at night. The jaw was drawn down to the left side, and the face looked swollen, from the constant tension of the masseters. She could not open her mouth to eat anything solid; the tongue, as far as could be seen, was covered with a black fur. Sometimes she had rigors and also tremors of the arms and hands. She had no marked hysterical symptoms. Every remedy was tried without avail.

Spasm of the facial muscles is also sometimes seen, but it is usually paroxysmal and intermittent. It is also called spasmodic tic. It is often associated with a similar state of the muscles of the neck and head. I know a gentleman who for some years has had this painful affection; his head is suddenly jerked round, and at the same time there is a violent twitching of the muscles of the face.

Spasm of the sterno-mastoid.—This may be paroxysmal. I occasionally see a gentleman, when I visit his sick children, who whilst conversing with me is constantly nodding his head forward. The chronic form is, however, more common, and is called *wryneck*, or *torticollis*. It usually affects the muscle on one side, whereby the head is drawn back towards that side, whilst the chin is carried upwards in the opposite direction. Cases are related where the muscles on both sides being affected, drag the head back, and others where the spasms are paroxysmal, causing a constant nodding of the head. One of the most frightful cases of spasmodic contortions I have ever seen began with wryneck; subsequently other muscles of the trunk became affected, and now the patient has sudden violent paroxysms, in which the body is twisted round in a most extraordinary shape. The following is an example of wryneck lately in the hospital :

Torticollis

(Reported by Mr. Knott)

CASE.—Alfred G—, æt. 37, admitted February 28th, and left June 23rd. Works at an oilcloth factory. Always enjoyed good health, and habits were steady and regular. Never had any severe illness. Exposed to cold. About a twelvemonth since he found his head becoming gradually drawn to one side, his face looking to the right shoulder, the muscles of the neck stiff, and some little difficulty in swallowing. This continued until admission.

Patient looks well, and whilst lying on his pillow his complaint is scarcely apparent, as he moves his head from side to side; but immediately he rises, the left sterno-mastoid contracts, and his head moves to the right side.

Faradization was used to the left sterno-mastoid muscle, and caused marked contraction. This was continued daily, and after some time the muscle seemed more readily affected by a less strong current. It seemed, indeed, as well as the trapezius, to have greater excitability than the muscle on the other side. The continuous current from twenty Daniell's cells was then tried, and had no effect on the left sterno-mastoid, although it caused contraction of the platysma, whilst it produced the usual effect on the sterno-mastoid and trapezius of the right or unaffected side. The two forms were tried for six weeks, but without any permanent benefit, and were, therefore, discontinued. He was then ordered injections of arsenic for nine days, but without any result. He subsequently took bromide of potassium and then Tr. Cannabis, and on some days he thought he was better and the muscle more supple; but he again became as bad as before. A consultation then took place with the surgeon about the propriety of an operation for dividing the muscle or the muscular branch of the spinal accessory nerve; but the patient not approving the proposition, he left the hospital unrelieved.

In reference to wryneck there is a fact worthy of observation and further investigation; that in young persons subject to this affection the head and face on the contracted side do not develop as on the other, and in consequence there is a want of symmetry in the countenance when narrowly examined from the front. One eye is slightly lower than the other, and the whole of that side of the face and head is smaller than on the other. In a lad lately in the hospital with heart disease a wryneck existed from infancy, and this remarkable want of symmetry was very evident. In a young lady patient, also, who is otherwise well grown, this disproportion of the two sides of the head and face is clearly shown. It may be asked whether this is due to some failure of nervous power having its foundation in the same cause which produced the wryneck, or whether the contracted muscle itself exerts an influence on growth, and, if so, whether the division of the sterno-mastoid would allow development again to proceed.

Cases are described where quite independently of wryneck the muscles of one side of the face have become atrophied. This has arisen no doubt from some affection of the nerve, whereby its influence over the muscles has been lessened. These cases have been designated by the name of " facial hematrophia."

Spasmodic Contraction of Jaw

A gentleman, æt. 28, has suffered for four years from a spasmodic affection of the jaw and tongue. When he speaks, and more especially when he reads aloud, he feels a strange sensation in his tongue ; this is thrust out of his mouth as his jaw opens. This occurs constantly while he reads ; he used to speak in public, but this affection has entirely prevented him. In quiet conversation it does not often occur, but in reading to himself it constantly happens. It seems to correspond to spasmodic affections witnessed in other parts of the body, and therefore very little more than a bad habit, as shown by the action of reading inducing an attack. He says : " Once I was addressing an audience, when wishing to refer to the Bible to read a passage, my tongue began to jump about so violently that I was obliged to sit down. Sometimes the sensation comes on suddenly, like a flash of lightning, and shoots my tongue forward so that I have no control over it for a few seconds, and if reading am obliged to miss a word. In bad attacks the lower jaw drops simultaneously with the impulses given to the tongue."

I have also notes of the case of a lady who is suddenly seized with spasm of the jaw which prevents her speaking. She opens her mouth but cannot utter a word.

Other muscles may be affected spasmodically, as the trapezius, together with the sterno-mastoid already mentioned, and the muscles of the shoulder ; sometimes also those of the limbs, in a rhythmical manner, as I shall presently mention. Spasm of the muscles of the abdomen generally implies some source of irritation in the organs within the cavity. Spasm of the diaphragm, or hiccup, is occasionally met with. We have had in the hospital a girl with this affection. The hiccup never ceased unless she was asleep. It continued for some weeks, and then departed. We had in the very same bed, immediately before this girl occupied it, another who never ceased to cough. It was thought that it might own some distant cause, and that the spasm was reflex, but this was never ascertained. I may observe that the aurists speak of a cough arising from the ear, and which they call ear cough ; this may sometimes depart when the wax is removed from the meatus. Cases of involuntary laughter might perhaps be included under spasmodic affections. Sneezing also will continue for a great length of time without any apparent source of irritation. This may sometimes be arrested by forcibly pressing the finger against the upper lip and septum of the nose, at the same time taking a deep breath.

Writers' cramp, or scriveners' palsy.—This is a spasmodic affection of the muscles of the arm and hand, arising from their long-continued use in an unequal or inordinate manner. Particular muscles being employed in one constant and special direction they become unduly fatigued. In writing we educate a set of muscles for a given movement, as indeed we do for any other special performance of the hands. This is shown by the fact that if the right hand becomes useless we cannot at once take up the pen with the other. In the act of writing we make use of the first two fingers and thumb to hold and guide the pen, whilst other muscles of the forearm are employed to steady the hand. After much use in this way these muscles are apt to become contracted, and cease to be controlled by the will; the fingers become stiff, and the pen can no longer be grasped by the hand. As soon as the person so affected begins to attempt to write a contraction of the muscles of the forearm takes place, the fingers become stiffly extended, and the pen falls out of the hand. In some forms, the great difficulty in holding the pen is owing to a crampy flexion of the thumb. Where the spasm has given way to actual feebleness of the muscles the hand affected may be assisted and steadied by the other one. This I have seen done by a gentleman, who in signing his cheques grasps the wrist of his right hand by the left, and so steadies it. A number of other methods, too, are often resorted to before the patient gives up his occupation. He will hold the pen between the two fingers, and dispense with the thumb altogether; or rest the whole arm on the table so as to move the muscles as little as possible, and then make the movement from the shoulder.

As regards the treatment of these cases, the complaint is sometimes so obstinate that nothing but perfect and absolute desistance from the usual employment will allow of a cure. In the case of a banker's clerk, where all treatment failed, he was obliged to leave his desk and take to an outdoor occupation. The best treatment is by the continuous galvanic current, after Dr Poore's plan of directing the current down the muscles of the forearm, and at the same time making the patient continually contract them by opening and closing the hand. The cold douche to the arm is also useful, and the adopting any plan by which a regular movement of all the muscles shall be maintained.

Any other movements besides those of writing which produce a continued strain on the muscles will cause the affection. It is not, however, met with in the ordinary class of workpeople who are using all their muscles equally and in a natural manner, but only in skilled persons who are employing some particular muscles inordidinately whilst others are kept in restraint. Thus, I have had as a patient a violin player, who has now been obliged to give up his profession from his inability to hold the bow. Of late, too, we have heard of telegraph clerks' cramp, a complaint where the fingers become stiffened from constant use. In France, it is said, the difficulty is overcome by setting the clerks to work on a different kind of instrument, whereby another set of muscles come into play. Then there is the dancer's cramp, arising from the constant and inordinate use of particular muscles of the leg, which keep the foot in a constrained position so as to enable the dancer to stand on the last phalanges, or toes. I have been reading lately also of the milk-maid's cramp. Duchenne described a case of a writer who struggled with the spasmodic difficulty in his arms so long that the complaint crept higher and higher, until it reached his body. When his arm had quite failed to be controlled he leaned his head on his shoulder until his head and neck became drawn down and contracted like his arm. One of the foreign journals lately contained the case of a tailor who, in consequence of sitting in a restrained posture, had not only spasmodic contraction of the arm but of the neck and upper part of the body. If he ceased to work and forcibly bent the arm the whole spasm would be arrested and he would become unlocked. He was at last obliged to give up work, but if at any time he put himself in a sewing attitude the whole spasm would return. The contraction would first begin in the arm, which would become rigidly flexed; then his body would bend forward until his head almost touched his knee, and the muscles of his face would under-go violent contortion. Under any circumstances where particular muscles are kept in a state of overstrain, then spasm and a feeling of fatigue will come on. I have lately been consulted by a vete-rinary surgeon, in whom I believe a cause of this kind existed to produce his complaint. He had cramps and pains in both his arms, which interfered with his professional work. I found that he was in the habit of driving a very powerful horse in a gig, and for several hours in a day he was holding him in. Besides this he was always trying fresh horses. Occasionally we may meet with cases where there is a tendency to spasmodic affections of the hands, preventing the patient from grasping, where no good cause can be arrived at for its occurrence. You must bear in mind the existence of a local disease produced by contraction of the tendons and fascia.

Dr Frank Smith has described, under the name of "hephæstic hemiplegia," or hammer palsy ('Ηφαιστος Vulcan), a very inter-esting form of paralysis occurring in the Sheffield workmen, who are constantly using the hammer in making files, knife-blades, &c. He reckons that the hammer weighs three pounds, and that a man during the day must make 28,000 strokes with it. He publishes several cases where, in consequence of this continued exertion, the right arm begins to waste, especially in the muscles of the forearm, and a painful contraction occurs in the flexors and pronators when an attempt is made to grasp any instrument like a hammer. The remarkable circumstance is, that although the complaint com-mences after the manner of those already mentioned, yet that, according to Dr F. Smith's observations, the higher centres sometimes become involved; consequently, not only does the arm become weak but subsequently the right leg, and even in exceptional cases the face, accompanied by aphasia. Together with this wasting and cramp, sensibility is impaired as regards touch and electro-action, but thermal sensation remains.

A man, a cooper by trade, was lately in the hospital, who could no longer grasp his hammer, as it produced pain and spasm; but he was effectually cured by galvanism.

Dr Hammond, of New York, has described a form of spasm of the fingers associated with continual movement, so that the patient is unable to retain his hand in any fixed position. To this affection he has given the name of *athetosis*. If, for example, the hand be placed on a table, the tendency is for the fingers to become firmly extended, and then to continue moving on over its surface, the thumb and little finger being mostly affected; the movement is quite involuntary, and it is only by a great effort that the fingers can be again flexed. Sometimes the movements are not so steady, are more of the chorcal kind, and occur in paroxysms. This is, however, the exception, for the motion of the hand is not jerky, but quite orderly, and it is only by a very strong effort of the will that it can for a time be controlled. From the almost continuous action the muscular development becomes increased. On investigating the case we often find evidence that the disease is not purely local. We may observe that the feet become paroxysmally flexed, and other nervous disorders are also present, suggesting a spinal or other central cause; in fact, the case might be one of commencing sclerosis of the anterior columns. Occasionally, also, the patient has been epileptic, and these movements constitute only a part of a more general nervous disorder. Dr Shaw has shown how in lunatic asylums, and especially amongst imbeciles, the hands are placed in all kinds of positions; they are constantly moving, and the patients have little control over them. I have seen examples of this.

CASE.—Samuel M—, a farmer, said that three months ago he had what the doctor called a low fever; after his recovery he observed that his right hand was continually shaking. Whilst he was talking to me his hand rested on his knee, and I noticed that there was no trembling movement, as in paralysis agitans, but it was continually travelling outwards until it fell off his leg, when he brought it back again, and the same process occurred over again. On asking him to write, I found that he could not grasp the pen well; but that when he had it fairly in his grasp it gradually ran away from him across the paper. There was no alteration of sensation and no wasting of the muscles. He said he was otherwise quite well, but I detected a slight quivering in his lips when he spoke. He was strictly temperate, had had no fits, and had not used his arm inordinately in his occupation or in sports.

Rhythmical spasms.—Under this name Sir J. Paget described, many years ago, cases where spasmodic contraction of a uniform kind came on in paroxysms, and sometimes became permanent. The tonic nature of the spasm distinguishes it from chorea, although there may occasionally be a difficulty in declaring to which class of malady it belongs. In true chorea the movements are irregular and ill-defined, and to a certain extent are under the control of the will, or are caused by any attempt at voluntary action, whilst the spasms of which I speak are regular and rhythmical. I once had the case of a girl in the clinical ward who was continually bending herself forward and bowing in a perfectly regular and methodical manner. If she were restrained, great distress and agitation were produced, and she expressed herself as feeling much worse than when in movement. A case, probably of the same nature, was that of a girl who had a most rapid breathing, each respiration accompanying a beat of the heart ; or of another girl, who every now and then repeated the same words with great rapidity. Dr Bright described a case of a girl who uttered a sound like "Heigh ho! Heigh ho!" at regular intervals of three seconds, so that the word was repeated twenty times in every minute. She could check it by great exertion, but it was immediately resumed when she ceased to converse. In men, too, and even amongst the aged, we meet with these strange paroxysmal spasmodic movements. Thus, lately we had a patient who was constantly jerking his head backward ; another who performed the same movement and raised his arm, putting himself in a fighting attitude, as if he were about to strike some one. I have also had an old man several times in the hospital who had paroxysmal spasmodic attacks, which lasted during two or three weeks, after which he was free from them for a considerable time. The paroxysms consisted in his raising his arm and violently beating his chest for about a minute, and then he would desist. This man subsequently died of pyæmia, but a careful examination failed to find any visible morbid alteration in his nervous centres. A lady patient of mine had for some time spasmodic twitches in the sterno-mastoid, and subsequently the whole head became turned round and violently shaken. Sometimes during these attacks the whole body would participate in the movements. I have already alluded, under the subject of spinal irritation, to the case of the lad who was thrown into convulsions by touching a particular spot in the neck.

In some cases there is probably a hyper-excitability of the cord, and so the body is thrown into spasm on any external stimulus ; as, for example, if the legs are touched. Thus, a patient of mine appeared very well whilst lying in bed, but immediately he put his feet to the ground his legs became perfectly rigid, so that he could with difficulty walk. The students used to feel the muscles of his thigh and calf whilst he was standing up, which were then found to be perfectly hard, but immediately he sat down the spasm was gone. It seemed as if he possessed a morbid excitability of the cord, which which was set in action by touching the ground. Sometimes an affection of this kind begins in one part of the leg. In the case of a patient, a gentleman, he would be walking along the road, when he would be suddenly seized with a spasm which would jerk him off the ground. This has continued in spite of treatment, and now his arm has commenced to jerk as well as his leg. If both legs are affected simultaneously the patient is forced suddenly to jump.

Where there is a continuous movement of one side of the body, I see Dr H. Jackson has suggested the term *hemikinesis*.

GENERAL REMARKS ON REMEDIES

I have already spoken of the treatment of individual diseases under their respective heads, but it is desirable to make a few general remarks with respect to the remedies most in vogue for nervous affections.

The remedies for nervous diseases are mostly of two kinds. There are those which act directly on the nervous system, and are supposed to cure either by setting up a counter-action, or by producing a temporary soothing effect until time works the result, and there are those which are styled the nervine tonics, consisting mostly of the metals.

It is remarkable how little has been accomplished with the first class of remedies—those which have a physiological action on the nerves. It does not seem to follow that a medicine which has a striking physiological action is of any value in a therapeutical point of view. It might be thought that strychnia was the remedy to rouse the dormant nerve centres, or opium to allay their excitability ; but the happy anticipation is not realised, for opium seems to have no curative influence on such diseases as chorea or tetanus. Far more efficacious remedies are to be found in simple tonics. There is, however, another class of remedies to be thought of before either of these, and which has no special relation to the nervous system. You must remember that affections of the nervous system need not originate therein, but may be altogether dependent on an external or independent cause, and in such instances our nervine medicines would be useless—as, for instance, in a convulsive attack arising from an intestinal worm. Hence the absurdity of any system which is founded on treating symptoms alone. Suppose a brain or spine disease arose from some affection of the skull or vertebræ whereby an inflammatory lymph or syphilitic deposit irritated the adjacent nerve structure, you would, of course, direct your efforts against the cause. Now, since it often happens that various nervous diseases have such an origin, I should recommend you in all doubtful cases to commence with such remedies as iodide of potassium or perchloride of mercury, for you may by these means actually cure your patient, whilst tonics would be useless. In cases of epilepsy and many obscure nervous affections I usually commence with this class of remedies, knowing that a curable disease has sometimes ended fatally because they have been overlooked. I have seen a case of epilepsy dependent on syphilitic disease treated ineffectually by zinc, and I have seen a case of painful affection of the leg ending in paraplegia treated in vain by strychnia, when, according to the post-mortem revelation, iodide of potassium would have been the effectual medicine. I remember some years ago seeing a case of severe epilepsy treated by Dr Rees with mercury, and evidently with the happiest result. Cases of the same disease apparently cured by the iodide of potassium are very numerous. In epilepsy I often like to try the iodide of potassium, although, as I have told you, there are other remedies, such as bromide, which are supposed to have a specific effect. You must not forget, then, the class of medicines to which I allude in reference to extraneous causes of disease.

Amongst the medicines which act directly on the nervous system there are few which I believe can be regarded as valuable remedies against its diseases. Thus *opium*, which, by its indirect influence on nutritive processes, is one of the most valuable remedies in the Pharmacopœia, is all but powerless in such diseases as mania, chorea, tetanus, and convulsions of all kinds. An all but poisonous dose may arrest the symptoms for a time, but only for them to recur with the same violence as before. *Belladonna*, again, may, through the nerves, control the disordered action of a particular part, but I think very little can be said in favour of its influence over diseases of the brain and spinal cord. I except a few cases where epilepsy has been apparently relieved by it. It seems to have more power over the vaso-motor system, and therefore has been found useful in exophthalmic goitre and in arterial disease. *Conium* is said by Harley to act on the motor centres or nerves, producing a paralysis, whilst the reflex action of the cord is preserved. *Jaborandi* produces sweating, and is therefore thought to be antagonistic to belladonna, which checks it. None of these remedies, however, are of much value in purely nervous affections. I should say the same of *strychnia*, a medicine the effects of which are slight, considering the extent to which it is administered. Its general effects on the nervous system are as disappointing as its direct effects on the stomach are encouraging, for I regard it as one of our very best tonics in some forms of dyspepsia. I would say the same of *aconite;* it is a drug which, acting powerfully on the nervous system, influences nutritive processes in various parts, but its direct operation on the centres so as to alter their morbid states appears to be very slight indeed. *Chloroform,* which as a temporary remedy promotes such a wonderful stillness of the nervous system, is productive of no permanent effect in its various lesions. In affections of other organs it is useful, as in allaying hiccup, pain in the bowels, &c., or arresting convulsions. *Chloral,* also, in gastralgia, in irritable bladder, &c., is very beneficial, but it cannot be called a remedy in any sense that opium can. It produces sleep, but has no power in arresting the morbid conditions which produce insomnia. We gain, therefore, but little by possessing it, and I am not at all sure that mankind has been much better off for the invention. *Cannabis indica,* which has so powerful an influence over the nervous centres, is a very poor medicine, although I have spoken of its benefit in migraine. *Physostigma* has been found useful occasionally in tetanus. *Phosphorus* is again coming into use in nervous affections. *Gelseminum* is our new remedy for neuralgia ; and *camphor* is a good thing to play with. *Bromide of potassium* is a very valuable remedy ; it is said, but I scarcely think on sufficient grounds, to cause contraction of the small blood-vessels, and so anemia and sleep.

Since in very many nervous diseases a disordered action of the centres has been of long duration, you can see how a temporary soothing or exciting remedy can be of little use compared with one which shall have a slower but more permanent effect. Thus we find that remedies which act indirectly, it may be, upon the blood-vessels of the centres, such as the metals, have contributed more than any other means to the cure of nervous disorders. Foremost stands *iron,* and then *zinc; silver* has been found useful in some cases, and in not a few *arsenic.* The most striking effects are seen in neuralgia, where iron and arsenic are often found to produce a cure without any possibility of doubt. In this class of affections I should say that arsenic is one of the most important medicines which we possess ; it is difficult to foretell a cure, but in tic of the face, sciatica, pleurodynia, gastralgia, and other nervous affections, its beneficial effect is often most marked. There is, again, *quinine,* which has cured more nervous disorders than all the physiological remedies combined. I have also given the Tinct. *Acteæ racemosæ,* and should say that in lumbago, sciatica, and some similar nervous affections, it seems to have some efficacy.

Moreover, I must not forget to mention the novel method of introducing medicine by the skin—the hypodermic method. A small syringe contains the solution, and, having a needle point, is inserted into the skin and the fluid forced in by gentle pressure or by means of a screw. Many remedies have been thus used, but more especially morphia. When first adopted it was thought to be eminently effi- cacious by acting directly on the painful part, but further experience has shown that an equally good result is obtained into whatever part of the body it is thrown. The advantages are that it acts speedily, and does not injuriously affect the system as when taken by the mouth. I have seen a gentleman who suffered agonies with spine disease take morphia in the usual way, and it produced sickness, parched mouth, and other unpleasant symptoms, long before the system responded to its influence; but when injected through the skin it speedily soothed the system, relieved the local pain, and no unpleasant consequences resulted.

Then, again, amongst the remedies for local nervous affections we have local remedies, and these are of various kinds. There is the class of soothing medicines already named, made into the form of liniments, ointments, &c. These are sometimes useful, but often less efficacious than applications of an altogether different kind, as blisters and hot applications. There are many instances where a blister is efficacious after every soothing remedy has failed, and, as regards hot applications, I cannot speak too highly of them. These are popular remedies, but nevertheless much less seldom used than a particular medicine which can be taken from a bottle, because, indeed, the latter practice entails less trouble ; but I know from ex- perience that there is many a sciatica or lumbago which can be cured in a few hours by a constant application of heat. Besides the heat, stimulating lotions are often highly efficacious, as, for example, the tincture of capsicum or mustard. I dislike to hear that a patient has failed to gain relief from the medicine of some eminent physician or surgeon, when some old woman or quack has effected a cure by a simple method. Amongst popular remedies is the Tinctura Arnicæ. I cannot say that my experience of it has been large, but I have seen enough of it not to ignore it, and consider it to be sometimes a useful remedy. In one case of a patient who had a violent neuralgic pain following shingles, we used the arnica, and the patient soon got relief, but at the same time an eruption came out, which is very usual after the use of this drug. The lotion was then discontinued, the erup- tion faded, and the pain returned. In this case it seemed to act as a counter-irritant.

I should say that just as hot applications are useful in many painful affections of the nerves, so is the cold douche in some para- lytic conditions. I have seen cases of writers' cramp and similar maladies much benefited by allowing a stream of cold water to run upon the weakened limb.

As regards medicated plasters, they may relieve directly by the influence of the drug which covers them, or by simply producing a new sensation in the place of the old one, or generally I believe, by the support they give to the part to which they are applied. If the pain be due to what is usually called muscular rheumatism, they prevent the movement of the muscle and its attachments.

Electricity.—After the discovery of electricity as one of the forces of nature, and its remarkable effects on the animal body, it was naturally thought that its services might be commanded for the alleviation of sickness; but it is only of late years that it has been applied in a scientific method. One reason for its neglect by physicians was no doubt the early meddling with it by charlatans, and consequently for a long time the only electricians were the most notorious quacks. There was the mountebank who travelled the country with his electrifying machine made out of an old glass vessel and a Leyden jar, consisting of a bottle with a nail inside, wherewith to "shock" the people, and the cures, of course, were numerous, as, for instance, that of a bishop long paralysed, who jumped out of his chair on the first application. After this, we heard of the wonderful properties of pulverised loadstones, and when the galvanic battery was invented, the effects of this in vivifying weak mortals were marvellous. We can now scarcely credit the fact that the celebrated quack Graham instituted at Leicester-square a temple of health, where, amongst the furniture, was a celestial bed provided with costly draperies, and standing on glass legs, so that married couples who slept in this couch were sure of being blessed with a beautiful progeny. For its use £100 a night was demanded, and many persons of rank were foolish enough to comply with the terms. When, shortly afterwards, Franklin dragged the lightning from the clouds, and showed its identity with electricity, we heard how an old woman, whilst at work in the fields, was struck with the flash, and how her uterine function was restored, and she was blessed with a second family. It can scarcely, then, be wondered at that respectable medical men up to the present day held aloof from the subject of electricity, and regarded it at the best as a pretty plaything for their patients. It has been quite of late years that the subject has been investigated in a scientific spirit; and I think we at Guy's may be proud that it was at this institution, under the auspices of the late Dr Golding Bird, that it began to be systematically used as a therapeutic agent. The instrument which you now see in our room was the same which this physician used for many years. His instrument was a simple cylindrical electrifying machine, and an insulated stool on which the patient sat. By this means the patient was charged, and sparks were drawn from his back or elsewhere. The Leyden jar was also sometimes put into use. At this time galvanism had not been employed for therapeutic purposes.

If you refer to the ' Guy's Hospital Reports,' you will see that a considerable amount of good was effected in cases of chorea and some forms of paralysis by Franklinism or frictional electricity. It seemed to act favorably in the same class of cases which we now find relieved by the simple current.

You may remember that it is now some eighty or ninety years ago since Galvani performed his experiments on a frog and made the discovery that electric currents ran through the animal's body ; this was supposed to be refuted by Volta, who, placing together a number of pieces of metal separated by wet cloths, gained the same result : the latter believed that the forces were generated in the metals, and that the animal body merely acted as a conductor. That Volta had a force developed by the chemical action of the metals was no doubt correct, but Galvani's surmise was also true that electric currents were continuously passing in the animal body. The well-known experiments made of late years by Matteucci, Du Bois-Reymond, Radcliffe, and others, have sufficiently confirmed his discovery that there are currents continually developed both in muscles and nerves. Just as in unscientific hands attempts had been made to use the electrifying machine as a therapeutic agent, so now the galvanic battery was thought to possess wonderful curative properties, on sending currents through the body. It was found, however, to be all but useless in the manner applied, and the machine, together with the galvanic bath, remained in the hands of charlatans almost up to the present time, and even now the bath is scarcely used in a scientific manner. The object proposed by it was to extract metals from the body which had been introduced as medicines or in various trades. The patient sat on a wooden stool in a bath containing some acid, and, by his holding one pole of the battery whilst the other was attached to the outside of the bath, the metals were said to be drawn out of him. The method is now being more thoroughly tested.

The subsequent discovery of electro-magnetism gave a new impulse to the use of this agent in medicine. You know how a current of galvanism in the conducting-wire of a battery induces a current in another wire, and how, if the latter be made into a coil and a piece of iron be inserted in its midst, the iron becomes a magnet, and how, by this means, if the current is applied or cut off, very rapid minute shocks are felt. You know also the counter discovery of Faraday of the magnet giving rise to an electric current whenever contact is made or unmade with one of its poles. Now, whether there be any difference in the therapeutical effect of the current induced by the galvanic battery and that induced by the magnet I cannot tell you. The magneto-electric machine has of late come more into use because more convenient. This induced or intermittent galvanism, when employed for the treatment of disease, is usually styled faradization, in distinction to the constant galvanic current from the simple battery.

On the discovery of this form of galvanism, and its striking effects on the muscles of the body when the poles were applied to different parts of the limbs, the method of treatment by faradization came at once into favour, and we are indebted especially to Duchenne for the stimulus which he gave to its use. This physician made long and careful experiments on healthy and diseased persons, and thus not only supplied us with new methods respecting the cure of disease, but with new facts as to the action of particular muscles in the body. If you read his works, you will see that if dry metallic points were applied to the surface, the skin was merely affected, but that if wet sponges were firmly pressed on a muscle it was excited to contraction, and more especially if the poles were applied to certain spots towards its edges. This was thought to be due to the nerves entering at these places. The electro-magnetic apparatus, owing more especially to Duchenne's writings, thus came into general use, and it is the instrument which we were formerly solely employing. Every ward had one, and if a patient was recommended galvanism this was used, the poles of the battery being applied to the muscles in the manner mentioned; its efficacy, however, was very uncertain when used indiscriminately in all cases. In those instances where a set of muscles were inactive from long disuse, its value was great. Thus, in the case of a girl who had hysterical paralysis of one leg, which, in consequence of her having been long bedridden, had become much smaller and weaker than the other, the galvanism effected an entire cure; so also in some cases of facial paralysis. In progressive muscular atrophy, also, as in the remarkable case of the girl already mentioned, it was very useful. On the other hand, we found it quite inefficacious in the infantile paralysis, and in a similar class of cases sometimes met with in the adult, where a limb, without any apparent cause, becomes wasted and useless.

Thus we went on until other observers, and especially Remak, informed us that in the supposed efficacy of the induced electric current we had overlooked the effects of the constant or continuous current as produced by the simple cell; and, moreover, that the effects of the two forms of galvanism were different on the human body, and that consequently they had their own special curative pro-

perties in different diseases; not only was faradization or the induction current of the magneto-electric and galvano-electric machine
useless in some forms of paralysis, but actually injurious; and that
diseases which could not be remedied by it could by the other.
Remak undertook many elaborate experiments on the human body,
in order to prove that the constant current was often the much
more useful agent. Its application produced also a soothing influence on the nervous system, although whilst being applied it
stimulated all the nerves of the body.

It was not long, as you may imagine, before a galvanic battery
was obtained for the use of the Hospital for Epilepsy and Paralysis, and, under the superintendence of Mr Radcliffe, the statements of Remak were confirmed. In consequence of what I heard,
I paid a visit to the institution, in company with Mr Branford
Edwards, when Mr Radcliffe was good enough to show us some of
his cases. There was one of a man who had been suffering for some
months with wasting and paralysis of the right arm. Faradization
was powerless in producing contraction of the muscles, and therefore valueless as a remedy, whilst the new machine was producing a
rapid cure. When the poles were applied a sudden contraction of
the muscles ensued, and in this way were daily growing in strength.
As we had always taken a great interest in the subject of electricity
as a means of cure at Guy's Hospital, we had a galvanic machine
at once fitted up in our room, consisting of a hundred cells, which
could be combined in any number; the results have fully
borne out all that was anticipated—in fact, some of the cures have
been most remarkable. Its value has been greatest where the faradisation had previously failed. Thus in lead paralysis, where very
little result had accrued from the use of the induced current, a most
marked effect was now obtained. This was not only seen in the
final cure, but in the greater susceptibility to its influence. Thus
in the man now in Stephen ward, who is recovering from lead palsy
of the arms, an action was produced by the combination of fifteen
cells, and a most marked result by twenty-five, whilst in a healthy
man there was no evident effect. In progressive muscular atrophy
the constant current has been recommended not in the course
of the muscles or nerves, but along the spine, and faradization
is said to be useless. Now, the first statement we have proved
to be true, but not the last. The case which I have already mentioned of the girl who was little more than a skeleton, and who quite
recovered under the use of faradization, is sufficient to show this;
at the same time we have already had cases which prove the assertion of Remak. Dr Fagge has had under his care the case
of a man with commencing progressive muscular atrophy, who

rapidly recovered by the use of the continuous current down the spine, one pole being placed over the nape of the neck and the other over the lumbar region.   One of the most remarkable cases I have seen proving its efficacy is that of the man who has just left Stephen ward.   I can give no other name to his complaint than partial paraplegia.   For six years he had been weak in his legs, so that they dragged when he walked, and he had great difficulty in raising them from the ground.   I ordered the continuous current to the spine, and he began to improve at once.   After each application he said his legs were more free, and at the end of two months he left the hospital well.   Of course there was no organic disease, but recovery after so great a length of time was most remarkable and encouraging.   I gave him no medicine, in order not to complicate the case.   That the whole nervous system is affected by its application is certain from the sensations which the patients experience— they almost always have a metallic taste in the mouth, sometimes flashes of fire in the eyes, and sometimes a more troublesome vertigo.

We do, then, find that there are different kinds of paralysis in which the induced current and the constant current have respectively their curative effects ; but much yet has to be learned as to the further application of these remedies, and perhaps some care is required in their management, for, if galvanism is a useful agent, we may suppose it also to be injurious if wrongly applied. In the experiments on frogs and other animals, if a current pass down a motor nerve, the function is increased, but an opposite effect produced if the poles are reversed. Whilst the current passes downwards the hind legs are moved ; if the poles are reversed, the front legs are moved, and the animal at the same time, cries out.

That the forms of galvanism used, and the direction of the current adopted are important, we should think from the different effects produced in apparently similar cases, so that at present our method is to a certain extent experimental. I have already said, in speaking of the treatment of different forms of paralysis, that the simple battery current is especially to be used in cases where we believe the nerve is at fault, as, for example, in the ordinary facial paralysis. Here, as a rule, faradization is powerless, the very opposite of what is seen in hemiplegia, where faradization excites the muscles well. But even in this marked instance some exceptions occur, as in a case I saw lately in our electrifying room, in which motility was obtained only by an inverse galvanic current, and subsequently by faradization. I have spoken of local paralysis of the limbs in which the simple galvanic current appears the most appropriate, and it may therefore be used as a test in feigned paralysis. Some of the cases which attend our room have arisen from the use of splints and other surgical apparatus, and in these cases faradization sometimes succeeds better than the continuous current. You have, indeed, to try both forms. I have spoken of the remarkable case of lead paralysis, where faradization appears to exert no influence over the contraction of the muscles, and yet it is a fact that when we had no other machine but the coil we succeeded sometimes in curing our patient. In infantile paralysis there is but little susceptibility to either form of galvanism, but both of them should be used in order to sustain the nutrition of the muscle.

I have already spoken of the use of galvanism in neuralgia, as tic or sciatica; also in lumbago, where I have seen the secondary current sometimes do as much good as the simple continuous one. Even in the pains of organic disease it is beneficial. In a case of spinal meningitis, with painful contraction of the legs, galvanism acted as a soother and enabled the patient to sleep.

I have spoken of the popular views respecting electricity, because very similar opinions are held about it now as in former times. You will be constantly asked as to its value in particular forms of complaint, and you will find they are generally those in which the profession knows nothing of its uses ; in fact, although we and the public both regard electricity as a very potent remedial agent, we look upon it in a totally different light. The popular notion of electricity is that it is a life restorer, and invigorates the system in all forms of depression ; it is therefore had recourse to by hypochondriacs and others, who seek from it some tonic influence for their weakened nerves ; or at least they think they are making use of it by wearing galvanic chains, belts, rings, and breast-plates. We know nothing of electricity in this sense. We are not even aware that a current continually passing through the system has any effect whatever. We employ it for producing a molecular change in the spinal centres, in the nerves, or the muscles, and we find by a succession of shocks of different lengths and intensities a commotion is set up in the organs, and in this way an alteration in their functions ensues.

I might also allude to the fact that when you enter upon practice you will find persons have very different susceptibilities as regards the action of medicines upon them. These idiosyncrasies, however, are probably not nearly so frequent as are supposed, since the patient feels a pleasure in asserting that he possesses some peculiarity of constitution which renders him or her more susceptible than other people ; and yet this is entirely imaginary. So often is this the case, that unless I have good evidence of it, I totally disregard the statements of the patient in this respect. What, however, is far more common is the extreme susceptibility to medicines of all kinds which we find in some persons. Just as there are those who with tolerably strong bodies always respond to treatment, so there are others who appear never to be benefited by it. This is due also to the mental attitude towards medicine which each class of patients presents; the one seeks the doctor's advice with confidence, intending to be better for it, whilst the other approaches him with scepticism, feeling sure that medicine will avail him nothing. The character of mind and body in the two persons being different, it becomes absolutely true that the incredulous patient has good grounds for his doubts. The doctor, therefore, soon finds that in treating his patient the practice of medicine is not only one of physic but of metaphysics, and that the effect of his drugs depends as much upon the constitution of the patient's mind as his body. I know several persons, amongst others, two notable examples in our profession, who say they cannot take physic; they mean that two or three grains of rhubarb will violently purge them, that a few drops of opium upset their livers and stomach for several days, that three grains of iodide of potassium will cause coryza and headache, and so on through the whole list of drugs. These very unpleasant people and unsatisfactory patients are counterbalanced by our old and steadfast adherents, who ask for a prescription with confidence, and declare that whatever you give them does them good. You therefore have much to learn when you get into practice as to the individual peculiarities of your patients.

INDEX

PRINTED BY J. E. ADLARD, BARTHOLOMEW CLOSE.